BIOCHEMISTRY RESEARCH TRENDS

HANDBOOK ON GALLIC ACID

NATURAL OCCURRENCES, ANTIOXIDANT PROPERTIES AND HEALTH IMPLICATIONS

Biochemistry Research Trends

Additional books in this series can be found on Nova's website under the Series tab.

Additional E-books in this series can be found on Nova's website under the E-book tab.

BIOCHEMISTRY RESEARCH TRENDS

HANDBOOK ON GALLIC ACID

NATURAL OCCURRENCES, ANTIOXIDANT PROPERTIES AND HEALTH IMPLICATIONS

MICHELLE A. THOMPSON
AND
PARKER B. COLLINS
EDITORS

nova publishers
New York

For permission to use material from this book please contact us:
Telephone 631-231-7269; Fax 631-231-8175
Web Site: http://www.novapublishers.com

Additional color graphics may be available in the e-book version of this book.

Library of Congress Cataloging-in-Publication Data

Handbook on gallic acid : natural occurrences, antioxidant properties and health implications / editors, Michelle A. Thompson and Parker B. Collins.

pages cm

Includes index.

ISBN 978-1-62618-921-8 (hardcover)

1. Antioxidants--Physiological effect. 2. Gallic acid--Health aspects. 3. Phenols--Health aspects. 4. Phytochemicals--Physiological effect. I. Thompson, Michelle A., editor of compilation. II. Collins, Parker B., editor of compilation.

QK898.A57H36 2013

613.2'86--dc23

2013016534

Published by Nova Science Publishers, Inc. † New York

CONTENTS

PREFACE

Gallic acid and its structurally related compounds are found widely distributed in fruits, plants, vegetables, and derivatives. Esters of gallic acid have a diverse range of industrial uses, as antioxidants in food, in cosmetics, and in the pharmaceutical industry. The authors in this book discuss the natural occurrences, antioxidant properties and health implications of gallic acid. Topics include gallic acid as a source to use for increasing functional properties in food products; gallic acid implications in health as a multi-therapeutic protective agent; the thermal, anti-inflammatory, and antioxidant properties of gallic acid; gallic acid extraction and its application in the prevention and treatment of cancer; application of spectroscopic techniques for the study of gallic acid autoxidation; gallic acid bioavailability in humans; and gallic acid and its derivatives and their occurrence and identification in high altitude edible and medicinal plants.

Chapter 1 - Gallic acid and its structurally related compounds are found widely distributed in fruits, plants, vegetables, and derivatives. Esters of gallic acid have a diverse range of industrial uses, as antioxidants in food, in cosmetics and in the pharmaceutical industry. It has drawn particular attention since it has been demonstrated to have many medicinally and therapeutically important properties such as antimicrobial, antiviral, and antitumor. Moreover it has the ability to scavenge peroxyl and DPPH radicals and antioxidant activity at stomach pH. In this chapter, studies of the effects of gallic acid and its derivatives on pathogens, enhancement of cell viability and functionality of probiotics, and their potential applications in food safety and human health are investigated.

Chapter 2 - Gallic acid is a naturally occurring polyphenol widely distributed in vegetables and fruits ubiquitous in diet. This trihydroxybenzoic acid is most frequently encountered in ester form as a basic structural unit of many health beneficial plant polyphenols. Due to the diverse range of biological effects, gallic acid and its derivatives have been extensively studied for implication in human health. Its well-known potent antioxidant and free radical scavenging properties are conventionally exploited for the therapy of diseases related with oxidative stress, whereas plant polyphenols are generally active against wide range of microbial infections. Moreover, recent experimental and epidemiological studies evidenced an even broader therapeutic potential of gallic acid such as for prevention and treatment of neurodegenerative processes or long-term anti-inflammatory effect for protection against different chronic disorders. In addition, gallic acid elicits chemopreventive effect by inhibiting, retarding and reversing the process of carcinogenesis without harming healthy cells. The present chapter summarises the relevance of gallic acid to human health emphasising the most recent advances in its application as a preventive and therapeutic agent for various diseases. New medical care strategies involving this polyphenol are also reviewed.

Chapter 3 - Gallic acid belongs to a wide class of naturally occurring phenolic compounds. It is mainly found in the fruits, leaves, twigs, bark and wood of several plants. It presents with a variety of biological functions, displaying anti-inflammatory, cardiovascular and anticarcinogenic properties, in addition to its antioxidant activity. Epidemiological evidence suggests that an antioxidant-rich diet plays an important role in cancer prevention. This antioxidant activity is considered as paramount in the inhibition of free radical-related diseases. The action of phenolic compounds, such as gallic acid, is carried out through the inhibition and/or delayed production of free radicals generated by cellular metabolism or exogenous sources. Gallic acid action aims at preventing the attacks of these free radicals on lipids, amino acids of the proteins, double bonds of the polyunsaturated fatty acids and/or the base-paring of DNA, thus avoiding the formation of lesions and also the loss of cellular integrity. Moreover, gallic acid displays scavenging properties that inhibit reactions of iron and copper via metal chelation. Gallic acid is also widely used as a starting material in the synthesis of drugs in pharmaceutical industries. Due to the widespread application of gallic acid, it is important to identify its main natural sources, determine the antioxidant potential and obtain specific knowledge concerning gallic acid's structural features and chemical properties. These are important benchmarks for the roles of antioxidant, since they evaluate the resistance to thermal decomposition and antioxidative power, inhibiting the free radical generation through donations of protons and/or metal chelation.

Chapter 4 - The oxidative stress on human health has been a serious issue. It may involve ageing, inflammation, metabolic disorders and cancer. Gallic acid is widely distributed in a number of of fruits and plants. It possesses antioxidant activity, thus could decrease the oxidative stress in human. It could also possess anti-cancer activity and other bioactivities related to human health.

The techniques used to determine the antioxidant activity of gallic acid included *in vitro* and *in vivo* methods. The available assays are based on different mechanisms (hydrogen atom transfer and electron transfer), bioavailability, metabolism and other factors, which can lead to varied results. Therefore, this chapter is intended to provide a brief update of recent literature on the *in vitro* and *in vivo* methods available for evaluating the antioxidant activities of gallic acid and the antioxidant values of pure compound or food/plants were also compared and discussed.

The chemical-based methods such as total radical trapping antioxidant potential (TRAP) assay, oxygen radical absorbance capacity (ORAC) assay, luminol-photochemiluminescence (PCL) assay, ferric reducing antioxidant power (FRAP) assay, Trolox equivalent antioxidant capacity (TEAC) assay and 1,1-diphenyl-2-picrylhydrazine (DPPH) radical scavenging assay are used to evaluate the antioxidant activities of phytochemicals in food. The cell-based models include the inhibition on the growth of cancer cells, the protective effect of oxidative stress in various cells exogenously exposed to external stimuli and the cellular antioxidant assay (CAA). The high antioxidant activity *in vitro* does not always lead to high *ex vivo* or *in vivo* antioxidant property. Some *in vivo* studies have also shown promising health effects of gallic acid and the food/plants rich in gallic acid. The animal and human experiments are parts of the methods to evaluate *in vivo* antioxidant potentials. The oxidative stress biomarkers in serum or tissue such as glutathione peroxidase (GSH-Px), glutathione (GSH), superoxide dismutase (SOD), catalase (CAT) and lipid peroxides (LPO) could be used to express the *in vivo* antioxidant activity.

Chapter 5 - Electrochemical DNA biosensors are promising tools for fast, inexpensive and simple in vitro analysis for the determination of free radicals and antioxidants. Reactive Oxygen Spieces (ROS) and especially hydroxyl radicals ($^{\bullet}OH$), are known to damage DNA by oxidation of the bases, which results in their destruction, release or attack of the deoxyribose moieties by strand breaking. These lesions are related to the aging process, cancer, Alzheimer's disease and atherosclerosis. In living systems, hydroxyl radicals are generated from the metal (ferrous or cupric ions) ion-dependent breakdown of hydrogen peroxide by Fenton' s reaction, which is an important mediator of oxidative damage in vivo. Antioxidants such as phenolic acids and plant extracts containing high amounts of them, act as free radical terminators and reduce the effect of the oxidative damage on dsDNA.

In the present work, an overview of natural occurrences, antioxidant properties and health implications is being thoroughly considered.

Chapter 6 - Gallic acid, a polyhydroxy phenolic compound, is found in various natural products such as tea leave, gallnuts, grapes, green tea, strawberries, lemons, pineapples and in red and white wines. Phenolic compounds are known to donate a proton from their hydroxyl (O–H) bond through hemolytic cleavage and form a stable phenoxy radical. Compounds possessing more than one phenolic hydroxyl group are further shown to generate an even more stabilized phenoxyl radical. As a result, gallic acid is regarded as an excellent antioxidant with a high free radical (reactive oxygen species or reactive nitrogen species-origin) scavenging effect. Among other properties, the anti-inflammatory, anti-bacterial, and antitumor properties are crucial. In this chapter, the authors mainly focus on the anti-oxidative and anti-inflammatory roles of gallic acid on transcriptional regulation of various cellular regulators. The authors will describe how gallic acid affects the suppression of anti-oxidative and anti-inflammatory regulators through the controlled regulation of key transcription factors. Since gallic acid and its esters can inhibit the TNFα-induced nuclear translocation of nuclear factor kappa-B (NF-κB) by way of a mechanism independent of inhibitors of kappa-B (IκB) degradation, the authors also deeply correlate the effect of gallic acid on the NF-κB-mediated transcriptional regulation of various early and late inflammatory mediators involved in related disease pathologies. In addition, the authors will focus on gallic acid's role on the epigenetic level since it exhibits potent p300/CBP-mediated anti-histone acetyl transferase (HAT) activity that inhibits the p65 acetylation-dependent NF-κB activation and production of inflammatory markers. Moreover, this review will show various gallic acid-mediated molecular targets, which affect signaling pathways within cells in different ways involving genetic or epigenetic mechanisms, and engage functional proteins in the formation of inhibiting complexes of strong affinities, ultimately perturbing the expression of oxidative and inflammatory molecules.

Chapter 7 - Gallic Acid (GA) is one of the most important compounds of the hydrolysable polyphenols group; and can be found in a variety of plant tissues. Extraction of this compound has been performed using optimized conditions such as temperature and solvent type used that ensure the best yields without affecting its properties. Polyphenols are associated with color and flavor of fruits and vegetables; however, the nutritional value of this compound is that it may promote a wide variety of biological activities. Also, it has been reported that Gallic acid participates in regulation of damage by inflammatory diseases, microbial infections, cardiovascular conditions as well as antitumor and anticancer protection.

These effects occur because it is a potent antioxidant, and can delay, inhibit, or prevent oxidation of oxidizable materials by scavenging free radicals and decreasing oxidative stress. For this reason several investigations are aimed at the evaluation of phenolic compounds like gallic acid for prevention or treatment of degenerative diseases such as cancer. Results have demonstrated that polyphenols affect some mechanisms of tumor developed in various stages, including initiation, promotion and progression of cancer, which triggers the metastasis process. The objective of the present chapter is to present a review about extraction process of gallic acid as well as the steps and factors implicates in this extraction, properties of this phenolic acid and its applications for cancer treatment as an alternative for the control of the one disease of high incidence, worldwide.

Chapter 8 - Gallic acid or trihydroxybenzoic acid belongs to the class of phenolic acids, has antioxidant and antimicrobial properties, and shows cytotoxicity against cancer cells. It occurs in different amounts in a great number of fruits and vegetables and, thus, in their food derivatives. Gallic acid can also arise from the breakdown of tannins. Berries supply gallic acid together with anthocyanins. A cup of hot chocolate contains higher amounts of this compound than beverages such as wine and tea. Other sources of gallic acid are walnuts, hops, apples, and flax seeds.

Agricultural practices, vegetative stage and processing parameters affect the phenolic content and profile of foods. Furthermore, processing operations including washing, cutting, size reduction, mechanical separations, and thermal treatments can modify the gallic acid content.

This review represents a survey of the gallic acid content of plants and plant-based foods and highlights the effects of agricultural practices, vegetative stage, and processing on the content of this phenolic acid.

Chapter 9 - Studies confirm that oxidative stress may be involved in the etiology of several chronic diseases, episodes genetic, metabolic and neurodegenerative processes such as cardiovascular, diabetes, cancer, Alzheimer's, among others, also working in the acceleration of aging and apoptosis. However it is known that consumption of fruit and vegetables have shown a positive correlation reducing the chances of developing pathologies.

The aim of this study is to relate the bioactive compounds contained in foods derived from plants, among these, the almond baru [Dipteryx alata Vog.], A fruit of savannah, the Cerrado and its relationship to antioxidant activity as a prevention of chronic degenerative diseases and their complications. Almonds were obtained in local trade Brasilia-DF, from three regions of the Cerrado (MT, MG and GO), which were previously selected and randomly assigned into two groups: raw almonds and almond raw film without film. Part of the kernels of these two groups was packaged in transparent polyethylene and stored at -80ºC.

The remainder of almonds with and without film was subjected to roasting, which was held in an oven without air circulation at 150°C/45 minutes. The raw almonds and toast were crushed and analyzed in triplicate.

Thus, knowing that the cellular components are not fully protected by endogenous mechanisms is well established that antioxidants obtained from nature, are indispensable for the proper defense against oxidation and thus begin to play an important role in maintaining health, as These, and contain essential nutrients and micronutrients, contribute to the compounds present in the leaves, fruits, films, and seeds, from its chemical structure, promote action varied in the body. Certain phenolic substances such as flavonoids and phenolic acids, are considerably more potent antioxidant than vitamin C and vitamin E. Gallic acid in a group of phenolic compounds simpler and is inserted in the group of hydroxybenzoic acids or non-flavonoids, being commonly in grapes, strawberries, nuts, cereals and leguminous seeds that act as potential antioxidants with proven preventive action degenerative processes in cardiovascular, and activity against some mutagenic chemicals, striking chemopreventive action, antiviral and antibacterial.

Recent studies show that daily consumption of almond baru (Dipteryx alata Vog), a seed of edible oilseed feature, reduces oxidative stress induced in rats.

It was confirmed these kernels, the presence of substances with a high content of bioactive compounds, wherein, of the eight phenolic compounds identified by liquid chromatography (HPLC), at concentrations ranging between 67.7 and 224.0 mg/100 acid, gallic was the predominant compound and found that a thermal process is not capable of reducing the content of phenol. Furthermore, the antioxidant activity involves a set of intrinsic factors that protects the body against harmful damage, being able to prevent and heal pre-existing diseases.

Chapter 10 - Previous studies have demonstrated a variety of biological activity of gallic acid (GA) including antitumoral, antioxidant and antiinflammatory properties. Among these numerous data, *in vitro* assays showed that the GA and its derivatives are active against several types of tumor cells. Particularly, the studies have shown that the antitumor activity seems to be related to the induction of apoptosis involving different signalling pathways. Apoptosis induced by GA may be associated with oxidative stress derived from reactive oxygen species (ROS) and mitochondrial dysfunction. Interestingly, GA has both pro-oxidant and antioxidant properties. Intracellular ROS levels were increased in cancer cells treated with GA and their derivatives and these results suggested that GA plays a central role as a pro-oxidant in cancer cell models.

It also has been reported that the cell death promoted by GA in different cell lines may be related with glutathione (GSH) depletion. Since the intracellular GSH has a decisive effect on anticancer drug-induced apoptosis, those apoptotic effects were inversely related to GSH content. The reduction of GSH levels might be related to the capability of GA to decrease the expression, as well as the activity of γ-glutamyl cysteine synthase (γ-GCS).

Regarding tumorigenesis, it has been widely reported that GA is responsible for suppression of tumor angiogenesis. The mechanisms underlying this event might be related to alterations in the interaction among the stromal cell-derived factor-1 (SDF-1), an alpha-chemokine, and CXCR4, a G-protein-coupled receptor. These changes are involved in angiogenesis, upregulating key angiogenic factors including vascular endothelial growth factor (VEGF). Furthermore, GA suppressed ADAM17 expression, a transmembrane protease, which may be associated with the inhibition of invasiveness through the inactivation of the well-known cell-survival signalling pathways PI3K/Akt and Ras/MAPK. Additionally, GA might inhibit the activity of matrix metalloproteinases (MMP) involved in tumor angiogenesis and metastasis, such as MMP-2 and MMP-9. These MMP are directly involved in degradation of extracellular matrix and play pivotal roles in cancer cell migration and invasion.

Finally, GA induced caspase-3 and -8 expression, alteration in the Bcl-2/Bax ratio, and an inhibition in tyrosine phosphorylation by BCR/ALB kinase, as well as a down- regulation in COX-2 levels in leukemia cell lines. The GA was also able to inhibit COX-1 and COX-2 activities in these cell lines, suggesting an antiinflammatory activity of the compound. Increased COX-2 expression seems to be involved in the development of cancer by triggering cell division and inhibiting apoptosis. These studies suggest that GA induces cell death mediating multivariant mechanisms, being a good candidate to clinical studies for cancer treatment.

Chapter 11 - Gallic acid (3, 4, 5-trihydroxybenzoic acid), is pytochemical compound occurring as a free molecule or as part of tannic acid molecule in plant tissue. Gallic acid has many significant biological activities such as antioxidant, antiinflmatory, antigungal,, and carcinogenic properties, it has attracted considerable interest Tannic acid biodegradation is the key condition for the gallic acid production. Biodegradation of tannic acid can be controlled by enzymatic or microbiological ways, giving the possibility for developing biotechnological processes for the production of gallic acid. Several studies on biotechnological production of gallic acid have generated significant achievements in recent years, interest in the search for new routes to produce more "*clean*" compounds has turned the interest of researchers for fermentation processes, biotransformation rich substrates tannins by microorganisms has led to high yields of gallic acid, however, these new areas of opportunity still present challenges which need studies involve more factors in the microbial production of gallic acid is necessary to make efforts to deepen scale up production studies. In addition, methods for higher recovery and purification of this important compound are necessary.

Chapter 12 - In terms of public health, drinking fruit and vegetable juices may well be as effective as consuming whole fruits and vegetables with regard to reducing the risk of chronic disorders. Many of their health benefits can be attributed to phenolic compounds. Phenolic compounds are aromatic metabolites of plants secondary metabolism that have a common structure with an aromatic ring with at least one hydroxyl group, which provides the ability to neutralize reactive species, helping the body to protect itself from oxidative stress. Phenols contribute to fruits' color and taste and have been described as possessing anticarcinogenic and antimutagenic activity. Furthermore, they have been shown to be good contributors to the total antioxidant capacity of foods, resulting in a beneficial effect to human health.

Due to the growing popularity of phenolic antioxidant over the past two decades, an increasing interest in determining the antioxidant activities exhibited by phenolic compounds and their derivatives should also be noted. In this line, special attention has been given to gallic acid. Gallic acid is a naturally abundant plant phenolic compound which is present in liquid foods derived from fruits and vegetables and has been found to exhibit antioxidative properties. Therefore, at this stage of development, there is a need to establish the relationship between phenolic compounds, including gallic acid, with the antioxidant capacity and total phenolic compounds of these food products.

Chapter 13 - Recent findings of the authors' group indicate that a small amount of gallic acid (GA 12 mg/P/d) protects humans against DNA damage and induces the activities of antioxidant enzymes. Furthermore, the authors saw in animal studies that similar doses prevent damage of the genetic material in inner organs of rats and the induction of preneoplastic hepatic foci by γ-radiation. These promising findings stimulated us to evaluate the current knowledge on ROS and cancer protective properties of the phenolic. Approximately a dozen *in vitro* studies have been published which concern the antioxidant properties of GA; with high concentrations sometimes pro-oxidant effects were observed. Also a few *in vivo* results which were obtained with laboratory rodents are available which indicate GA induces ROS protective enzymes possibly *via* induction of the transcription factor Nrf2. In regard to the cancer protective properties of the compound, it is notable that *in vitro* experiments showed that GA inhibits cell cycle progression at G0/G1 and also at G2/M, (via modulation of the transcription of cyclines) and induces apoptosis in various human cancer cell lines (possible modes of action are inhibition of the synthesis of DNA bases, induction of NfκB, depletion of glutathione and release of ROS). Furthermore, was found that the phenolic decreases the motility and adhesion of cancer cells, inhibits metastasis and prevents angiogenesis *via* inhibition of VEGF. Also data from animal experiments (xenograft models with mice) show that the compound prevents the tumour growth. Most of these findings were obtained with doses which are substantially higher as the daily intake of GA *via* the diet which is in Middle Europe in the rage of 18 mg/P/d; however some findings were seen with lower levels that can be easily reached after increase of the consumption by dietary measures. Taken together the current data indicate that GA is an extremely potent antioxidant in humans and may protect against malignant diseases.

Chapter 14 - Gallic acid is a polyphenolic compound widely distributed in plants ether in the free form or as a part of more complex polyphenolic compounds. It has a range of biological activities but is best known for its antioxidative activity. All the biological effects of gallic acids in living cells may be connected to the ease of its oxidation or/and autoxidation. Various methods may be used for the study of such processes but spectroscopic methods seem to be especially well suited for such purpose. Out of many spectroscopic methods used for the study of oxidation/autoxidation of various phenolic compounds UV/Vis spectrophotometry and electron spin resonance (ESR) have some particular advantages. UV/Vis spectrophotometry enables direct study of the oxidation/autoxidation of polyphenolic compounds but it lacks the specificity of other spectroscopic techniques. This disadvantage may be overcomed by the application of various chemometric techniques for spectra processing. UV/Vis spectrophotometry is also one of the most economic spectroscopic techniques found in virtually all analytical laboratories and is readily used for detection in liquid chromatography, usually as a HPLC-DAD (high performance liquid chromatography with diode array detection). ESR spectroscopy allows direct detection and characterization of transient free radical species formed in the course of polyphenolic compounds transformations under various conditions. In this chapter the authors present the results of the spectroscopic studies of the gallic acid autoxidation under various conditions with special emphasis on the results obtained by the UV/Vis spectrophotometry with chemometric spectral processing and ESR spectroscopy and their application for a better understanding of mechanisms of gallic acid transformation and its action upon living cells.

Chapter 15 - Polyphenols are extensively metabolized either in tissues, once they are absorbed through the gut barrier or, for the non-absorbed fraction and the fraction re-excreted in the bile, by the colonic microflora. All polyphenols are conjugated in our body to form O-glucuronides, sulphate esters and O-methyl ethers. The formation of anionic derivatives by conjugation with glucuronides and sulphate groups facilitates their urinary and biliary excretion and explains their rapid elimination.

Animal studies first revealed the possible metabolic fate of gallic acid. In rats, rabbits and chickens the major urinary metabolite is 4-O-methylgallic acid, followed by pyrogallol (conjugated and unconjugated). Small amounts of conjugated 2-O-methylpyrogallol were also detected in rats. In sheep, resorcinol glucuronide is the major product of gallic acid metabolism and unconjugated pyrogallol and resorcinol comprise minor urinary metabolites.

Moreover, recent *in vitro* studies have reported phenolic acids to be main metabolites of anthocyanins after fecal fermentation. However, fragmentation of anthocyanins to phenolic acids in humans has not been studied in detail. One recently identified anthocyanin metabolite is 3-O-methylgallic acid, which is presumably the metabolite of petunidin-3-glucoside, but possibly also a demethylation product of malvidin-3-glucoside.

The few existing studies addressing the bioavailability of gallic acid in humans revealed that, compared to other polyphenols, this phenolic compound is extremely well absorbed. The most common metabolite of gallic acid identified in human biological fluids is 4-O-methylgallic acid, which was found to increase significantly in human plasma within the first 4 h after consumption of a single dose of red wine or dealcoholized red wine. Plasma concentrations of free and glucuronidated forms of gallic acid and 4-O-methylgallic acid also increased after ingestion of 50 mg pure gallic acid, while gallic acid was rapidly absorbed after human administration with tea or acidum gallicum. Concerning the excretion of gallic acid, many studies have identified gallic acid and its metabolites in urinary samples. For example, after acute consumption of 3 cups of black tea, gallic acid was not detected in urine of subjects. However, three gallic acid methyl ethers, 4-O-methylgallic acid, 3-O-methylgallic acid and 3,4-O-dimethylgallic acid, were identified. A recent study on the bioavailability of phenolic compounds present in Corinthian raisins (*Vitis vinifera*) –containing 115.7 ng gallic acid per g- indicated that gallic acid is bioavailable in humans, as its concentrations in plasma increased one hour after raisins consumption. Overall, gallic acid concentrations in plasma were found to range between 2.2±0.9 to 9.9±4.6 (SEM) μg/L.

Chapter 16 - Phenolic compounds are molecules that have one or more hydroxyl groups bound to an aromatic ring. Together with vitamins, phenolic compounds are considered important dietary antioxidants. Thousands of phenolic compounds are found in plants, and they are classified into different types of functional groups. Phenolic compounds play a number of metabolic roles in plants, in growth and reproduction, and in protection against pathogens and external stress, UV radiation and predators for instance.

Gallic acid is trihydroxybenzoic acid, a phenolic acid, also known as 3,4,5-trihydroxybenzoic acid. The presence of this phenolic compound and its derivatives is extensive in nature and it can be found in many plants and foods like fruits, vegetables and cereals.

Bolivia is currently considered among few countries with a high bio-diversity in the world. This diversity is evident from many species of food plants found in so contrasting geographical regions like the Andean mountains or the Amazon flood plain. In particular, the Andean region has a variety of endemic and adapted foods like roots, fruits, cereals and pseudo-cereals all growing at high altitude (2000 to 4200 m.a.s.l.). The presence of ancient native populations such as the Aymara, Quechua and Guarani ethnic groups contributed through the centuries to present with their own and particular *savoir-faire* to the adaptation and use of these plants.

As a sample, this part of the world has given to the world's kitchen two of the most celebrated foods in mankind's history -potato and corn- in their many varieties. Endemic Bolivian non-conventional crops have increased dramatically their importance during recent years due to the introduction of Andean foods onto the international market, never as widely open as nowadays. A typical example is quinoa -*Chenopodium quinoa*-, a protein-rich Andean grain currently available and a very common product in European markets.

Bolivia offers a unique environment for researches on high solar UV radiation levels and related phenolics' contents in plants. The altiplano region particularly, is a *sui generis* site (because of its high altitude), where populations and food plants grow in an abundant variety and quantity with a non-negligible percentage in the per-year national statistics.

This chapter presents the results of gallic acid and its derivatives found in food and high altitude plants. These results include an EIMS and NMR spectroscopic approach for gallic acid.

In: Handbook on Gallic Acid
Editors: M. A. Thompson and P. B. Collins
ISBN: 978-1-62618-921-8

Chapter 1

GALLIC ACID AS A SOURCE TO USE FOR INCREASING FUNCTIONAL PROPERTIES OF FOOD PRODUCTS

Irene Dini*

Dipartimento di Chimica delle Sostanze Naturali, Università di Napoli "Federico II", Naples, Italy

ABSTRACT

Gallic acid and its structurally related compounds are found widely distributed in fruits, plants, vegetables, and derivatives. Esters of gallic acid have a diverse range of industrial uses, as antioxidants in food, in cosmetics and in the pharmaceutical industry. It has drawn particular attention since it has been demonstrated to have many medicinally and therapeutically important properties such as antimicrobial, antiviral, and antitumor. Moreover it has the ability to scavenge peroxyl and DPPH radicals and antioxidant activity at stomach pH. In this chapter, studies of the effects of gallic acid and its derivatives on pathogens, enhancement of cell viability and functionality of probiotics, and their potential applications in food safety and human health are investigated.

INTRODUCTION

Pure gallic acid [3,4,5-trihydroxybenzoic acid, 4 ($C_6H_2(OH)_3COOH$)], is a colorless crystalline organic powder (Figure1).

Gallic acid molecule is essentially planar and has two intramolecular hydrogen bonds between hydroxyl groups, the hydrogen atoms of the three hydroxyl groups are oriented in the same direction around the ring and form intra-and intermolecular hydrogen bonds. The crystal structure is stabilized by all available intermolecular hydrogen bonds (Okabe, Kyoyama and Suzuki; 2001). The two adjacent hydroxyl groups are engaged in a complex (Fatima, Maqsood and Kazmi; 2002) and the remaining hydroxyl group is suggested to form hydrogen

* Email: irdini@unina.it.

bonding with COO^-. Gallic acid is a strong chelating agent and forms complexes of high stability with iron (Li, Bandy and Tsang; 2000). It is most frequently encountered in plants in ester form. Gallate may be classified into several broad categories (Haslama and Caib; 1994):

(1) Simple esters.
(2) Depside metabolites (syn-gallo tannins).
(3) Hexahydroxydiphenoyl and dehydrohexahydroxydiphenoyl esters (syn-ellagitannins)based upon
    (a) $^4C_1$ conformation of D-glucose ;
    (b) $^1C_4$ conformation of D-glucose ;
    (c) 'open-chain' derivatives of D-glucose.
(4) 'Dimers' and 'higher oligomers' formed by oxidative coupling of 'monomers', principally those of class above.
(5) Condensed proanthocyanidins.

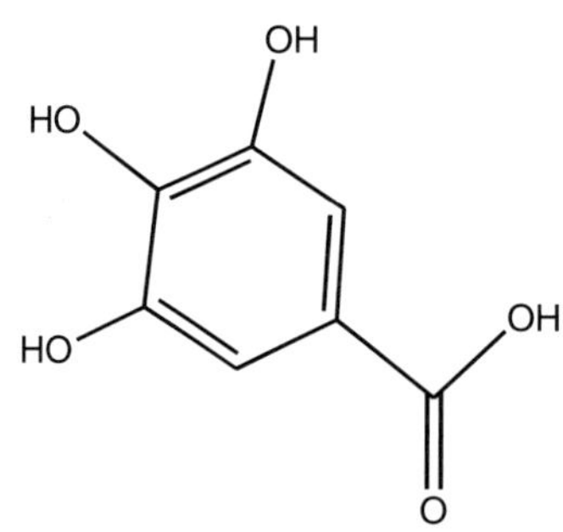

Figure 1. Gallic acid.

Gallate is an essential precursor for many plant secondary metabolites, particularly hydrolyzable tannins such as gallo- and ellagitannins (Grundhofer, Niemetz, Schilling and Gross; 2001). Both plants and fungi produce high concentrations of gallate (Werner, Rossmann, Schwarz, Bacher, Schmidt and Eisenreich; 2004), while bacteria are known to produce small amounts. Gallic acid containing metabolites are not universally distributed in higher plants. They are found in blueberries, walnuts, apples, flax seed, tea, sumac, hazel, watercress, oak bark, and a variety of other plants and herbs. They occur within clearly defined taxonomic limits in both woody and herbaceous dicotyledons. Ellagitannins are widely distributed in the lower *Hamamelidae*, *Dilleniidae*, and *Rosidae* (the HDR complex) and have been used as prominent chemotaxonomic markers. It has been suggested that the low degree of diversification in gallate-dominated taxa may be a result of the electron scavenging properties of these metabolites which, in turn, inhibit oxidation (Haslama and Caib; 1994).

BIOSYNTHESIS OF GALLIC ACID

Biosynthesis of gallic acid has been narrowed to two possible routes. Oxidation of 3-dehydroshikimic acid (DHS) to a diketo intermediate followed by spontaneous aromatization might lead to gallic acid. Alternatively, gallic acid may result from the dehydration of DHS by the action of shikimate dehydrogenase, a multifunctional enzyme that also catalyzes the reversible reduction of 3-dehydroshikimate to shikimate, followed by hydroxylation of the intermediate protocatechuic acid (Figure 2) (Kambourakis and Frost; 2000).

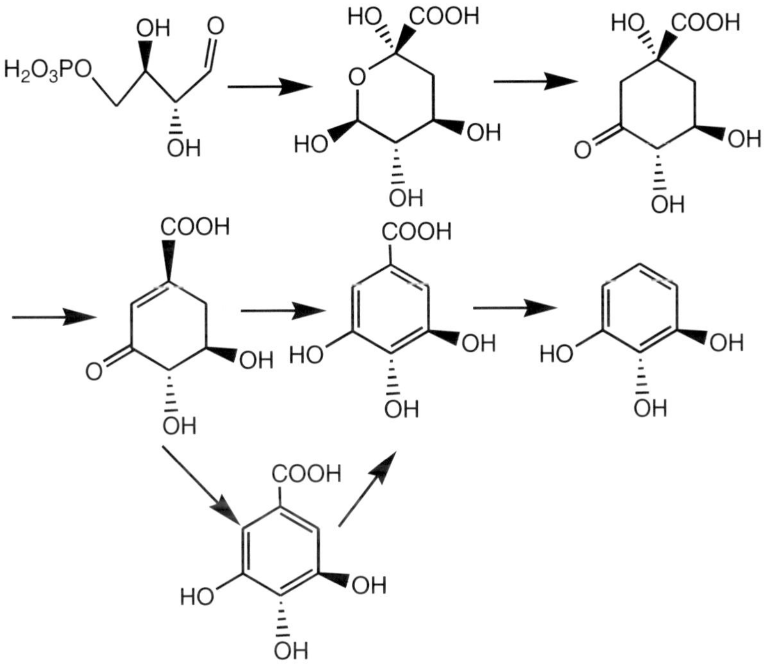

Figure 2. Biosynthesis of gallic acid.

Gallic Acid Flavour

Gallic acid is more astringent than bitter (Robichaud and Noble; 1990) and could help to decrease the perception of bitterness. The amount of flavour released depends on the retention of flavour components in the food matrix. Consequently, the nature of the mechanism of binding of flavour components with nonvolatile components is of importance in the perception of the food products. The systematic investigation of complexation between gallic acid, and three aromatic flavours, vanillin, 2-methylpyrazine, and ethyl benzoate, indicates that gallic acid interacts strongly with the flavours. The mechanism involved in the interaction is an intricate combination of weak non covalent interactions, of which hydrogen bonding makes a major contribution to the specific conformational geometry (Da-Mi, de Ropp and Ebeler; 2000). Instead, the astringent taste, may depend on its ability to dissociate into ions in solution. The negatively charged portion of chlorogenic acid would have affected the positively charged lipids to lower the membrane potentials. Moreover the astringent substances precipitate some proteins producing the widely recognized tactile sensation. Therefore, gallic acid will has two kinds of sensation, namely a tactile sensation and a gustatory sensation.

Absorption and Metabolism

Recent *in vivo* data from rat experiments suggest that orally administered gallic acid is directly metabolized into 4-*O*-methyl gallic acid in the liver and then excreted via the kidneys in urine (Zong, Inoue, Nose, Kojima, Sakaguchi, Isuzugawa, Takeda and Ogihara; 1999). In a study on the absorption and metabolism of gallic acid (gallic acid; 50 mg per *os*) in a single volunteer, gallic acid and its metabolite 4-O-methylgallic acid appeared rapidly in the plasma and urine. Plasma levels for free gallic acid and 4-O-methylgallic acid 90 min. after dosing, were 305 and 422 μgl -respectively, and after deconjugation these values rose to 3.8 and 4.4 $\mu g l^{-1}$. These two compounds and their conjugates in urine accounted for 55.6% of the dose. The metabolism of n-propyl gallate has been investigated in mice and a no-effect level of 1.5 g kg^{-1} BW (body weigth) day^{-1} established. *In vitro* studies have been performed on n-propyl, n-octyl and n-dodccyl gallatc using homogenates of liver, mucosa of the small intestine and the contents of the caecum/colon as a source of the intestinal microflora. All test substances were extensively metabolised by the homogenate of the intestinal mucosa, and the microflora also showed a high metabolic activity, especially to n-propyl gallate (Toma´s-Barbera´and Clifford; 2000).

Effect of the Cooking Process and pH on the Gallic Acid Content in Fruits and Vegetables

In recent years, there has been an increasing trend towards the exploration of functional ingredients from natural dietary sources like fruits, vegetables, oilseeds, cereals, grains and herbs. During vegetable processing, qualitative changes, breakdown and leaching into surrounding water may influence the phytochemicals content of the vegetables (Hunter and Fletcher; 2002). In addition, operations such as cutting and slicing may provoke a rapid enzymatic depletion of several naturally occurring antioxidants as a result of cellular disruption, which allows contacts of substrates and enzymes (Davey, Van Montagu, Inze, Sanmartin, Kanellis and Smirnoff; 2000). These processes would bring about a number of changes in physical characteristics and chemical composition of vegetables (Rehman, Islam and Shah; 2003). Several studies have reported both positive and negative effects of thermal treatment depending upon differences in the processing conditions, as well as the morphological and nutritional characteristics of the food. Some research demonstrates that the phenolic compound can decrease after dry heat treatment, for example in moth bean (11%); dark brown seed coated cowpea (16%); and Bauhinia vahlii (20%); a drastic loss of 48% was also reported in light brown colored cowpea, instead the gallic acid content increases significantly in boiled, in roasted chestnuts and in toasted and microwaved white beans (Barros, Nunes, Gonçalves, Bennett and Silva; 2011). This finding may be due to conversion of tannins to gallic acid during various technological process (Hassan and Youssef; 2012). On the other hand gallic acids are not stable to high pH and the transformations are not reversible. Therefore if exposure to high pH is used to recover proteins from cereals and legumes, to induce the formation of fiber-forming meat analogue vegetable protein, for preparing peeled fruits and vegetables, and for destroying microorganisms, nevertheless this treatment may cause destruction of gallic acid (Friedman and Jurgens; 2000). Finally gallic acid could have an important role in food technological processes to preserve the functional properties of the vegetable after cooking, because it significantly decreases the loss of flavonoids from vegetable due to the addition of salt (Takei and Ioku; 2004).

PAST USES OF GALLIC ACID

About 2000 years ago, Hippocrates correctly emphasized "Let food be your medicine and medicine be your food". Today, we know that the choice of comestible goods is an important contributor to our well-being and that the targeted use of specific nutrients and other bioactive molecules (e.g. secondary plant products) significantly helps in the regulation of metabolic processes in a variety of situations of particular requirements. Many of the foods containing gallic acid have been used for years as natural remedies, and were relied upon by various cultures for their medicinal properties. The ancient cultures may not have understood the role of gallic acid, or even that gallic acid existed, in the foods and herbs they used for medicinal purposes. They did, however, believe that their medicines worked. Blueberries, for example, were used by Europeans for treatment of diabetes and by Native Americans were consumed as a good tonic for purifying the blood (Schatz; 2009). Another plant (also rich in gallic acid) that has long been used for its medicinal properties is hazel. Native American tribes used hazel leaves and bark in haemorrhages, congestion, inflammation of eyes and skin burns, haemorrhoids (Darshan and Doreswamy; 2004).

Gallic acid content in blackberry and kernel

Fruit	Gallic acid	reference
Blackberry	8-67 mg/Kg	Mosel and Herrmann; 1974
Hazelnut kernel (with skin)	127 mg/Kg	Shahidi, Alasalvar and Chandrika; 2007

MODERN USE OF GALLIC ACID

1. Cosmetic Use

Gallic acid can be used in cosmetic as an acidizer to instead of citric acid. It can be added into toothpaste to restrain bleeding and it can be utilized in hair dye as pigmentation aid. Gallic acid powerfully absorbs ultraviolet radiation and can be added into anti-UV lotions. Finally it is used to clarify skin and us skincare. There are some evidence about high skin, spot and speckle clarifying effect of the complex comprising gallic acid obtained from *Shinus sp.* and the linoleic acid extracted from passion flower (Da Luz, Ferrari, Teixeira Tage, Siqueira and Arroteia; 2011). Moreover gallic acid contributes to protect skin from wrinkles, leading to glowing and healthy younger skin because it scavenges free radicals from skin cells and prevent trans-epidermal water loss (Pulok, Mukherjee, Niladri Maity, Neelesh, Nema and Birendra; 2011).

2. Use as a Component for Deodorizing Air or Gas

Gallic acid is a strong deodorizing agent. When a mixed acid containing tannic acid and/or gallic acid in ascorbic acid, citric acid, tartaric acid and/or gluconic acid, are allowed to react with Fe, Mn, Cr, Ni, Zn, Al and Cu or a powdery mixture of two or more of these metals, and then dried in the air, the reaction product of the metal and galic acid exhibits a strong air cleaning ability through a reaction with sulfur type compounds, nitrogen type compounds and lower fatty acids, toxic gases, and objectionable odor gases such as SOx, NOx, O_3, CO_2, CO, in the presence of oxygen and moisture in the air (Tamio and Yozo; 1993).

3. Use in Biodegradable Active Packaging

Biodegradable active packaging is the main focus of current food packaging research and developments. Extensive research has been conducted to employ some natural antioxidants such as phenolic compounds as alternatives to synthetic antioxidants into packaging materials. Incorporation of antioxidants has become popular since oxidation is a major problem affecting the food quality. Currently, the most frequently used antioxidants in active packaging are butylated hydroxyanisole and butylated hydroxytoluene. Although these synthetic antioxidants can effectively be used in active food packaging because of high stability, low cost and efficiency, there are significant concerns related to their toxicological aspects. Instead, the addition of gallic acid improved mechanical, water vapor barrier and antioxidant properties of the resulting films without toxicological aspect related to use of synthetic antioxidants (Siripatrawan and Harte; 2010). Moreover gallic acid was found to potentially act like a plasticizer, therefore films containing gallic acid became more elastic (Hager, Vallons and Arendt; 2012).

4. Food Additives (Antioxidant Additives)

Gallic acid and its esters [E-310 (propyl gallate), E-311 (octyl gallate) and E-312 (dodecyl gallate)), in turn, are used as antioxidant additives in both food and pharmaceutical industry. They are known to protect against oxidative damage induced by reactive oxygen species, as hydroxyl radicals or hydrogen peroxide, and reactive sulfur species. Synthetic galloyl esters showed to be potent and selective enzyme inhibitors, as well as synergistic protectors against peroxyl radical damage in membranes.

5. Antifungal Activity

Fungi, especially *Aspergillus flavus* and *A. parasiticus*, during plant defense responses can induce rapid generation of reactive oxygen species, production of hydrogen peroxide and generate a harmful toxin known as aflatoxin that can contaminate foods such as nuts, peanuts, corn, wheat and cottonseed. Aflatoxins are well known for their possible role as carcinogen affecting different human systems. They have been also reported to enhance the generation of free radicals. Hence, prevention of fungal growth, aflatoxin secretion as well as lipid peroxidation by using a single measure will be a novel and economical strategy to combat food losses during storage and transit. Recently it has been observed that gallic acid is able to alleviate oxidative stress and it is also potent inhibitors of aflatoxin biosynthesis in *Aspergillus flavus* (Kim, Campbell, Mahoney, Chan, Molyneux, Bhatnagar and Cleveland; 2005).

6. Use in Probiotics Additional Food

There are very few people who have not heard about the "new" health miracles called probiotics. They claim to help with health problems, ranging from constipation to diarrhea, and prevent colds or fight them once you already have one. Probiotics are showing up in foods, beverages, and supplements. The root of the word probiotic comes from the Greek word pro, meaning "promoting" and biotic, meaning "life". There is some debate about how to define probiotics. The Food and Agriculture Organization of the United Nations (FAO) defines probiotics as "live microorganisms, which, when administered in adequate amounts, confer a health benefit on the host. According to the German definition, probiotics are defined viable microorganisms, sufficient amounts of which reach the intestine in an active state and thus exert positive health effects. Our digestive system normally has what we would call "good" bacteria and "bad" bacteria. Maintaining the correct balance between the "good" bacteria and the "bad" bacteria is necessary for optimal health. Things like medications, diet, diseases, and environment can upset that balance (de Vrese and Schrezenmeir; 2008). Numerous probiotic microorganisms (e.g. *Lactobacillus rhamnosus GG*, *L. reuteri*, *bifidobacteria* and certain strains of *L. casei* or the *L. acidophilus*-group) are used in probiotic food, particularly fermented milk products, or have been investigated, as well as *Escherichia coli* strain Nissle 1917, certain enterococci (*Enterococcus faecium SF68*) and the probiotic yeast *Saccharomyces boulardii*, with regard to their medicinal use. Among the numerous purported health benefits attributed to the probiotic bacteria, the (transient) modulation of the intestinal microflora of the host and the capacity to interact with the immune system directly or mediated by the autochthonous microflora, are basic mechanisms. Plant species have been known to contain a wide variety of functional components, among these, antioxidants present in plant, have shown to stimulate the growth of probiotics. The impact of these components on gallic acid play an important role in the functionality of many biological systems microorganisms can be either direct or indirect. Most of these compounds have shown antimicrobial properties against foodborne pathogens. These probiotics in turn produce organic acids, which act as antimicrobial agents. On the other hand, mineral nutrients present in plants indirectly induce growth-specific metabolites of probiotics to promote the production of antimicrobial compounds. Hence, the production of these organic acids and antimicrobial compounds could contribute to improve food safety and human health. The data demonstrate that the selection of *Streptococcus thermophilus* strain in dairy products and formulations supplemented with polyphenolics may present a nutritional benefit as a promising probiotic starter candidate as well as a potential source of dietary antioxidants (Rowaida and Khalil; 2010). In addition studies on probiotic ice cream produced using *Lactobacillus casei Shirota* and gallic acid have shown, that gallic acid could be utilized for increasing functional properties of ice cream without causing a negative effect on survival of

L. casei Shirota and on organoleptic properties (Sagdic, Ozturk, Cankurt and Tornuk; 2012). Moreover indian researchers have found that consuming probiotics supplemented with the micronutrients quercetin and gallic acid may double the beneficial effect to the intestinal tract by improving zinc and iron status and by enhancing the balance of microbes in the colon (Supriya, Snehal, Smita and Vaishali; 2011).

7. Use in Prebiotics Additional Food

A prebiotic is "a selectively fermented ingredient that allows specific changes, both in the composition and/or activity in the gastrointestinal microflora that confers benefits upon host well being and health". Today, only bifidogenic, non-digestible oligosaccharides (particularly inulin, its hydrolysis product oligofructose, and (trans) galactooligosaccharides), fulfil all the criteria for prebiotic classification. They are dietary fibers with a well-established positive impact on the intestinal microflora. Other health effects of prebiotics (prevention of diarrhoea or obstipation, modulation of the metabolism of the intestinal flora, cancer prevention, positive effects on lipid metabolism, stimulation of mineral adsorption and immunomodulatory properties) are indirect, i.e. mediated by the intestinal microflora, and therefore less-well proven (de Vrese and Schrezenmeir; 2008). Moderate amounts of prebiotics and natural antioxidants incorporated into a main food had an influence on several parameters of the immune system and play an important role in the maintenance of gastrointestinal health. For example Lee et al. have been shown that tea phenolics (including 3-O-methyl gallic acid and gallic acid) exert significant effects on the intestinal environment by modulation of the intestinal bacterial population. Phenols significantly repressed the growth of certain pathogenic bacteria such as *Clostridium perfringens*, *Clostridium difficile* and *Bacteroides spp.*. While they less severely changed commensal anaerobes like *Clostridium spp.*, *Bifidobacterium spp.* and probiotics such as *Lactobacillus sp.* (Lee, Jenner, Low and Lee; 2006).

8. Health Benefits

Gallic acid has been shown to have immense health benefits, and may prove to be an effective treatment for certain cancers and neural disorders. It has been shown to prevent cellular mutations and to be toxic to cancer cells, while having no negative effect on healthy cells (Elvira, Chandra, Vinicio and Wenyi; 2006). Studies have shown that gallic acid has anti-cancer properties against leukemia, certain prostate, colon and lung cancer cells. Gallic acid also has therapeutic applications for inflammatory and allergic diseases, such as: asthma, allergic rhinitis, sinusitis due to its ability to inhibit histamine release and the expression of pro-inflammatory cytokine. Moreover gallic acid is a powerful antioxidant (Jittawan and Siriamornpun; 2008) and it can be used to treat albuminuria, diabetes (Lan, Charn, Chin and Yin; 2007) and as a remote astringent (it constrict tissues and stop bleeding). Administered internally, it has shown to be useful in the treatment of uterine, pulmonary, and nephritic hemorrhages, as well as all hemorrhages of a passive character (Hurrell, Reddy and Cook; 1999). Gallic acid has also shown to cease the occurrence of menorrhagia (unusually heavy and prolonged menstrual periods), when the individual takes 5 grains of gallic acid in pill-form 3 – 4 times per day. Finally it has anti-viral and anti-fungal properties (Kinjo, Nagao, Tanaka, Nonaka, Okawa, Nohara and Okabe; 2002).

9.1. Anti-neoplastic Effects

Cancer occurs after a single cell in a tissue is progressively genetically damaged to produce a cancer stem cell possessing a malignant phenotype. These cancer stem cells are able to undergo uncontrolled abnormal mitosis, which serves to increase the total number of cancer cells at that location. When the area of cancer cells at the originating site becomes clinically detectable, it is called a primary tumor. Some cancer cells also acquire the ability to penetrate and infiltrate surrounding normal tissues in the local area, forming a new tumor. The newly formed "daughter" tumor in the adjacent site within the tissue is called a local metastasis. All cancers share have six common traits ("hallmarks") that govern the transformation of normal cells to cancer (malignant or tumor) cells. Those hallmarks are:

(1) cancer cells stimulate their own growth;
(2) they resist inhibitory signals that might otherwise stop their growth;
(3) they resist their own programmed cell death (apoptosis);
(4) they stimulate the growth of blood vessels to supply nutrients to tumors (angiogenesis);
(5) they can multiply forever;
(6) they invade local tissue and spread to distant sites (metastasis).

(Hanahan and Weinberg; 2000).

Gallic acid plays antitumor action either by increasing apoptosis of cancer cells as well as it happens *in vitro* for lung cancer cell (Kawada, Ohno, Ri, Yunmo, Ikoma, Yuugetu, Asai, Watanabe, Yasuda, Akao, Takemura, Minatoguchi, Gotoh, Fujiwara and Fukuda; 2001), prostate cancer cells (Agarwal, Tyagi, and Agarwal; 2006) and human leukemia cells (Yeh, Chen, Lai, Yang, Yu, Chiang, Lu, Yang, Yu, Chang, Lin and Chung; 2011), either by anti metastases effect as has been found in gastric (Ho, Chang Ho, Liao Wu and Wang; 2010) and liver tumors (Ohno, Inoue and Ogihara; 2001).

9.1.1. Apoptosis

Apoptosis is a form of programmed cell death (cell suicide), the mechanism by which cells are programmed to die in the event they become damaged. Cancer cells characteristically are able to bypass this mechanism. Apoptosis can be triggered by an overexpressed oncogene, and this may be the primary means by which such mutant cells are continuously removed. Conversely, cancer cells must overcome apoptosis to progress. Gallic acid derivatives are known to cause apoptosis in tumor cell lines and to inhibit lymphocyte proliferation. The mechanism by which gallic acid analogues induce apoptosis in some cell lines is yet not completely understood, probably involving the paradoxical generation of reactive oxygen species (ROS), which interfere with the homeostatic redox balance of the cell. It is also a fact that these compounds are excellent inhibitors of protein tyrosinase kinases. The recognized cytotoxic activity of phenolic derivatives depends on their rate of incorporation into cells, directly related to their lipophilicity, as much as to their antioxidant activity. Both these properties are affected by the presence of the ring substituent hydroxyl groups and by the length of the ester moiety, in polyhydroxylated phenolic esters (Fiuza, Gomes, Teixeira, da Cruz, Cordeiro, Milhazes, Borges and Marques; 2004). In the investigation of the effects of gallic acid on various human oral squamous carcinoma cell lines (HOSCC), including UM1, UM2, SCC-4, and SCC-9, researchers showed that gallic acid up-regulated pro-apoptotic genes such TNF-α, TP53BP2, GADD45A, and down-regulated the anti-apoptotic genes (Chia, Rajbanshi, Calhoun and Chiu; 2010).

Other researchers in the investigation of the apoptotic activity induced by gallic acid on chronic myeloid leukemia (CML) cell line-K562 and the underlying mechanism, found that gallic acid inhibited BCR/ABL tyrosine kinase (BCR-ABL play critical roles in the pathogenesis of CML, thus representing potential therapeutic targets revealed from several clinical studies) and NF-κB, (nuclear factor kappa-light-chain-enhancer of activated B cells is a protein complex that controls the transcription of DNA. NF-κB is found in almost all animal cell types and is involved in cellular responses to stimuli such as stress, cytokines, free radicals, ultraviolet irradiation, oxidized LDL, and bacterial or viral antigens), therefore they concluded that gallic acid induced apoptosis in K562 cells involves death receptor and mitochondrial-mediated pathways by inhibiting BCR/ABL kinase, NF-κB activity and COX-2 (enzyme responsible for inflammation and pain) (Chandramohan, Bharat, Aparna, Arunasree, Gupta, Achari, Reddy, Lakshmipathi, Subramanyam and Reddanna; 2012; Ohno, Fukuda, Takemura, Toyota, Watanabe, Yasuda, Xinbin, Maruyama, Akao, Gotou and Fujiwara; 1999).

9.1.2. Antimetastasis Effect

Metastatic cancer is cancer that has spread from the place where it first started to another place in the body. A tumor formed by metastatic cancer cells is called a metastatic tumor or a metastasis. The process by which cancer cells spread to other parts of the body is also called metastasis. Metastatic cancer has the same name and the same type of cancer cells as the original, or primary, cancer. For example, breast cancer that spreads to the lungs and forms a metastatic tumor is metastatic breast cancer, not lung cancer. Metastasis is a multistep process made up of a cascade of interrelated, sequential steps including invasion, migration, adhesion, infiltration, colonization at a distant site, and the subsequent formation of new capillaries. To successfully metastasize, invasive tumor cells must overcome three barriers: first, the attachment to basement membrane or extracellular matrices; second, protease activity must induce local degradation of the matrix and third, tumor cells must migrate through the modified matrix. Recently, a new step has been proposed; the creation of a ''premetastatic niche'' at the target site before the first tumor cells arrive at this distant location. The precise timing of the establishment of the premetastatic niche cannot be determined (Weng and Yen; 2012). Gallic acid has been reported to have anti-invasive and anti-metastatic activities in various cancer cells. Gallic acid antimetastasis effect is induced by matrix metalloproteinase (MMP) downregulation. The molecular mechanism of gallic acid in genetically regulating the expression of MMPs is not fully elucidated. Increasing evidence suggests that MMPs promote tumor progression. The main role of MMPs in angiogenesis, tumor growth, and metastasis is degradation of the extracellular matrix and release and/or activation of growth factors through their degradative activity. MMP-2 and MMP-9 are presumed to be especially important for cell transmigration because these proteinases act on type IV collagen and other basement membrane components. Ho et al. suggested that suppression of NF-B pathway and

PI3K/Akt/small GTPase pathways may lead to MMP- 2 and MMP-9 downregulation in gallic acid treated AGS cells (cells extract from an adenocarcinoma of the stomach of a Caucasian female) (Chen and Chang; 2012).

In Gastric Adenocarcinoma

Gallic acid has potent inhibitory effects on AGS cells migration by suppressing the expression of MMP-2/-9 and cytoskeletal F-actin. The antimigratory effect of gallic acid may involve the inhibition of NF-jB activity and multiple proteins related to metastasis and cytoskeletal reorganization signal pathways, including Ras, Cdc42, Rac1, RhoA, RhoB, PI3K, and p38.

In Mastocytoma

Gallic acid has an inhibitory effect on P815 cell (mouse lymphoblast-like mastocytoma cell line) metastasis. In fact DBA/2 mice in which P815 murine mastocytoma cells have been injected intravenously, when treated with gallic acid, had a decreased number of nodules in the liver.

In Glioma

Gallic acid-induced decrease of U87 cell (human primary glioblastoma cell line) invasiveness by the suppression of ADAM metallopeptidase domain 17 (ADAM17), also called tumor necrosis factor-α-converting enzyme (TACE), and the down-regulation of PI3K/Akt (phosphatidylinositol 3-kinase/Akt signalling) and *ERK* (Extracellular signal-regulated kinases)/*MAPK* (Mitogen-activated protein kinases) signalling pathways.

9.2. Anti-HIV Effects

The most common secondary immunodeficiency is Acquired Immunodeficiency Syndrome or AIDS, which results from infection with the Human Immunodeficiency Virus 1 (HIV1). HIV preferentially infects T-cells

(T cells or T lymphocytes belong to a group of white blood cells known as lymphocytes, and play a central role in cell-mediated immunity), attacking the very system that protects us from viruses. Each T-cell has its own type of T-cell receptor, which recognizes its own type of peptide. Anti-HSV-1 (Anti-Herpes simplex virus) activity of gallic acid was attributed to the inhibition of virus attachment and penetration into cells. Gallic acid and gallates with no more than five carbons in the alkyl moiety presented higher percentages of viral replication inhibition (78%). A similar association between the size of the alkyl moiety and the antiviral activity was found for HSV-1, suggesting that the lipophilicity of these molecules may be involved in their biological properties (Kratz Jadel, Andrighetti-Frohner, Kolling Deise, Leal, Cirne-Santos, Yunes Rosendo, Nunes, Trybala, Bergstrom and Frugulhetti; 2008).

9.3. Antidiabetic Effects

Type II diabetes mellitus (DM) is one of the fastest growing epidemics. Type II DM occurs predominantly in adults more than 30 years old. It is characterized by decreased insulin sensitivity leading to insulin resistance in its target tissues (mainly liver, skeletal muscle, and adipose tissues). On the other hand, impaired glucose induced insulin secretion with a decrease in pancreatic β cell mass will eventually lead to chronic hyperglycemia. The overall protective effects of gallic acid in type II diabetic rats might be due to decrease in blood glucose, hepatic lipid peroxides, hepatic glycoprotein components, and hepatic lipids and increase in hepatic glycogen, plasma insulin, and hepatic antioxidant systems. Thus, gallic acid protect the liver tissue by maintaining all the biochemical parameters studied in diabetic rats (Punithavathi, Stanely Mainzen Prince, Kumar, and Selvakumari; 2011). Moreover gallic acid is beneficial for the treatment of myocardial damage associated with type-1 diabetes. It significantly lower fasting glucose, and prevent loss of body weight, polyphagia, and polydypsia in diabetic rats. Gallic acid also prevent hyperlipidemia, hypertension, bradycardia, structural alterations in cardiac tissue such as increase in force of contraction, left ventricular weight to body weight ratio, collagen content, protein content, serum lactate dehydrogenase, and creatinine kinase levels in a dose-dependent manner. Further, treatment also produce reduction in lipid peroxidation and the increase in antioxidant parameters in heart of diabetic rats (Patel and Goyal; 2011). Further studies are needed to find out the exact mechanism of action of gallic acid in diabetes mellitus (Punithavathi, Stanely Mainzen Prince, Kumar, and Selvakumari; 2011).

9.4. Antiobesity Effects

Obesity is associated with high blood cholesterol and a high risk for developing diabetes and cardiovascular disease. Therefore, the management of body weight and obesity are increasingly considered important to maintain healthy cholesterol profiles and to reduce cardiovascular risk. Several drugs are used for the therapy of obesity-related metabolic diseases, and frequently the possibility of preventing body fat accumulation with these drugs is discussed. Globally, one in three of the World's adults is overweight and one in 10 is obese. By 2015, World Health Organization (WHO) estimates the number of chubby adults will balloon to 2.3 billion—Equal to the combined populations of China, Europe and the United States. The discovery of bioactive compounds from herbs is one possible way to control obesity and to prevent or reduce the risks of developing various obesity-related diseases. Gallic acid administration was found to reduce the food intake (at a statistically non-significant level) in treating mice (Glick; 1981). Administration of gallic acid, also suppressed the postprandial elevation of the blood triglyceride in the single oral administration test and contributes to the antiobesity effect by inhibiting pancreatic lipase activity (Yasuyuki, I-Ching, Hiroyuki, and Kazunaga; 2012). Pancreatic lipase, also known as pancreatic triacylglycerol lipase, is secreted from the pancreas, and is the primary lipase (enzyme) that hydrolyzes (breaks down) dietary fat molecules in the human digestive system, converting triglyceride substrates found in ingested oils to monoglycerides and free fatty acids. Hsu and Yen showed that the supplement of gallic acid on high fat diet-decreases body weight gain, the weights of liver and adipose tissue, serum parameters (TAG, phospholipid, total cholesterol, LDL cholesterol, insulin and leptin), hepatic steatosis and reduced oxidative stress (reduced GSSG and enhanced GSH, glutathione peroxidase, glutathione reductase and glutathione S-transferase) (Hsu and Yen; 2007). Finally, studies on gallic acid derivatives showed that ethyl gallate (methyl ester of 3,4,5-trihydroxybenzoic acid) give lipid inhibition (Changhyun Roh and Sung; 2012) and epigallocatechin-3-gallate inhibits extracellular signal-related kinases, activates AMP-activated protein kinase (enzyme that plays a role in cellular energy homeostasis), modulates adipocyte marker proteins, and down-regulates lipogenic enzymes as well as other potential targets (Moon, Lee, Choi, Kim and Cho; 2007).

9.5. Carbonic Anhydrase Inhibition

Gallic acid is a weak carbonic anhydrase inhibitor. The carbonic anhydrases (or carbonate dehydratases) form a family of enzymes that catalyze the rapid interconversion of carbon dioxide and water to bicarbonate and protons (or vice versa), a reversible reaction that occurs rather slowly in the absence of a catalyst (Badger and Price, 1994). The active site of most carbonic anhydrases contains a zinc ion; they are therefore classified as metalloenzymes. Carbonic anhydrase inhibitors (CAIs) are a class of pharmaceuticals used as antiglaucoma agents, diuretics, antiepileptics, in the management of mountain sickness, gastric and duodenal ulcers, neurological disorders or osteoporosis (Supuran, Scozzafava and Conway; 2004; Supuran and Scozzafava; 2000; Supuran, Scozzafava and Casini; 2003). CAs, of which 16 isoforms are presently known in mammals, are inhibited by three main mechanisms:

1. coordination of the inhibitor to the Zn(II) ion (critical for catalysis) from the enzyme active site by replacing the zinc-bound water/hydroxide ion and leading to a tetrahedral geometry of Zn(II), or by addition of the inhibitor to the metal coordination sphere, when the Zn(II) ion is in a trigonal bipyramidal geometry
2. by anchoring of the inhibitor to the Zn (II)- bound solvent molecule, that is, a water or hydroxide ion.
3. by occlusion of the entrance to the active site cavity, when the inhibitors bind in the activator binding region from the CA active site

Gallic acid is anchored at the Zn(II)-coordinated water molecule and bind more externally within the active site cavity, making contacts with various amino acid residues (Innocenti, Ozturk Sarıkaya, Gulcin and Supuran; 2010).

9.6. Antioxidant and Pro-oxidant Activity

Antioxidants are compounds capable to either delay or inhibit the oxidation processes which occur under the influence of atmospheric oxygen or reactive oxygen species. They are used for the stabilization of polymeric products, of petrochemicals, foodstuffs, cosmetics and pharmaceuticals. Antioxidants are involved in the defence mechanism of the organism against the pathologies associated with the attack of free radicals. The essential features of oxidation via a free radical-mediated chain reaction are initiation, propagation, branching and termination steps (Pisoschi & Negulescu; 2011). When a hydrogen atom is abstracted from an unsaturated fatty acid forming an alkyl radical, lipid oxidation is initiated. In order to stabilize, the alkyl radical usually undergoes a shift in the position of the double bond (cis to trans) and production of a conjugated diene system. The initiating oxidizing radical can react with O_2 to form a high-energy peroxyl radical. The peroxyl radical can then abstract a hydrogen atom from another unsaturated fatty acid forming a hydroperoxide and a new, free alkyl radical. This process then propagates to another fatty acid. Lipid hydroperoxides are the primary products of lipid oxidation. They are tasteless and odorless; however, in the presence of heat, metal ions, and/or light, they can decompose to compounds that contribute off-odors and offtastes. Alkoxy radicals can also abstract $H^{\cdot}$ from unsaturated fatty acids continuing the chain reaction. Hydroxyl radicals can react with conjugated systems continuing the oxidation process. This chain reaction terminates when two radical species combine to form a nonradical species (Figure 3).

Initiation

$LH + R^{\cdot} \rightarrow L^{\cdot} + RH$

Propagation

$L^{\cdot} + O_2 \rightarrow LOO^{\cdot}$

$LOO^{\cdot} + LH \rightarrow L^{\cdot} + LOOH$

Branching

$LOOH \rightarrow LO^{\cdot} + HO\cdot$

$2\ LOOH \rightarrow LOO^{\cdot} + LO^{\cdot} + H_2O$

Termination

$LO^{\cdot} + LO^{\cdot}$

$L^{\cdot} + AH \rightarrow LH + A^{\cdot}$

$LOO^{\cdot} + LOO^{\cdot}$

$LO^{\cdot} + LOO^{\cdot}$

LH = substrate molecule (for example, a lipid)
R = the initiating oxidizing radical.
$L^{\cdot}$ = allyl radical (highly reactive that can rapidly react with oxygen)
$LOO^{\cdot\cdot}$=lipid peroxyl radical
(LOOH) = lipid hydroperoxides
($LO^{\cdot}$) = alkoxyl radical

Figure 3. Lipid oxidation.

The oxygen-related products can also interact in the presence of suitable transition metal catalysts to form highly toxic hydroxyl radicals and other oxidizing species. As a result, NADH, GSH and ATP are depleted, whereas calcium ion is increased, inducing cell damage and causing diseases such as atherosclerosis, cancer, and ischemia (Gow-Chin, Pin-Der and Hui-Ling; 2002). In human cells exist antioxidant defenses able to remove $O_2\cdot-$ and H_2O_2. Superoxide dismutases remove $O_2 -$ by greatly accelerating its conversion to H_2O_2. Catalases, in peroxisomes, convert H_2O_2 into water and O_2 and help to dispose of H_2O_2 generated by the action of the oxidase enzymes that are located in these organelles. Other important H_2O_2-removing enzymes in human cells are the glutathione peroxidases. When produced in excess, ROS can cause tissue injury. However, tissue injury can itself cause ROS generation (e.g., by causing activation of phagocytes or releasing transition metal ions from damaged cells), which may (or may not, depending on the situation) contribute to a worsening of the injury (Aruoma; 1998). Gallic acid received much attention because of its potent ability to scavenge ROS, such as superoxide anions, hydrogen peroxide, hydroxyl radicals and hypochlorous acid (Hansi and Stanely; 2009; Kim; 2007; Polewsk, Kniat and Slawinska; 2002). Gallic acid is able to inhibit pro-oxidant enzymes activity such us tyrosinase and xanthine oxidase

(Rangkadilok, Sitthimonchai, Worasuttayangkurn, Mahidol, Ruchirawat and Satayavivad; 2007; Leopoldini, Russo and Toscano; 2011). The pro-oxidant action of gallic acid at lower concentrations may be due to its weak metal-chelating effects and its strong electron-donating effects (reducing ability) (Gow-Chin, Pin-Der and Hui-Ling; 2002).

9.7. Anti-inflammatory Effect

Inflammation (Latin, īnflammō, "I ignite, set alight") is part of the complex biological response of vascular tissues to harmful stimuli, such as pathogens, damaged cells, or irritants (Ferrero-Miliani, Nielsen, Andersen and Girardini; 2007), characterized by redness, heat, swelling, loss of function and pain. Redness and heat result from an increase in blood flow, swelling is associated with increased vascular permeability, and pain is the consequence of activation and sensitization of primary afferent nerve fibers. Inflammation can be classified as either acute or chronic. Acute inflammation is the initial response of the body to harmful stimuli and is achieved by the increased movement of plasma and leukocytes (especially granulocytes) from the blood into the injured tissues. A cascade of biochemical events propagates and matures the inflammatory response, involving the local vascular system, the immune system, and various cells within the injured tissue. Prolonged inflammation, known as chronic inflammation, leads to a progressive shift in the type of cells present at the site of inflammation and is characterized by simultaneous destruction and healing of the tissue from the inflammatory process. The lists below show the difference between chronic and acute inflammation regarding the causative agents, which major cells are involved, features regarding onset, duration, and outcomes:

Acute Inflammation

- *Causative agents* - harmful bacteria or injury to tissue
- *Major cells involved* - mainly neutrophils, basophils (in the inflammatory response), and eosinophils (response to parasites and worms), and mononuclear cells (macrophages, monocytes)
- *Primary mediators* - eicosanoids, vasoactive amines
- *Onset* (when does the inflammation start) - straight away
- *Duration* - short lived, only a few days
- *Outcomes* - the inflammation either gets better (resolution), develops into an abscess, or becomes a chronic inflammation

Chronic inflammation

- *Causative agent* - non-degradable pathogens that cause persistent inflammation, infection with some types of viruses, persistent foreign bodies, overactive immune system reactions
- *Major cells involved* - Macrophages, lymphocytes, plasma cells (these three are mononuclear cells), and fibroblasts
- *Primary mediators* - reactive oxygen species, hydrolytic enzymes, IFN-γ and other cytokines, growth factors
- *Duration* - from several months to years

- *Outcomes* - the destruction of tissue, thickening and scarring of connective tissue (fibrosis), death of cells or tissues (necrosis)

Anti-inflammatory activity is important for screening compounds for possible health benefits because inflammation is associated with a vast variety of human diseases, including cancer, heart disease, asthma et al. Several mechanisms of action have been proposed to explain the anti-inflammatory actions of phytoconstituents, it consists broadly in:

(1) Antioxidative and radical scavenging activities;
(2) Modulation of cellular activities of inflammation-related cells (mast cells, macrophages, lymphocytes, and neutrophils);
(3) Modulation of proinflammatory enzyme activities such as phospholipase A2, cyclooxygenase, and lipoxygenase and the nitric oxide producing enzyme, nitric oxide synthase;
(4) Modulation of the production of other proinflammatory molecules;
(5) Modulation of proinflammatory gene expression.

Gallic acid possesses significant anti-inflammatory properties and prevents the expression of inflammatory chemicals including cytokines and histamines. It interferes with the functioning of polymorphonuclear leukocytes, determines the scavenge of superoxide anions, influence the inhibition of myeloperoxidase release and activity, and mediates the assembly of active NADPH-oxidase (Kroes, van den Berg, Quarles van Ufford, van Dijk, H. and Labadie; 1992).

9.8. Anti-allergic Effects

Allergy is considered as a disorder of the immune system in which an exaggerated response occurs when a person is exposed to normally harmless environmental substances, such as animal dander, house dust mites, foods, pollen, insects, and chemical agents. The initial event responsible for the development of allergic diseases is the generation of allergen-specific CD^{4+} T_h2 cells (helper cells). T_h2 cells are a sub-group of lymphocytes, a type of white blood cell, that play an important role in the immune system, particularly in the adaptive immune system. They help the activity of other immune cells by releasing T cell cytokines. They are essential in B cell antibody class switching, in the activation and growth of cytotoxic T cells, and in maximizing bactericidal activity of phagocytes such as macrophages. Mature Th cells express the surface protein CD^4 and are referred to as CD^{4+} T cells. CD^{4+} T cells are generally treated as having a pre-defined role as helper T cells within the immune system. Once generated, effector Th2 cells produce interleukins: IL-4, IL-5, IL-9, and IL-13 which cause the production of allergen-specific IgE by B cells. Subsequently, allergic reactions are induced upon binding of allergen to IgE, which is tethered to the high-affinity IgE receptor (FcεRI) on the surface of mast cells. Following the aggregation of cell-surface receptors are a cascade of intracellular events, including the increase of intracellular Ca^{2+} level, the release of preformed inflammatory mediators from secretory granules such as histamine and β-hexosaminidase, the generation and secretion of the newly synthesized substances such as leukotrienes, prostaglandins, and cytokines. Gallic acid has been determined to be a strong inhibitor against histamine release and pro-inflammatory cytokine production in mast cells (Thanh-Sang, Dai-Hung and Se-Kwon; 2012).

9.9. Cardiovascular Protective Effect

Gallic acid had been widely investigated for its cardiovascular protective properties which is likely to be related to its antioxidant effects, inhibition of platelet aggregation, its interaction with leukocytes, and anti-dyslipidemic activities. Platelet aggregation and activation is a primary contributor to a variety of atherosclerotic diseases, including coronary artery disease, transplant vasculopathy, and carotid artery disease. Platelets are essential for primary hemostasis and the repair of endothelium, but they also play a key role in the development of acute coronary syndromes and contribute to cerebrovascular events. Platelet activation triggered by inflammation is the critical component of atherothrombosis. In addition, platelets participate in the process of forming and extending atherosclerotic plaques. When activated, platelets coaggregate with circulating leukocytes via P-selectin glycoprotein ligand-1 (PSGL-1) and P-selectin interactions. These interactions trigger autocrine and paracrine activation processes leading to the recruitment of the leukocytes into the vascular wall, which is important in the formation of atherothrombosis. In a large-scale prospective human study, the risk of future cardiovascular events increased with increasing levels of plasma platelet-leukocyte aggregation. Two factors shown to play an indispensable role in arterial thrombogenesis: platelet P-selectin and Ca^{2+}. Platelet P-selectin should be able to form large stable platelet-leukocyte aggregates. Ca^{2+} is a potent stimulus of platelet granule secretion. Recent reports demonstrate that gallic acid, as a structure-like molecule to P-selectin, can inhibit P-selectin-mediated adhesion, moreover pre-incubation of gallic acid with mast cells decreased the intracellular calcium level after provocation. The inhibition of Ca^{2+} release may lead to inhibition of platelet granule secretion and P-selectin expression (Chang, Lee, Tseng, Chang, Chen, Chen, and Li; 2012). Moreover gallic acid, is shown to inhibit pancreatic cholesterol esterase which may enhance control of the bioavailability of dietary cholesterol derived from cholesterol esters and determine the limitation of absorption of free cholesterol into blood circulation. In general, pancreatic cholesterol esterase plays an important role in hydrolyzing dietary cholesterol esters which liberates free cholesterol in the lumen of the small intestine. Furthermore, it enhances the incorporation of cholesterol into mixed micelles and aids transport of free cholesterol to the enterocyte. Therefore, the inhibition of cholesterol esterase is expected to limit the absorption of dietary cholesterol, resulting in reduced cholesterol absorption (Ngamukote, Mäkynen, Thilawech and Adisakwattana; 2011).

9.10. Hepatoprotective Effect

Liver is a vital organ playing a major role in metabolism and excretion of xenobiotics from the body. Liver injury or liver dysfunction is a major health problem that challenges not only health care professionals but also the pharmaceutical industry and drug regulatory agencies. Liver cell injury should be caused by various toxic chemicals, anti-biotic, chemotherapeutic agents, carbon tetrachloride (CCL_4), thioacetamide (TAA), excessive alcohol consumption and microbes. Gallic acid may protect the liver from the harmful effects of free radicals that are formed as a result of various metabolic processes in the body (Mahaboob Khan, Evan Prince, Segu, Pranatharthiharan, Smita, Niharika, Punya, and Jaisy; 2010). Moreover certain animal studies indicate that gallic acid and caffeine have a strong hepatoprotective effect against CCl_4 induced hepatotoxicity. Carbon tetrachloride is a classical hepatotoxicant which causes acute and reversible liver injury characterized by centrilobular necrosis, followed by hepatic regeneration and tissue repair (Orfila, Lepert, Alric, Carrera, Beraud, Vinel and Pipy; 1999).

9.11. Kidney Protective Effect

Administration of gallic acid protects the kidney function from oxytetracycline damage. Nephrotoxicity is one of the major side effects of drug therapy in clinical practice frequently leading to acute renal failure. The main mechanism of nephrotoxicity is vasoconstriction, altered intra glomerular hemodynamics, tubular cell toxicity, interstitial nephritis, crystal deposition, thrombotic microangiopathy and osmotic nephrosis. Oxytetracycline administration in rats determines renal damage (increase levels of serum creatinine, urea and decrease levels of creatinine clearance). It is well documented that tetracycline inhibit the incorporation of amino acid into protein causing an increase in urea level. Administration of gallic acid protects the kidney function from oxytetracycline as indicated by a significant decrease of serum creatinine and restoration of creatinine clearance levels (Balagangadharan; 2012).

9.12. Neuroprotective Properties Effect

Neurodegenerative diseases are thought to be related to the free radical mediated reactions. Parkinson disease is characterized by a selective degeneration of dopaminergic neurons in the substantia nigra pars compacta resulting in a reduction of dopamine level in the striatum. Oxidative stress is implicated, where the oxidation of dopamine generates reactive oxygen species and unbalanced over production of reactive oxygen species induces neuronal damage, ultimately leading to neuronal death via apoptosis or necrosis. Therefore the neuroprotective effect depends on both antioxidant capabilities and hydrophobicity of drugs. The acetyl derivatives of gallic acid derivatives facilitates dopaminergic transmission or decreases cholinergic transmission and hence exhibit antiparkinson activity. None of this compound exhibited any toxic effect on grass behaviour. The penetration into brain indicated with rapid onset of action can considered for future drug development (Kasture, Kasture, Sahu1, Kumar and Musmade; 2012).

9.13. Attenuation of Lead Toxicity

The detection and prevention of lead toxicity have been a major international public health priority. The central nervous system is a primary target for lead toxicity, and exposure to this metal is associated with several neurobehavioral and psychological alterations besides cognitive and motor deficits. Moreover, elevated lead levels cause neurodegeneration leading to impairment of neuromuscular coordination and motor control. The possible molecular mechanism involved in lead toxicity is oxidative stress. Lead toxicity causes oxidative stress by two different pathways:

1. Generation of reactive oxygen species, as singlet oxygen and hydrogen peroxide,
2. Depletion of the antioxidant defences like superoxide dismutase, catalase, glutathione peroxidase and reduced glutathione. In addition, δ-aminolevulinic acid dehydratase – an enzyme of heme biosynthesis that is highly sensitive to lead– is inhibited by low blood lead levels. δ-aminolevulinic acid dehydratase inhibition contributes to reactive oxygen species generation and is used as one of the most reliable indicators of lead intoxication

Currently, the therapeutic approach to lead poisoning is to increase lead excretion using chelating agents (EDTA), 2,3-dimercaptopropanol, and meso 2,3-dimercaptosuccinic acid (DMSA). However, new therapies are needed because most chelating agents exhibit many side effects, such as haematopoietic disorders, impairment of cellular metabolism and of synthesis of DNA, RNA and proteins, as well as an imbalance of trace elements. A new experimental approach to decrease the concentration of metals in the body use chelating agents together with antioxidants. Gallic acid could be considered as a supplementary agent useful in lead poisoning treatment due to its high antioxidant potential, which can add to its relative ability to remove lead from brain (Reckziegela, Tironi Diasc, Benvegnúa, Boufleura, Barcelosa, Segatc, Pasec, dos Santosb, Floresb and Bürger; 2011; Testino; 2008).

9.14. Preventive Role of Gallic Acid on Acute Alcohol Intoxication

High levels of alcohol consumption are associated with an increased risk of alcoholism, malnutrition, chronic pancreatitis, alcoholic liver disease, and cancer. In addition damage to the central nervous system and peripheral nervous system can occur from chronic alcohol abuse (Müller, Koch, von Specht, Völker. and Münch; 1985). Long-term use of alcohol in excessive quantities is capable of damaging nearly every organ and system in the body (Caan and Belleroche; 2002). The developing adolescent brain is particularly vulnerable to the toxic effects of alcohol (Mellon, Simone and Rappaport, 2007). Four distinct pathways for ethanol degradation have been described in the human - three oxidative pathways and one non-oxidative pathway. Each of the oxidative pathways starts with the oxidation of ethanol to acetaldehyde, which is then oxidized to acetate for subsequent extra-hepatic activation to acetyl- CoA. The three oxidative pathways can be differentiated on the basis of the enzyme and the mechanism by which ethanol is oxidized to acetaldehyde. The first pathway utilizes cytoplasmic alcohol dehydrogenase, the second oxidative pathway uses the endoplasmic reticulum Microsomal Ethanol Oxidizing System (MEOS) and the third pathway uses peroxisomal catalase. MEOS is better known as Cytochrome P450 2E1. The nonoxidative pathway for ethanol metabolism is less well characterized but produces fatty acid ethyl esters as primary end products. Acetaldehyde is the first metabolic product of alcohol metabolism regardless of whether oxidation is affected by alcohol dehydrogenase, catalase, or MEOS. It has attracted considerable interest in the past few years and has been held responsible for many, if not most, of the tissue-damaging effects of chronic alcohol abuse. Acetaldehyde modifies the lysine residues of LDL, and the product rapidly disappears from the plasma. Paraoxonase is believed to prevent the oxidation of LDL and to destroy the oxidized LDL to biologically inactive products. Heavy alcohol feeding in rats significantly reduces both serum and liver paraoxonase activities with a corresponding down-regulation of paraoxonase mRNA in the liver (Marks; 1983). Among the multiple-organ damages due to excessive alcohol consumption, alcoholic liver disease (ALD) and alcohol-mediated acute and chronic pancreatic tissue damage are the most common disease. Alcohol causes a wide range of hepatic damages ranging from steatosis to minimal injury to advanced fibrosis or cirrhosis. The etiology of ALD is well known, but its pathogenesis is quite complex. Acetaldehyde, alcohol dehydrogenase, reactive oxygen and nitrogen species, inflammatory mediators, immunity and genetic factors and their interactions appear to play prominent roles in the development of ALD. Gallic acid therapy significantly reduces the increase in serum alanine aminotransferase, aspartate aminotransferase, lactate dehydrogenase activities and rejuvenates arylesterase and paraoxonase activities which are indicators of liver injury after acute ethanol consumption in a dose-dependent way (Kartkaya, Oğlakç, Şentürk, Bayramoğlu, Canbek and Kanbak; 2012). Moreover excessive ethanol consumption is a principal cause of the common human disease pancreatitis. Studies using cultured pancreatic acinar cells and isolated pancreatic acini have reported that pancreas can metabolize ethanol via the oxidative pathway enzymes alcohol dehydrogenase and possibly cytochrome P4502E1 (CYP2E1) and by non-oxidative pathway producing fatty acid ethyl esters (FAEEs). Non oxidative fatty acid metabolites, rather than ethanol itself, are responsible for the marked elevations of [Ca^{2+}] that mediate toxicity in the pancreatic acinar cell and that these compounds act primarily by releasing Ca^{2+} from the endoplasmic reticulum. The administration of gallic acid might be due to maintaining lysosomal membrane integrity and protecting lysosomal enzymes, against lipid peroxide mediated radicals generated in the lysosomal membrane by isoproterenol (Prince, Priscilla and Devika; 2009).

9.15. Use of Gallic Acid Esters to Increase Bioavailability of Orally Administered Pharmaceutical Compounds

The cytochromes P450 are a superfamily of hemoproteins that catalyze the metabolism of a large number of xenobiotics and endobiotics. Genetic polymorphisms within Cytochrome P450 (CYP) enzymes mainly affect pharmacokinetics of drugs that are substrates for those enzymes. Those differences in drug metabolism, exposition and elimination lead to differences in drug response as well as to an altered risk for adverse drug reactions. Accumulation of a metabolite which is not further metabolized because of lacking enzyme activity might lead to differential pharmacological effects or even to toxic events. Cytochrome P450 enzymes, however, do not only play a role in drug metabolism. Also physiological substrates are metabolized by those enzymes as for example most steroid hormones or arachidonic acid, and genetic polymorphisms also play a role in the biotransformation of those. This is the reason why genetic polymorphisms within CYP enzymes also have been shown to be associated with risk for certain diseases like hypertension and cancer (Kirchheiner and Seeringer; 2007). The CYP3A locus includes all the known members of the 3A subfamily of the cytochrome P450 superfamily of genes. A growing number of preclinical and clinical studies have demonstrated that the oral bioavailability of many CYP3A and/or P-glycoprotein substrate drugs can be increased by concomitant administration of CYP3A inhibitors and/or P-glycoprotein inhibitors. These genes encode monooxygenases which catalyze many reactions involved in drug metabolism and synthesis of cholesterol, steroids and other lipids. Recent patent applications of use of gallic acid esters to increase bioavailability are described. Gallic acid ester (octyl gallate, propyl gallate, lauryl gallate, and methyl gallate) are able to increase bioavailability by inhibition of Cytochrome P450 and CYP3A activity in gut which leads to a total increase in drug bioavailability in the serum (Wacher and Vincent; 2001).

9.16. Supplements

Gallic acid is commonly used in the pharmaceutical industry because many in *vivo* and in *vitro* studies in humans, animals, and cell culture have provided evidence for the following actions of gallic acid. For pharmaceutical purposes, gallic acid is employed as a reducing agent (Ritzer and Sundermann; 1981) and as a homeopathic drug (*Acidum gallicum*) (Singleton; 1981; Keller, Greiner and Stockebrand; 1995). Its supplements are available as capsules, ointments and liquid extracts, and have been used traditionally to treat a variety of ailments. There are not supplements specifically of gallic acid. However, there are plenty of readily available vitamins and nutritional supplements that contain gallic acid and will help to give the benefits associated with gallic acid. Most of these supplements are specifically designed to be rich in antioxidants. Some examples of pill form supplements that include gallic acid are grape seed extract, Rooibos tea extract, Amla and Livotide, among others. Finnaly, supplementation of dietary gallic acid and linoleic acid mixture or their synthetic salt can improve the antioxidative potential of egg and the fatty acid quality of egg yolk while lowering the cholesterol level (Samooel, Byung, Kichang, Dong, Heon and Cheorun; 2011).

9.17. Side Effects of Gallic Acid

The ultimate goal for any toxicological testing is to establish safe levels of the compounds to which humans are exposed. Gallic acid is non-toxic up to a level of 5000 mg/kg body weight, when given orally. Subacute administration of gallic acid at a dose of 1000 mg/kg body weight did not cause any significant measurable toxicological side-effects (Rajalakshmi, Devarajb and Devaraj; 2001). However, oral intake of gallic acid cause toxicity to rabbits with the LD50 of 5g/kg. It was also found that intraperitoneal injection of gallic acid led to liver injury for CD-1 mice with a dosage of 500 mg/kg. Furthermore, in vitro experiments showed some gallic acid toxicities to mouse spermatogonia, spermatocytes, and sertoli cells as well as in normal rat liver epithelial cells. In addition a single oral intake of gallic acid can induce rat metabolic alterations in multiple biological matrices including plasma, liver extract, urine, and feces even when gallic acid dosage is well below no-obvious adverse-effect-levels. These gallic acid -induced changes are probably associated with alterations to liver functions resulting from some mild oxidative stresses, thus pro-oxidative activities. Metabolic changes associated with gallic acid intake include promotion of glycogenolysis, glycolysis, and TCA cycle, some effects on metabolism of nucleotides, choline, bile acids, amino acids, and gut microbiota. Such effects are both dose- and time-dependent. Gallic acid effects on liver metabolism were more severe than on the other matrices and metabolic changes in plasma were only minor at 120 mg/kg gallic acid. Within 80 h after a single oral dose of both 120 and 600 mg/kg, such metabolic effects were completely recoverable for urinary, plasma, and fecal metabonomes. In contrast, gallic acid - induced metabolic changes in liver were not completely recoverable within 144 h (Shi, Xiao, Wang and Tang; 2013). Finally, in animal studies, gallic acid has triggered contractile responses and inhibited the decrease of vascular pressure in the thoracic aorta. Gallic acid has also been shown to interfere with medications taken for high blood pressure (Sanae, Miyaichi and Hayashi; 2002).

REFERENCES

Agarwal, C.; Tyagi, A. & Agarwal R. (2006). Gallic acid causes inactivating phosphorylation of cdc25A/cdc25C-cdc2 via ATM-Chk2 activation, leading to cell cycle arrest, and induces apoptosis in human prostate. *Molecular cancer therapeutics,* 5, 3294-3302.

Aruoma O. I. (1998). Free radicals, oxidative stress, and antioxidants in human health and disease. *Journal of the American Oil Chemists' Society,* 75, 199-212

Badger, M.R. & Price, G.D. (1994). The role of carbonic anhydrase in photosynthesis. *Annual Review of Plant Physiology and Plant Molecular Biology,* 45, 369-392.

Balagangadharan, K. (2012). Efficacy of gallic acid on oxytetracycline induced neprotoxicity in ratsinternational. Journal of Pharmacology and Bio Sciences, 3 (2), B-393-B401.

Barros, A.I.R.N.A.; Nunes, F.N.; Gonçalves, B.; Bennett , R.B. & Silva, A.P. (2011). Effect of cooking on total vitamin C contents and antioxidant activity of sweet chestnuts (Castanea sativa Mill.). Food Chemistry, 128, 165-172.

Caan, W. & Belleroche, J. (2002). *Drink, Drugs and Dependence: From Science to Clinical Practice* (1st ed.). Routledge, 19-20.

Chandramohan, T.R.; Bharat R.D.; Aparnaa, A.; Arunasreea, K.M.; Guptaa, G.; Acharia, C.; Reddya, G.V.; Lakshmipathia, V.; Subramanyamc, A. & Reddanna, P. (2012). Anti-leukemic effects of gallic acid on human leukemia K562 cells: Downregulation of COX-2, inhibition of BCR/ABL kinase and NF-κB inactivation. *Toxicology in Vitro,* 26 (3), 396-405.

Chang, S.S.; Lee, V.S.Y.; Tseng, Y.L.; Chang, K.C. Chen, KB., Chen, Y.L. & Li C.Y. (2012). Gallic acid attenuates platelet activation and platelet-leukocyte aggregation: involving pathways of Akt and GSK3*β*. *Evidence-based complementary and alternative medicine,* [Article ID 683872], Volume 2012.

Changhyun Roh, U.J. & Sung,-K.J. (2012). Screening of anti-obesity agent from herbal mixtures. *Molecules,* 17, 3630-3638.

Chen, Y.-J. & Chang L.-S. (2012) Gallic acid downregulates matrix metalloproteinase-2 (MMP-2) and MMP-9 in human leukemia cells with expressed Bcr/Abl. *Molecular Nutrition & Food Research,* 56 (9), 1398-1412.

Chia, Y.C.; Rajbanshi, R.; Calhoun, C. & Chiu, R.H. (2010). Anti-neoplastic effects of gallic acid, a major component of Toona sinensis leaf extract, on oral squamous carcinoma cells. *Molecules,* 15 (11), 8377-89.

Da Luz M.P.; Ferrari, C.R.; Teixeira Tage B.R.M.; Siqueira J.A.T.; Arroteia, K.F. (2011). *Sk*in clarifying complex comprising gallic acid and linoleic acid, cosmetic or pharmaceutical composition comprising it *Patent Cooperation Treaty (PCT) international application.*

Da-Mi Jung, J. de Ropp, S. and Ebeler S.E. (2000). Study of Interactions between food phenolics and aromatic flavors using one- and two-dimensional ^{1}H NMR Spectroscopy. *Journal of Agricultural and Food Chemistry,* 48 (2), 407-412.

Darshan, S. & Doreswamy, R. (2004). Patented antiinflammatory plant drug development from traditional medicine. *Phytotherapy Research,* 18, 343-357.

Davey, M.W.; Van Montagu, M.; Inze, D., Sanmartin, M., Kanellis, A. & Smirnoff, N. (2000). Plant L-ascorbic acid: chemistry, function, metabolism, bioavailability and effects of processing. *Journal of the Science of Food and Agriculture,* 80, 825-860.

de Vrese, M. & Schrezenmeir, J. (2008). Probiotics, prebiotics, and synbiotics. *Advances in Biochemical Engineering/Biotechnology*, 111, 1-66.

Elvira, G.M.; Chandra, S.; Vinicio, R.M.M. & Wenyi, W. (2006). Catalytic inhibition of human DNA topoisomerase by phenolic compounds in *Ardisia compressa* extracts and their effect on human colon cancer cells. *Food & Chemical Toxicology,* 44, 1191-1203.

Fatima, N; Maqsood, Z.T.; Kazmi, S.A. (2002). Complex formation of vanadium ions with various catechols. *Journal of Chemical Society of Pakistan,* 24, 49-57.

Ferrero-Miliani, L.; Nielsen, O. H.; Andersen, P.S.; Girardini, S. E. (2007). Chronic inflammation: importance of NOD2 and NALP3 in interleukin-1beta generation. *Clinical & Experimental Immunology,* 147 (2), 227-235.

Fiuza, S.M.; Gomes, C.; Teixeira, L.J.; Girao da Cruz, M.T.; Cordeiro, M.N. D.S.; Milhazes, N.; Borges, F. & Marques, M.P.M. (2004). Phenolic acid derivatives with potential anticancer properties–a structure–activity relationship study. Part 1: Methyl, propyl and octyl esters of caffeic and gallic acids. *Bioorganic & Medicinal Chemistry,* 12, 3581-3589.

Friedman, M. & Jurgens, H.S. (2000). Effect of pH on the stability of plant phenolic compounds. *Journal of Agricultural and Food Chemistry,* 48, 2101-2110.

Glick, Z. (1981). Modes of action of gallic acid in suppressing food intake of rats. Journal of Nutrition, 111, 1910-1916.

Gow-Chin, Y.; Pin-Der, D. & Hui-Ling, T. (2002). Antioxidant and pro-oxidant properties of ascorbic acid and gallic acid. *Food Chemistry,* 79, 307-313

Grundhofer, P.; Niemetz, R.; Schilling, G. & Gross G.G. (2001). Biosynthesis and subcellular distribution of hydrolyzable tannins. *Phytochemistry,* 57, (6); 915-27.

Hager, A.S.; Vallons, K. J. R. & Arendt, E. K. (2012). Influence of Gallic acid and Tannic Acid on the Mechanical and Barrier Properties of Wheat Gluten Films. *Journal of Agricultural and Food Chemistry*, 60, 6157-6163.

Hanahan D. & Weinberg R. A. (2000). The Hallmarks of Cancer. *Cell,* 100 (1): 57-70.

Hansi, P.D. & Stanely, P.P. (2009). Cardioprotective effect of gallic acid on cardiac troponin-T, cardiac marker enzymes, lipid peroxidation products and antioxidants in experimentally induced myocardial infarction in Wistar rats. *Chemico-Biological Interactions*, 179, 118–124.

Haslama E. & Caib, Y. (1994). Natural product report Plant Polyphenols (Vegetable Tannins*): Gallic acid Metabolism. *Natural Product Report,* 11, 41-66.

Hassan M.A.M. & Youssef A.M.K. (2012). Effect of toasting and microwaving on gross chemical composition, total phenolics, antioxidant activity and phenolic acids fractionation of white beans flour (Phaseolus vulgaris L.). *World Applied Sciences Journal,* 18 (6), 731-736.

Ho, H.H.; Chang C.S.; Ho, W.C.; Liao S.Y.; Wu, C.H. & Wang, C.J. (2010). Anti-metastasis effects of gallic acid on gastric cancer cells involves inhibition of NF-kappaB activity and downregulation of PI3K/AKT/small GTPase signals. *Food Chemical & Toxicology,* 48 (8-9), 2508-2516.

Hsu, C.L. & Yen, G. C. (2007). Effect of gallic acid on high fat diet-induced dyslipidaemia, hepatosteatosis and oxidative stress in rats. *British Journal of Nutrition,* 98, 727-735.

Hunter, K.J. & Fletcher, J.M. (2002). The antioxidant activity and composition of fresh, frozen, jarred and canned vegetables. *Innovative Food Science & Emerging Technology,* 3, 399-406.

Hurrell, R.F.; Reddy, M. & Cook J.D. (1999). Inhibition of non-haem iron absorption in man by polyphenolic-containing beverages. *British Journal of Nutrition,* 81, 289-295.

Innocenti, A.; Beyza Ozturk Sarıkaya, S. ;Gulcin, I. & Supuran C.T. (2010). Carbonic anhydrase inhibitors. Inhibition of mammalian isoforms I–XIV with a series of natural product polyphenols and phenolic acids. *Bioorganic & Medicinal Chemistry,* 18, 2159-2164.

Jittawan, K. & Siriamornpun, S. (2008). Phenolic contents and antioxidant activities of bitter gourd (Momordica charantia L.) leaf, stem and fruit fraction extracts in vitro. *Food Chemistry,* 110, 881-890.

Kambourakis, S.K.M & Frost, D.J.W. (2000). Synthesis of gallic acid and pyrogallol from glucose: replacing natural product isolation with microbial catalysis. *Journal of American Chemical Society,* 122, 9042-9043.

Kartkaya, K.; Oğlakç, A.; Şentürk, H.; Bayramoğlu, G.; Canbek, M. & Kanbak, G. (2012). Investigation of the possible protective role of gallic acid on paraoxanase and arylesterase activities in livers of rats with acute alcohol intoxication. *Cell Biochemistry & Function,* doi:10.1002/cbf.2874.

Kasture, V.S.; Kasture, S.B.; Sahu1, R.K.; Kumar, A. & Musmade, D.S. (2012). Synthesis and antiparkinson activity of Gallic acid derivatives. *Medicinal Chemistry & Drug Discovery,* 3(2), 134-139.

Kawada, M.; Ohno, Y.; Ri, Yunmo1; Ikoma, T.; Yuugetu, H.; Asai, T.; Watanabe, M.; Yasuda, N.; Akao, S.; Takemura, G.; Minatoguchi, S.; Gotoh, K.; Fujiwara, H. & Fukuda, K. (2001). Anti-tumor effect of gallic acid on LL-2 lung cancer cells transplanted in mice. *Anti-Cancer Drugs,* 12 (10), 847-852.

Keller, K.; Greiner,S. & Stockebrand, P.; 1995. *Homeopathic medicines: materials for evaluation Arzneimittel Govi-Verlag, Eschborn.* DM 198,-. (German)

Kim, J.H.; Campbell, B.C.; Yu, J.; Mahoney, N.; Chan, K L.; Molyneux, R.J.; Bhatnagar, D. & Cleveland, T.E. (2005). Examination of fungal stress response genes using Saccharomyces cerevisiae as a model system: targeting genes affecting aflatoxin biosynthesis by Aspergillus flavus. *Microbiology and Biotechnology,* 67(6), 807-815.

Kim, Y. (2007). Antimelanogenic and antioxidant properties of gallic acid. *Biological & Pharmaceutical Bulletin,* 30, 1052-1055.

Kinjo, J.; Nagao, T.; Tanaka, T.; Nonaka, G.; Okawa, M.; Nohara, T. & Okabe, H. (2002). Activity-guided fractionation of green tea extract with antiproliferative activity against human stomach cancer cells. *Journal of Biological & Pharmaceutical Bullettin,* 25 1238-1240.

Kirchheiner, J. & Seeringer, A. (2007). Clinical implications of pharmacogenetics of cytochrome P450 drug metabolizing enzymes. *Biochimica et Biophysica Acta,* 1770 (3), 489-494.

Kratz Jadel, M.; Andrighetti-Frohner, C.R.; Kolling Deise, J.; Leal, P.C.; Cirne-Santos, C.C.; Yunes Rosendo, A.; Nunes, R.J.; Trybala, E.; Bergstrom, T. & Frugulhetti, I. (2008). Anti-HSV-1 and anti-HIV-1 activity of gallic acid and pentyl gallate. *Memorias do Instituto Oswaldo Cruz,* 103, (5), 437-42.

Kroes, B.H.; van den Berg, A.J.; Quarles van Ufford, H. C.; van Dijk, H. & Labadie, R.P. (1992). Anti-inflammatory activity of gallic acid. *Planta Medica,* 58(6), 499-504.

Lan, H.C.; Charn, L.Y.; Chin, Y.G. & Yin, C.H. (2007). Preventive effects of guava (*Psidium guajava* L.) leaves and its active compounds against α-dicarbonyl compounds-induced blood coagulation. *Food Chemistry,* 103, 528-535.

Lee, H.C.; Jenner, A.M.; Low, C.S & Lee, Y.K. (2006). Effect of tea phenolics and their aromatic fecal bacterial metabolites on intestinal microbiota. *Research in Microbiology,* 157, 876-84.

Leopoldini, M.; Russo, N. & Toscano, M. (2011). The molecular basis of working mechanism of natural polyphenolic antioxidants. *Food Chemistry,* 125, 288-306.

Li, A.S; Bandy, B.; Tsang, S.S.; Davison, A.J. (2000). DNA-Breaking versus DNA-protecting activity of four phenolic compounds in vitro. *Journal of Free Radical Research,* 33, 551-566.

Mahaboob Khan, R.; Evan Prince S.; Segu R.R.; Pranatharthiharan, P.; Smita, P.; Niharika, M., Punya, P.M. & Jaisy S. (2010). Hepatoprotective and antioxidant effects of gallic acid in paracetamol-induced liver damage in mice. *Journal of Pharmacy and Pharmacology,* 62, 638-643.

Marks, V. (1983). Clinical pathology of alcohol. *Journal of Clinical Pathology,* 36, 365-378.

Mellon, R.D.; Simone, A.F. & Rappaport, B.A. (2007). Use of anesthetic agents in neonates and young children. *Anesthesia & Analgesia.* 104 (3), 509-20.

Moon, H.S.; Lee, H.G.; Choi, Y.J.; Kim, T.G. & Cho, C.S. (2007). Proposed mechanisms of (-)-epigallocatechin-3-gallate for anti-obesity. *Chemico-Biological Interactions,* 167 (2), 85-98.

Mosel, H.D. & Herrmann K. (1974). Phenolics of fruits. IV. The phenolics of blackberries and raspberries and their changes during development and ripeness of the fruits. *Z Lebensm Untersuch Forsch, 154,* 324-327.

Müller, D.; Koch, R.D.; von Specht, H.; Völker, W. & Münch , E.M. (1985). Neurophysiologic findings in chronic alcohol abuse. *Psychiatrie, Neurologie und medizinische Psychologie (Leipz),* 37 (3), 129-32.

Ngamukote, S.; Mäkynen, K.; Thilawech, T. & Adisakwattana, S. (2011). Cholesterol-Lowering activity of the major polyphenols in grape seed. *Molecules,* 16, 5054-5061.

Ohno, T.; Inoue, M. & Ogihara, Y. (2001). Cytotoxic activity of gallic acid against liver metastasis of mastocytoma cells P-815. *Anticancer Research,* 21(6A), 3875-3880.

Ohno, Y.; Fukuda, K.; Takemura, G.; Toyota, M; Watanabe, M.; Yasuda, N.; Xinbin, Q.; Maruyama, R.; Akao, S.; Gotou, K.; Fujiwara, T. & Fujiwara, H. (1999). Induction of apoptosis by gallic acid in lung cancer cells. *Anticancer Drugs,* 10 (9). 845-51.

Okabe, N.; Kyoyama, H. & Suzuki, M. J. (2001). 3,4-Dihydroxyphenylacetic acid. *Acta Crystallographica Section*, E57, 764-766.

Orfila, C.; Lepert, J.C.; Alric, L.; Carrera, G.; Beraud, M.; Vinel, J.P. & Pipy, B. (1999). Expression of TNF-alpha and immunohistochemical distribution of hepatic macrophage surface markers in carbon tetrachloride-induced chronic liver injury in rats. *The Histochemical Journal.* 31, 677-685.

Patel, S.S. & Goyal, R.K. (2011). Cardioprotective effects of gallic acid in diabetes-induced myocardial dysfunction in rats. *Pharmacognosy Research,* 3(4), 239-245.

Pisoschi, A.M. & Negulescu G.P. (2011). *Methods for Total Antioxidant Activity Determination: A Review Pisoschi and Negulescu, Biochemestry & Analitical Biochemistry* 1:1 http://dx.doi.org/10.4172/2161-1009.1000106.

Polewski, K.; Kniat, S. & Slawinska D. (2002). Gallic acid, a natural antioxidant, in aqueous and micellar environment: spectroscopic studies. *Current Topics in Biophysics,* 26, 217-227.

Prince, M.S.P.; Priscilla, H. & Devika, P.T. (2009). Gallic acid prevents lysosomal damage in isoproterenol induced cardiotoxicity in wistar rats. *European Journal of Pharmacology.* 139, 139-143.

Pulok, K.; Mukherjee, Niladri, M.; Neelesh K.; Nema, Birendra S. K. (2011). Bioactive compounds from natural resources against skin aging. *Phytomedicine,* 19, 64-73.

Punithavathi, V.R.; Stanely Mainzen Prince, P.; Kumar, M. R. & Selvakumari, C.J. (2011). Protective effects of gallic acid on hepatic lipid peroxide metabolism, glycoprotein components and lipids in streptozotocin-induced type II diabetic wistar rats. *Journal of Biochemical and Molecular Toxicology,* 25, 68-76.

Rajalakshmi, K.; Devarajb, H. & Devaraj, S. N. (2001). Assessment of the no-observed-adverse-effect level (NOAEL) of gallic acid in mice. *Food and Chemical Toxicology,* 39, 919-922.

Rangkadilok, N.; Sitthimonchai, S.; Worasuttayangkurn, L.; Mahidol, C.; Ruchirawat, M. & Satayavivad, J. (2007). Evaluation of free radical scavenging and antityrosinase activities of standardized longan fruit extract. *Food Chemistry and Toxicology,* 45, 328-336.

Reckziegela, P.; Tironi Diasc, V.; Benvegnúa, D.; Boufleura, N; Barcelosa, R. C.S.; Segatc, H.J.; Pasec, C. S.; dos Santosb, C.M.M.; Floresb, E.M.M. & Bürger, M.E. (2011). Locomotor damage and brain oxidative stress induced by lead exposure are attenuated by gallic acid treatment. *Toxicology Letters.* 203 (1), 74-81.

Rehman, Z.U.; Islam, M. & Shah, W.H. (2003). Effect of microwave and conventional cooking on insoluble dietary fiber components of vegetables. *Food Chemistry,* 17, 237-240.

Ritzer, E. & Sundermann, R. (1981), *Ullmann's Encyclopedia of Industrial Chemistry,* A 13, 519-526.

Robichaud, J.L. & Noble, A.C. (1990). Astringency and bitterness of selected phenolics in wine. *Journal of the Science of Food and Agriculture,* 53(3), 343-353.

Rowaida, K. & Khalil S. (2010). Influence of gallic acid and catechin polyphenols on probiotic properties of Streptococcus thermophilus CHCC 3534 strain. *World Journal of Microbiology & Biotechnology,* 26, 2069-2079.

Sagdic, O.; Ozturk, I.; Cankurt, H. & Tornuk, F. (2012). Interaction between some phenolic compounds and probiotic bacterium in functional ice cream production. *Food Bioprocess Technology,* 5, 2964-2971.

Samooel, J.; Byung, H.H.; Kichang, N.; Dong, A.U.; Heon, L. J. & Cheorun, J. (2011). Effect of dietary supplementation of gallic acid and linoleic acid mixture or their synthetic salt on egg quality. *Food Chemistry,* 129, 822-829.

Sanae, F.; Miyaichi, Y. & Hayashi, H. (2002). Potentiation of vasoconstrictor response and inhibition of endothelium-dependent vasorelaxation by gallic acid in rat aorta. *Planta medica,* 68 (8), 690-693.

Schatz, H. (2009). Herbal remedies and other complementary and alternative treatments for diabetes mellitus. *Experimental and Clinical Endocrinology & Diabetes,* 117, 1-2.

Shahidi, F.; Alasalvar, C. & Chandrika M.L.P.C.M. (2007). Antioxidant phytochemicals in Hazelnut Kernel (Corylus avellana L.) and hazelnut byproducts. *Journal Agricultural and Food Chemistry,* 55 (4), 1212-1220.

Shi, X.; Xiao, C.; Wang, Y. & Tang, H. (2013). Gallic acid intake induces alterations to systems metabolism in rats. *Journal of Proteome Research,* 12 (2), 991-1006.

Singleton, V.L (1981). Naturally occurring food toxicants: phenolic substances of plant origin common in foods. *Advances in Food Research,* 27, 149-242.

Siripatrawan, U. & Harte, B.R. (2010). Physical properties and antioxidant activity of an active film from chitosan incorporated with green tea extract. *Food Hydrocolloids,* 24, 770-775.

Supriya, Y.; Snehal, G.; Smita, N. & Vaishali A. (2011). Effect of supplementation of micronutrients and phytochemicals to fructooligosaccharides on growth response of probiotics and E. coli. *BioFactors,* 37, 58-64.

Supuran, C.T.; Scozzafava, A. & Conway J, (2004). *Carbonic anhydrase: its inhibitors and activators.* Boca Raton: CRC Press. New York, London, 1-364.

Supuran, C.T. & Scozzafava, A. (2000). Carbonic anhydrase inhibitors and their therapeutic potential. *Expert Opinion on Therapeutic Patents,* 10 (5), 575-600.

Supuran, C.T.; Scozzafava, A. & Casini, A. (2003). Carbonic anhydrase inhibitors. *Medicinal Reserche Reviews, 23* (2). 146-89.

Tamio, N. & Yozo, T. (1993). *Manufacture of agents for air purification* Jpn. Kokai Tokkyo Koho Jun 29, Patent: 052232301

Takei Y. and Ioku K. (2004). Effect of the addition of salt in a cooking process on the flavonoid content in vegetables. *Urakami Foundation Memoirs* (12), 1-12.

Thanh-Sang, V.; Dai-Hung, N. & Se-Kwon, K. (2012). Gallic acid-grafted chitooligosaccharides suppress antigen-induced allergic reactions in RBL-2H3 mast cells. *European Journal of Pharmaceutical Sciences,* 27, 527-533.

Testino, G. (2008). Alcoholic diseases in hepato-gastroenterology: a point of view. *Hepatogastroenterology,* 55 (82-83), 371-7.

Toma´s-Barbera´n, F.A. & Clifford, M.N. (2000). Dietary hydroxybenzoic acid derivatives – nature, occurrence and dietary burden. *Journal of the Science of Food and Agriculture,* 80, 1024-1032.

Yasuyuki, O.; I-Ching, H.; Hiroyuki, F. & Kazunaga, Y. (2012). Antiobesity effects of chinese black tea (pu-erh tea) extract and gallic acid. *Phytotherapy Research,* 26, 4, 475-481.

Yeh, R. D; Chen, J.C.; Lai, T.Y.; Yang, J.S.; Yu, C.S.; Chiang, J.H.; Lu, C.C.; Yang, S.T.; Yu, C.C.; Chang, S.J.; Lin, H.-Y. and Chung, J.G. (2011). Gallic acid induces G0/G1 phase arrest and apoptosis in human leukemia HL-60 cells through inhibiting Cyclin D and E, and activating mitochondria-dependent pathway. *Anticancer Researche,* 31 (2821-2832.

Wacher, B.L.Z. & Vincent J. (2001). Use of gallic acid esters to increase bioavailability of orally administered pharmaceutical compounds. *Patent Drawings:* 6180666 I.

Weng, C.J. & Yen, G.C. (2012). Chemopreventive effects of dietary phytochemicals against cancer invasionand metastasis: Phenolic acids, monophenol, polyphenol, and their derivatives. *Cancer Treatment Reviews,* 38, 76-87.

Werner, R.A.; Rossmann, A.; Schwarz, C.; Bacher, A.; Schmidt., H.L. & Eisenreich, W. (2004). Biosynthesis of gallic acid in Rhus typhina: discrimination between alternative pathways from natural oxygen isotope abundance. *Phytochemistry,* 65 (20), 2809-13.

Zong, L.; Inoue, M.; Nose, M.; Kojima, K.; Sakaguchi, N.; Isuzugawa, K.; Takeda, T. and Ogihara, Y. (1999). Metabolic fate of gallic acid orally administered to rats. *Biological and Pharmaceutical Bulletin,* 22, 326-329.

In: Handbook on Gallic Acid
Editors: M. A. Thompson and P. B. Collins
ISBN: 978-1-62618-921-8

Chapter 2

Gallic Acid Implications in Health: Multi-Therapeutic and Protective Agent

Antonio Francesko, Margarida M. Fernandes and Tzanko Tzanov [*]
Group of Molecular and Industrial Biotechnology, Department of Chemical Engineering, Universitat Politècnica de Catalunya, Terrassa, Spain

Abstract

Gallic acid is a naturally occurring polyphenol widely distributed in vegetables and fruits ubiquitous in diet. This trihydroxybenzoic acid is most frequently encountered in ester form as a basic structural unit of many health beneficial plant polyphenols. Due to the diverse range of biological effects, gallic acid and its derivatives have been extensively studied for implication in human health. Its well-known potent antioxidant and free radical scavenging properties are conventionally exploited for the therapy of diseases related with oxidative stress, whereas plant polyphenols are generally active against wide range of microbial infections. Moreover, recent experimental and epidemiological studies evidenced an even broader therapeutic potential of gallic acid such as for prevention and treatment of neurodegenerative processes or long-term anti-inflammatory effect for protection against different chronic disorders. In addition, gallic acid elicits chemopreventive effect by inhibiting, retarding and reversing the process of carcinogenesis without harming healthy cells. The present chapter summarises the relevance of gallic acid to human health emphasising the most recent advances in its application as a preventive and therapeutic agent for various diseases. New medical care strategies involving this polyphenol are also reviewed.

1. Introduction

Bioactive compounds extracted from plants have long been used as active agents for medical purposes. Polyphenols in particular have been attracting a great deal of attention due

[*] Corresponding author: tzanko.tzanov@upc.edu.

to their abundance in plants along with their pharmacological properties, such as antioxidant, antimicrobial, antiviral and antitumoral activities [1]. Among them, gallic acid (GA, 3,4,5-trihydroxybenzoic acid, $C_6H_2(OH)_3COOH$), a natural triphenol present as an endogenous product in plants [2, 3], has been increasingly studied for its therapeutic potential (Figure 1). Rich sources of GA are fruits, nuts, red wine and, particularly, tea leaves, where GA is encountered as a free molecule or esterified to glucose, catechins or proanthocyanidins [3-6]. The well-known GA derivatives are those in the form of methylated gallic acid, galloyl conjugates of catechin derivatives and polygalloyl esters of glucose [7]. GA is, therefore, a typical representative of a large family of plant secondary phenolic metabolites [8]. Other GA derivatives can be synthesised by the esterification of the GA carboxylic group and these are widely used as antioxidant supplements in food, pharmaceutical and cosmetic industries due to their superior physicochemical, pharmacokinetic and pharmacodynamic properties. These industries also use a non-derivatized GA as a supplement. As such, GA is often obtained by alkaline or acid hydrolysis of tannins or from hydrolysis of spent broths from *Penicillium glaucum* or *Aspergillus niger.*

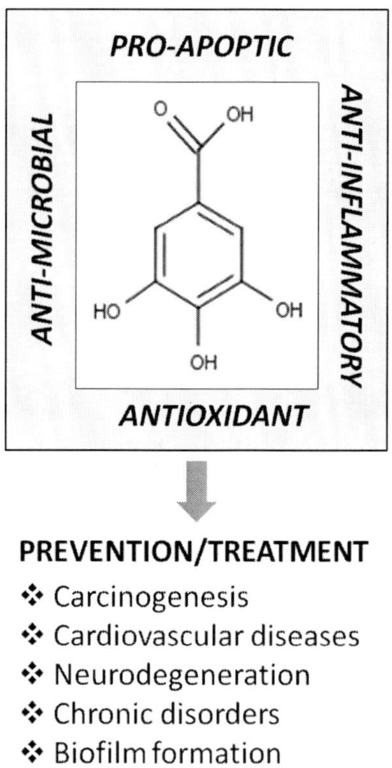

Figure 1. Gallic acid structure, biological effects and implications in health.

Little is known about the absorption and metabolism of this triphenol. Nevertheless, recent studies addressing the bioavailability of polyphenols in humans revealed that GA is considerably better absorbed compared with other phenolic nutraceuticals [9]. Furthermore, 4-*O*-methylgallic acid was found to be the main metabolite of GA in both animals [10, 11] and humans [12], and it was recently shown to be involved in the prevention of chronic inflammatory responses [13]. Large therapeutic potential coupled to high absorption related to the concentration of an active metabolite in target tissues makes GA a promising candidate for many therapeutic applications. Indeed, GA possesses remarkable properties for medical purposes including antitumoral [14], antimicrobial and antiviral [15], as well as being a potent antioxidant, acting as scavengers of reactive oxygen species (ROS) [16]. Logically, GA and its derivatives are lately receiving increased attention for their therapeutic potential, especially if administrated through dietary intake or as nutritional supplements for prevention of many chronic conditions and diseases. This chapter provides a comprehensive overview of the biological effects of GA and related disease preventive and therapeutic potential.

2. Biological Effects of Gallic Acid

Like other natural polyphenols, gallic acid is a subject of increasing scientific interest due to the multiple beneficial effects on human health. Numerous epidemiological studies and clinical trials performed at the end of 20th century strongly suggested that long term consumption of diets rich in plant polyphenols, including gallic acid, offered significant protection against the development of conditions and diseases related to chronic inflammation. The most studied biological activities underlying these preventive effects include antioxidant, anti-inflammatory and anti-microbial properties.

Antioxidant Functions

Reactive oxygen and reactive nitrogen species such as hydroxyl radical ($^{\bullet}OH$), hydrogen peroxide (H_2O_2), superoxide ($O_2^{\bullet-}$), nitric oxide ($NO^{\bullet}$) and peroxynitrite ($ONOO^-$), are the main causative agents of oxidative stress in cells inducing damage of several cellular components including proteins, lipids and DNA [17]. These species are generated in biological systems either as by-products of oxygen reduction or through the xenobiotic catabolic processes. In pharmacology the term "antioxidant activity" is used to describe the capability to prevent the damage of cell components induced by these species.

One of the best studied properties of GA is its antioxidant activity, being capable of intercepting a free radical before it can react with a substrate [18]. Several mechanisms by which GA prevent oxidative stress in cells have been proposed. One of the most accepted is associated to GA scavenging properties of deleterious reactive species [19]. In this radical scavenging mechanism GA "sacrificially" reduces ROS/RNS, preventing damage to biomolecules or formation of more reactive species. The process simply involves an electron donation from GA to the free radical which results in its reduction and consequent inactivation [20]. This property enables GA to effectively limit the breakdown of membrane lipids by preventing their peroxidation by free radicals, in addition preventing DNA damage [21].

The capacity to bind iron is another property of GA that explains its antioxidant activity. Multiple pathways result on the formation of free radicals in biological systems such as the decomposition of peroxynitrous acid [22] or the metal- mediated reduction of peroxides, a common product from the cellular respiration [23] (reaction 1).

$$Fe^{2+} \text{ or } Cu^{+} + H_2O_2 + H^{+} \rightarrow Fe^{3+} \text{ or } Cu^{2+} + {}^{\bullet}OH + H_2O \quad (1)$$

The underlying antioxidant mechanism of GA involves the chelation of metal ions (Fe^{2+} or Cu^{+}) hindering these Fenton-type reactions. In this mechanism, GA prevents the reduction of biological peroxides and, consequently, the radical formation and induced cell damage/death [24]. In fact, GA has been reported to be an efficient chelating agent that forms high stability complexes with iron (III) [25, 26].

Anti-Inflammatory Effects

The molecular mechanisms involved in the anti-inflammatory activity of polyphenols include: i) inhibition of enzymes that mediate inflammatory processes [27], ii) inhibition of ROS generating transcription factors linked to inflammation [28], iii) activation of antioxidant detoxifying enzymes [29], and iv) modulation of cell survival/cell-cycle genes [30]. Some of these mechanisms have been identified in GA anti-inflammatory activity suggesting the multiple sites of action that this phenolic compound possesses. *In vitro* studies indicated that GA interferes with the function of the polymorphonuclear leukocytes by scavenging superoxide anions, thus suppressing the release of oxidative myeloperoxidase (MPO), in addition to inhibiting MPO and NADPH-oxidase activity [31]. Other studies performed in mast cells assigned the anti-inflammatory effects of GA to the inhibition of transcription factors related to pro-inflammatory cytokine expression [32]. A recent work reported the agonist activity of GA at GPR35, a G protein-coupled receptor and the target of anti-inflammatory drugs, suggesting another mechanism of anti-inflammatory action [33].

Anti-Microbial Effect

The antibacterial properties of plant-derived compounds such as polyphenols have been attracting great attention lately, mainly due to the increased antibiotic resistance both in hospital- and community-acquired infections. However, not all the polyphenols are efficient antimicrobials: the antimicrobial activity varies depending on their chemical structure being the flavonoids more efficient than the non-flavonoids [34]. In line with this, flavonoids are being extensively investigated due to their bactericidal effect and the successful synergistic interaction with antibiotics which increase its activity against resistant bacterial strains [35, 36]. As a non-flavonoid, GA induces weaker antimicrobial function. Nevertheless, in several studies GA induced antibacterial action against both Gram-positive (*Staphylococcus aureus* and *Listeria monocytogenes*) and Gram-negative bacteria (*Escherichia coli* and *Pseudomonas aeruginosa*) through the cell enlargement, and cell wall or cell membrane disruption [37-39]. In fact, different extracts of *Caesalpinia mimosoides*, which were found to possess high gallic acid content, inhibited the growth of several pathogenic bacteria [40].

Cell Apoptosis Inducing Effect

Recent research studies have shown that the polyphenolic antioxidants in green tea possess cancer chemopreventive effects [41]. The aromatic nature of polyphenols makes them potential targets of pro-inflammatory oxidants, such as hypochlorous acid (HOCl) and peroxynitrite, associated to the first stages of carcinogenesis. Cell cycle deregulation and overexpression of growth promotion kinases are accepted to be indicators of carcinogenesis [42]. Nuclear factor-kB (NF-kB), a transcription factor that regulates the expression of various genes involved in inflammation and carcinogenesis, is normally inactive in the cytosol. Its degradation by kinases activates the expression of carcinogenic proliferative genes. GA was reported to interfere with this pathway acting as a chemopreventive agent [43]. Moreover, recent studies have shown that GA and its derivatives induce apoptosis in tumoral cell lines and inhibit lymphocyte proliferation [44].

3. Gallic Acid and its Derivatives in Prevention Of Diseases

The prevention of a number of diseases is tightly connected mainly with dietary intake of polyphenols due to the wide range of their biological effects. The regular consumption of these nutraceuticals has been associated with a reduced risk of chronic conditions, including cancer, cardiovascular disease, neurodegenerative disorders, diabetes, gastric injury and others. Phenolic acids, with GA as a typical representative, account for only about one-third of the total daily intake of polyphenolic nutraceuticals in a normal human diet [27]. For this reason the following section exemplifies the preventive effects on various disorders of polyphenols in general, further summarising available *in vitro* and *in vivo* investigations related to GA. The latter is also outlined in Tables 1 and 2.

Chemoprevention

The development of efficient chemopreventive strategies is a priority in public health since the conventional therapeutic approaches are not able to control the incidence of most cancer types. Chemoprevention comprises multiple intervention to prevent, arrest or reverse the carcinogenesis process at various stages. In fact, chemoprevention is a protection from cancer by the administration of pharmacological drugs or naturally occurring constituents of the diet [45]. As the uncontrolled cellular proliferation is a main characteristic of cancers, the application of agents capable to inhibit cancer cell proliferation represents an efficient strategy. In addition, a chemopreventive agent should also intervene in the early process of carcinogenesis, exerting a protection from cancer by enhancing elimination of initiated, pre-malignant cells.

Table 1. Effect of GA, GA-rich extracts and GA derivatives related to disease preventive effects in relevant *in vitro* models

Physiological disorder	Experimental model	Physiological/pathological associated effects	References
Cancer	Human prostate cancer cell lines	Growth inhibition Mitochondria-mediated apoptosis Cell cycle arrest Overexpression of pro-apoptotic proteins	57, 58, 59, 60, 62
	Human leukaemia cell lines	ROS production Elevation of intracellular Ca^{2+} Mitochondrial dysfunction Inhibition of DNA synthesis	63, 64, 65, 66, 68
	Human skin cancer cell lines (oral squamous and melanoma)	Cell apoptosis/necrosis Up-regulation of pro-apoptotic genes	69, 70
	Human cervical cancer cell lines	Cell apoptosis/necrosis ROS production	71
Cardiovascular diseases	Animal and human aortic smooth muscle cell line	Angiogenesis and atherosclerosis inhibition Cell migration inhibition MMP inhibition Downregulation of proteins and vascular endothelial growth factor expression	90, 99
	Human blood platelets	Improved endothelial function Attenuation of platelet activation Inhibition of platelet-leukocyte aggregation	104, 105
Neurodisorders	Liposomes and human bone marrow cell line	Radical scavenging Inhibition of cell apoptosis	133

Table 2. Effect of GA and GA-rich extracts related to disease preventive effects *in vivo*

Physiological disorder	Type of disorder	Experimental model	Effects	References
Cancer	Prostate tumour	Transgenic adenocarcinoma and human xenograft tumour in mice	Inhibition of cancer progression Suppression of cell proliferation Cell cycle arrest and apoptosis Reduction of tumour microvessel density	61, 62
	Leukaemia	Mouse injected with acute leukaemia cells	Delayed tumour incidence Suppressed spleen and liver enlargement Leukocyte activity promotion	67, 68
	Lung non-small carcinoma	Human xenograft tumour in mice	Inhibition of tumour growth Cell cycle arrest and apoptosis	72
	Liver carcinoma	Chemically-induced carcinoma in rat	Suppression of cell proliferation	73
	Colon carcinoma	Chemically-induced carcinoma in rat	Decreased tumour incidence Modulation of xenobiotic-metabolizing enzymes	74
Cardiovascular diseases	Coronary artery disease	Diagnosed patients	Improved endothelial function Improved vasodilation Inhibition of LDL oxidation	79, 98
		Healthy volunteers; Rabbit aorta	Improved endothelial function Improved vasodilation Anti-ischemic activity	85, 97, 103

Table 2. (Continued)

Physiological disorder	Type of disorder	Experimental model	Effects	References
Neurodisorders	Neural damage	Chemically-induced oxidative stress in rats	Neuroprotection Improved memory performance/ cerebral antioxidant potential	135
	Parkinson's disease	Drug-induced catalepsy in mice	Antiparkinson activity Improved dopaminergic transmission	138
	Brain degeneration	Senescence-accelerated mice	Improved memory performance Reversed cognitive impairment	139
	Alzheimer's disease	Transgenic mouse model	Delayed disease onset Decreased amyloid load	140
	Mood disorders	Mice subjected to chronic stress	Anti-depressant Anti-anxiety	141, 142
Diabetes	Type II diabetes	Chemically-induced diabetic rats	Decreased biochemical parameters in blood Inhibition of lipid peroxidation	144, 145
Gastric injury		Drug-induced ulcers in rats	Ulcer-preventive properties Inhibition of *H. pylori* growth Anti-apoptosis of gastric cells Blockage of oxidative damage	147, 148
Chronic kidney disease		Drug/chemically-induced disease in rats	Long-term protective effect Suppressed oxidative stress MMP inhibition	153, 154

Several case-control studies indicated that practicing a polyphenols-rich diet, including GA abundant foods and beverages, may decrease the risk for cancer development [46-48], further demonstrated in different gender and geographical cohort investigations [49-51]. Recent studies on animal models and human cell lines enabled investigation at the molecular level to elucidate the underlying mechanisms of chemoprevention. Nowadays it is known that these compounds exert anti-carcinogenic properties through multiple mechanisms combining: i) removal of carcinogenic agents, ii) modulation of cell cycle signalling by altering the activity of signalling proteins [52], iii) inhibition of regulatory proteins to arrest cell cycle progression [53], iv) induction of apoptosis [54], and v) alteration of the enzymatic activities in the cancer cell [55]. In addition, antioxidant and anti-inflammatory properties of phenolics may account for decreased incidence of cell mutations and cancer development by lowering the oxidative damage of genomic DNA of proliferating cells caused by continuous oxidative stress in the site of inflammation [56].

Both practicing GA-rich diet and GA supplementation are reported to induce chemopreventive effect on many types of cancer cell lines and cancer models in animals. Majority of the reports investigate GA influence on prostate cancer, the most frequently diagnosed malignancy in elderly man in developed countries. For example, two studies identified GA as one of the major anti-cancer constituents of phenolic extracts from different geographical regions and ascribe its cytotoxic activity on prostate cancer cell lines to its anti-proliferative, pro-apoptotic and anti-angiogenic activities [57, 58]. Agarwal et al. [59] further evidenced that GA causes cell cycle arrest and consequent apoptotic death in human prostate carcinoma cells in a dose- and time-dependent manner. A more recent study demonstrated the effect of GA on malignant transformation and prostate cancer development primarily investigating survival rate and mechanisms of cancer cell apoptosis [60]. A strong suppression of cell cycle progression and cell proliferation and an increase in apoptosis in presence of GA have been further confirmed *in vivo* using mouse models [61, 62], supporting the *in vitro* studies and providing a rationale for future preclinical trials with GA for its efficacy against prostate cancer. Regarding haematological malignancies, GA and its derivatives are found to induce Ca^{2+} dependent apoptosis in leukaemia cells [63]. Additional biochemical targets for chemotherapy in human premyelocytic leukaemia were identified in mitochondrial and rate limiting enzymes of *de novo* DNA synthesis [64-66]. Later *in vivo* studies supported these findings by the GA-induced concentration dependent apoptosis of leukaemia cells in mouse models [67, 68]. Anti-neoplastic effects of GA and apoptosis induction in human skin cancer cells have been also recently demonstrated [69, 70]. However, although the authors envision the possible modes of action, in depth *in vivo* investigations are necessary to elucidate the exact mechanisms of GA-induced chemopreventive effect on skin cancer development. The GA capacity to inhibit the initiation of malignant transformations is further exemplified in studies conducted *in vitro* on cervical cancer cells [71] and lung cancer cells, *ex vivo* on lung xenograft tumor [72], and *in vivo* on primary liver cancer [73] and colon carcinogenesis [74].

Prevention of Cardiovascular Diseases

Cardiovascular diseases remain one of the principal causes of death in both developed and developing countries. Growing evidence indicates that majority of cardiovascular conditions share the common etiological base related to the presence of oxidative stress [75]. For example, chronic and acute overproduction of ROS under pathophysiologic conditions is integral in the development of atherosclerotic lesions, the most common cardiovascular condition [76].

Prevention of cardiovascular diseases through dietary intervention has recently received an increasing interest, and nutraceuticals are nowadays recognised as important preventive and therapeutic agents. Larrosa et al. [77] reviewed a considerable amount of *in vitro*, *in vivo* (animal models) and human intervention studies on phenolic nutraceuticals associated with cardiovascular health benefits. The authors also observe discordance between *in vitro* and *in vivo* results on the cardiovascular disease prevention by phenolic compounds. For example, while the GA-rich pomegranate juice extracts showed relevant activity *in vitro* towards molecular mechanisms underlying vascular health benefits, the low bioavalability of phenolics led to the much scarcer evidences on animal models or humans. Furthermore, the latter studies also lack in proofs that the phenolic compounds solely account for strong cardio-protective effects, since most are carried out using foods and beverages that in addition contain sugars, organic acids and other compounds that might be partially responsible for the observed effects. Another review pays attention to the variable effect of phenolics in the context of cardiovascular protection, especially if obtained from different sources [1].

However, in spite of inconsistency in the existing data, in general there is a growing body of evidences supporting the short- and long-term benefits of phenolics-rich food and beverages consumption with regards to decreased risk for development of cardiovascular conditions [78-85]. These compounds, including GA and its derivatives, exert the anti-atherogenic, anti-thrombotic, anti-inflammatory and anti-angiogenic activity in the vascular system. Traditionally, these effects were only attributed to their antioxidant activity, combining direct ROS scavenging activity and transient metal chelation. However, recent studies showed that cardiovascular effects of these compounds were not caused only by the direct antioxidant activity [75]. Other potentially preventive effects on the development of atherosclerotic lesions involve: i) inhibition of platelet aggregation [86, 87] and low density lipoprotein oxidation [88, 89], ii) reduced inflammatory responses by direct inhibition of endothelial enzymes [90] iii) vasodilation [80, 91-93], and iv) improved endothelial function [94-96].

Red wine, rich in GA and related compounds, showed multiple cardiovascular disease prevention properties by improving endothelial function [79, 97, 98] and vasodilatation [85]. In addition, wine extracts prevent the formation of the atherosclerotic plaque by inhibiting the release of the vascular endothelium growth factor from human aortic smooth muscle cells [99]. This potent growth factor is an inducer of angiogenesis, an important process during atherosclerosis [100]. Another GA-rich source, pomegranate juice, has shown significant anti-atherosclerotic and anti-hypertensive activities in both human subjects and mouse models [101]. GA and derivatives from mulberry extract are shown to impede the migration of rat thoracic smooth muscle cells, in addition inhibiting matrix metalloproteinases (MMPs) activity [90]. Proteins of the MMP family are involved in the breakdown of extracellular matrix and remodelling of the vascular wall. However, their elevated expression is associated to vascular smooth cells migration and thus linked to the atherosclerotic processes [102]. *Ex vivo* studies on isolated rabbit aortic rings demonstrated endothelium-dependent vasorelaxation in the presence of extracts containing GA and derivatives [103]. Grape extracts have also been reported to exert anti-thrombotic effects by inhibiting platelets aggregation [104]. The anti-thrombotic effect of solely GA was also confirmed in a recent study revealing GA dose-dependent attenuation of platelet activation and platelet-leukocyte aggregation [105]. Overall, single GA, its derivatives and related components present in the plant extracts show a strong preventive effect on the development of atherosclerotic lesions and cardiovascular diseases.

Protection from Neurodegeneration

Many studies pointed out an important role of oxidative stress with induced neuroinflammation in pathogenesis of neurodegenerative diseases. Neurodisorders, such as Parkinson's and Alzheimer's diseases, represent an emerging problem in developed societies due to their increased prevalence with age [106, 107]. In most of these several markers of oxidative stress including MPO are increased in brain tissue and traces of lipid peroxidation are found in cerebral cortex and brain neurons [108]. Furthermore, evidences show that many neurodegenerative conditions are also tightly connected with the lower antioxidant capacity in the organism [109, 110].

Thus, the investigation on the dietary prevention of the neurodegenerative diseases has started on the basis of the antioxidative potential of phenolic nutraceuticals. Indeed, epidemiological studies linked the regular dietary intake of phenolics-rich food and beverages with preservation of cognitive performance with aging [111, 112], reduced risk of dementia [113], and the delay in onset of developing Alzheimer's [114] and Parkinson's disease [115]. *In vivo* animal studies suggested that consuming food supplemented with phenolic compounds reverse age-related deficits in spatial working memory [116-118] and improve object recognition memory [119]. Randomised control trials on humans further confirmed improvement of mood, memory and motor skills of older adults with consumption of these compounds [120-123]. It is very likely that phenolic nutraceuticals account for at least some portion of these neuropreventive effects since some are reported to successfully permeate the brain barrier in relevant *in vitro* models [124] and various *in vivo* studies regarding fast access and bioavailability of these compounds in animal brain [125-127].

Phenolic compounds preserve brain function via: i) protection of vulnerable neurons against oxidative stress and neurotoxicity [128, 129], ii) stimulation of neuronal regeneration [130], and iii) interactions with neuronal signalling pathways to interfere with the cell death mechanisms [131]. Moreover, they also enhance the expression of detoxifying enzymes enhancing the overall antioxidant capacity in the brain [132]. However, the neurorescue capacity of phenolic compounds does not only depend on their antioxidant potential but also on their structure and molecular conformation. For example, the neurorpotective capacity of GA derivatives increases with their hydrophobicity, related to ability to cross through the lipidic membrane of neuroblasts and induce the effect [133]. This effect was also observed in the treatment of other diseases, with GA analogues obtained by the esterification of its carboxylic group with substituents containing eight or more carbons, widely used as antioxidant supplements in food, pharmaceutical and cosmetic industries. Such chemical changes in the GA molecule are reported to modify their physicochemical, pharmacokinetic and pharmacodynamic properties, improving its antiviral, antifungal, antioxidant and anticancer activities [16]. The improvement of these biological activities have been correlated to the hydrophobic moiety to increase the transfer through lipid bilayers in cell membranes [134].

In animal models GA and its derivative epigallocatechin gallate (EGCG) have been shown to exert neuroprotection via enhancement of cerebral antioxidant defence [135] and neurorescue against neurotoxic inducer of Parkinson's disease [136]. EGCG may also protect neurons by activating signalling enzymes fundamental for cell survival. Its ability to chelate iron contributes to the overall antioxidant defence of the body by preventing redox-active transition metal from catalysing free radicals formation in the onset of the Parkinson's disease [137]. On the base of these studies many other GA derivatives were found to exert antiparkinson activity [138] and the neuroprotective effect in general.

The beneficial effects of GA-rich sources, tea and pomegranate, on psychomotor activity in older animals have also been reported [139, 140]. Showing the beneficial effects in reversing neuronal and behavioural aging was of particular importance in those studies. Finally, GA effects have been recently reported in animal models subjected to stress, suggesting GA therapeutic potential for treatment of depression and anxiety [141, 142].

Prevention from Developing Other Chronic Conditions

Besides of the above described pathological conditions, there is a significant amount of evidences showing that dietary intake or other means of GA administration may exert protective effects against other diseases, especially the ones related with chronic inflammation. Diabetes is a chronic metabolic disorder characterised with hyperglycaemia and associated with a relative or absolute deficiency in the insulin secretion. Oxidative stress in diabetes coexists with a reduction in the antioxidant status, which can increase the deleterious effects of free radicals damaging the pancreas and altering the insulin secretion. Supplementation with free radical scavengers and antioxidants may protect pancreas [143]. Two recent studies evaluated the protective effects of GA on laboratory rats with chemically induced diabetes [144, 145]. It has been found that GA induces antihyperglycaemic, antilipid peroxidative and antioxidant effects suggesting its benefits in treatment of diabetic patients.

Long term protective effects against other chronic diseases and conditions are reported, especially using animal models with chemically-induced tissue injuries. For example, it is known that GA induces hepatoprotective activity through inhibition of lipid peroxidation and triglycerides [146]. GA also inhibits apoptosis in gastric mucosal cells and prevents gastric injury [147, 148]. Gastric ulcers are associated with the use of non-steroidal anti-inflammatory drugs for the treatment of pain, rheumatic disorders, and osteoarthritis. Long term use of these drugs promotes ROS production and induces gastric mucosal apoptosis, which is thought to be the main cause of gastropathy [149, 150]. The oxidative stress in stomach is mainly induced by the generation of $^{\bullet}OH$ through the release of free iron. GA thus prevents gastrophaty mainly due to its iron-chelating properties [151], but also antioxidant and antimicrobial activity against *Helicobacter pylori*, pathogenic bacteria existing in human stomach and known to cause ulcers [152]. Furthermore, GA showed a nephroprotective effect in a model study showing that daily administration of GA suppresses a chemically-induced oxidative stress in rat kidney [153]. In fact, due to its superior renal protective properties without inducing nephron damage, GA is the most recommended phenolic acid in long term treatment of chronic kidney disease [154].

Prevention of Biofilm Formation

Biofilm is a complex community of bacterial cells that commonly grows on indwelling medical devices and provokes difficult-to-treat infections due to its innate resistance to the immune system and antibiotics [155-157]. For example, patients with contaminated urinary catheters need longer hospitalisations and possibly surgeries. If biofilms are not treated properly, in a worse scenario, the severe infections may even lead to death [158, 159]. Antimicrobial activity of polyhenols may account for their potential to prevent biofilm formation. For example, cranberry polyphenolic extracts were shown to inhibit the initial adhesion of *Escherichia coli* and *Enterococcus faecalis* and thereby biofilm formation on various biomaterials [160], while the same extract was found to efficiently suppress the adherence of the specific uropathogenic isolates of P-fimbriated *Escherichia coli* [161]. Purified GA promoted a reduction to above 70 % of *Escherichia coli, Pseudomonas aeruginosa, Staphylococcus aureus* and *Listeria monocytogenes* biofilms together with the inhibition of the microorganisms motility [162].

The biofilm growth is now believed to be regulated by a quorum-sensing (QS) system, a unique mechanism of communication that bacterial cells use through the secretion and uptake of small hormone-like molecules, called autoinducers [163]. Quenching of the QS signals by interfering with their "command language" is an emerging alternative to the relatively ineffective conventional biofilm control strategies. The first evidences that GA derivatives, such as EGCG, interfere with the bacterial quorum-sensing were attained by *Huber et al.* [164]. Since then, studies regarding the use of polyphenols as quorum quenching agents have been rapidly increasing [165, 166].

4. Advanced Treatment Strategies Involving Gallic Acid

A substantial amount of evidences from human, animal, and cell culture studies have shown a potential of phenolic nutraceuticals for prevention of numerous diseases. These compounds were found to exert antiproliferative and cytotoxic effects towards tumour cells, thus blocking or delaying the progression of malignancies in animal and *in vitro* models. They are also toxicity-specific with pro-apoptotic ability towards malignant and pre-malignant, but not normal cells. Relative to the way of administration, it is clear that the circulating concentrations of phenolics after e.g. intravenous injection are much higher than those achieved after dietary intake, and therefore the expected effect as well. In line with this, almost exclusively discrepancies are observed between *in vitro* and *in vivo* experimental findings on chemopreventive effect of phenolics, usually due to the low bioavailability of the food antioxidants after dietary intake (with the exception of their bioavailability in gastrointestinal tract) [6]. Thus, a clear distinction between the cancer treatment at pharmacological doses and cancer prevention at dietary levels should be made when discussing the experimental results. One of the most effective strategies for cancer treatment involves the pharmacological administration of drugs to individuals with higher risk of cancer development and already affected patients. In the case of phenolic compounds the advantage of pharmacological administration is the use of purified agents in much higher concentrations than those achieved by intake, i.e. in doses that *in vivo* are not available to the targeted cells, tissues and organs. Phenolic acids, such as GA, possess significant inhibitory effects on cell proliferation and induced apoptosis in a series of cancer cell lines, showing the selective cytotoxicity against tumour cells with higher sensitivity than normal cells [167]. Gallic acid also inhibits cancer cell migration by suppressing the expression or inhibiting multiple proteins related to metastasis [168], such as MMPs [169]. *In vivo* studies documented that administration of GA for twenty consecutive weeks to the mice with prostate tumour partially suppressed cancer cell cycle progression and proliferation, and increased apoptosis [61].

An ideal chemotherapeutic agent should combine: i) multiple mechanisms of chemoprevention to increase the effectiveness, ii) selectivity for recognition of damaged or transformed cells, and iii) significant bioavailability in the target region. It is hard to believe that there exists a single agent combining such multifunctionality. In general, regardless of the drug administrated, a single-agent intervention has failed to produce the expected outcome for the respective cancer treatments in clinical trials. Without exception, the intervention based on phenolic nutraceuticals is often insufficient to control the progress of most types of malignancies [170]. Thus, there is an urgent need to develop novel, advanced treatment strategies in order to achieve this goal.

One strategy combining conventional approaches with administration of phenolic compounds recently indicated a progress in this field. For example, GA synergises with doxorubicin, a drug used in cancer chemotherapy, to more effectively suppress the growth of a prostate cancer [57]. Another example is the administration of protective agents prior to or during the cancer radiation treatment [171]. The radiation is non-selective, affecting both cancer and normal cells during the treatment. The damage induced on the normal cells is irreparable and frequently physicians are forced to discontinue the treatment. In such situations, an agent that may selectively protect the normal cells from the deleterious effects of radiation (radioprotector), and thus render a therapeutic potential is highly beneficial [48]. GA has shown such potential by inhibiting the radiation-induced damage to DNA and lipid peroxidation both *in vitro* and *in vivo* [21]. Furthermore, phenolics administration can prevent the radiation-induced decline in antioxidant enzyme levels [172]. These findings suggest that radioprotection is mediated by synergistic mechanisms related to antioxidant activity.

On the other hand, polymer-drug combinations have become promising tools to improve therapeutic strategies. The efficacy of small chemotherapeutic molecules may be limited by poor pharmacokinetics and undesired toxicity with side effects. Numerous chemotherapeutic drugs used today have a relatively low therapeutic index. Their conjugation to suitable polymers, by e.g. grafting using a radical initiator system, can greatly enhance therapeutic efficacy by altering their bioavailability [173]. Using this approach GA has been combined with gelatine into a conjugate suitable for anticancer treatment strategies [174, 175]. After covalent immobilisation onto the biopolymer, GA retained its functional properties and the obtained conjugate demonstrated the potential anticancer activity for the treatment of prostate and renal carcinoma. The same approach could be used for grafting nutraceuticals onto other polymers [176] to expand the potential applications. Likewise, the multifunctional devices containing phenolic compounds can be used for the treatment of many other chronic diseases. Our group recently reported functionalisation of several biopolymers and blends thereof with phenolic compounds, including GA, targeting controlled attenuation of the overexpressed chronic wound enzymes, MPO and MMPs [177-179]. By controlling the activities of these enzymes it is possible to suppress the persistent inflammation, restore a protease/antiprotease imbalance and avoid the excessive breakdown of the extracellular matrix in the ulcers.

Depending on the functionalisation approach used, e.g. enzymatically-assisted covalent immobilisation or simple incorporation, phenolic compounds can act from the platform or be released into the wound environment in controllable manner. GA in the functional materials retained its functional properties including antioxidant and antibacterial activity, in addition to enzymatic inhibition. This approach is advantageous comparing to the topical application of active agents that might lead to undesired side effects due to the accumulation of immunoreactive compounds at the wound site, i.e. overdoses.

Conclusion

During the last decades a significant amount of data has emerged with regards to the relevance of gallic acid to human health. Biological effects that this compound exerts account for suppressing of various pathogenic processes related to chronic disease progression. Along with this, reasonable understandings of the GA bioavailability and the mechanisms by which it engage multiple health benefits *in vivo* are now available. However, there are no evidences to claim undisputed therapeutic effects of GA itself for already progressed diseases, especially in the treatment of cancer. Initial attempts are currently being made to increase the efficiency of the existing strategies and expand the application potential of the polyphenols in general, through the combination with other pharmacological drugs or other suitable molecules. Meanwhile, a more comprehensive view on the molecular events underlying biological activities of GA upon all bioentities, including healthy cells, is essential for evaluation of the overall impact of both prevention and treatment strategies. For example, risks inherent to the possible prooxidant activities in the pharmaceutical doses are still unanswered. Additional investigations are, therefore, required before GA can be routinely adopted as a pharmacological agent in clinical practice.

References

[1] Vauzour, D; Rodriguez-Mateos, A; Corona, G; Oruna-Concha, MJ; Spencer, JPE, Polyphenols and Human Health: Prevention of Disease and Mechanisms of Action. *Nutrients*, 2010 2, 1106-1131.

[2] Shahrzad, S; Bitsch, I, Determination of some pharmacologically active phenolic acids in juices by high-performance liquid chromatography. *Journal of Chromatography A*, 1996 741, 223-231.

[3] Kuhr, S; Engelhardt, UH, Determination of flavanols, theogallin, gallic acid and caffeine in tea using HPLC ; Bestimmung von Flavanolen, Theogallin. *Zeitschrift für Lebensmittel-Untersuchung und Forschung*, 1991 192, 526-529.

[4] Gu, L; Kelm, MA; Hammerstone, JF; Beecher, G; Holden, J; Haytowitz, D; Prior, RL, Screening of Foods Containing Proanthocyanidins and Their Structural Characterization Using LC-MS/MS and Thiolytic Degradation. *Journal of Agricultural and Food Chemistry*, 2003 51, 7513-7521.

[5] Shahrzad, S; Bitsch, I, Determination of gallic acid and its metabolites in human plasma and urine by high-performance liquid chromatography. *Journal of chromatography B, Biomedical sciences and applications*, 1998 705, 87-95.

[6] Landete, JM, Ellagitannins, ellagic acid and their derived metabolites: A review about source, metabolism, functions and health. *Food Research International*, 2011 44, 1150-1160.

[7] Tang, HR; Covington, AD; Hancock, RA, Structure–activity relationships in the hydrophobic interactions of polyphenols with cellulose and collagen. *Biopolymers*, 2003 70, 403-413.

[8] Locatelli, C; Filippin-Monteiro, FB; Creczynski-Pasa, TB, Alkyl esters of gallic acid as anticancer agents: A review. *European Journal of Medicinal Chemistry*, 2013 60, 233-239.

[9] Manach, C; Williamson, G; Morand, C; Scalbert, A; Rémésy, C, Bioavailability and bioefficacy of polyphenols in humans. I. Review of 97 bioavailability studies. *The American Journal of Clinical Nutrition*, 2005 81, 230S-242S.

[10] Zong, L; Inoue, M; Nose, M; Kojima, K; Sakaguchi, N; Isuzugawa, K; Takeda, T; Ogihara, Y, Metabolic fate of gallic acid orally administered to rats. *Biological & Pharmaceutical Bulletin*, 1999 22, 326-329.

[11] Potter, DK; Fuller, HL, Metabolic fate of dietary tannins in chickens. *The Journal of nutrition*, 1968 96, 187-191.

[12] Shahrzad, S; Bitsch, I, Determination of gallic acid and its metabolites in human plasma and urine by high-performance liquid chromatography. *Journal of Chromatography B: Biomedical Sciences and Applications*, 1998 705, 87-95.

[13] Lee, G; Na, H-J; Namkoong, S; Jeong Kwon, H; Han, S; Ha, K-S; Kwon, Y-G; Lee, H; Kim, Y-M, 4-O-methylgallic acid down-regulates endothelial adhesion molecule expression by inhibiting NF-κB-DNA-binding activity. *European Journal of Pharmacology*, 2006 551, 143-151.

[14] Inoue, M; Suzuki, R; Sakaguchi, N; Li, Z; Takeda, T; Ogihara, Y; Jiang, BY; Chen, Y, Selective induction of cell death in cancer cells by gallic acid. *Biological & Pharmaceutical Bulletin*, 1995 18, 1526-1530.

[15] Ow, Y-Y; Stupans, I, Gallic acid and gallic acid derivatives: effects on drug metabolizing enzymes. *Current drug metabolism*, 2003 4, 241-248.

[16] Savi, LA; Leal, PC; Vieira, TO; Rosso, R; Nunes, RJ; Yunes, RA; Creczynski-Pasa, TB; Barardi, CRM; Simões, CMO, Evaluation of anti-herpetic and antioxidant activities, and cytotoxic and genotoxic effects of synthetic alkyl-esters of gallic acid. *Arzneimittel-Forschung*, 2005 55, 66-75.

[17] Orrenius, S; Gogvadze, V; Zhivotovsky, B, Mitochondrial Oxidative Stress: Implications for Cell Death. *Annual Review of Pharmacology and Toxicology*, 2007 47, 143-183.

[18] Halliwell, B, Antioxidants in human health and disease. *Annual review of nutrition*, 1996 16, 33-50.

[19] Perron, N; Brumaghim, J, A Review of the Antioxidant Mechanisms of Polyphenol Compounds Related to Iron Binding. *Cell Biochemistry and Biophysics*, 2009 53, 75-100.

[20] Pozharitskaya, ON; Ivanova, SA; Shikov, AN; Makarov, VG; Galambosi, B, Separation and evaluation of free radical-scavenging activity of phenol components of green,

brown, and black leaves of Bergenia crassifolia by using HPTLC-DPPH* method. *Journal of Separation Science*, 2007 30, 2447-2451.
[21] Gandhi, N; Nair, C, Protection of DNA and membrane from gamma radiation induced damage by gallic acid. *Molecular and Cellular Biochemistry*, 2005 278, 111-117.
[22] Beckman, JS; Beckman, TW; Chen, J; Marshall, PA; Freeman, BA, Apparent hydroxyl radical production by peroxynitrite: implications for endothelial injury from nitric oxide and superoxide. *Proceedings of the National Academy of Sciences of the United States of America*, 1990 87, 1620-1624.
[23] Pryor, WA, Reactive oxygen species in biological systems: An interdisciplinary approach. *Free Radical Biology and Medicine*, 2005 39, 426.
[24] Flora, SJS, Structural, chemical and biological aspects of antioxidants for strategies against metal and metalloid exposure. *Oxidative medicine and cellular longevity*, 2009 2, 191-206.
[25] Perron, NR; Wang, HC; Deguire, SN; Jenkins, M; Lawson, M; Brumaghim, JL, Kinetics of iron oxidation upon polyphenol binding. *Dalton transactions*, 2010 39, 9982-9987.
[26] Sroka, Z; Rzadkowska-Bodalska, H; Mazol, I, Antioxidative effect of extracts from Erodium cicutarium L. *Zeitschrift für Naturforschung. C, Journal of biosciences*, 1994 49, 881-884.
[27] Fresco, P; Borges, F; Diniz, C; Marques, MPM, New insights on the anticancer properties of dietary polyphenols. *Medicinal Research Reviews*, 2006 26, 747-766.
[28] Kundu, JK; Surh, Y-J, Molecular basis of chemoprevention by resveratrol: NF-κB and AP-1 as potential targets. *Mutation Research/Fundamental and Molecular Mechanisms of Mutagenesis*, 2004 555, 65-80.
[29] Yu, R; Jiao, JJ; Duh, JL; Gudchithlu, K; Tan, TH; Kong, AN, Activation of mitogen-activated protein kinases by green tea polyphenols: potential signaling pathways in the regulation of antioxidant-responsive element-mediated phase II enzyme gene expression. *Carcinogenesis*, 1997 18, 451-456.
[30] Kim, HP; Son, KH; Chang, HW; Kang, SS, Anti-inflammatory Plant Flavonoids and Cellular Action Mechanisms. *Journal of Pharmacological Sciences*, 2004 96, 229-245.
[31] Kroes, BH; van den Berg, AJ; Quarles van Ufford, HC; van Dijk, H; Labadie, RP, Anti-inflammatory activity of gallic acid. *Planta medica*, 1992 58, 499-504.
[32] Kim, S-H; Jun, C-D; Suk, K; Choi, B-J; Lim, H; Park, S; Lee, SH; Shin, H-Y; Kim, D-K; Shin, T-Y, Gallic Acid Inhibits Histamine Release and Pro-inflammatory Cytokine Production in Mast Cells. *Toxicological Sciences*, 2006 91, 123-131.
[33] Deng, H; Fang, Y, Anti-inflammatory gallic Acid and wedelolactone are G protein-coupled receptor-35 agonists. *Pharmacology*, 89, 211-219.
[34] Daglia, M, Polyphenols as antimicrobial agents. *Current Opinion in Biotechnology*, 2012 23, 174-181.
[35] Eumkeb, G; Sakdarat, S; Siriwong, S, Reversing β-lactam antibiotic resistance of Staphylococcus aureus with galangin from Alpinia officinarum Hance and synergism with ceftazidime. *Phytomedicine*, 2010 18, 40-45.
[36] D Arrigo, M; Ginestra, G; Mandalari, G; Furneri, PM; Bisignano, G, Synergism and postantibiotic effect of tobramycin and Melaleuca alternifolia (tea tree) oil against Staphylococcus aureus and Escherichia coli. *Phytomedicine*, 2010 17, 317-322.

[37] Saavedra, MJ; Borges, A; Dias, C; Aires, A; Bennett, RN; Rosa, ES; Simoes, M, Antimicrobial activity of phenolics and glucosinolate hydrolysis products and their synergy with streptomycin against pathogenic bacteria. *Medicinal Chemistry*, 2010 6, 174-183.
[38] Bisignano, C; Filocamo, A; Faulks, RM; Mandalari, G, *In vitro* antimicrobial activity of pistachio (Pistacia vera L.) polyphenols. *FEMS Microbiology Letters*, 2013 DOI: 10.1111/1574-6968.12091.
[39] Al-Habib, A; Al-Saleh, E; Safer, A-M; Afzal, M, Bactericidal effect of grape seed extract on methicillin-resistant Staphylococcus aureus (MRSA). *The Journal of toxicological sciences*, 2010 35, 357-364.
[40] Chanwitheesuk, A; Teerawutgulrag, A; Kilburn, JD; Rakariyatham, N, Antimicrobial gallic acid from Caesalpinia mimosoides Lamk. *Food Chemistry*, 2007 100, 1044-1048.
[41] Ahmad, N; Mukhtar, H, Green Tea Polyphenols and Cancer: Biologic Mechanisms and Practical Implications. *Nutrition Reviews*, 1999 57, 78-83.
[42] Semczuk, A; Jakowicki, JA, Alterations of pRb1-cyclin D1-cdk4/6-p16^{INK4A} pathway in endometrial carcinogenesis *Cancer Letters*, 2004 203, 1-12.
[43] Fiuza, SM; Gomes, C; Teixeira, LJ; Girão da Cruz, MT; Cordeiro, MNDS; Milhazes, N; Borges, F; Marques, MPM, Phenolic acid derivatives with potential anticancer properties- a structure-activity relationship study. Part 1: Methyl, propyl and octyl esters of caffeic and gallic acids. *Bioorganic & Medicinal Chemistry*, 2004 12, 3581-3589.
[44] Serrano, A; Palacios, C; Roy, G; Cespón, C; Villar, ML; Nocito, M; González-Porqué, P, Derivatives of Gallic Acid Induce Apoptosis in Tumoral Cell Lines and Inhibit Lymphocyte Proliferation. *Archives of Biochemistry and Biophysics*, 1998 350, 49-54.
[45] Morse, MA; Stoner, GD, Cancer chemoprevention: principles and prospects. *Carcinogenesis*, 1993 14, 1737-1746.
[46] Franceschi, S; Parpinel, M; Vecchia, CL; Favero, A; Talamini, R; Negri, E, Role of Different Types of Vegetables and Fruit in the Prevention of Cancer of the Colon, Rectum, and Breast. *Epidemiology*, 1998 9, 338-341.
[47] La Vecchia, C; Chatenoud, L; Franceschi, S; Soler, M; Parazzini, F; Negri, E, Vegetables and fruit and human cancer: Update of an Italian study. *International Journal of Cancer*, 1999 82, 151-152.
[48] Baliga, MS, Anticancer, chemopreventive and radioprotective potential of black plum (Eugenia jambolana lam.). *Asian Pacific journal of cancer prevention*, 2011 12, 3-15.
[49] Feskanich, D; Ziegler, RG; Michaud, DS; Giovannucci, EL; Speizer, FE; Willett, WC; Colditz, GA, Prospective Study of Fruit and Vegetable Consumption and Risk of Lung Cancer Among Men and Women. *Journal of the National Cancer Institute*, 2000 92, 1812-1823.
[50] Zhang, SM; Hunter, DJ; Rosner, BA; Giovannucci, EL; Colditz, GA; Speizer, FE; Willett, WC, Intakes of Fruits, Vegetables, and Related Nutrients and the Risk of Non-Hodgkin's Lymphoma among Women. *Cancer Epidemiology Biomarkers & Prevention*, 2000 9, 477-485.
[51] Benetou, V; Orfanos, P; Lagiou, P; Trichopoulos, D; Boffetta, P; Trichopoulou, A, Vegetables and Fruits in Relation to Cancer Risk: Evidence from the Greek EPIC Cohort Study. *Cancer Epidemiology Biomarkers & Prevention*, 2008 17, 387-392.

[52] Khan, N; Afaq, F; Saleem, M; Ahmad, N; Mukhtar, H, Targeting Multiple Signaling Pathways by Green Tea Polyphenol (−)-Epigallocatechin-3-Gallate. *Cancer Research*, 2006 66, 2500-2505.

[53] Wang, W; Heideman, L; Chung, CS; Pelling, JC; Koehler, KJ; Birt, DF, Cell-Cycle Arrest at G2/M and Growth Inhibition by Apigenin in Human Colon Carcinoma Cell Lines. *Molecular Carcinogenesis*, 2000 28, 102-110.

[54] Mantena, SK; Baliga, MS; Katiyar, SK, Grape seed proanthocyanidins induce apoptosis and inhibit metastasis of highly metastatic breast carcinoma cells. *Carcinogenesis*, 2006 27, 1682-1691.

[55] Ko, C-H; Shen, S-C; Lee, TJF; Chen, Y-C, Myricetin inhibits matrix metalloproteinase 2 protein expression and enzyme activity in colorectal carcinoma cells. *Molecular Cancer Therapeutics*, 2005 4, 281-290.

[56] Multhoff, G; Molls, M; Radons, J, Chronic inflammation in cancer development. *Frontiers in immunology*, 2011 2, 98.

[57] Chen, H-M; Wu, Y-C; Chia, Y-C; Chang, F-R; Hsu, H-K; Hsieh, Y-C; Chen, C-C; Yuan, S-S, Gallic acid, a major component of Toona sinensis leaf extracts, contains a ROS-mediated anti-cancer activity in human prostate cancer cells. *Cancer letters*, 2009 286, 161-171.

[58] Veluri, R; Singh, RP; Liu, Z; Thompson, JA; Agarwal, R; Agarwal, C, Fractionation of grape seed extract and identification of gallic acid as one of the major active constituents causing growth inhibition and apoptotic death of DU145 human prostate carcinoma cells. *Carcinogenesis*, 2006 27, 1445-1453.

[59] Agarwal, C; Tyagi, A; Agarwal, R, Gallic acid causes inactivating phosphorylation of cdc25A/cdc25C-cdc2 via ATM-Chk2 activation, leading to cell cycle arrest, and induces apoptosis in human prostate carcinoma DU145 cells. *Molecular Cancer Therapeutics*, 2006 5, 3294-3302.

[60] Reddivari, L; Vanamala, J; Safe, SH; Miller, JC, The Bioactive Compounds α-Chaconine and Gallic Acid in Potato Extracts Decrease Survival and Induce Apoptosis in LNCaP and PC3 Prostate Cancer Cells. *Nutrition and Cancer*, 2010 62, 601-610.

[61] Raina, K; Rajamanickam, S; Deep, G; Singh, M; Agarwal, R; Agarwal, C, Chemopreventive effects of oral gallic acid feeding on tumor growth and progression in TRAMP mice. *Molecular Cancer Therapeutics*, 2008 7, 1258-1267.

[62] Kaur, M; Velmurugan, B; Rajamanickam, S; Agarwal, R; Agarwal, C, Gallic Acid, an Active Constituent of Grape Seed Extract, Exhibits Anti-proliferative, Pro-apoptotic and Anti-tumorigenic Effects Against Prostate Carcinoma Xenograft Growth in Nude Mice. *Pharmaceutical Research*, 2009 26, 2133-2140.

[63] Sakaguchi, N; Inoue, M; Ogihara, Y, Reactive Oxygen Species and Intracellular Ca^{2+}, Common Signals for Apoptosis Induced by Gallic Acid. *Biochemical Pharmacology*, 1998 55, 1973-1981.

[64] Zhang, W; Hashimoto, K; Yu, GY; Sakagami, H, Decline of superoxide dismutase activity during antioxidant-induced apoptosis in HL-60 cells. *Anticancer research*, 2002 22, 219-224.

[65] Madlener, S; Illmer, C; Horvath, Z; Saiko, P; Losert, A; Herbacek, I; Grusch, M; Elford, HL; Krupitza, G; Bernhaus, A; Fritzer-Szekeres, M; Szekeres, T, Gallic acid inhibits ribonucleotide reductase and cyclooxygenases in human HL-60 promyelocytic leukemia cells. *Cancer letters*, 2007 245, 156-162.

[66] Yeh, R-D; Chen, J-C; Lai, T-Y; Yang, J-S; Yu, C-S; Chiang, J-H; Lu, C-C; Yang, S-T; Yu, C-C; Chang, S-J; Lin, H-Y; Chung, J-G, Gallic Acid Induces G0/G1 Phase Arrest and Apoptosis in Human Leukemia HL-60 Cells through Inhibiting Cyclin D and E, and Activating Mitochondria-dependent Pathway. *Anticancer research*, 2011 31, 2821-2832.

[67] Ho, C-C; Lin, S-Y; Yang, J-S; Liu, K-C; Tang, Y-J; Yang, M-D; Chiang, J-H; Lu, C-C; Wu, C-L; Chiu, T-H; Chung, J-G, Gallic Acid Inhibits Murine Leukemia WEHI-3 Cells In Vivo and Promotes Macrophage Phagocytosis. *In Vivo*, 2009 23, 409-413.

[68] Huang, P-J; Hseu, Y-C; Lee, M-S; Senthil Kumar, KJ; Wu, C-R; Hsu, L-S; Liao, J-W; Cheng, IS; Kuo, Y-T; Huang, S-Y; Yang, H-L, *In vitro* and *in vivo* activity of gallic acid and Toona sinensis leaf extracts against HL-60 human premyelocytic leukemia. *Food and Chemical Toxicology*, 2012 50, 3489-3497.

[69] Chia, Y-C; Rajbanshi, R; Calhoun, C; Chiu, RH, Anti-Neoplastic Effects of Gallic Acid, a Major Component of Toona sinensis Leaf Extract, on Oral Squamous Carcinoma Cells. *Molecules*, 2010 15, 8377-8389.

[70] Lo C, LT, Yang JH, Yang JS, Ma YS, Weng SW, Chen YY, Lin JG, Chung JG, Gallic acid induces apoptosis in A375.S2 human melanoma cells through caspase-dependent and -independent pathways. *Int J Oncol*, 2010 37, 377-385.

[71] You, BR; Moon, HJ; Han, YH; Park, WH, Gallic acid inhibits the growth of HeLa cervical cancer cells via apoptosis and/or necrosis. *Food and Chemical Toxicology*, 2010 48, 1334-1340.

[72] Ji, B-C; Hsu, W-H; Yang, J-S; Hsia, T-C; Lu, C-C; Chiang, J-H; Yang, J-L; Lin, C-H; Lin, J-J; Suen, L-JW; Gibson Wood, W; Chung, J-G, Gallic Acid Induces Apoptosis via Caspase-3 and Mitochondrion-Dependent Pathways in Vitro and Suppresses Lung Xenograft Tumor Growth in Vivo. *Journal of Agricultural and Food Chemistry*, 2009 57, 7596-7604.

[73] Jagan, S; Ramakrishnan, G; Anandakumar, P; Kamaraj, S; Devaki, T, Antiproliferative potential of gallic acid against diethylnitrosamine-induced rat hepatocellular carcinoma. *Molecular and Cellular Biochemistry*, 2008 319, 51-59.

[74] Giftson Senapathy, J; Jayanthi, S; Viswanathan, P; Umadevi, P; Nalini, N, Effect of gallic acid on xenobiotic metabolizing enzymes in 1,2-dimethyl hydrazine induced colon carcinogenesis in Wistar rats – A chemopreventive approach. *Food and Chemical Toxicology*, 2011 49, 887-892.

[75] Mladěnka, P; Zatloukalová, L; Filipský, T; Hrdina, R, Cardiovascular effects of flavonoids are not caused only by direct antioxidant activity. *Free Radical Biology and Medicine*, 2010 49, 963-975.

[76] Singh, U; Jialal, I, Oxidative stress and atherosclerosis. *Pathophysiology*, 2006 13, 129-142.

[77] Larrosa, M; García-Conesa, MT; Espín, JC; Tomás-Barberán, FA, Ellagitannins, ellagic acid and vascular health. *Molecular Aspects of Medicine*, 2010 31, 513-539.

[78] Arts, IC; Hollman, PC, Polyphenols and disease risk in epidemiologic studies. *The American Journal of Clinical Nutrition*, 2005 81, 317S-325S.

[79] Stein, JH; Keevil, JG; Wiebe, DA; Aeschlimann, S; Folts, JD, Purple Grape Juice Improves Endothelial Function and Reduces the Susceptibility of LDL Cholesterol to Oxidation in Patients With Coronary Artery Disease. *Circulation*, 1999 100, 1050-1055.

[80] Grassi, D; Necozione, S; Lippi, C; Croce, G; Valeri, L; Pasqualetti, P; Desideri, G; Blumberg, JB; Ferri, C, Cocoa Reduces Blood Pressure and Insulin Resistance and Improves Endothelium-Dependent Vasodilation in Hypertensives. *Hypertension*, 2005 46, 398-405.

[81] Heiss, C; Kleinbongard, P; Dejam, A; Perré, S; Schroeter, H; Sies, H; Kelm, M, Acute Consumption of Flavanol-Rich Cocoa and the Reversal of Endothelial Dysfunction in Smokers. *Journal of the American College of Cardiology*, 2005 46, 1276-1283.

[82] Engler, MB; Engler, MM; Chen, CY; Malloy, MJ; Browne, A; Chiu, EY; Kwak, H-K; Milbury, P; Paul, SM; Blumberg, J; Mietus-Snyder, ML, Flavonoid-Rich Dark Chocolate Improves Endothelial Function and Increases Plasma Epicatechin Concentrations in Healthy Adults. *Journal of the American College of Nutrition*, 2004 23, 197-204.

[83] Duffy, SJ; Keaney Jr, JF; Holbrook, M; Gokce, N; Swerdloff, PL; Frei, B; Vita, JA, Short- and Long-Term Black Tea Consumption Reverses Endothelial Dysfunction in Patients With Coronary Artery Disease. *Circulation*, 2001 104, 151-156.

[84] Hodgson, JM; Burke, V; Puddey, IB, Acute effects of tea on fasting and postprandial vascular function and blood pressure in humans. *Journal of Hypertension*, 2005 23, 47-54.

[85] Hashimoto, M; Kim, S; Eto, M; Iijima, K; Ako, J; Yoshizumi, M; Akishita, M; Kondo, K; Itakura, H; Hosoda, K; Toba, K; Ouchi, Y, Effect of acute intake of red wine on flow-mediated vasodilatation of the brachial artery. *The American Journal of Cardiology*, 2001 88, 1457-1460.

[86] Keevil, JG; Osman, HE; Reed, JD; Folts, JD, Grape Juice, But Not Orange Juice or Grapefruit Juice, Inhibits Human Platelet Aggregation. *The Journal of Nutrition*, 2000 130, 53-56.

[87] Erlund, I; Koli, R; Alfthan, G; Marniemi, J; Puukka, P; Mustonen, P; Mattila, P; Jula, A, Favorable effects of berry consumption on platelet function, blood pressure, and HDL cholesterol. *The American Journal of Clinical Nutrition*, 2008 87, 323-331.

[88] Priscilla, DHea, Gallic acid pretreatment ameliorates lipids, lipoproteins and HMG-CoA reductase in normal and isoproterenol induced myocardial infarction in male Wistar rats. *Journal of Pharmacy Research*, 2010 3, 960-964.

[89] Mathur, S; Devaraj, S; Grundy, SM; Jialal, I, Cocoa Products Decrease Low Density Lipoprotein Oxidative Susceptibility but Do Not Affect Biomarkers of Inflammation in Humans. *The Journal of Nutrition*, 2002 132, 3663-3667.

[90] Chan, K-C; Ho, H-H; Huang, C-N; Lin, M-C; Chen, H-M; Wang, C-J, Mulberry Leaf Extract Inhibits Vascular Smooth Muscle Cell Migration Involving a Block of Small GTPase and Akt/NF-κB Signals. *Journal of Agricultural and Food Chemistry*, 2009 57, 9147-9153.

[91] Park, YK; Kim, J-S; Kang, M-H, Concord grape juice supplementation reduces blood pressure in Korean hypertensive men: double-blind, placebo controlled intervention trial. *BioFactors (Oxford, England)*, 2004 22, 145-147.

[92] Taubert D; Roesen R.; E., S, Effect of cocoa and tea intake on blood pressure: A meta-analysis. *Archives of Internal Medicine*, 2007 167, 626-634.

[93] Desch, S; Schmidt, J; Kobler, D; Sonnabend, M; Eitel, I; Sareban, M; Rahimi, K; Schuler, G; Thiele, H, Effect of cocoa products on blood pressure: systematic review and meta-analysis. *American journal of hypertension*, 2010 23, 97-103.

[94] Schroeter, H; Heiss, C; Balzer, J; Kleinbongard, P; Keen, CL; Hollenberg, NK; Sies, H; Kwik-Uribe, C; Schmitz, HH; Kelm, M, (–)-Epicatechin mediates beneficial effects of flavanol-rich cocoa on vascular function in humans. *Proceedings of the National Academy of Sciences of the United States of America*, 2006 103, 1024-1029.

[95] Grassi, D; Mulder, TP; Draijer, R; Desideri, G; Molhuizen, HO; Ferri, C, Black tea consumption dose-dependently improves flow-mediated dilation in healthy males. *Journal of Hypertension*, 2009 27, 774-781.

[96] Widlansky, ME; Hamburg, NM; Anter, E; Holbrook, M; Kahn, DF; Elliott, JG; Keaney, JF; Vita, JA, Acute EGCG Supplementation Reverses Endothelial Dysfunction in Patients with Coronary Artery Disease. *Journal of the American College of Nutrition*, 2007 26, 95-102.

[97] Agewall, S; Wright, S; Doughty, RN; Whalley, GA; Duxbury, M; Sharpe, N, Does a glass of red wine improve endothelial function? *European Heart Journal*, 2000 21, 74-78.

[98] Karatzi, K; Papamichael, C; Aznaouridis, K; Karatzis, E; Lekakis, J; Matsouka, C; Boskou, G; Chiou, A; Sitara, M; Feliou, G; Kontoyiannis, D; Zampelas, A; Mavrikakis, M, Constituents of red wine other than alcohol improve endothelial function in patients with coronary artery disease. *Coronary Artery Disease*, 2004 15, 485-490.

[99] Oak, MH; Bedoui, JE; Madeira, SVF; Chalupsky, K; Schini-Kerth, VB, Delphinidin and cyanidin inhibit PDGFAB-induced VEGF release in vascular smooth muscle cells by preventing activation of p38 MAPK and JNK. *British Journal of Pharmacology*, 2006 149, 283-290.

[100] Holm, PW; Slart, RHJA; Zeebregts, CJ; Hillebrands, JL; Tio, RA, Atherosclerotic plaque development and instability: A dual role for VEGF. *Annals of Medicine*, 2009 41, 257-264.

[101] Basu, A; Penugonda, K, Pomegranate juice: a heart-healthy fruit juice. *Nutrition Reviews*, 2009 67, 49-56.

[102] Watanabe, N; Ikeda, U, Matrix metalloproteinases and atherosclerosis. *Current Atherosclerosis Reports*, 2004 6, 112-120.

[103] Beretta, G; Rossoni, G; Santagati, NA; Facino, RM, Anti-ischemic activity and endothelium-dependent vasorelaxant effect of hydrolysable tannins from the leaves of Rhus coriaria (Sumac) in isolated rabbit heart and thoracic aorta. *Planta medica*, 2009 75, 1482-1488.

[104] De Lange, DW; Verhoef, S; Gorter, G; Kraaijenhagen, RJ; Van De Wiel, A; Akkerman, J-WN, Polyphenolic Grape Extract Inhibits Platelet Activation Through PECAM-1: An Explanation for the French Paradox. *Alcoholism: Clinical and Experimental Research*, 2007 31, 1308-1314.

[105] Chang, S-S; Lee, VSY; Tseng, Y-L; Chang, K-C; Chen, K-B; Chen, Y-L; Li, C-Y, Gallic Acid Attenuates Platelet Activation and Platelet-Leukocyte Aggregation: Involving Pathways of Akt and GSK3². *Evidence-Based Complementary and Alternative Medicine*, 2012 2012, 1-8.

[106] Van Den Eeden, SK; Tanner, CM; Bernstein, AL; Fross, RD; Leimpeter, A; Bloch, DA; Nelson, LM, Incidence of Parkinson's Disease: Variation by Age, Gender, and Race/Ethnicity. *American Journal of Epidemiology*, 2003 157, 1015-1022.

[107] Gao S; Hendrie HC; Hall KS; S, H, The relationships between age, sex, and the incidence of dementia and alzheimer disease: A meta-analysis. *Archives of General Psychiatry*, 1998 55, 809-815.
[108] Green, PS; Mendez, AJ; Jacob, JS; Crowley, JR; Growdon, W; Hyman, BT; Heinecke, JW, Neuronal expression of myeloperoxidase is increased in Alzheimer's disease. *Journal of Neurochemistry*, 2004 90, 724-733.
[109] Maes, M; Mihaylova, I; Kubera, M; Uytterhoeven, M; Vrydags, N; Bosmans, E, Increased 8-hydroxy-deoxyguanosine, a marker of oxidative damage to DNA, in major depression and myalgic encephalomyelitis/chronic fatigue syndrome. *Neuroendocrinology Letters*, 2009 30, 715-722.
[110] Chan, A; Rogers, E; Shea, TB, Dietary deficiency in folate and vitamin e under conditions of oxidative stress increases phospho-tau levels: Potentiation by ApoE4 and alleviation by S-adenosylmethionine. *Journal of Alzheimer's Disease*, 2009 17, 483-487.
[111] Letenneur, L; Proust-Lima, C; Le Gouge, A; Dartigues, J; Barberger-Gateau, P, Flavonoid Intake and Cognitive Decline over a 10-Year Period. *American Journal of Epidemiology*, 2007 165, 1364-1371.
[112] Morris, MC; Evans, DA; Tangney, CC; Bienias, JL; Wilson, RS, Associations of vegetable and fruit consumption with age-related cognitive change. *Neurology*, 2006 67, 1370-1376.
[113] Commenges, D; Scotet, V; Renaud, S; Jacqmin-Gadda, H; Barberger-Gateau, P; Dartigues, JF, Intake of flavonoids and risk of dementia. *European Journal of Epidemiology*, 2000 16, 357-363.
[114] Dai, Q; Borenstein, AR; Wu, Y; Jackson, JC; Larson, EB, Fruit and Vegetable Juices and Alzheimer's Disease: The Kame Project. *The American journal of medicine*, 2006 119, 751-759.
[115] Checkoway, H; Powers, K; Smith-Weller, T; Franklin, GM; Longstreth, WT; Swanson, PD, Parkinson's Disease Risks Associated with Cigarette Smoking, Alcohol Consumption, and Caffeine Intake. *American Journal of Epidemiology*, 2002 155, 732-738.
[116] Joseph, JA; Shukitt-Hale, B; Denisova, NA; Prior, RL; Cao, G; Martin, A; Taglialatela, G; Bickford, PC, Long-term dietary strawberry, spinach, or vitamin E supplementation retards the onset of age-related neuronal signal-transduction and cognitive behavioral deficits. *The Journal of neuroscience : the official journal of the Society for Neuroscience*, 1998 18, 8047-8055.
[117] Joseph, JA; Shukitt-Hale, B; Denisova, NA; Bielinski, D; Martin, A; McEwen, JJ; Bickford, PC, Reversals of age-related declines in neuronal signal transduction, cognitive, and motor behavioral deficits with blueberry, spinach, or strawberry dietary supplementation. *The Journal of neuroscience : the official journal of the Society for Neuroscience*, 1999 19, 8114-8121.
[118] Casadesus, G; Shukitt-Hale, B; Stellwagen, HM; Zhu, X; Lee, H-G; Smith, MA; Joseph, JA, Modulation of hippocampal plasticity and cognitive behavior by short-term blueberry supplementation in aged rats. *Nutritional neuroscience*, 2004 7, 309-316.
[119] Goyarzu, P; Malin, DH; Lau, FC; Taglialatela, G; Moon, WD; Jennings, R; Moy, E; Moy, D; Lippold, S; Shukitt-Hale, B; Joseph, JA, Blueberry supplemented diet: effects

on object recognition memory and nuclear factor-kappa B levels in aged rats. *Nutritional neuroscience*, 2004 7, 75-83.

[120] How, PS; Ellis, JA; Spencer, JPE; Williams, C, The impact of plant-derived flavonoids on mood, memory and motor skills in UK adults. *Appetite*, 2008 51, 754.

[121] Macready, AL; Kennedy, OB; Ellis, JA; Williams, CM; Spencer, JPE; Butler, LT, Flavonoids and cognitive function: a review of human randomized controlled trial studies and recommendations for future studies. *Genes & nutrition*, 2009 4, 227-242.

[122] Krikorian, R; Boespflug, EL; Fleck, DE; Stein, AL; Wightman, JD; Shidler, MD; Sadat-Hossieny, S, Concord Grape Juice Supplementation and Neurocognitive Function in Human Aging. *Journal of Agricultural and Food Chemistry*, 2012 60, 5736-5742.

[123] Krikorian, R; Shidler, MD; Nash, TA; Kalt, W; Vinqvist-Tymchuk, MR; Shukitt-Hale, B; Joseph, JA, Blueberry supplementation improves memory in older adults. *Journal of Agricultural and Food Chemistry*, 2010 58, 3996-4000.

[124] Youdim, KA; Qaiser, MZ; Begley, DJ; Rice-Evans, CA; Abbott, NJ, Flavonoid permeability across an in situ model of the blood–brain barrier. *Free Radical Biology and Medicine*, 2004 36, 592-604.

[125] Passamonti, S; Vrhovsek, U; Vanzo, A; Mattivi, F, Fast Access of Some Grape Pigments to the Brain. *Journal of Agricultural and Food Chemistry*, 2005 53, 7029-7034.

[126] Kalt, W; Blumberg, JB; McDonald, JE; Vinqvist-Tymchuk, MR; Fillmore, SAE; Graf, BA; O'Leary, JM; Milbury, PE, Identification of Anthocyanins in the Liver, Eye, and Brain of Blueberry-Fed Pigs. *Journal of Agricultural and Food Chemistry*, 2008 56, 705-712.

[127] Milbury, PE; Kalt, W, Xenobiotic Metabolism and Berry Flavonoid Transport across the Blood−Brain Barrier†. *Journal of Agricultural and Food Chemistry*, 2010 58, 3950-3956.

[128] Wang, Q; Sun, AY; Simonyi, A; Miller, DK; Smith, RE; Luchtefeld, RG; Korthuis, RJ; Sun, GY, Oral administration of grape polyphenol extract ameliorates cerebral ischemia/reperfusion-induced neuronal damage and behavioral deficits in gerbils: comparison of pre- and post-ischemic administration. *The Journal of Nutritional Biochemistry*, 2009 20, 369-377.

[129] Vauzour, D; Corona, G; Spencer, JPE, Caffeic acid, tyrosol and p-coumaric acid are potent inhibitors of 5-S-cysteinyl-dopamine induced neurotoxicity. *Archives of Biochemistry and Biophysics*, 2010 501, 106-111.

[130] Youdim, KA; Joseph, JA, A possible emerging role of phytochemicals in improving age-related neurological dysfunctions: a multiplicity of effects. *Free Radical Biology and Medicine*, 2001 30, 583-594.

[131] Vauzour, D; Vafeiadou, K; Rice-Evans, C; Williams, RJ; Spencer, JPE, Activation of pro-survival Akt and ERK1/2 signalling pathways underlie the anti-apoptotic effects of flavanones in cortical neurons. *Journal of Neurochemistry*, 2007 103, 1355-1367.

[132] Aquilano, K; Baldelli, S; Rotilio, G; Ciriolo, M, Role of Nitric Oxide Synthases in Parkinson's Disease: A Review on the Antioxidant and Anti-inflammatory Activity of Polyphenols. *Neurochemical Research*, 2008 33, 2416-2426.

[133] Lu, Z; Nie, G; Belton, PS; Tang, H; Zhao, B, Structure–activity relationship analysis of antioxidant ability and neuroprotective effect of gallic acid derivatives. *Neurochemistry International*, 2006 48, 263-274.

[134] Saeki, K; Yuo, A; Isemura, M; Abe, I; Seki, T; Noguchi, H, Apoptosis-inducing activity of lipid derivatives of gallic acid. *Biological & Pharmaceutical Bulletin*, 2000 23, 1391-1394.
[135] Mansouri, MT; Farbood, Y; Sameri, MJ; Sarkaki, A; Naghizadeh, B; Rafeirad, M, Neuroprotective effects of oral gallic acid against oxidative stress induced by 6-hydroxydopamine in rats. *Food Chemistry*, 2013 138, 1028-1033.
[136] Levites, Y; Amit, T; Mandel, S; Youdim, MBH, Neuroprotection and neurorescue against Aβtoxicity and PKC-dependent release of non-amyloidogenic soluble precursor protein by green tea polyphenol (-)- epigallocatechin-3-gallate. *The FASEB Journal*, 2003.
[137] Pandey, KB; Rizvi, SI, Plant polyphenols as dietary antioxidants in human health and disease. *Oxidative medicine and cellular longevity*, 2009 2, 270-278.
[138] Kasture V.S., KSB, Kumar Sahu R., Kumar A., Musmade D.S., Synthesis and antiparkinson activity of Gallic acid derivatives. *Medicinal Chemistry & Drug Discovery*, 2012 3, 134-139.
[139] Chan, Y-C; Hosoda, K; Tsai, C-J; Yamamoto, S; Wang, M-F, Favorable effects of tea on reducing the cognitive deficits and brain morphological changes in senescence-accelerated mice. *Journal of nutritional science and vitaminology*, 2006 52, 266-273.
[140] Hartman, RE; Shah, A; Fagan, AM; Schwetye, KE; Parsadanian, M; Schulman, RN; Finn, MB; Holtzman, DM, Pomegranate juice decreases amyloid load and improves behavior in a mouse model of Alzheimer's disease. *Neurobiology of Disease*, 2006 24, 506-515.
[141] Chhillar, R; Dhingra, D, Antidepressant-like activity of gallic acid in mice subjected to unpredictable chronic mild stress. *Fundamental & Clinical Pharmacology*, 2012, no-no.
[142] Dhingra, D; Chhillar, R; Gupta, A, Antianxiety-Like Activity of Gallic Acid in Unstressed and Stressed Mice: Possible Involvement of Nitriergic System. *Neurochemical Research*, 2012 37, 487-494.
[143] Coskun, O; Kanter, M; Korkmaz, A; Oter, S, Quercetin, a flavonoid antioxidant, prevents and protects streptozotocin-induced oxidative stress and β-cell damage in rat pancreas. *Pharmacological Research*, 2005 51, 117-123.
[144] Punithavathi, VR; Stanely Mainzen Prince, P; Kumar, MR; Selvakumari, CJ, Protective effects of gallic acid on hepatic lipid peroxide metabolism, glycoprotein components and lipids in streptozotocin-induced type II diabetic wistar rats. *Journal of Biochemical and Molecular Toxicology*, 2011 25, 68 76.
[145] Punithavathi, VR; Prince, PSM; Kumar, R; Selvakumari, J, Antihyperglycaemic, antilipid peroxidative and antioxidant effects of gallic acid on streptozotocin induced diabetic Wistar rats. *European Journal of Pharmacology*, 2011 650, 465-471.
[146] Anand, KK; Singh, B; Saxena, AK; Chandan, BK; Gupta, VN; Bhardwaj, V, 3,4,5-Trihydroxy benzoic acid (gallic acid), the hepatoprotective principle in the fruits ofterminalia belerica-bioassay guided activity. *Pharmacological Research*, 1997 36, 315-321.
[147] Nanjundaiah, SM; Annaiah, HNM; Dharmesh, SM, Gastroprotective Effect of Ginger Rhizome (*Zingiber officinale*) Extract: Role of Gallic Acid and Cinnamic Acid in H^+, K^+-ATPase/*H. pylori* Inhibition and Anti-Oxidative Mechanism. *Evidence-based complementary and alternative medicine : eCAM*, 2011 2011, 249487.

[148] Pal, C; Bindu, S; Dey, S; Alam, A; Goyal, M; Iqbal, MS; Maity, P; Adhikari, SS; Bandyopadhyay, U, Gallic acid prevents nonsteroidal anti-inflammatory drug-induced gastropathy in rat by blocking oxidative stress and apoptosis. *Free Radical Biology and Medicine*, 2010 49, 258-267.

[149] Wallace, JL, Pathogenesis of NSAID-induced gastroduodenal mucosal injury. *Best practice & research. Clinical gastroenterology*, 2001 15, 691-703.

[150] Maity, P; Bindu, S; Dey, S; Goyal, M; Alam, A; Pal, C; Mitra, K; Bandyopadhyay, U, Indomethacin, a Non-steroidal Anti-inflammatory Drug, Develops Gastropathy by Inducing Reactive Oxygen Species-mediated Mitochondrial Pathology and Associated Apoptosis in Gastric Mucosa: a novel role of mitochondrial aconitase oxidation. *Journal of Biological Chemistry*, 2009 284, 3058-3068.

[151] Fazary, AE; Taha, M; Ju, Y-H, Iron Complexation Studies of Gallic Acid. *Journal of Chemical & Engineering Data*, 2008 54, 35-42.

[152] Kurata, JH; Nogawa, AN, Meta-analysis of Risk Factors for Peptic Ulcer: Nonsteroidal Antiinflammatory Drugs, Helicobacter pylori, and Smoking. *Journal of Clinical Gastroenterology*, 1997 24, 2-17.

[153] Nabavi, S; Habtemariam, S; Nabavi, S; Sureda, A; Daglia, M; Moghaddam, A; Amani, M, Protective effect of gallic acid isolated from Peltiphyllum peltatum against sodium fluoride-induced oxidative stress in rat's kidney. *Molecular and Cellular Biochemistry*, 2013 372, 233-239.

[154] Peng, C-C; Hsieh, C-L; Wang, H-E; Chung, J-Y; Chen, K-C; Peng, RY, Ferulic acid is nephrodamaging while gallic acid is renal protective in long term treatment of chronic kidney disease. *Clinical Nutrition*, 2012 31, 405-414.

[155] Davies, D, Understanding biofilm resistance to antibacterial agents. *Nature Reviews Drug Discovery*, 2003 2, 114-122.

[156] Fux, CA; Costerton, JW; Stewart, PS; Stoodley, P, Survival strategies of infectious biofilms. *Trends in Microbiology*, 2005 13, 34-40.

[157] Vreuls, C; Zocchi, G; Garitte, G; Archambeau, C; Martial, J; Van de Weerdt, C, Biomolecules in multilayer film for antimicrobial and easy-cleaning stainless steel surface applications. *Biofouling*, 2010 26, 645-656.

[158] Gabriel, M; Pukacki, F; Checinski, P; Oszkinis, G; Stanisic, M; Zielinski, M; Pawlaczyk, K, Current options in prosthetic vascular graft infection: comparative analysis of 63 consecutive cases. *Langenbeck's Archives of Surgery*, 2004 389, 272-277.

[159] Hall-Stoodley, L; Costerton, JW; Stoodley, P, Bacterial biofilms: from the Natural environment to infectious diseases. *Nat Rev Micro*, 2004 2, 95-108.

[160] Eydelnant, IA; Tufenkji, N, Cranberry Derived Proanthocyanidins Reduce Bacterial Adhesion to Selected Biomaterials. *Langmuir*, 2008 24, 10273-10281.

[161] Foo, LY; Lu, Y; Howell, AB; Vorsa, N, The structure of cranberry proanthocyanidins which inhibit adherence of uropathogenic P-fimbriated Escherichia coli in vitro. *Phytochemistry*, 2000 54, 173-181.

[162] Borges, A; Saavedra, MJ; Simões, M, The activity of ferulic and gallic acids in biofilm prevention and control of pathogenic bacteria. *Biofouling*, 2012 28, 755-767.

[163] Bassler, BL; Losick, R, Bacterially Speaking. *Cell*, 2006 125, 237-246.

[164] Huber, B; Eberl, L; Feucht, W; Polster, J, Influence of polyphenols on bacterial biofilm formation and quorum-sensing. *Zeitschrift für Naturforschung. C, Journal of biosciences*, 2003 58, 879-884.

[165] Riedel, K; Köthe, M; Kramer, B; Saeb, W; Gotschlich, A; Ammendola, A; Eberl, L, Computer-aided design of agents that inhibit the cep quorum-sensing system of Burkholderia cenocepacia. *Antimicrobial agents and chemotherapy*, 2006 50, 318-323.

[166] Brackman, G; Hillaert, U; Van Calenbergh, S; Nelis, HJ; Coenye, T, Use of quorum sensing inhibitors to interfere with biofilm formation and development in Burkholderia multivorans and Burkholderia cenocepacia. *Research in Microbiology*, 2009 160, 144-151.

[167] Faried, A; Kurnia, D; Faried, LS; Usman, N; Miyazaki, T; Kato, H; Kuwano, H, Anticancer effects of gallic acid isolated from Indonesian herbal medicine, Phaleria macrocarpa (Scheff.) Boerl, on human cancer cell lines. *International Journal of Oncology*, 2007 30, 605-613.

[168] Ho, H-H; Chang, C-S; Ho, W-C; Liao, S-Y; Wu, C-H; Wang, C-J, Anti-metastasis effects of gallic acid on gastric cancer cells involves inhibition of NF-κB activity and downregulation of PI3K/AKT/small GTPase signals. *Food and Chemical Toxicology*, 2010 48, 2508-2516.

[169] Lo, C; Lai, T-Y; Yang, J-S; Yang, J-H; Ma, Y-S; Weng, S-W; Lin, H-Y; Chen, H-Y; Lin, J-G; Chung, J-G, Gallic acid inhibits the migration and invasion of A375.S2 human melanoma cells through the inhibition of matrix metalloproteinase-2 and Ras. *Melanoma Research*, 2011 21, 267-273.

[170] Shukla, Y; George, J, Combinatorial strategies employing nutraceuticals for cancer development. *Annals of the New York Academy of Sciences*, 2011 1229, 162-175.

[171] Hosseinimehr, SJ, Trends in the development of radioprotective agents. *Drug Discovery Today*, 2007 12, 794-805.

[172] Joy, J; Nair, CKK, Protection of DNA and membranes from gamma-radiation induced damages by Centella asiatica. *The Journal of pharmacy and pharmacology*, 2009 61, 941-947.

[173] Geng, J; Mantovani, G; Tao, L; Nicolas, J; Chen, G; Wallis, R; Mitchell, DA; Johnson, BRG; Evans, SD; Haddleton, DM, Site-Directed Conjugation of "Clicked" Glycopolymers To Form Glycoprotein Mimics: Binding to Mammalian Lectin and Induction of Immunological Function. *Journal of the American Chemical Society*, 2007 129, 15156-15163.

[174] Cirillo, G; Kraemer, K; Fuessel, S; Puoci, F; Curcio, M; Spizzirri, UG; Altimari, I; Iemma, F, Biological Activity of a Gallic Acid−Gelatin Conjugate. *Biomacromolecules*, 2010 11, 3309-3315.

[175] Spizzirri, UG; Iemma, F; Puoci, F; Cirillo, G; Curcio, M; Parisi, OI; Picci, N, Synthesis of Antioxidant Polymers by Grafting of Gallic Acid and Catechin on Gelatin. *Biomacromolecules*, 2009 10, 1923-1930.

[176] Curcio, M; Puoci, F; Iemma, F; Parisi, OI; Cirillo, G; Spizzirri, UG; Picci, N, Covalent Insertion of Antioxidant Molecules on Chitosan by a Free Radical Grafting Procedure. *Journal of Agricultural and Food Chemistry*, 2009 57, 5933-5938.

[177] Antonio, F; Guillem, R; Sonia, T; Clara, M; Piergiorgio, G; Valeria, C; Gianluca, C; Tzanov, T, Cross-linked collagen sponges loaded with plant polyphenols with

inhibitory activity towards chronic wound enzymes. *Biotechnology Journal*, 2011 6, 1208-1218.

[178] Francesko, A; Soares da Costa, D; Reis, RL; Pashkuleva, I; Tzanov, T, Functional biopolymer-based matrices for modulation of chronic wound enzyme activities. *Acta Biomaterialia*, 2013 9, 5216-5225.

[179] Rocasalbas, G; Francesko, A; Touriño, S; Fernández-Francos, X; Guebitz, GM; Tzanov, T, Laccase-assisted formation of bioactive chitosan/gelatin hydrogel stabilized with plant polyphenols. *Carbohydrate Polymers*, 2013 92, 989-996.

In: Handbook on Gallic Acid
Editors: M. A. Thompson and P. B. Collins

ISBN: 978-1-62618-921-8

Chapter 3

GALLIC ACID: THERMAL AND ANTIOXIDANT PROPERTIES

***A. M. T. M. Cordeiro*[1], *S. S. Damasceno*[2], *J. G. M. Costa*[2], *R. Rosenhaim*[3], *L. E. B. Soledade*[4], *A. G. Souza*[5] and *N. A. Santos*[6,7]**

[1]Campus Manaus-Zona Leste, Instituto Federal de Educação Ciência e Tecnologia do Amazonas, Manaus, AM, Brazil
[2]Departamento de Química Biológica, CCBS, Universidade Regional do Cariri, Crato, CE, Brazil
[3]Departamento de Engenharia Química, CT, Universidade Federal da Paraíba, João Pessoa, PB, Brazil
[4]Campus de Pinheiro, Universidade Federal do Maranhão, Pinheiro, MA, Brazil
[5]Departamento de Química, CCEN, Universidade Federal da Paraíba, João Pessoa, PB, Brazil
[6]Departamento de Tecnologia Sucroalcooleira, CTDR, Universidade Federal da Paraíba, João Pessoa, PB, Brazil
[7]Instituto UFPB de Desenvolvimento da Paraíba, Universidade Federal da Paraíba, João Pessoa, PB, Brazil

ABSTRACT

Gallic acid belongs to a wide class of naturally occurring phenolic compounds. It is mainly found in the fruits, leaves, twigs, bark and wood of several plants. It presents with a variety of biological functions, displaying anti-inflammatory, cardiovascular and anticarcinogenic properties, in addition to its antioxidant activity. Epidemiological evidence suggests that an antioxidant-rich diet plays an important role in cancer prevention. This antioxidant activity is considered as paramount in the inhibition of free radical-related diseases. The action of phenolic compounds, such as gallic acid, is carried out through the inhibition and/or delayed production of free radicals generated by cellular metabolism or exogenous sources. Gallic acid action aims at preventing the attacks of these free radicals on lipids, amino acids of the proteins, double bonds of the polyunsaturated fatty acids and/or the base-paring of DNA, thus avoiding the formation of lesions and also the loss of cellular integrity. Moreover, gallic acid displays scavenging properties that inhibit reactions of iron and copper via metal chelation. Gallic acid is also widely used as a starting material in the synthesis of drugs in pharmaceutical industries. Due to the widespread application of gallic acid, it is important to identify its main natural sources, determine the antioxidant potential and obtain specific knowledge concerning gallic acid's structural features and chemical properties. These are important benchmarks for the roles of antioxidant, since they evaluate the resistance to thermal decomposition and antioxidative power, inhibiting the free radical generation through donations of protons and/or metal chelation.

1. INTRODUCTION

The formation of free radicals from reactive oxygen species (ROS) and reactive nitrogen species (RNS), among other reactive species, is supposedly the factor for unchaining diseases such as cancer and coronary diseases. Free radicals are formed in physiological conditions in proportions controlled by cell defense mechanisms. However, in pathology situations, the production of free radicals can increase substantially and attack biomolecules, among which lipids, proteins and DNA should be detached (Halliwell, 1997), but these biomolecules can be preserved by the action of antioxidants.

The thesis that plants and their active principles have an important role in the prevention of chronic and degenerative diseases has been increasingly sustained by the scientific point of view. Various and powerful action mechanisms make secondary metabolites, especially phenolic compounds, subjects for the investigation of health-beneficial phytochemicals, as well as sources of natural antioxidants, with the purpose of minimizing the oxidative damage of living cells and to prevent the deterioration of food, thus increasing its shelf life (Lia et al., 2008; Yanishlieva and Marinova, 2001; André et al., 2010; Maganha et al., 2010).

1.1. Biological Effect of Free Radicals

Free radicals are atoms, molecules or ions with unpaired electrons, highly active to chemical reactions with other molecules. In biological systems, free radicals are frequently derived from molecules of oxygen, nitrogen and sulfur (Lü et al., 2010). These reactive oxygen species (ROS) include oxygen radicals, such as superoxide ($O_2\bullet^-$), hydroxyl ($\bullet OH$), hydroperoxyl ($HO_2\bullet$), nitric oxide (NO) and other species such as hydrogen peroxide (H_2O_2), hypochlorous acid (HClO) and singlet oxygen ($^1O_2$) (Bayir, 2005). The reactive nitrogen species (RNS) are derived from nitric oxide, by a reaction with the superoxide radical, in order to form the peroxynitrite radical ($ONOO^-$) (Batinic–Haberle et al., 2010). As for the reactive sulfur species (RSS), they are easily formed from thiols from ROS reactions ROS. These radical species are produced during cellular metabolism and functional activities, playing an important role in cell signaling, apoptosis (via inactivation of mitochondrial cytochrome c), gene expression and ion transport. The ROS provide signals that can trigger mitogen-activated protein kinases and modulate neutrophil adhesion to the substrate, initiating its activation and stimulating hormonal responses. Furthermore, ROS also activate cells of the immune system to remove pathogens (Kohlmeier, 2003).

However, excessive amounts of ROS may lead to harmful consequences. ROS may attack biomolecules, including proteins, lipids and nucleic acid, by reacting with nucleophilic groups (thiols and amines), abstracting hydrogen-atoms from biomolecules, and forming dimers. The interaction of free radicals with cellular components can lead to the formation of secondary radicals derived from proteins, lipids and nucleic acids. These, in turn, react with other cellular macromolecules and thus initiate a chain reaction (Younes, 1999).

1.1.1. Interactions of Free Radicals with Nucleic Acids

Studies indicate that free radicals can damage nucleic acid in two different ways: by the reaction of hydroxyl radicals with the sugar phosphate structure of the nucleic acid, resulting in the chain scission and chromosome breakage, or by oxidation of the sugar moiety (e.g. deoxyribose, in the case of DNA) and different bases. Examples of oxidized nucleobases are: thymine glycol, 5 - hydroxymethyluracil, and 8- hydroxyguanine (Younes, 1999). These oxidative modifications in DNA represent a process of tumor initiation. In addition, oxidative damage implies a cancer etiology involving carcinogens, serving as a biomarker of the oxidative DNA damage (Tarlovsky, 2013). The reactive oxygen species act in all stages of the carcinogenesis (Figure 1), a process characterized by the action of events in series, described in three stages:

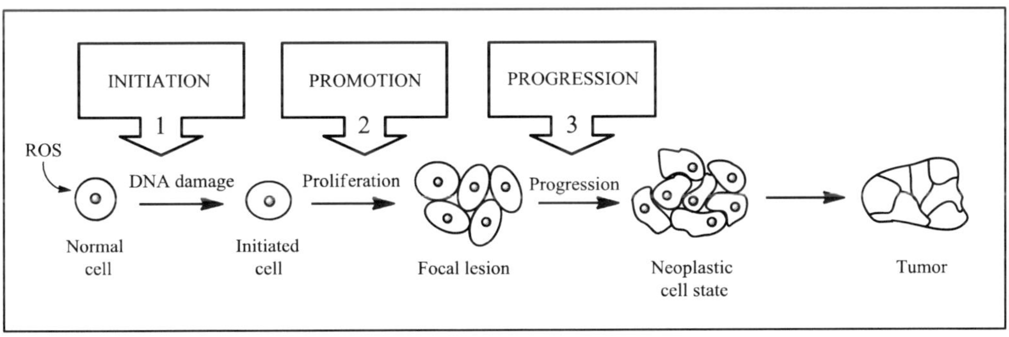

Figure 1. Development of carcinogenesis is described in three steps. (1) initiation, (2) promotion and (3) progression. (1) The initiation phase corresponds to the first step of carcinogenis, when the cell can undergo the action of ROS that provoke DNA damages. (2) Promotion, phase in which the genetically damaged cell can suffer the action of promoting agents that stimulate its multiplication, transforming the initiated cell into a malignant cell. (3) Progression, at that stage, reveals that the altered cells multiply uncontrollably and irreversibly (adapted from Tarlovsky, 2013).

1.1.2. Interactions of the Free Radicals with Polyunsaturated Fatty Acids

The polyunsaturated fatty acids contained in the cell membranes (lipids) make them powerful generators of the free radicals alkoxyl (LO•) and peroxyl (LOO•) by means of lipid peroxidation (Barbosa et al., 2010). Consequently, there is a loss of selectivity in ion exchange and in the release of the contents of organelles, such as hydrolytic enzymes, from the lysosomes and the formation of cytotoxic products (such as malonaldehyde), culminating in cell death.

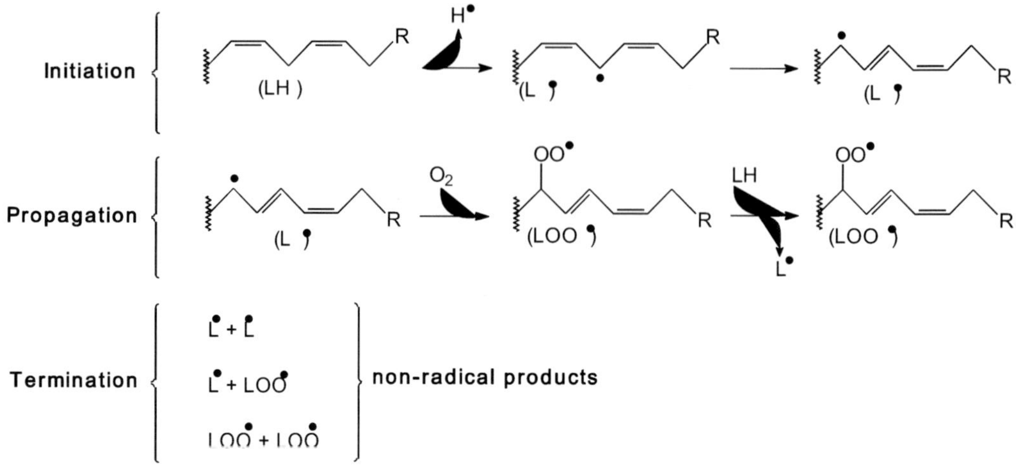

Figure 2. Schematic presentation of lipid oxidation. The peroxidation of the phospholipid membrane is initiated by the abstraction of hydrogen from the allylic C-H bond of polyunsaturated fatty acids, resulting in the formation of lipid alkyl radical (L•). The lipid alkyl radical, added to molecular oxygen, leads to the formation of the lipid peroxyl radical (LOO•), which abstracts a hydrogen atom from the neighboring fatty acid. A lipid hydroperoxide and a new alkyl radical are formed. The chain reaction is maintained and its end takes place by the recombination of radicals (adapted by Schaich, 2005).

Lipid peroxidation may also be associated with the mechanisms of aging, cancer and the exacerbation of the toxicity of xenobiotics (Ferreira and Matsubara, 1997). Such processes constitute a chain reaction, represented by the stages of initiation, propagation and termination (Figure 2) (Barbosa et al., 2010).

Polyunsaturated fatty acids (represented as LH) (Figure 2) generally display double bonds. In the initiation step, the C-H allylic bond is "activated" and is therefore susceptible to oxidation reactions. Subsequently, free radicals are formed by the removal of hydrogen from the allylic/bis-allylic carbon from the lipid molecule (Younes, 1999). The initially formed peroxyl radicals transfer unpaired electrons to neighbor lipid molecules through the abstraction of a hydrogen in an allylic position or from a hydroperoxide. The process is repeated indefinitely, until the chain is intercepted. In the termination step, free radicals form non radical-radical products from their recombination (Schaich, 2005).

The radical •OH, by removing one hydrogen atom from polyunsaturated fatty acids of the cell membrane, plays an important role in lipid peroxidation, and it is considered the major initiator of this process (Welch, 2002). Studies on the mechanisms of oxidative injury have gradually confirmed the catalytic action of metals in the reactions that lead to these injuries (Ferreira, 1997; Won et al., 2012).

The role of metals on the formation of ROS is confirmed by the reactions of Fenton and Haber-Weiss. These oxidative processes are evident in the reaction between Fe2+ and H_2O_2 (Fenton-type reaction, Eq. 1-2), generating the •OH radical with high oxi-reduction potential, indistinctly attacking all the species present in the reaction medium. When Fe3+ (instead of Fe2+) is utilized, other radicals, of smaller oxidation potential than hydroperoxyl (HO_2•), and the superoxide anion (O_2•-) are also formed. The pH of the reaction medium is a determining factor for the proportion at which these radicals are formed, once the protonation of the superoxide anion takes place in the acid medium (Aguiar et al., 2007).

$$Fe^{2+} + H_2O_2 \rightarrow {}^{-}OH + {}^{\bullet}OH + Fe^{3+}$$ (Eq. 1)

$$Fe^{3+} + H_2O_2 \rightarrow Fe^{3+} + HO_2^{\bullet}$$ (Eq.2)

It is suggested that ROS form from a Fenton-type mechanism from traumatic brain injury. The release of intracellular iron, the low cerebrospinal fluid capacity of the iron-protein bond and the deficiency of antioxidant enzymes in the central nervous system extend the risk of injury induced by the trauma (Halliwell, 1992).

On the other hand, the Haber-Weiss reaction (Eq. 3-5) involves the ions (Fe2+/Fe3+) that catalyze the reaction between H_2O_2 and the radical O_2•, both generating the radical •OH (Barbosa et al., 2010; Ferreira and Matsubara, 1997).

$$Fe^{3+} + O_2^{\bullet} \rightarrow Fe^{2+} + O_2$$ (Eq. 3)

$$Fe^{2+} + H_2O_2 \rightarrow Fe^{3+} + {}^{-}OH + {}^{\bullet}OH$$ (Eq. 4)

$$O_2^{\bullet} + H_2O_2 \rightarrow O_2 + {}^{-}OH + {}^{\bullet}OH$$ (Eq. 5)

In vivo experimental data suggests that the syndrome of postischemic reperfusion injury in rat hearts subjected to the overload of Fe3+ may be related to ROS production via the Haber-Weiss reaction. The excess of Fe3+, catalyzing the Haber-Weiss reaction, and of •OH stimulates the lipid peroxidation of the membranes, and is responsible for the decreased myocardial contractility (Ferreira and Matsubara, 1997).

Copper also acts as a catalyst for the H_2O_2 reaction (cuprous Fenton reaction), similarly to iron. Both transition metals react with H_2O_2 yielding intermediate complexes, which, subsequently, decompose to form the radical •OH (Aguiar et al., 2007; Won et al., 2012).

1.1.3. Defense System

Generally, the reducing environment within the cells, being maintained by the action of antioxidant enzymes and substances, helps to prevent the damage caused by free radicals (Halliwel, 1997). This defense system may act in two lines. One of them acts as a detoxifier of the agent before it causes damage. This line comprises reduced glutathione (GSH), superoxide dismutase (SOD), catalase, glutathione peroxidase (GSH-Px) and vitamin E. The other defense line has the function of repairing the lesion, being constituted of ascorbic acid, glutathione reductase (GSH-red) and by GSH-Px, among others. With the exception of vitamin E (α-tocopherol), which is a structural membrane antioxidant, most of the antioxidants are placed in the intracellular medium.

A better understanding of the involvement of oxidative stress in many pathological conditions has occurred in parallel with the development of synthetic antioxidants and the usage of natural compounds with potential antioxidant activity. According to the literature, there is evidence associated with the protective activity of antioxidants in biological systems.

Recently, researchers identified antioxidants, including resveratrol, genistein, and baicalein, with the ability to induce DNA damage (cancer cells), resulting in the cellular death of dividing cancer cells (Fox et al., 2012).

Researchers evaluated the cytotoxic property of gallic acid and its derivatives via cells of human adenocarcinom and theoretical calculations, using the *ab initio* method. According to the results obtained, the molecular structure of the compounds fully correlates with the anticancer activity. The same authors pointed out that the effectiveness of these compounds in biological systems occurs by means of their ability to interact in different environments, lipophilic and hydrophilic (Fiuza et al., 2004).

Dairam and coworkers (2008) reported the importance of the phenolic substituents as chelant agents, scavengers of metallic ions. Fiuza and coworkers (2004) also reported the capacity of the antioxidant gallic acid in causing apoptosis in tumor cell lines and inhibiting lymphocyte proliferation.

1.2. Antioxidants

Antioxidants can be defined as substances that, at small concentrations as compared to the oxidizable substrate, significantly delay or prevent the beginning or the propagation of oxidation chain reactions (Halliwell et al., 1995).

Among several classes of antioxidant substances of natural sources, phenolic compounds, in recent years, have received much attention, mainly as they inhibit lipid peroxidation and *in vitro* lipoxygenase (Soares, 2002). The main sources of phenolic compounds are citrus fruits (lemon, orange and mandarin), cherries, grapes, plums, pears, apples, tomatos and papayas, as well as green peppers, broccoli, red cabbage, onions and garlic (Angelo and Jorge, 2007).

Antioxidant compounds of vegetable origin have demonstrated in *in vitro* experimentsthat they protect against oxidation damages, chiefly by their redox properties, which act in the absorption and neutralization of free radicals, chelating singlet and triplet oxygen, or decompounding peroxides (Zheng and Wang, 2001).

Natural antioxidants occur in all higher plants and in all plant parts, as in husks, stalks, wood, leaves, fruits, roots, flowers, pollen and seeds.

The protection that vegetables provide against diseases is ascribed to the various antioxidants therein contained, as tocopherols, carotenoids and phenolic compounds or polyphenols, including phenolic acids, flavonoids and isoflavones, hydrolyzable tannins, lignins, coumarins and anthocyanins (Zheng and Wang, 2001; Kahkonen et al., 1999). These compounds are secondary metabolites of plants and undergo variations in their contents, depending on the different stages of plant development.

Secondary metabolites are chemically divided into three distinct groups: terpenes, nitrogen-containing compounds and phenolic compounds (Dewick, 2002).

Phenolic compounds are ubiquitous constituents of higher plants and they are in the forms of either free radicals or in sugar and protein complexes. They are found in a wide range of plant foods such as fruits, vegetables, cereals and legumes; and also in beverages of plant origin such as as wine, tea and coffee (Cheynier, 2005; Manach et al., 2004). Phenolic compounds constitute a large and complex group of phytochemicals, with more than 6,000 known structures (Lee et al., 2008; Croft, 1998).

Phenolics are essential to the growth and reproduction of vegetables, in addition to acting as anti-pathogenic agents and contributing to pigmentation, being responsible for the color, astringency, flavor and oxidative stability in food (Angelo and Jorge, 2007).

Table 1. Classes of phenolic compounds in plants

Class	Structure
Simple phenolics, benzoquinones	C6
Hydroxybenzoic acids	C6-C1
Phenylacetic acids, acetophenol	C6-C2
Hydroxycinnamic acids, phenylpropanoids	C6-C3
Naphthoquinones	C6-C4
Xanthones	C6-C1-C6
Stilbenes, anthraquinones	C6-C2-C6
Flavonoids	C6-C3-C6
Lignanes, neolignanes	(C6-C3)2
Lignins	(C6-C3)*n*
Condensed tannins	(C6-C3-C6)*n*

Chemically, the so-called phenolic compounds are substances that display at least one benzene/aromatic ring, in which at least one hydrogen is replaced by a hydroxyl group, thus conferring antioxidant properties to the plants (Soares, 2002).

The structural diversity of polyphenols is due to the great variety of combinations that take place in nature, allowing to group them in different classes according to their basic structure (Table 1), in association with carbohydrates and polymerized forms (Manach et al., 2004; Farah and Donangelo, 2006; Angelo and Jorge, 2007). Polyphenols can be classified according to the type of the main skeleton (C6), which will constitute the benzene ring and the substituent chain (C*n*) (Table 1) (Angelo and Jorge, 2007).

Phenolic compounds can also be classified according to their occurrence in the plant kingdom and are classified as: scarcely distributed (simple phenols), polymers (lignins and tannins) and widely distributed (Ribéreau-Gayon, 1968).

In the widely distributed group are the phenolics found in all the plant kingdom. Among them, flavonoids and phenolic acids should be highlighted (Soares, 2002).

The expression "phenolic acids" generally refers to phenols that display a functional carboxylic acid (Figure 3). According to Soares (2002), they are divided in three groups, in which the first one is composed of benzoic acids and derivatives. Benzoic acid is the most simple found in nature, with seven carbon atoms and the carboxylic group bond to the aromatic ring. In this first group, emphasis should be given to protocatequinic acid, vanillic acid, syringic acid, gentisic acid, salicylic acid and gallic acid as the most common. The second group is formed by the cinnamic acids (derivatives of the hydroxycinnamic acid), which present nine carbon atoms (C6-C3), of which the acids p-coumaric, caffeic, ferulic and sinapic are the most commonly found in the plant kingdom. The third group is composed of coumarins that are derived from cinnamic acid by cyclization of the side chain of o-coumaric acid.

Phenolic acids	Class	R1	R2	R3	R4
Gallic acid	Benzoic	na	na	OH	OH
Protocatequinic acid	Benzoic	na	na	H	OH
Chlorogenic acid	Cinnamic	OH	*	na	na
Vanillic acid	Benzoic	na	na	H	OCH_3
Caffeic acid	Cinnamic	OH	H	na	na
Syringic acid	Benzoic	na	na	OCH_3	OCH_3
p-coumaric acid	Cinnamic	H	H	na	na
Ferulic acid	Cinnamic	OCH_3	H	na	na

Legend: na= not applied; *Quinic acid.

Figure 3. Structure of phenolic acids.

1.2.1. Classification and Action Mechanisms of the Antioxidants

Antioxidants represent a class of chemical substances with varied structures and diverse action mechanisms. They may be present as natural constituents of foods or they may be intentionally added to preserve lipid components from deterioration (Wanasundara and Shahidi, 2005).

According to the action mechanism, the antioxidants may be classified as either primary or secondary.

Primary antioxidants are phenolic compounds that promote the removal or inactivation of the free radicals formed during the initiation or propagation of the reaction, by means of donating hydrogen atoms, thus interrupting the chain reaction. Equations (6-11) illustrate possible events that may take place with primary antioxidants (AH is an antioxidant molecule), according the mechanism of the primary antioxidant activity proposed by Wanasundara and Shahidi (2005). However, these primary antioxidants cannot inhibit photo oxidation or scavenge the singlet oxygen.

The active hydrogen atom from the antioxidant is more easily removed by the free radicals than the allylic hydrogens of the unsaturated molecules. The phenolic derivatives stabilize the free radical by electronic delocalization of the aromatic ring (resonance effect), thereby interrupting the propagation of the oxidative radical reactions in the medium (Litwinienko et al., 1999).

$$AH + R\bullet \rightarrow A\bullet + RH \quad \text{(Eq. 6)}$$

$$AH + ROO\bullet \rightarrow A\bullet + ROOH \quad \text{(Eq. 7)}$$

$$A\bullet + ROO\bullet \rightarrow ROOA \quad \text{(Eq. 8)}$$

$$AH + RO\bullet \rightarrow ROH + A\bullet \quad \text{(Eq. 9)}$$

$$RO\bullet + A\bullet \rightarrow ROA \quad \text{(Eq. 10)}$$

$$A\bullet + A\bullet \rightarrow AA \quad \text{(Eq. 11)}$$

The antioxidant radicals are relatively stable, in such a way that they react with the peroxide radicals (ROO•) in order to interrupt the chain propagation, thus they inhibit the formation of peroxides. Furthermore, the reaction with alkoxyl radicals (RO•) diminish the decomposition of hydroperoxides (considered as harmful products).

Secondary antioxidants delay the oxidation rate by several mechanisms (Wanasundara and Shahidi, 2005). Among these mechanisms are:

- Metal chelating agents - are antioxidants that complex metals and act in a preventive way, once the interaction with radicals does not take place, and they inhibit the peroxide decomposition. They complex metal ions, mainly copper and iron, which catalyze the lipid oxidation. Multiple carboxylic acid compounds, such as citric acid, ethylenediaminetetraacetic acid (EDTA), and phosphoric acid derivatives (polyphosphates and phytic acid) are commonly used to increase the shelf life of foods containing lipids, due to their metal-chelating properties.
- Oxygen removers - As oxygen is essential for the oxidation process and it is a catalyst in the auto-oxidation process, substances such as ascorbic acid and ascorbyl palmitate act capturing the oxygen present in the medium, by means of stable chemical reactions, thereby making it unavailable to act as an auto-oxidation propagator.
- Compounds that regenerate primary antioxidants – as ascorbic acid that regenerates α- tocopherol.

The antioxidants can act, inhibiting the oxidation reactions that lead to complex processes of auto-oxidation, photo-oxidation and/or enzymatic oxidation.

In the auto-oxidation reaction, the antioxidants work by preventing the oxidative reactions or as blockers of the chain reaction; they may be electron donors or electron acceptors. Electron donors compete with the lipid for the peroxyl radical, resulting in a decrease of the reaction rate. On the other hand, electron acceptors compete with triplet oxygen, reducing the formation of the peroxyl radical (Pokorny et al., 2001).

In the photo-oxidation, carotenoids and tocopherols can inactivate photo-activated sensors, physically absorbing their energy, thus converting singlet oxygen to triplet oxygen. On the other hand, in the enzymatic oxidation, flavonoids, phenolic acids and gallates inhibit the action of lipoxygenase (Pokorny et al., 2001).

Some antioxidants exhibit more than one action mechanism; therefore they are referred to as multifunctional antioxidants. The antioxidants from natural sources are present many times in combinations involving many different compounds. Each compound may be present together with its precursor/precursors. Thus, the mode of action of the antioxidant from natural sources can vary and thereby involve several action mechanisms (Shahidi, 2000).

1.3. Gallic Acid: Natural Occurrence and Chemical Properties

1.3.1. Natural Occurrence and Applications

Gallic acid (3,4,5-trihydroxybenzoic acid) (Figure 4) and its derivatives are known as secondary metabolites of plants and are widely distributed in the plant kingdom, both in free form and in the form of methylated gallic acid (e.g. syringic acid) or galloyl conjugates of catechin derivatives (Lu et al., 2006). Gallic acid (GA) is present in the composition of many fruits and vegetables such as: gallnuts, oak bark, sumac, grapes, and tea leaves as one of the chief phenolic components. Gallic acid can be isolated from the plant *Galega officinalis*, an herbaceous plant from the Faboideae subfamily (Zhao et al., 2011). Pure, GA is a colorless crystalline powder, whose salts and esters are named as gallates (Masoud et al., 2012).

According to the literature, the beneficial effects of GA in human health are associated with its antioxidant properties (Fernandes et al., 2009). Studies consistently indicate that the balance between oxidative stress and the antioxidant functions of living organisms play a role in the carcinogenesis (Machado at al., 2008). GA, with its antioxidant activity is capable of preventing or inhibiting carcinogenesis at different levels, thereby reducing the risk and/or propagation of determined cancer types. It thus represents an inhibitor agent of cancer diseases by means of blocking free radicals. It is emphasized that gallic acid may cause apoptosis in tumor cell lines and inhibit lymphocyte proliferation (Fiuza et al. 2004). Fox and co-workers (2012) also mentioned that GA does not cause mutagenesis, the main side effect of anticancer drugs, and is effective in multidrug-resistant cancer cells.

In addition to its antioxidant action, gallic acid exhibits significant properties: anti-inflammatory, antibacterial (Masoud et al., 2012), antiangiogenic, antidiabetic (Liao et al., 2012), antifungal (Zhao et al., 2011), cardioprotective and vasodilatory (Ferreres et al., 2011). All these properties make gallic acid a multifunctional compound, including use in the food and cosmetic industries for preventing rancidity induced by lipid peroxidation and deterioration (Pasanphan and Chirachanchai, 2008). It is also utilized in the pharmaceutical industry; it is found in many phytotherapics, with many biological factors and pharmacological activities (Lu et al., 2006). In chemical research, GA has been used as a standard for the determination of phenolics in various analytes by the Folin-Ciocalteau method.

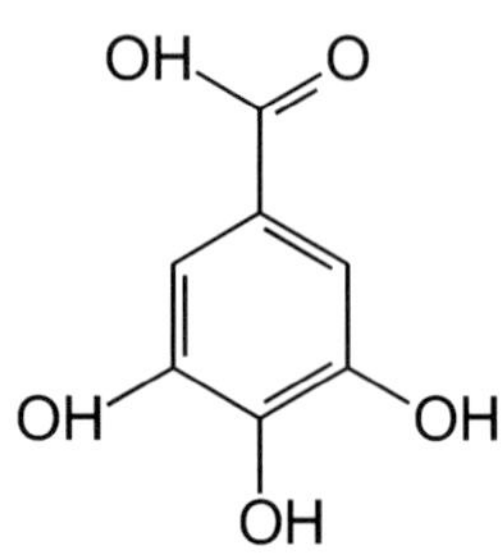

Figure 4. Gallic acid.

1.3.2. Structure-Activity Relationship

The understanding of the structural characteristics of gallic acid has a unique importance in understanding the structure-activity relationship of this compound. From the chemical point of view, the most relevant parameters to evaluate the activity are: dissociation energy of the phenolic hydrogen atom bond (O-H BDE), ionization potential (IP), and energy-eigenvalue of the highest occupied molecular orbital (E_{HOMO}).

The literature states that the pKa value of GA is approximately 4, and therefore the anions derived from the proton dissociation are dominant in the neutral system.

According to the action mechanism of the phenolic compounds mentioned in the present chapter, the predominant mechanism is elucidated by the donation of phenolic hydrogen to the free radical. This bond rupture is characterized by a heterolytic breakage (Eq. 12), in which the more favorable the hydrogen transfer, the lower the dissociation energy of the O-H bond. The second mechanism (Eq. 13) for the GA performance is reported as an electron donation to the radical (homolytic bond breakage), to form the ArOH• radical, which is ionized to form the radical ArO•. In both cases, the ionization potential (IP) is crucial in the chelating activity (Lu et al., 2006; Ji et al., 2006).

$$RO^{\bullet} + ArOH \rightarrow ROH + ArO^{\bullet} \quad \text{(Eq. 12)}$$

$$RO^{\bullet} + ArOH \rightarrow RO^{-} + ArOH^{\bullet +} \rightarrow ROH + ArO^{\bullet} \quad \text{(Eq. 13)}$$

The molecular structure of GA (Figure 4) displays three phenolic hydroxyls at positions 3, 4 and 5 of the benzene ring and the carboxyl group. Theoretical studies on the GA conformation suggested the preference for a planar geometry. This planarity is predictable, taking into account that the electron delocalization is favored by an expanded π system (Fiuza et al., 2004). The significant efficiency of gallic acid is also evaluated by its ability to form catechol after a donation of the phenolic hydrogen; this fact evidences its good performance as a chelating agent.

1.3.3. Chelating Properties

A plethora of free metals in the diet can catalyze the oxidation of biomolecules. This is effectively prevented by the complexation of metals with proteins. The role of these proteins can be efficiently complemented by the antioxidant system.

The versatility of the oxidant system allows for profiting the diverse antioxidants supplied by the diet. It has been reported that flavonoids and phenolic acids play an important role in the antioxidant system of the organism. Their ability to reduce the effect of a free radical attack is partly due to the fact that they are able to provide specific chelating agents, which are able to establish a chemical bond with any iron available, thus reducing their bioavailability (Hynes and Coinceanainn, 2001).

Several studies showed the antioxidant properties of gallic acid (Fernandes et al., 2009; Pasanphan and Chirachanchai, 2008). Studies showed that gallic acid displays diverse properties: anti-inflammatory (Hsiang et al., 2013), antibacterial (Nakamura et al., 2012) and anticarcinogenic (Ho et al., 2013). Other studies demonstrated the effects of gallic acid in systems containing metal ions (Piker et al., 2012; Strlic et al., 2002; Rajendran et al., 2012).

The interactions between transition metal ions and phenolic compounds are common in nature and involve the complexation of metal ions with phenols or their respective products of oxidation, redox or polymerization. Generally, complexation occurs in biological systems in redox reaction processes (Piker et al., 2012).

Since iron is the most abundant transition metal in the human body, and most of the intake of iron is in ferric form (Hynes and Coinceanainn, 2001), interactions between Fe (III) and GA are of primary importance.

Gallic acid is a strong chelating agent and forms high stability complexes with iron. The complex is formed at a pH range from 3 to 9. The degree of chelation increases as the pH increases. The iron is bound to gallic acid by means of two adjacent hydroxyl groups present in the aromatic ring (Masoud et al., 2012).

The interactions between catechol and iron (III) have been investigated. The reactions involve a variety of different mechanisms. (Hynes and Coinceanainn, 2012; Kazmi et al., 1987). The metal - ligand proportion varies depending on the concentration of the precursors, the temperature and, more importantly, the pH of the reaction medium (Rajendran et al., 2012).

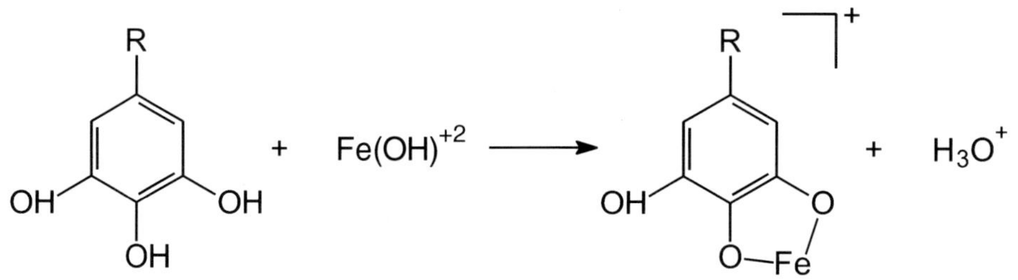

Figure 5. Gallic acid-iron complex (Adapted from Hynes and Coinceanainn, 2001).

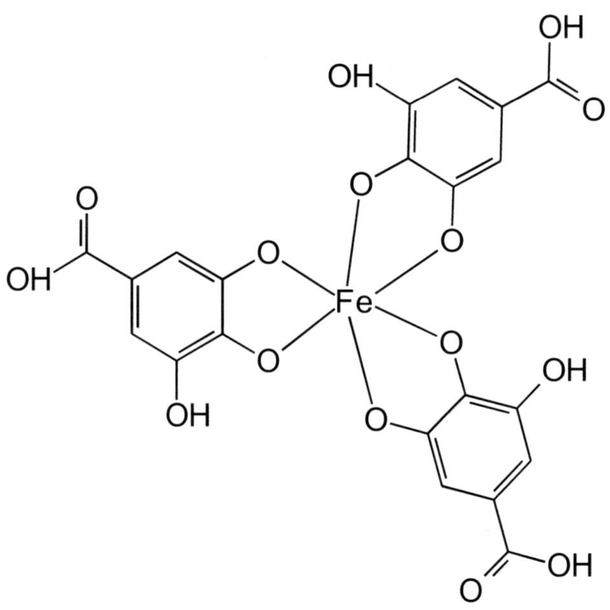

Figure 6. Structure of a gallic acid-iron complex. (Adapted from Rajendran et al., 2012).

Hynes and Coinceanainn (2001) studied the kinetics and mechanisms for the reaction of iron (III) with gallic acid. In this study, when an excess of iron (III) is added to solutions of gallic acid, a dark blue color firstly appears and then quickly disappears. The appearance of the blue color is attributed to the formation of a 1:1 complex (Figure 5), wherein the metal ion is coordinated by two adjacent hydroxyl groups (the third hydroxyl group remains protonated due to its high pKa).

Rajendran and co-workers (2012) demonstrated that a gallic acid-iron (III) complex is easily formed by means of phenolic groups; the carboxylic acid group of gallic acid was not involved in the coordination of the metal ion. Figure 6 shows the structure of the gallic acid – iron (III) complex proposed by Rajendran and co-workers (2012), in which the metal/ligand ratio is 1:3 (ML3 complex).

Studies were also performed with the system copper/gallic acid. In the redox reactions between Cu (II) and molecules of polyphenols, Cu (II) is reduced to Cu (I) and hydroquinone (H_2Q) is oxidized to semiquinone (HQ) (Eq. 14). In the second oxidation step (Eq. 15), semiquinone (HQ) is oxidized to quinone (Q) by Cu (II) (Piker et al., 2012).

$$Cu(II) + H_2Q \rightarrow Cu(I) + HQ \qquad \text{(Eq. 14)}$$

$$Cu(II) + HQ \rightarrow Cu(I) + Q \qquad \text{(Eq. 15)}$$

Piker et al. (2012) evaluated the influence of the pH on the reactions of copper (II) in the presence of gallic acid. The work reports that the formation of chelates with Cu (II) is favored in alkaline pH. Another study was performed by Hynes and Coinceanainn (2001) that evaluated the kinetics and mechanisms of the reactions involving aluminum (III) and gallic acid. Therefore, gallic acid is considered a multifunctional antioxidant, which, in addition to acting as a primary antioxidant, displays chelant properties.

1.4. Evaluation Methods of the Antioxidant Activity

There are many evaluation methods for the antioxidation activity and many different ways of expressing their results. In turn, the applications are countless (physiology, pharmacology, nutrition, agrochemical, cosmetics and other), making it difficult to choose the most appropriate method in order to avoid misunderstandings in the interpretation of the results.

Therefore, the utilization of distinct methods, with different substrates and components is required, in such a way to evaluate the different antioxidant effects (Oliveira et al., 2009). Thus, among the different existing methods, the most suitable method should be evaluated for a specific study, based on its advantages and disadvantages, thus generating a profile of the antioxidant, without losing sight of the relationship with the potential application of the product.

According to Oliveira et al. (2009), it is necessary to stress that the comparison between the antioxidant capacities of different methods is not performed in terms of absolute values, because each method has its own scale of values.

They also notice that the study of the total antioxidant capacity is preferable, instead of the analysis of isolate antioxidants, since it is difficult to measure the performance of each individual antioxidant, chiefly due to the interaction among them.

The antioxidant activity can be directly measured by its capacity of scavenging free radicals, or indirectly by the effects of the antioxidant on the oxidation rate. However, there are differences in the tests that measure the radical scavenging capacity, considering the free radicals generated, chosen oxidation indicators and methods used for detection and quantification. The results of these tests can be expressed as an inhibition percentage, 50% reduction of free radical concentration, the time required to achieve such a 50% reduction in concentration, a Trolox equivalent or complexation with iron (Antolovich et al., 2002; Sánchez-Moreno, 2002).

The indirect measurements are applied tests that evaluate the state of the lipid oxidation in samples with and without antioxidants; and the results are expressed as inhibiting effects, and as rates of reagent consumption or product formation. The antioxidant capacity can also be indirectly measured by the increase of oxidative stability expressed as an antioxidant index or as a protection factor (Antolovich et al., 2002).

Different methods are used to characterize the antioxidant capacity of foods, but no one is complete (Prior et al., 2005). The literature reports numerous studies upon the antioxidant properties of vegetable species carried out utilizing different test methods (Zheng and Wang, 2001; Ninfali et al., 2005; Miura et al., 2002; Wootton-Beard et al., 2011; Sánchez-Moreno, 2002). Nevertheless, the big variety of oxidation systems and methods for measuring the antioxidant activity makes it difficult to compare of the results of several studies (Shan et al., 2005).

Among the several methods and techniques used for the identification and/or quantification of the performance of these natural antioxidants, the following can be cited: the determination of the total phenolic compounds, High-Performance Liquid Chromatography (HPLC), Gas Chromatography Mass Spectrometry (GC/MS), the methods for capturing the peroxyl radical (ORAC - oxygen radical absorbance capacity), of Fe^{+3} reduction FRAP (Ferric Reducing Antioxidant Power), of capture of organic radical (ABTS - 2,2'-azino-bis(3-ethylbenzothiazoline-6-sulphonic acid) and DPPH• - 2,2-diphenyl-1-picrylhydrazy), of capturing the hydroxyl radical - $HO^{\bullet}$ (deoxyribose method) and the methods quantifying the products formed during the lipid peroxidation (β -carotene/linoleic acid co-oxidation system). Among these, emphasis should be given to the DPPH•, β-carotene/linoleic acid system and FRAP.

1.4.1. DPPH• Method

This test, developed by Brand-Williams et al. (1995), measures the antioxidant activity of a substance by quantifying its capacity of scavenging the free radical DPPH•, 2,2-diphenyl-1-picryl-hydrazyl (DPPH•), reducing it to hydrazine (Figure 7). DPPH• is classified as a stable free radical, due to the delocalization of the unpaired electron throughout the molecule. This delocalization confers to DPPH•, a violet color; its alcoholic solution is characterized by an absorption band centered at around 515 nm.

When a substance that acts as a donor of hydrogen atoms is added to a DPPH• solution, hydrazine is obtained with the simultaneous change in color from violet to pale yellow.

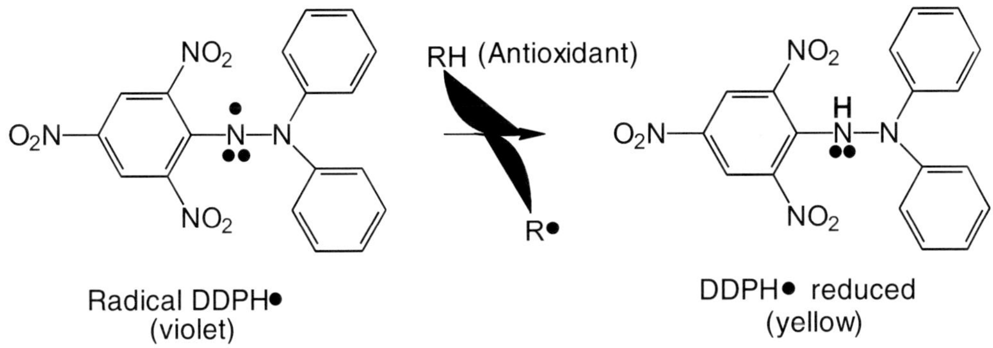

Figure 7. Structure of the DPPH• radical and its stabilization reaction with an antioxidant.

The absorbance values obtained allows for the determination of a percentage total antioxidant activity (%TAA) or the free radical scavenging activity and/or the percentage DPPH• remaining in the reaction medium, related to the amount of DPPH• consumed by the antioxidant.

The utilization of several sample concentrations is required to determine the amount of antioxidant required to decrease the initial DPPH• concentration in 50%. The concentration thus determined is called effective concentration (EC_{50}), which is also known as the inhibitory concentration (EC_{50}). The higher the DPPH• consumption of a sample, the lower is its EC_{50} and the higher is its antioxidant activity (Brand-Williams et al. 1995).

This test is widely used because of its simplicity, speed, and its reproducibility (Silva et al., 2005; Capecka et al., 2005; Lee et al., 2008; Ghasemzadeh et al., 2011).

1.4.2. FRAP Method

The FRAP (Ferric Reducing Antioxidant Power) assay is based on the capacity of an antioxidant to reduce Fe^{3+} to Fe^{2+}. When this reduction takes place in the presence of 2,4,6-tri(2-pyridyl)-1,3,5-triazine (TPTZ) in acidic conditions, the reduction is accompanied by the formation of a colored complex (intense blue) with Fe^{2+}, displaying its maximum absorption at 593 nm (Figure 8).

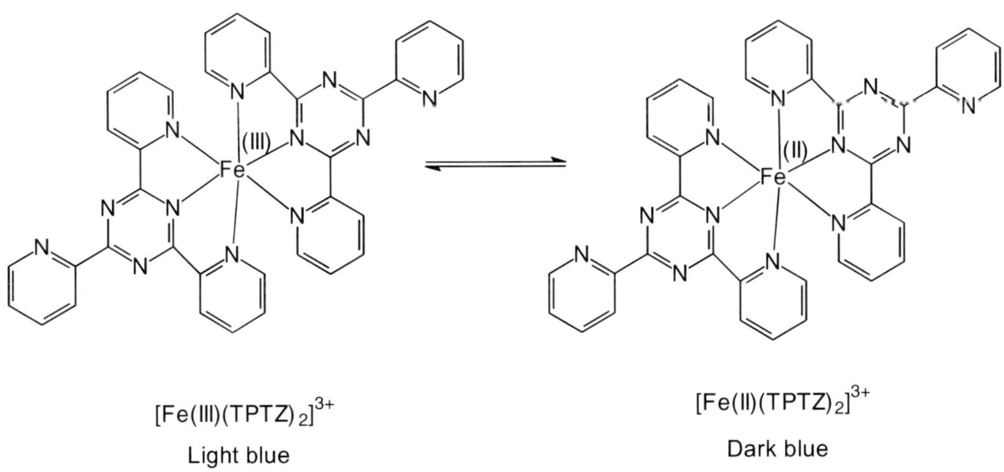

Figure 8. TPTZ complex reduction by Fe^{2+}.

1.4.3. Method of the β-Carotene/Linoleic Acid System

The co-oxidation of the β-carotene /linoleic acid system, originally described by Marco (1968), allows for assessing the ability of a determined substance in preventing the oxidation of β-carotene, protecting it from the free radicals generated during the peroxidation of linoleic acid. The reaction can be monitored spectrophotometrically by the color loss of β-carotene at 470 nm, with an immediate reading and another reading after 120 minutes. The emulsifier Tween - 40 is used to obtain a stable emulsion from the β-carotene/linoleic acid mixture in chloroform. After the chloroform removal, the contents are homogenized in water, previously saturated with atmospheric oxygen. The result is compared with synthetic standards, as Trolox, or with natural standards as gallic acid or quercetin.

This method has been widely used to evaluate the antioxidation activity, either of compounds isolated from plant extracts, or from fruits and beverages. This method is simple, sensitive, and, as it does not use high temperatures, it allows for the determination of the antioxidant activity (AA) of thermosensitive substances. However, this method has some drawbacks, as with the interference in the absorbance values, due to use of the emulsified medium, causing low reproducibility, and interaction of β-carotene with oxygen from the environment, complicating the interpretation of the results.

1.5. Thermal Stability by Thermogravimetry

The thermal stability of an oil, drug, antioxidant or any sample is defined as the ability of the substance to keep its properties, during the thermal processing, as close as possible to its initial characteristics.

Thermogravimetry - TG is an analytical technique used to determine the thermal stability of a material and its fraction of volatile components, by means of monitoring the mass change that takes place when the sample is heated.

The mass change is recorded as a function of the temperature increase. The plot of the mass derivative against temperature (DTG) points out the different steps in the thermal decomposition process (Mothé and Azevedo, 2009).

This technique is based on the mass variation of a sample resulting from a physical (sublimation, evaporation, condensation) or chemical transformation (degradation, decomposition, oxidation), as a function of time or temperature (Mothé and Azevedo, 2009). It has been used to characterize oil matrices (Neto et al., 2009; Diniz et al., 2008) with the objective of investigating the thermal stability (Fonseca and Yoshida, 2009), decomposition kinetics (Santos et al., 2002), effects of antioxidants (Arora et al., 2010, Van Aardt et al., 2004), and product adulterations (Torrecilla et al., 2011), as well as to characterize and determine the thermal stability, degradation kinetics and purity of drugs (Rodrigues et al., 2005, Bazzo and Silva, 2005) and antioxidant substances (Santos et al., 2012), among other applications.

The differential thermal analysis (DTA) shows changes, such as melting, solidification and crystallization, in samples recorded in the form of peaks. Changes in the heat capacity of the sample are noted as a shift of the baseline. The main use of the DTA curve is to detect the initial temperature of the thermal processes and qualitatively characterize them as endothermic/exothermic; reversible/or irreversible and also or first/second order transitions (Mothé and Azevedo, 2009).

2. Materials and Methods

2.1. Materials

The reagents 2 2'-diphenyl-1-picrylhydrazyl ($DPPH^{\bullet}$), 6-hydroxy-2,5,7,8-tetramethylchroman- 2- carboxylic acid (TROLOX), 2,4,6-tris(2-pyridyl)-s-triazine) (TPTZ), 6-hydroxy-2,5,7,8-tetramethylchroman-2-carboxylic acid) (TWEEN ® 40) and linoleic acid were supplied by Sigma-Aldrich and ß-carotene by Merck. Gallic acid (3,4,5-trihydroxybenzoic acid) monohydrate, ferric chloride hexahydrate ($FeCl_3 \cdot 6H_2O$) , ferrous sulfate heptahydrate ($FeSO_4 \cdot 7H_2O$) and sodium acetate trihydrate ($CH_3COONa \cdot 3H_2O$) were provided by Vetec Química Fina Ltda, and the remaining reagents and analytical grade solvents by FMAIA.

2.2. Methods

2.2.1. Methods for the Determination of the Antioxidant Activity

2.2.1.1. Assay for the Radical-Scavenging Activity of 2,2'-Diphenyl-2-Picrylhydrazyl ($DPPH^{\bullet}$) Radical

The DPPH assay was carried out according to the methodology described by Brand-Williams and co-workers (1995), with some modifications. Based on a preliminary screening study, appropriate amounts of DPPH• solution (23.6 $\mu g.mL^{-1}$ in EtOH) were added to the ethanolic sample solutions, aiming at obtaining final concentrations ranging from 0.5 to 7.5 $\mu g/mL^{-1}$. In triplicate, aliquots of samples from different concentrations were mixed with ethanol and 2,700 μL of the DPPH• solution. After 30 minutes, the reading was performed at 517 nm in a Shimadzu 2550 model UV-UV-vis spectrophotometer. The readings allowed for plotting a graph of concentration ($\mu g.ml^{-1}$) against %TAA (Eq. 16):

$$\%TAA = 100 \times \left(\frac{Abs_{control} - Abs_{sample}}{Abs_{control}}\right) \quad \text{(Eq. 16)}$$

in which, $Abs_{control}$ is the absorbance of the DPPH• radical ethanol solution and Abs_{sample} is the sample absorption, in the presence of the DPPH• radical.

The results were expressed in terms of EC_{50} and the calculation was performed from the standard line equation, replacing the value of Y by 50%, for determining the sample concentration with the capacity to reduce 50% of DPPH•.

2.2.1.2. Ferric Reducing Antioxidant Power (FRAP) Assay

The ferric reducing antioxidant power (FRAP) was performed according to Benzie and Strain (1996) with modifications. For the preparation of the FRAP reagent solution, 25 mL of 0.3 M acetate buffer solution was added to a 2.5-mL aliquot of a 10-mM TPTZ solution and to 2.5 mL of a 20-mM aqueous ferric chloride solution. 90 μL of each sample at three different concentrations were homogenized with a 2.7-mL FRAP reagent solution (freshly prepared) and 270μL of distilled water. The solutions were stirred in a temperature-controlled water bath at 37 ° C for 30 minutes. For the absorbance readings, the spectrophotometer was calibrated with the FRAP reagent solution and they were performed at 595 nm in a UV-2550 model Shimadzu UV-vis spectrophotometer.

A control curve was prepared using a ferrous sulfate solution (500-2000 μM). Total antioxidant activity can be calculated from the linear relationship of absorbance *versus* concentration (mg/L^{-1}) (Y *vs* X), substituting in the line equation the absorbance equivalent to 1,000 μM of the ferrous sulfate standard solution. The value obtained for the term X corresponds to sample dilution (mg/L^{-1}) equivalent to a 1,000 μM of ferrous sulfate solution. The result is expressed in a μM ferrous sulfate g^{-1} of sample.

2.2.1.3. β-Carotene/Linoleic Acid System Method

This experiment was conducted using the methodology described by Marco (1968), with modifications. For the preparation of the system solution, 40 μL of linoleic acid, 530 mL of Tween-40, 50 μL of β-carotene solution (20 mg/mL^{-1} in chloroform) and 1 mL of chloroform were added together in a 250 ml Erlenmeyer flask. After homogenizing the mixture, chloroform was evaporated with the aid of the oxygenator. Subsequently, water previously oxygenated for 30 minutes was added to the Erlenmeyer flask contents. The absorbance, at 470 nm, was read in the aforementioned spectrophotometer and the water addition continued until an absorbance between 0.6 and 0.7 was achieved.

After the addition of 0.4 mL of each sample concentration (0.5, 0.75 and 1.0 mg/mL^{-1}), 5 mL of the system solution (β-carotene/linoleic acid) were added, in triplicate. Trolox (200 mg/mL^{-1}) was used for the control, utilizing the same conditions employed for the samples. The tubes were stirred and kept in a water bath at 40 °C. For the absorbance measurements, the spectrometer was calibrated with water. The readings were made 2 minutes and 120 minutes after the mixing.

The absorbance decrease of the samples correlates with the system and establishes the proportion of oxidation (Eq. 17). The results were expressed as a percentage of protection, by subtracting the percentage of oxidation of each sample from 100 (Eq. 18):

$$\%\,Oxidation = \frac{[(Absbance\ Reduction)_{sample}\ x\ 100]}{(Absorbance\ Reduction)_{system}} \tag{17}$$

$$\%Protection\ = 100 - (\%Oxidation) \tag{18}$$

in which % Oxidation is the percentage of oxidation, Absorbance Reduction is the difference between the initial absorbance and the final absorption, both for the samples and for the system without antioxidants.

2.2.2. Thermal Profile

The thermogravimetric curves (TG/DTA) of gallic acid were obtained in a TA Instruments, model SDT 2960, TG-DTA simultaneous thermal analyzer, coupled to an Ametek mass spectrometer, model Dymaxion Dycor Mass Spectrometer. The operating conditions were: synthetic air and nitrogen atmospheres, flow rate of 100 mL/min^{-1}, heating rate of 10 °C/min^{-1}, temperature range from room temperature up to 1,000 °C, alumina crucible and sample mass of approximately 10 mg.

3. RESULTS

3.1. Total Antioxidant Activity Determination

According to Rice-Evans (1996), for a substance to be considered an antioxidant and it to exert its biological role, it is necessary, in low concentrations, to be able to block, delay or prevent auto-oxidation or mediated oxidation by free radicals and the created product must be stable after the reaction.

The spectrophotometric quantification of phenolic compounds is accomplished by a variety of techniques, however, those using the Folin-Ciocalteau reagent, developed by Singleton and Rossi (1965) are among the most extensively used (Lee et el., 2008; Sousa et al., 2007; Shan et al., 2005; Omoni and Aluko, 2005; Roginsky and Lissi, 2005; Naczk and Shahidi, 2004).

The antioxidant activity of gallic acid was determined by FRAP and ß-Carotene linoleic acid bleaching assay methods. The synthetic antioxidant Trolox was used as a reference standard. The results are showed in Table 2.

DPPH• is a stable free radical that accepts an electron or hydrogen radical to become a stable diamagnetic molecule. The reduction in DPPH• radical was determined by the decrease in its absorbance at 517 nm induced by antioxidants. Hence, DPPH• radical is usually used as a substrate to evaluate the antioxidative action of antioxidants (Yen et al., 2002).

The antioxidant activity determined by the DPPH essay was expressed in EC_{50} ($\mu g/mL^{-1}$) ± SD corresponding to the sample concentration required to reduce the initial amount of DPPH• radical by 50%. The smaller the EC_{50}, the higher is the efficiency, and, therefore, the better is the antioxidant capacity of the sample.

The efficacy of gallic acid as an antioxidant is confirmed by the method of scavenging the DPPH• radical with antioxidant activity three times superior than Trolox control (Table 2).

Similar values were reported by Lu and co-workers (2006) who found values of CE_{50} for gallic acid at 6.00 μM. These authors evaluated gallic acid derivatives and found values ranging from 7.2 to 13.2 μM, as well as no sequestrated effect.

The antioxidant activity can also be measured by the reduction of metallic ions. Gallic acid was tested in the FRAP assay, which, according to Benzie and Strain (1996), is a fast method, reproducible and easy to realize. In this method, the antioxidant activity is determined based on the ability to reduce Fe (III) to Fe (II) (Benzie and Strain, 1996).

Table 2. Antioxidant activity of Gallic acid and Trolox by methods DPPH•, FRAP and β-carotene/linoleic acid system[a]

Antioxidant	Antioxidant capacity DPPH• EC_{50} ($\mu g/mL^{-1}$)[b]	FRAP (mmol Fe/g^{-1})	ß-carotene bleaching %O.I.[c]
Gallic acid	1.00 ±0.003	130.77 ±6.46	18.30 ±0.034
Trolox	3.09 ±0.082	35.12 ±7.83	51.50 ±0.003

[a] Mean value ± standard deviation; n=3.

[b] Concentration of antioxidant required to reduce the original amount of free radicals by 50%.

[c] Oxidation inhibition.

Although this assay was originally developed to measure plasma antioxidant capacity, it can be used to quantify the antioxidant capacity from a wide variety of biological samples from pure compounds to fruits, wines, and animal tissues (Katalinic et al., 2004). The results were expressed as mmol ferrous iron (II) equivalent per g of sample ± SD.

The results of the antioxidant activity by the FRAP assay (Table 2) show that gallic acid showed a high antioxidant capacity, as compared with the Trolox antioxidant, with values of 130.17 ± 6.46 and 35.15 ±7.83 mmol of Fe^{2+}/g^{-1} sample, respectively.

The antioxidant capacity was also determined from the ability of samples to inhibit □-carotene bleaching caused by free radicals generated during linoleic acid peroxidation. In the present study, the antioxidant activity of gallic acid determined in terms of percent oxidation inhibition is presented in Table 2. The antioxidant capacity was classified as high (>70%), intermediate (40-70%) or low (<40%) levels of oxidation inhibition (O.I.) (Hassimotto et al., 2005).

Gallic acid scored below 40% of O.I., protecting only 18.30 ±0.034% O.I., whereas the Trolox positive control presented an intermediary classification, controlling the oxidation by 51.50 ±0.003%.

Several studies report the use of gallic acid as a positive control in assays to determine the antioxidant activity (Sanchez-Moreno et al., 1998; Yen et al., 2002; Oliveira et al., 2009; Queirós et al., 2013). An association between these methods is essential for an efficient evaluation of the protective effect of a possible antioxidant. However, a large amount of diverse literature data was obtained in experimental conditions, making it difficult to compare the results.

For this reason, the methodology standardization for assessing the antioxidant potential, even by the capacity of scavenging free radicals, or indirectly through the antioxidant effects in oxidation control, is necessary.

3.2. Thermal Analysis

Thermogravimetry is a widely used technique to characterize degradation profiles of substances. The exposure to high temperature can, sometimes, change the chemical structure and, consequently, the substance's chemical and physical properties. Therefore, thermogravimetry is an important tool for quality control, assessing the substance's thermal stability when under processes that involve high temperatures.

The thermogravimetric curve of gallic acid in atmospheric air showed four stages of thermal decomposition (Figure 9a).The first stage of weight loss is an endothermic process and it is associated with loss of hydration water (Table 3). The second stage is an endothermic process associated with the decomposition of gallic acid, releasing CO_2, H_2O and OH (Figure 9b). The third and the fourth stages in atmospheric air are associated with exothermic events. In these stages, there is a release of CO_2 and H_2O (Table 3), products of the organic matter combustion reaction.

The thermal profile of gallic acid in a nitrogen atmosphere also showed four stages of thermal decomposition (Figure 10a). The first two steps presented results similar to those obtained in the air atmosphere (Table 3).

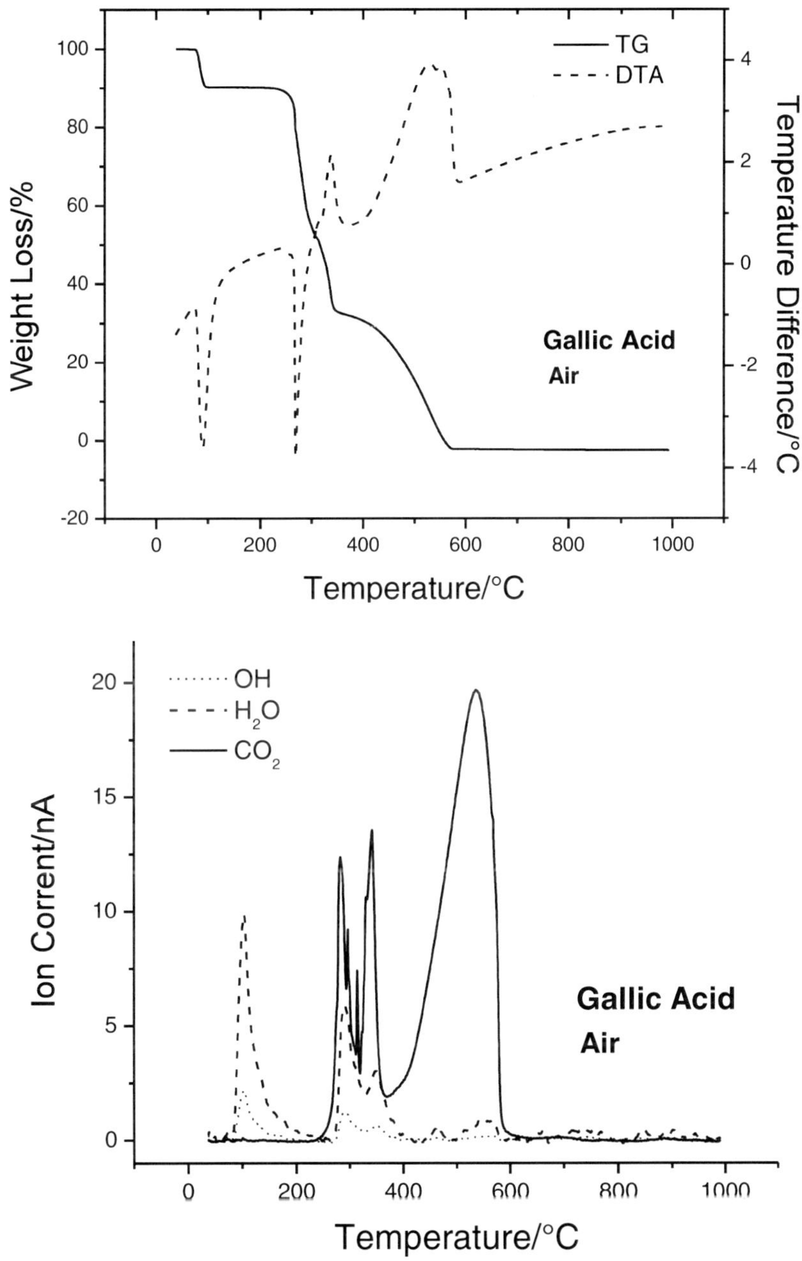

Figure 9. (a) TG and DTA curves of gallic acid in air atmosphere (b) Evolution of the signals m/z 17 (OH); 18 (H_2O) and 44 (CO_2).

The third and the fourth stages of weight loss are endothermic process and are related to the decomposition of gallic acid molecules, releasing CO_2, H_2O and OH in the third event and CO_2, H_2O in the last event (Figure 10b). At the end of the process, a temperature of 1,000 °C and a residue of 11.91% was quantified.

Therefore, gallic acid is a chemical constituent, thermally stable, and can be applied in foods which received higher temperatures, without losing its proprieties up to a temperature of approximately 200 °C (Table 3).

Table 3. Gallic acid thermal analysis

Atmosphere	Events	Temp. Range (°C)	Weight Loss (%)	T_p DTG (°C)	T_p DTA (°C)	Trans.	Assignment	T_p MS (°C)	Weight Loss (%)
Air	1	66.7 - 104.6	9.74	84.6	90.0	Endo	H_2O	102.7	8.09
							OH	103.2	1.65
	2	206.7 - 310.2	38.37	269.5	269.0	Endo	H_2O	290.3	14.39
							OH	290.7	2.41
							CO_2	282.3	21.57
	3	310.2 - 366.9	19.51	335.5	337.2	Exo	H_2O	349.6	3.69
							OH	347.4	0.79
							CO_2	341.3	15.02
	4	366.9 - 580.1	32.38	529.7	529.1	Exo	H_2O	567.7	0.83
							OH	563.3	0.16
							CO_2	535.6	31.38
Nitrogen	1	63.0 - 109.6	9.69	83.5	88.7	Endo	H_2O	99.6	7.68
							OH	99.8	2.01
	2	199.2 - 305-1	32.82	268.9	243.4 and 268.9	Endo	H_2O	290.4	9.62
							OH	289.2	2.81
							CO_2	278.4	20.39
	3	305.1 – 378.3	28.38	326.8	323.1	Endo	H_2O	508.0	20.33
							OH	513.0	0.41
							CO_2	328.1	7.64
	4	378.3 – 1,000.0	17.20	464.9	784.7	Endo	H_2O	826.3	14.89
							CO_2	384.8 and 751.2	2.31
	Residue						Carbon	-	11.91

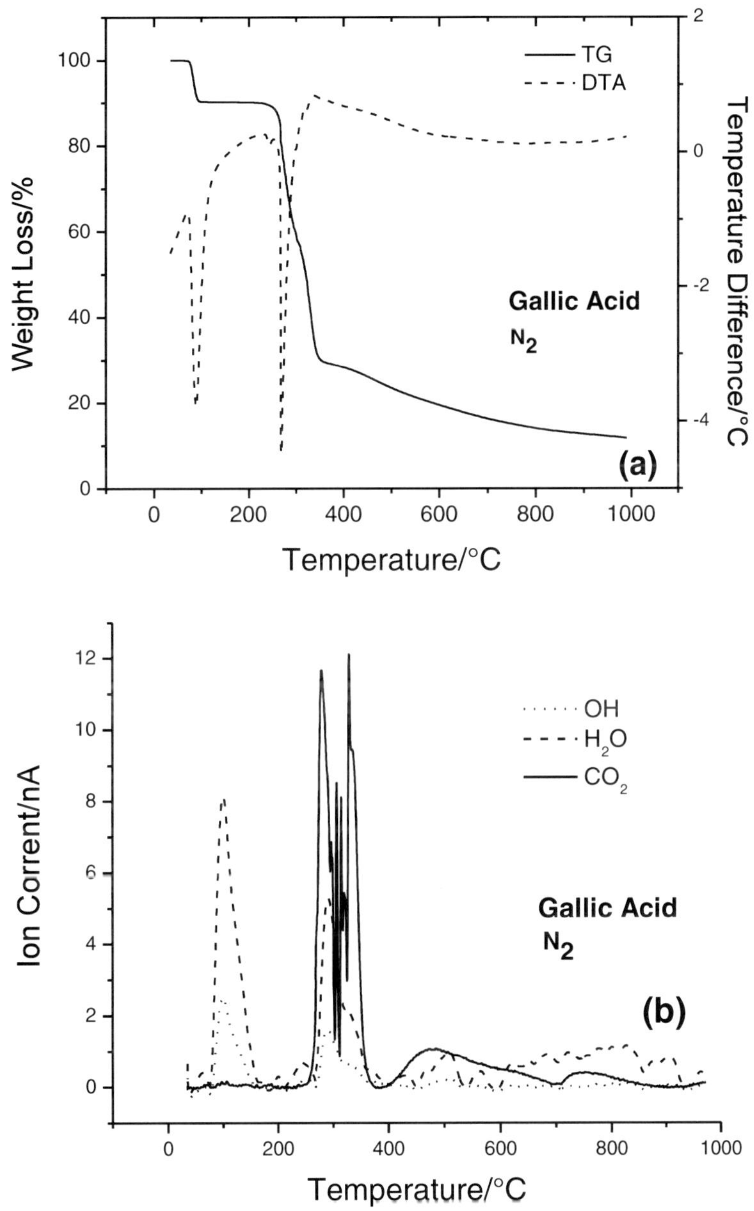

Figura 10. TG and DTA curves of gallic acid in N_2 atmosphere (b) Evolution of the m/z signals 17 (OH); 18 (H_2O) and 44 (CO_2).

CONCLUSION

It is well known that the generation of free radicals is naturally produced during cellular metabolism and functional activities and that oxidative stress is involved in the pathology of inflammatory diseases, cardiovascular diseases, diabetes, cancer, Alzheimer's disease, atherosclerosis, and rheumatoid arthritis and could play a role in neurodegeneration and ageing processes. The protection that fruits and vegetables provide against several diseases has been attributed to several antioxidants therein contained, such as tocopherols, carotenoids and phenolic compounds or polyphenols, including phenolic acids. Several studies have shown that phenolic compounds inhibit lipids *in vitro* peroxidation and lipoxygenase, particularly the phenolics with multiple hydroxyl groups, which are generally the most efficient in minimizing the oxidative damage of living cells and in preventing the deterioration of food. It is considered that the defense system is positioned against damage caused by free radicals, and keeps the oxidative process within physiological limits. It is also considered that the mechanism of action of antioxidants is accomplished by means of the inhibition of reactions with iron and copper, as well as by the donation of the phenolic hydrogen, preventing the attack of ROS on unsaturated fatty acids and nucleic bases. Studies consistently indicate that the beneficial effects of GA in human health are associated with antioxidant properties. It thus represents an inhibitor agent of cancer diseases by means of blocking free radicals. A large number of plants are rich sources of gallic acid, both in the free form and in the form of methylated gallic acid or galloyl conjugates of catechin derivatives. GA occurs naturally in all higher plants and in all plant parts. In addition to its antioxidant action, gallic acid exhibits significant properties: anti-inflammatory, antibacterial, antiangiogenic, antidiabetic, antifungal, cardioprotective and vasodilatory. All these properties make gallic acid a multifunctional compound, including having uses in food and cosmetic industries by preventing rancidity induced by lipid peroxidation and deterioration. The molecular structure of GA and its planar geometry demonstrates its good performance as a chelating agent and forms high stability complexes with transition metals.

Thus, the antioxidant activity assays performed in the present study showed that gallic acid showed a high antioxidant activity and was shown to be more efficient, by the DPPH• and FRAP techniques, than the synthetic antioxidant Trolox. Moreover, the GA thermogravimetric analysis showed that it is a thermostable chemical constituent, and it can be applied to foods processed at high temperatures (approximately 200 °C), because its multifunctional properties are not degraded, thus ensuring its wide application.

References

Aguiar, A.; Ferraz, A.; Contreras, D.; Rodríguez, J. Mechanism and applications of the Fenton reaction assisted by iron-reducing phenolic compounds. *Quim. Nova*. 2007, 30, 623-628.

André, C.; Castanheira, I.; Cruz, J. M.; Paseiro, P.; Sanches-Silva, A. Analytical strategies to evaluate antioxidants in food: a review. *Trends. Food Sci. Tech*. 2010, 21, 229-246.

Angelo, P. M.; Jorge, N. Phenolic compounds in foods – A brief review. *Rev. Inst. Adolfo Lutz*, 2007, 66, 232-240.

Antolovich, M.; Prenzler, P.D.; Patsalides, E.; Mcdonald, S.; Robards, K. Methods for testing antioxidant activity. *Analyst*. 2002, 127, 183-198.

Arora, S.; Bagoria, R.; Kumar, M. Effect of alpha-Tocopherol (Vitamin E) on the Thermal Degradation Behavior of Edible Oils. *JTAC*. 2010, 102, 375-381.

Barbosa, K. B. F.; Costa, N. M. B.; Alfenas, R. C. G.; De Paula, S. O.; Minim, V. P. R.; Bressan, J. Oxidative stress: concept, implications and modulating factors. *Rev. Nutr.* 2010, 23, 629-643.

Batinic–Haberle, I.; Rebouças, J.S.; Spasojevic, I. Superoxide Dismutase Mimics: Chemistry, Pharmacology, and Therapeutic Potential. *Antioxid. Redox. Sign.* 2010, 13, 877-918.

Bayir, H.M.D. Reactive oxygen species. *Crit. Care Med.* 2005, 33:498-501.

Bazzo, G. C.; Silva, M. A. S. Thermal analysis study of captopril coated tablets by thermogravimetry (TG) and differential scanning calorimetry (DSC). *Braz. J. Pharm. Sci.* 2005, 41, 315-322.

Benzie, I. F. F.; Strain, J. J. The ferric reducing ability of plasma (FRAP) as a measure of ''antioxidant power'': The FRAP assay. *Anal. Biochem.* 1996, 239, 70–76.

Brand-Williams, W.; Cuvelier, M. E.; Berset, C. Use of free radical method to evaluate antioxidant activity. *LWT-Food Sci. Technol.* 1995, 28, 25-30.

Capecka, E.; Mareczek, A.; Leja, M.. Antioxidant activity of fresh and dry herbs of some Lamiaceae species. *Food Chem.* 2005, 93, 223–226, 2005.

Cheynier, V. Polyphenols in foods are more complex than often thought. *Am. J. Clin. Nutrit.* 2005, 81, 223-229.

Croft, K. D. The chemistry and biological effects of flavonoids and phenolic acids. *Ann. N. Y. Acad. Sci.,* 1998, 854, 435-442.

Dairam, A.; Fogel, R.; Daya, S.; Limson, J.L. Antioxidant and iron-binding properties of curcumin, capsaicin, and S allylcysteine reduce oxidative stress in rat brain homogenate. *J. Agric. Food Chem.* 2008, 56, 3350–3356.

Dewick, P. M. Secundary metabolism: the building blocks and construction mechanisms. Medicinal Natural Products: A Biosynthetic Approach. 2nd ed. John Wiley and Sons: England, 2002.

Diniz, Z. N.; Bora, P. S.; Neto, V. Q.; Cavalheiro, J. M. O. *Sterculia striata* Seed Kernel Oil: Characterization and Thermal Stability. *Grasas Aceites*, 2008, 59, 160-165.

Farah, A.; Donangelo, C. M. Phenolic compounds in coffe. *Braz. J. Plant. Physiol.* 2006, 18, 23-36.

Fernandes, A.; Fernandes, I.; Cruz, L.; Mateus, N.; Cabral, M.; Freitas, V. Antioxidant and Biological Properties of Bioactive Phenolic Compounds from *Quercus suber* L. *J. Agr. Food. Chem.* 2009, 57, 11154-11160.

Ferreira, A. L. A.; Matsubara, L. S. Free radicals: concepts, related diseases, defense system and oxidative stress. *Rev. Assoc.Med. Bras.* 1997, 43, 61-68.

Ferreres, F.; Figueiredo, R.; Bettencourt, S.; Carqueijeiro, I.; Juliana, O.; Gil-Izquierdo, A.; Pereira, D. M.; Valentão, P.; Andrade, P. B.; Duarte, P.; Barcelo, A. R.; Sottomayor, M. Identification of phenolic compounds in isolated vacuoles of the medicinal plant Catharanthus roseus and their interaction with vacuolar class III peroxidase: an H_2O_2 affair? *J. Exp. Bot.* 2011, 62, 2841-2854.

Fiuza, S.M.; Gomes, C.; Teixeira, L.J.; Girão da Cruz, M.T.; Cordeiro, M.N.; Milhazes N.; Borges, F.; Marques, M.P. Phenolic acid derivatives with potential anticancer properties –a structure–activity relationship study. Part 1: Methyl, propyl and octyl esters of caffeic and gallic acids. *Bioorg. Med. Chem.* 2004, 12, 3581–3589.

Fonseca, M. M.; Yoshida, M. I. Thermal analysis of raw and aged linseed oil. *Vértices.* 2009, 11, 61-75.

Fox, J. T.; Sakamuru S.; Huang R.; Teneva N.; Simmons S. O.; Xia M.; Tice R. R.; Austin C. P.; Kyungjae Myung K. High-throughput genotoxicity assay identifies antioxidants as inducers of DNA damage response and cell death. *PNAS*, 2012, 109, 5423-5428.

Ghasemzadeh, A.; Jaafar, H. Z. E.; Rahmat, A. Effects of solvent type on phenolics and flavonoids content and antioxidant activities in two varieties of young ginger (*Zingiber officinale* Roscoe) extracts. *J. Med. Plants. Res*. 2011, 5, 1147-1154.

Halliwell B. Reactive oxygen species and the central nervous system. *J. Neurochem*. 1992, 59, 1.609-1623.

Halliwell, B. Antioxidants and humam disease: a general introduction. *Nutr. Rev*. 1997, 55, 44-52.

Halliwell, B.; Aeschbach, R.; Löliger, J.; Aruoma, O. I. The characterization of antioxidants. *Food Chem. Toxicol*. 1995, 33, 601-617.

Hassimotto N. M.; Genovese. M. I.; Lajolo, F.M. Antioxidant activity of dietary fruits, vegetables, and commercial frozen fruit pulps. *J. Agric. Food Chem*. 2005, 53, 2928-2935.

Ho, H.; Chang, C.; Ho, W.; Liao, S.; Lin, W.; Wang, C. Gallic acid inhibits gastric cancer cells metastasis and invasive growth via increased expression of RhoB, downregulation of AKT/small GTPase signals and inhibition of NF-κB activity. *Toxicol. Appl. Pharmacol*. 2013, 266, 76–85.

Hsiang, C.; Hseu Y.; Chang Y.; Kumar K. J. S.; Ho, T., Yang H. *Toona sinensis* and its major bioactive compound gallic acid inhibit LPS-induced inflammation in nuclear factor-κB transgenic mice as evaluated by in vivo bioluminescence imaging. *Food Chem*. 2013, 136, 426–434.

Hynes, M. J.; Coinceanainn, C.O. The kinetics and mechanisms of the reaction of iron (III) with gallic acid, gallic acid methyl ester and catechin. *J. Inorg. Biochem*. 2001, 85, 131–42.

Ji, H. F.; Zhang, H. Y.; Shen, L. Proton dissociation is important to understanding structure–activity relationships of gallic acid antioxidants. *Bioorg. Med. Chem. Lett*. 2006, 16, 4095–4098.

Kahkonen, M. P.; Hopia, A. I.; Vuorela, H. J.; Rauha, J. P.; Pihlaja, K.; Kujala, T. S.; Heinonen, M. Antioxidant activity of plant extracts containing phenolic compounds. *J. Agric. Food Chem*. 1999, 47, 3954-3962.

Katalinic, V.; Milos, M.; Modum, D.; Music, I.; Boban, M. Antioxidant effectiveness of selected wines in comparison with (+)-cathechin. *Food Chem*. 2004, 86, 593–600.

Kazmi, S. A.; Qureshi M. S.; Maqsood Z. Reactivity of an Iron (II1) Complex of Gallic Acid. *Inorg. Chim. Acta*. 1987, 137, 151-154.

Kohlmeier, M. Fat-soluble Vitamins and Non-nutrients. Nutrient Metabolism. Food Science and technology, International Series. Elsevier: London, 2003.

Lee, Y. C.; Chuah, A. M.; Yamaguchi, T.; Takamura, H.; Matoba, T. Antioxidant Activity of Traditional Chinese Medicinal Herbs. *Food Sci. Technol. Res.*, 2008, 14, 205-210.

Lia, H. B.; Wonga, C. C.; Chenga, K. W.; Chen, F. Antioxidant properties in vitro and total phenolic contents in methanol extracts from medicinal plants. *LWT-Food Sci. Technol*. 2008, 41, 385–390.

Liao C. L.; Lai, K. C.; Huang A.C.; Yang J. S.; Lin, J. J; Wu S. H.; Wood, W. G.; Lin, J. G.; Chung, J. G. Gallic acid inhibits migration and invasion in human osteosarcoma U-2 OS cells through suppressing the matrix metalloproteinase-2/-9, protein kinase B (PKB) and PKC signaling pathways. *Food Chem. Toxicol*. 2012, 50, 1734–1740.

Litwinienko, G.; Kasprzsycka-Guttman, T.; Jamanek, D. DSC study of antioxidant properties of dihydroxyphenols, *Thermochim. Acta*, 1999, 331, 79-86.

Lü, J.M.; Lin, P.H.; Yao, Q.; Chen, C. Chemical and molecular mechanisms of antioxidants: Experimental approaches and model systems. *J. Cell. Mol. Med.* 2010, 14, 840-860.

Lu, Z.; Nie, G.; Belton, P. S.; Tang, H.; Zhao, B. Structure–activity relationship analysis of antioxidant ability and neuroprotective effect of gallic acid derivatives. *Neurochem. Int.* 2006, 48, 263–274.

Machado, N. F. L.; Calheiros, R.; Gaspar, A.; Garrido, J.; Borges, F. Marques, M. P. M. Antioxidant phenolic esters with potential anticancer activity: solution equilibria studied by Raman spectroscopy. *J. Raman Spectrosc.* 2008, 1, 1-6.

Maganha, E. G.; Halmenschlager, R. C.; Rosa, R. M.; Henriques, J. A. P.; Ramos, A. L. L. P.; Safi, J. Pharmacological evidences for the extracts and secondary metabolites from plants of the genus *Hibiscus*. *Food Chem.* 2010, 118, 1–10.

Manach, C.; Scalbert, A.; Morand, C.; Rémésy, C.; Jimenez, L. Polyphenols: food sources and bioavailability. *Am. J. Clin. Nutrit.* 2004, 79, 727-747.

Marco, G. J. A Rapid Method for Evaluation of Antioxidants. *J. Am. Oil Chem. Soc.* 1968, 45, 594-598.

Masoud M. S.; Hagagg, S. S.; Ali, A. E; Nasr, N. M. Synthesis and spectroscopic characterization of gallic acid and some of its azo complexes. *J. Mol. Struct.* 2012, 1014, 17–25.

Miura, K.; Kikuzaki, H.; Nakatani, N. Antioxidant activity of chemical components from sage (*Salvia officinalis* L.) and thyme (*Thymus vulgaris* L.) measured by the oil stability index method. *J. Agric. Food Chem.* 2002, 50, 1845-1851.

Mothé, C. G.; Azevedo, A. D. Análise Térmica de Materiais. Artliber: São Paulo, 2009.

Naczk, M.; Shahidi, F. Extractions and analysis of phenolics in food. *J. Chromatography A.*, 2004, 1054, 95- 111.

Nakamura, K.; Yamada Y.; Ikai H., Kanno T., Sasaki K., Niwano Y. Bactericidal Action of Photoirradiated Gallic Acid via Reactive Oxygen Species Formation. *J. Agric. Food Chem.* 2012, 60, 10048−10054.

Neto, V. Q.; Bora, P. S.; Diniz, Z. N.; Cavalheiro, J. M. O.; Queiroga, K.F. Dipteryx lacunifera Seed Oil: Characterization and Thermal Stability. *Ciênc. Agrotec.* 2009, 33, 1601-1607.

Ninfali, P.; Mea, G.; Giorgini, S.; Rocchi, M.; Bacchiocca, M. Antioxidant capacity of vegetables, spices, and dressings relevant to nutrition. *Brit. J. Nutr.*, 2005, 93, 257-266.

Oliveira, A. C. De; Valentim, I. B.; Goulart, M. O. F. Vegetals as natural sources of antioxidants. *Quím. Nova*, 2009, 32, 689-702.

Omoni, A. O.; Aluko, R. E. The anti-carcinogenic and anti-atherogenic effects of lycopene: A review. *Trends Food Sci Tech*, 2005, 16, 344-350.

Pasanphan, W.; Chirachanchai, S. Conjugation of gallic acid onto chitosan: An approach for green and water-based antioxidant. *Carbohydr. Polym.* 2008, 72, 169–177.

Pirker, K. F.; Baratto, M. C.; Basosi, R.; Goodman, B. A. Influence of pH on the speciation of copper(II) in reactions with the green tea polyphenols, epigallocatechin gallate and gallic acid. *J. Inorg. Biochem.* 2012, 112, 10–16.

Pokorny, J.; Yanishlieva-Maslarova, N.V.; Gordon, M.H. Antioxidants in food – Practical applications. CRC Press:Cambridge, 2001.

Prior, R.L.; Wu, X.; Schaich, K. Standardized methods for the determination of antioxidant capacity and phenolics in foods and dietary supplements. *J. Agric. Food Chem.* 2005, 53, 4290-4303.

Queirós R. B.; Tafulo, P. A. R.; Sales, M. G. F. Assessing and Comparing the Total Antioxidant Capacity of Commercial Beverages: Application to Beers, Wines, Waters and Soft Drinks Using TRAP, TEAC and FRAP Methods. *Comb. Chem. High T. Scr.* 2013, 16, 22-31.

Rajendran N.; Pemiah, B.; Sekar, R. K.; Krishnan, U. M.; Sethuraman, S.; Krishnaswamy, S. Role of gallic acid in the preparation of an iron–based indian traditional medicine – lauha bhasma. *Int. J. Pharm. Pharm. Sci.* 2012, 4, 45-48.

Ribèreau-Gayon, P. Les composés phénoliques des végétaux. Dunod: Paris, 1968.

Rice-Evans, C.; Miller, N. J.; Papanga, G. Structure-antioxidant activity relationships of flavonoids and phenolics acids. *Free Radic. Biol. Med.*1996, 20, 933-956.

Rodrigues, P. O.; Cardoso, T. F. M.; Silva, M. A. S.; Matos, J. R. Application of Thermal Analytical Techniques on Characterization, Purity Determination and Degradation Kinetic of Zidovudine (AZT). *Acta Farmacéutica Bonaerense*. 2005, 24, 383-387.

Roginsky, V.; Lissi, E. A. Review of methods to determine chain-breaking antioxidant activity in food. *Food Chem.* 2005, 92, 235-254.

Sánchez-Moreno, C. Review: methods used to evaluate the free radical scavenging activity in foods and biological systems. *Food Sci. Technol. Int.* 2002, 8, 121-137.

Sanchez-Moreno, C.; Larrauri, J. A.; Saura-Calixto, F. A Procedure to Measure the Antiradical Efficiency of Polyphenols. *J. Sci. Food Agric.* 1998, 76, 270-276.

Santos, J. C. O; Santos, I. M. G.; Souza, A. G.; Prasad, S.; Santos, A. V. Thermal Stability and Kinetic Study on Thermal Decomposition of Commercial Edible Oils by Thermogravimetry. *J. Food Sci.* 2002, 67, 1393-1398.

Santos, N. A.; Cordeiro, A. M. T. M.; Damasceno, S. S.; Aguiar, R. T.; Rosenhaim, R.; Carvalho Filho, J. R.; Santos, I. M. G.; Maia, A. S.; Souza, A. G. Commercial antioxidants and thermal stability evaluations. *Fuel*, 2012, 97, 638-643.

Schaich, K. M. Lipid Oxidation: Theoretical Aspects. In Bailey's Industrial Oil and Fat Products. 6th ed. Fereidoon Shahidi, Ed.; John Wiley and Sons: New York, 2005.

Shahidi, F. Antioxidants in food and food antioxidants. *Nahrung*, 2000, 44, 158-163.

Shan, B.; Cai, Y. Z.; Sun, M.; Corke, H. Antioxidant capacity of 26 spice extracts and characterization of their phenolic constituents. *J. Agric. Food Chem.* 2005, 53, 7749-7759.

Silva, C.G.; Herdeiro, R.S.; C.J.; Mathias, A.D.; Panek, C.S.; Silveira, V.P.; Rodrigues, M.N.; Rennó, D.Q.; Falcão, D.M.; Cerqueira, A.B.M.; Minto, F.L.P.; Nogueira, C.H.; Quaresma, J.F.M. Silva, F.S. Menezes, E.C.A. Eleutherio. Evaluation of antioxidant activity of Brazilian plants. *Pharmacol Res*. 2005, 52, 229–233.

Singleton V. L.; Rossi J. A. Colorimetry of total phenolics with phosphomolybdic-phosphotungstic acid reagents. *Am. J. Enol. Vitic*., 1965, 16,144-158.

Soares, S. E. Phenolic acids as antioxidants. *Rev. Nutr.* 2002, 15, 71-81.

Sousa, C. M. M.; Silva, H. R. Vieira Jr., G. M.; Ayres, M. C. C.; Costa, C. L. S.; Araújo, D. S.; Cavalcante, L. C. D.; Barros, E. D. S.; Araújo, P. B. M.; Brandão, M. S.; Chaves, M. H. Total phenolics and antioxidant activity of five medicinal plants. *Quim. Nova*, 2007, 30, 351-355.

Strlic, M.; Radovic, T.; Kolar, J.; Pihlar, B. Anti- and Prooxidative Properties of Gallic Acid in Fenton-Type Systems. *J. Agric. Food Chem.* 2002, 50, 6313-6317.

Tarlovsky V.F. Role of antioxidants in cancer therapy. *Nutrition*. 2013, 29, 15–21.

Torrecilla, J. S.; Garcia, J.; Garcia, S.; Rodriguez, F. Application of lag-k autocorrelation coefficient and the TGA signals approach to detecting and quantifying adulterations of extra virgin olive oil with inferior edible oils. *Anal. Chim. Acta*. 2011, 688, 140-145.

Van Aardt, M.; Duncan, S. E.; Long, T. E.; O'keefe, S. F.; Marcy, J.E.; Sims, S. R. Effect of Antioxidants on Oxidative Stability of Edible Fats and Oils: Thermogravimetric Analysis. *J. Agric. Food Chem*. 2004, 52, 587-591.

Wanasundara, P. K. J. P. D.; Shahidi, F. Antioxidants: Science, Technology, and Applications. In Bailey's Industrial Oil and Fat Products. 6th ed. Fereidoon Shahidi, Ed.; John Wiley and Sons: New York, 2005.

Welch, K. D.; Davis T. Z.; Eden, M. E. V.; Aust, S. D. Deleterious iron-mediated oxidation of biomolecules. *Free Radic. Biol. Med*. 2002, 32, 577-583.

Won, E. J.; Rhee, J. S.; Kim, R. O.; Ra, K.; Kim, K. T.; Shin, K. H.; Lee, J. S. Susceptibility to oxidative stress and modulated expression of antioxidant genes in the copper-exposed *polychaete Perinereis nuntia. Comp. Biochem. Physiol*. 2012, 155, 344–351.

Wootton-Beard, P. C.; Moran, A.; Ryan, L. Stability of the total antioxidant capacity and total polyphenol content of 23 commercially available vegetable juices before and after in vitro digestion measured by FRAP, DPPH, ABTS and Folin–Ciocalteu methods. *Food Res. Int*., 2011, 44, 217–224.

Yanishlieva, N. V.; Marinova, E. M. Stabilisation of edible oils with natural antioxidants. *European Journal Lipid Science Technology*. 2001, 103, 752-767.

Yen, G. C.; Duh, P. D.; Tsai, H. L. Antioxidant and pro-oxidant properties of ascorbic acid and gallic acid. *Food Chem*. 2002, 79, 307–313.

Younes, M. Free Radicals and Reactive Oxygen Species. Toxicology. Elsevier: Geneva, 1999.

Zhao, J.; Khana, I. A.; Fronczek F. R. Gallic acid. *Acta Crystallogr*. 2011, 67, 316-317.

Zheng, W.; Wang, S. Y. Antioxidant activity and phenolic compounds in selected herbs. *J. Agric. Food Chem*. 2001, 49, 5165-5170.

In: Handbook on Gallic Acid
Editors: M. A. Thompson and P. B. Collins

ISBN: 978-1-62618-921-8

Chapter 4

AN OVERVIEW OF THE TECHNIQUES USED TO ESTIMATE THE ANTIOXIDANT ACTIVITY OF GALLIC ACID

Hongyan Li and Ze-Yuan Deng*
State Key Laboratory of Food Science and Technology, Nanchang University, Nanchang, Jiangxi, China
Institute for Advanced Study, Nanchang University, Nanchang, Jiangxi, China

ABSTRACT

The oxidative stress on human health has been a serious issue. It may involve ageing, inflammation, metabolic disorders and cancer. Gallic acid is widely distributed in a number of of fruits and plants. It possesses antioxidant activity, thus could decrease the oxidative stress in human. It could also possess anti-cancer activity and other bioactivities related to human health.

The techniques used to determine the antioxidant activity of gallic acid included *in vitro* and *in vivo* methods. The available assays are based on different mechanisms (hydrogen atom transfer and electron transfer), bioavailability, metabolism and other factors, which can lead to varied results. Therefore, this chapter is intended to provide a brief update of recent literature on the *in vitro* and *in vivo* methods available for evaluating the antioxidant activities of gallic acid and the antioxidant values of pure compound or food/plants were also compared and discussed.

The chemical-based methods such as total radical trapping antioxidant potential (TRAP) assay, oxygen radical absorbance capacity (ORAC) assay, luminol-photochemiluminescence (PCL) assay, ferric reducing antioxidant power (FRAP) assay, Trolox equivalent antioxidant capacity (TEAC) assay and 1,1-diphenyl-2-picrylhydrazine (DPPH) radical scavenging assay are used to evaluate the antioxidant activities of phytochemicals in food. The cell-based models include the inhibition on the growth of cancer cells, the protective effect of oxidative stress in various cells exogenously exposed

* Corresponding author. Tel./fax: +86 791 88304402. E-mail address: zeyuandeng@ hotmail. com.

to external stimuli and the cellular antioxidant assay (CAA). The high antioxidant activity *in vitro* does not always lead to high *ex vivo* or *in vivo* antioxidant property. Some *in vivo* studies have also shown promising health effects of gallic acid and the food/plants rich in gallic acid. The animal and human experiments are parts of the methods to evaluate *in vivo* antioxidant potentials. The oxidative stress biomarkers in serum or tissue such as glutathione peroxidase (GSH-Px), glutathione (GSH), superoxide dismutase (SOD), catalase (CAT) and lipid peroxides (LPO) could be used to express the *in vivo* antioxidant activity.

1. INTRODUCTION

Gallic acid (3,4,5-trihydroxybenzoic acid) is a trihydroxybenzoic acid and constitutes the basic building block of gallotannins. Gallic acid is obtained by alkaline or acid hydrolysis of tannins and known to have antioxidant, anti-inflammatory, antimutagenic and anticancer activities [1-3]. Studies in recent years have shown the important roles of gallic acid and the food/plants rich in gallic acid in the prevention of chronic diseases and promotion of human health [4-6]. Increasing evidence has pointed to gallic acid is extremely safe even at high doses. The epidemiology safety and efficacy of gallic acid makes it a potential compound for the prevention or treatment of human diseases [7-10].

Gallic acid is found widely in numerous plant species and is a naturally occurring plant phenolics with antioxidant activity [11]. Antioxidant activity is a key criterion for the evaluation of health benefit [5]. So far, there were a number of methods of evaluating the antioxidant activity and they were classified as *in vitro* and *in vivo* methods. However, there is still lack of standardized methods in the assessment of the antioxidant activity. Different *in vitro* chemical-based antioxidant assays rely on different chemical backgrounds, the analysis of the same sample by different methods may provide quite different results [12-14]. That is why the antioxidant activity of a sample should not be concluded based on a single analysis method [15, 16]. There are many *in vitro* chemical and cell based models available for the assessment of antioxidant activities; however, none of them can represent the "true" antioxidant status *in vivo* [17-19]. The antioxidant activities of phytochemicals such as gallic acid may be changed during the human digestive tract and cell biochemical processes once absorbed [20]. It is therefore imperative to develop analytical methods to access the antioxidant activity *in vivo* [21].

In this chapter, an overview of relevant methods (*in vitro* and *in vivo*) used in the evaluation of antioxidant activity of gallic acid and the food/plants rich in gallic acid are provided. Readers, who are interested in detailed information and in-depth discussion on the principle and procedures of antioxidant test models, are referred to comprehensive reviews by others [22-25].

2. *IN VITRO* CHEMICAL-BASED METHODS

In practice various assays are commonly carried out for investigating the antioxidant activity of gallic acid. Generally the available *in vitro* chemical-based assays are simple, reproducible and relatively straightforward to perform. The main reported *in vitro* chemical-based antioxidant assays and activities of gallic acid are listed in Table 1. They include methods based on hydrogen atom transfer (HAT) and electron transfer (SET) mechanisms [26]. Most of the assays can be carried out in a 96-well plate [27]. Trolox (6-Hydroxy-2,5,7,8-tetramethylchroman-2-carboxylic acid) is a water soluble analogue of vitamin E. The *in vitro* antioxidant activities are mostly expressed as Trolox or ascorbic acid equivalent values [28].

2.1. Methods Based on Hydrogen Atom Transfer (HAT)

HAT-based methods detect the ability of gallic acid to quench free radicals by hydrogen donation. In HAT-based assays, the gallic acid and substrate compete for thermally generated stable or short-lived radicals through the decomposition of azo compounds [29]. Gallic acid could inhibit or retard the oxidation of the oxidizable molecular probes.

Methods such as total radical trapping antioxidant potential (TRAP) assay, oxygen radical absorbance capacity (ORAC) assay and photochemi- luminescence (PCL) assay are the major HAT methods used to evaluate the antioxidant activity (Table 1).

Table 1. Reported major *in vitro* chemical-based antioxidant activities of gallic acid

Mechanisms	Method	Value	Reference
Hydrogen atom transfer (HAT)	TRAP assay	3.9 TE 3.3 AAE	[12]
	ORAC assay	5300 μM TE/g	[36]
	PCL assay	3.8 TE	[39]
Single electron transfer (SET)	FRAP assay	0.4 AAE 3 TE 30.17 ferrous sulfate equivalent	[12] [39] [45]
	TEAC assay	2 TE 3.48 TE 179 μM TE/mg	[39, 50]
Others	DPPH radical scavenging assay	EC_{50}: 8 μM 4.17 mM: 43.9% 5 μg/mL: 49% 37 μg/mL: 75%	[60] [61-63]

Abbreviations used: TRAP, Total radical trapping antioxidant potential; TE, Trolox Equivalent; AAE, Ascorbic acid equivalent; ORAC, Oxygen radical absorbance capacity; PCL, Photochemiluminescence; FPAP, Ferric reducing antioxidant power; TEAC, Trolox equivalent antioxidant capacity; DPPH, 2,2-diphenyl-1-picrylhydrazyl; EC_{50} is the amount of gallic acid necessary to decrease by 50% the initial DPPH concentration.

2.1.1. Total Radical Trapping Antioxidant Potential (TRAP) Assay

The TRAP method is based on the ability of gallic acid to inhibit the reaction between peroxyl radicals and a target probe during a controlled peroxidation reaction [13]. Peroxyl radicals are generally provided from thermal decomposition of AAPH (2,20-azobis-(2-methylpropionamidine) dihydrochlo- ride) or ABAP (2,2'-azo-bis(2-amidino-propane) hydrochloride). In addition, the TRAP assay uses R-phycoerythrin (R-PE) or dichlorofluorescin diacetate (DCFH-DA) as a fluorescent probe. The reaction progress of R-PE with AAPH or ABAP is monitored fluorometrically (excitation wavelength 495 nm, emission wavelength 575 nm).

The TRAP assay is automated, and serves as a useful tool for measuring antioxidant activity of gallic acid. According to Ghiselli *et al.* (1995), 120 μL of diluted sample is added to 2.4 mL of phosphate buffer (pH 7.4), 375 μL of bidistilled water, 30 μL of diluted R-PE and 75 μL of ABAP. The reaction kinetics at 38°C is recorded for 45 min by a luminescence spectrometer [30]. The antioxidant activity is evaluated by measuring the decay in decolouration. TRAP values are calculated as the lag time of the kinetic curve in the presence of the gallic acid compared to corresponding lag time for a standard (such as Trolox). The results are expressed as standard equivalent.

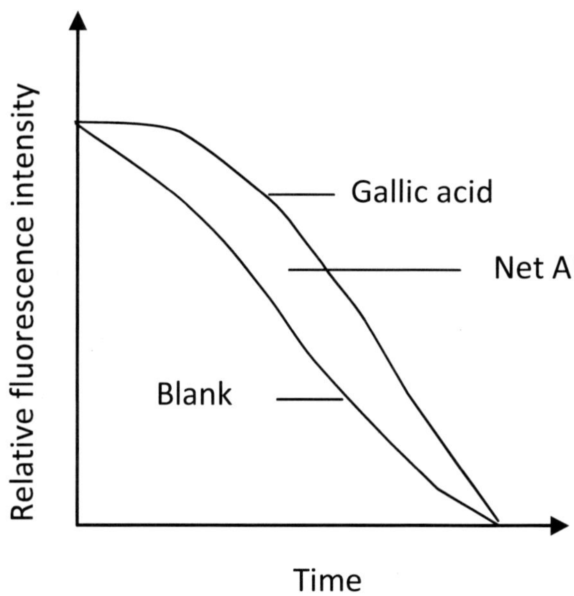

Figure 1. ORAC value of gallic acid expressed as the net area under the curve (AUC).

It was reported that based on the same TRAP value, 1 μM gallic acid was equivalent to 3.9 μM Trolox and 3.3 μM ascorbic acid (Table 1) [12]. Gallic acid is the main phytochemical in some food/plants and is also the main contributor to their good antioxidant activities (Table 2). Thus, the gallic acid content and the antioxidant activities in some typical food/plants were reviewed and listed in Table 2. For example, in extracts from seeds of lupin species *(Lupinus albus*, *Lupinus luteus* and *Lupinus angustifolius* cultivars), the gallic acid contents varied from 0.62 to 4.21 mg/kg DW (dry weight) and the TRAP value were 0.33-1.78 μM TE (Trolox equivalent)/g DW [31].

Table 2. The *in vitro* chemical-based antioxidant activities of typical food/plants rich in gallic acid

Food/plants	Gallic acid content	Antioxidant activity	Reference
Lupin seed	0.62 to 4.21 mg/kg DW	TRAP value: 0.33-1.78 μM TE/g DW	[31]
Longan seed	23.04 mg/g DW	ORAC value: 1400 μM TE/g DW	[36]
Staghorn sumac	5.97 mg/g DW	PCL value: 4513 μmol TE/g DW	[40]
Litchi pulp	11.51-105.52 μg/100 g FW	FRAP value: 121.53-220.96 mg TE/100 g FW	[46]
Habenaria edgeworthii	143.63 mg/100 g DW	FRAP value: 2.96 mM AAE/100g DW	[47]
Potherb mustard	2.0 μg/g DW	TEAC value: 47 mg TE/g DW	[52]
Raisin grape	24.5 μg/g DW	TEAC value: 60.9 μmol TE/g DW	[53]
Mango kernel	43.57 mg/100 g DW	TEAC value: 1.25 mol AAE/g DW	[51]
Toona sinensis leave	0.39 mg/g FW	DPPH value: 0.084 g TE/g FW	[64]
Red grape pomaces	18.68 mg/100 g DW	DPPH value: 361.12 μM TE/g DW	[65]
Portuguese wine	302.58 mg/L	DPPH scavenging: 93.5%	[66]

Abbreviations used: TRAP, Total radical trapping antioxidant potential; TE, Trolox Equivalent; DW, Dry weight, AAE, Ascorbic acid equivalent; ORAC, Oxygen radical absorbance capacity; PCL, Photochemiluminescence; FPAP, Ferric redu- cing antioxidant power; FW, Fresh weight; TEAC, Trolox equivalent antioxidant capacity; DPPH, 2,2-diphenyl-1-picrylhydrazyl.

2.1.2. Oxygen Radical Absorbance Capacity (ORAC) Assay

The OARC method is based on the radical chain breaking ability of gallic acid to inhibit the peroxyl radicals. Gallic acid can rapidly react with peroxyl radicals and inhibit the loss of fluorescence intensity. This inhibition is proportional to the antioxidant activity. AAPH, a water-soluble azo compound, was used as peroxyl radical generator. The peroxyl radical reacts with a fluorescent probe (fluorescein) resulting in the loss of fluorescence, which is technically much more sensitive to small variations in concentration.

The ORAC assay is considered to be a good method because of its biological relevance to the *in vivo* antioxidant efficacy [32]. According to Ou *et al.* (2001), 25 μL of blank, Trolox standard or sample were mixed with 200 μL fluorescein (0.0868 nM) solution and incubated for 30 min at 37°C before injection of 25 μL AAPH (153 mM) [33]. The fluorescence was measured every minute for about 120 min until it reached zero (excitation wavelength 485 nm, emission wavelength 528 nm) in a fluorescence spectrophotometer. ORAC values are calculated as the area under the fluorescein decay curve between the blank and the gallic acid (Net AUC, Figure 1). The results were expressed as Trolox equivalent [34].

It was shown that the ORAC value of 0.9 mg/mL gallic acid was 111 μM Trolox equivalent [35]. Based on the same ORAC value, 1 g gallic acid was equivalent to 5300 μM

Trolox [36]. Furthermore, dried longan seed extracts contained a high level of gallic acid (23.04 mg/g DW) and also exhibited a high ORAC value (1400 μM TE/g DW) [36].

2.1.3. Photochemiluminescence (PCL) Assay

The PCL method is based on the gallic acid generation of free radicals and the chemiluminescent detection. It does not require oxidizing reagent for the production of peroxyl radical. Antioxidant activity is determined as the disappearance of the radicals caused by hydrogen donation from gallic acid. Luminol is used as a photosensitizer which generates superoxide radicals and a chemiluminogenic probe for free radicals. It does not require high temperature to generate free radicals and is reported to have a high sensitivity (nanomolar range) [37].

PCL assay is a quick and cost effective method for the evaluation of antioxidant activity. It was performed on a Photochem® system [38]. Complete reagent kits were purchased from the manufacturer. The assay was a mixture of 2.3 mL of Reagent 1 (sample solvent), 0.2 mL of Reagent 2 (reaction buffer), 25 μL of diluted Reagent 3 (luminol, photosensitizer) and different concen- trations of Reagent 4 (Trolox) for the calibration curve. To measure the antioxidant activity of gallic acid, reagent 4 was simply replaced by gallic acid. The antioxidant activity is quantified based on their inhibitory effect on luminescence generation, and is expressed as Trolox equivalent.

It was reported that the PCL values of gallic acid were almost 2.5 Trolox and 2.9 ascorbic acid equivalent [39]. Wu *et al.* (2013) showed that the gallic acid content and PCL value of Staghorn sumac (*Rhus hirta* L.) were 5.97 mg/g DW and 4513 μmol TE/g DW [40]. This high antioxidant activity was related to the gallic acid content.

2.2. Methods Based on Single Electron Transfer (SET)

SET-based assays measure the ability of gallic acid to transfer one electron to reduce oxidants [29]. This ability is accompanied by colour change and the degree of this change is proportional to the antioxidant activity. Feric reducing antioxidant power (FRAP) and Trolox equivalent antioxidant capacity (TEAC) method represent SET-based methods used to evaluate the antioxidant activity of gallic acid (Table 1).

2.2.1. Feric Reducing Antioxidant Power (FRAP) Assay

The FRAP assay measures the ability of gallic acid to reduce ferric tripyridyltriazine (Fe^{3+}-TPTZ) complex to the intensely blue coloured ferrous form (Fe^{2+}) (Figure 2) [41]. This assay is carried out in acidic medium and the reduction is monitored by measuring the change in absorption at 593 nm [42]. It helps maintain iron solubility and increase the redox potential, which could cause a shift in dominant reaction mechanism.

FRAP assay is simple, fast and cost-effective and does not require specialized equipment. According to Khanizadeh *et al.* (2008), standard (such as ascorbic acid) or gallic acid (10 μL) was mixed with 300 μL of ferric-TPTZ reagent (prepared by mixing 300 mM acetate buffer, pH 3.6, 10 mM TPTZ in 40 mM HCl and 20 mM $FeCl_3 \cdot 6H_2O$ in the ratio of 10:1:1 (v/v/v)) and added to a 96 well plate [43]. The plate was incubated at 37°C for the duration of the reaction. The absorbance readings were taken at 593 nm using a UV-visible microplate kinetic reader. The FRAP value was expressed as ascorbic acid equivalent or ferrous sulfate ($Fe_2SO_4{\cdot}7H_2O$) equivalent concentration [44].

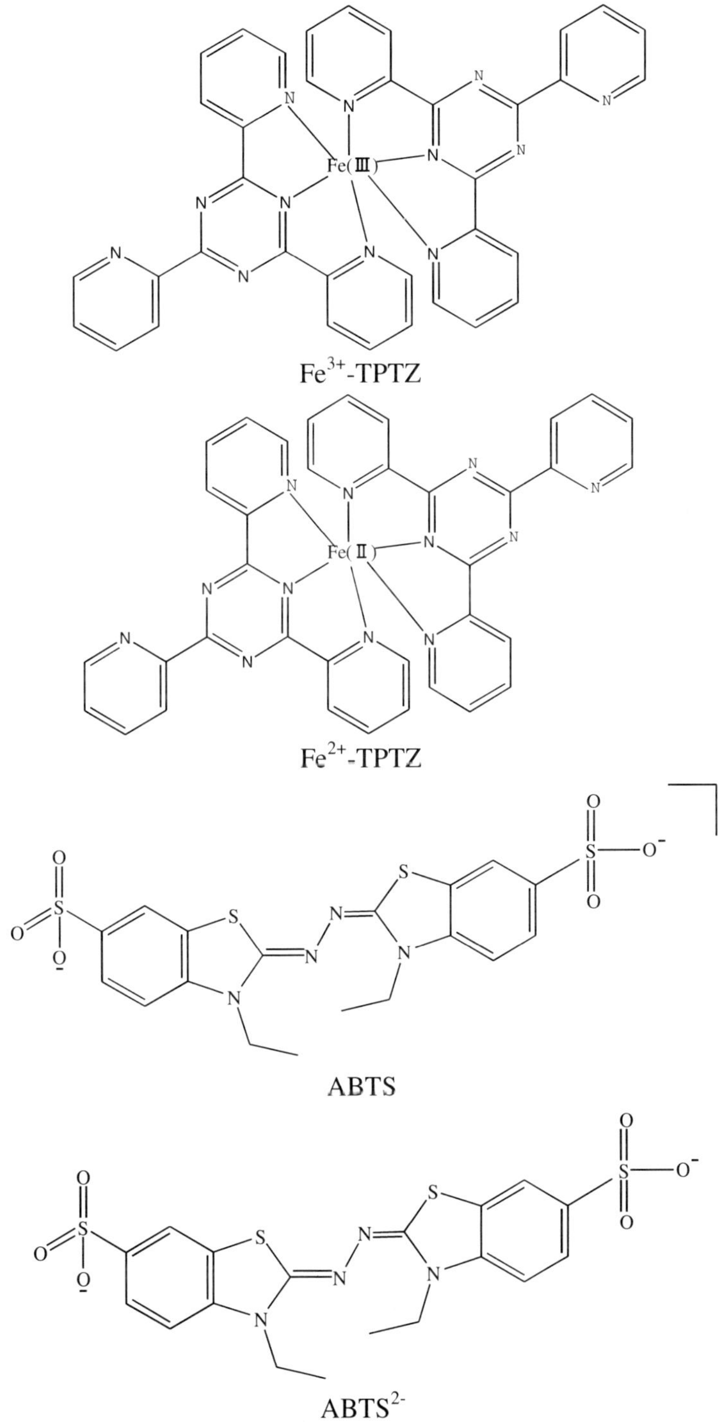

Figure 2. Structures of compounds used in the antioxidant reaction (FRAP and TEAC assays).

It was reported that the FRAP value of gallic acid was 30.17 ferrous sulfate ($Fe_2SO_4 \cdot 7H_2O$) equivalent [45]. Based on the same FRAP value, 1 μM gallic acid was equivalent to 1.1 μM Trolox and 0.4 μM ascorbic acid [12]. Kranl *et al.* (2005) also showed that the 1 mM gallic acid was equivalent to 6.3 mM ferrous ions formed, which is much higher than Trolox (2.1 ferrous ions formed), caffeic acid (2.2 ferrous ions formed) and ascorbic acid (2.4 ferrous ions formed) [39]. The contents of gallic acid in the pulp of 11 litchi cultivars were reported to 11.51-105.52 μg/100 g FW (fresh weight) and the FRAP values ranged from 121.53 to 220.96 mg TE/100 g FW [46]. The callus suspension cultures of *Habenaria edgeworthii* (A rare Himalayan medicinal orchid) also showed the high gallic acid content (143.63 mg/100 g DW) and FRAP value (2.96 mM AAE/100g DW) [47].

2.2.2. Trolox Equivalent Antioxidant Capacity (TEAC) Assay/ABTS Radical Cation Decolourization Assay

The TEAC assay is based on the capacity of gallic acid to inhibit the radical compared with a reference standard (Trolox) [48]. The oxidant $ABTS^{\bullet -}$ was generated by oxidation of $ABTS^{2-}$ using potassium persulfate ($K_2S_2O_8$) and the radical anion $ABTS^{\bullet -}$ can be reduced by gallic acid to release the colourless product $ABTS^{2-}$ (Figure 2). $ABTS^{\bullet -}$ is intensely coloured (blue-green chromophore) and shows maximum absorption at 734 nm [49].

The TEAC assay is simple in operation and can be automated. According to the method described by Zulueta *et al.* (2009), 25 mL ABTS (7 mM) was spiked with 440 μL of $K_2S_2O_8$ (140 mM) and allowed to stand in darkness at room temperature for 12-16 h (the time required for formation of the radical) [35]. The solution was diluted in ethanol until its absorbance at 734 nm was 0.70. The measurement was carried out in a spectrophotometer. 2 mL of the solution and 100 μL of sample or standard were added to the tube and absorbance is read at 734 nm after 6 min. Gallic acid could inhibit the radical, produce a reduction in absorbance and the results were expressed by comparison with the amountTrolox or ascorbic acid to give rise to the TEAC.

The TEAC value of 0.9 mg/mL gallic acid was 161 μM Trolox equivalent [35]. Kranl *et al.* (2005) and Apak *et al.* (2007) showed that the TEAC value of gallic acid was almost 2 and 3.48 Trolox equivalent, respectively [39, 50]. The gallic acid content of mango (*Mangifera indica* L.) kernel was 43.57 mg/100 g DW and the TEAC value was 1.25 mol AAE (ascorbic acid equivalent)/g DW [51]. Similarly, the gallic acid content and TEAC value in potherb mustard (*Brassica juncea*, Coss.) was 2.0 μg/g DW and 45 mg TE/g DW [52]. In ten new raisin grape selections, A95-27 had higher gallic acid content (24.5 μg/g DW) and greater TEAC value (60.9 μmol TE/g DW) than other cultivars [53].

2.3. Others

Methods such as DPPH (2, 2-diphenyl-1-picrylhydrazyl) are based on the free radical scavenging ability of gallic acid. However, the mechanism of DPPH assay is not clear [54].

DPPH is a stable free radical which gives a deep purple colour and it does not need to be generated prior to the assay. In the presence of antioxidant such as gallic acid, which could donate a hydrogen atom, DPPH radicals are reduced and the reduction reaction results in decolourization of the solution. Moreover, the delocalization of electron also gives rise to the deep violet colour. The DPPH assay is considered to be mainly based on SET, and HAT is a marginal reaction pathway in this assay [55]. The loss of colour can be monitored at 517 nm using a spectrophotometer. Results were reported as percent scavenging, EC_{50} or T_{EC50}. EC_{50} is the amount of gallic acid or sample necessary to decrease by 50% the initial DPPH concentration [56]. The time taken to reach the steady state to EC_{50} concentration is T_{EC50} [57].

The DPPH radical scavenging assay is a simple technique and requires only a UV spectrophotometer to perform. According to Wu *et al.* (2011), 100 µL of a methanolic solution of DPPH (0.065 mM) was mixed with 20 µL of gallic acid or a standard solution in a 96 well plate. The plate was allowed to stand for 30 min at room temperature before the absorbance was recorded at 517 nm using the microplate kinetic reader [58]. The radical scavenging activity of gallic acid was calculated as follows:

$$Percent\ scavenging\ (\%) = \frac{A_0 - (A_1 - A_s)}{A_0} \times 100$$

where A_0 is the absorbance of DPPH alone, A_1 is the absorbance of DPPH and gallic acid and A_S is the absorbance of gallic acid only.

The EC_{50} value of gallic acid was reported to be 8 µM, which was 90 times higher than chitosan-gallic acid (740 µM) [59]. The EC_{50} and T_{EC50} values of 1 g catechin:gallic acid (1:1) was 0.067 g DPPH and 19.7 min [60]. Yen *et al.* (2002), Cho *et al.* (2011) and Sroka *et al.* (2003) showed that gallic acid exhibited 43.9%, 49.03% and 75% scavenging effects on DPPH radicals at concentrations of 4.17 mM, 5 µg/mL and 37 µg/mL, respectively [61-63]. Moreover, it was shown that the gallic acid content in *Toona sinensis* leave was 0.39 mg/g FW and the DPPH value of 1 g sample was equivalent to 0.084 g Trolox [64]. Gallic acid was present in high concentration (18.68 mg/100 g DW) in red grape (*Vitis labrusca* L.) pomaces and the DPPH value was 361.12 µM TE/g DW [65]. Portuguese wine brandy aged in chestnut wood presents higher antioxidant activity (DPPH scavenging: 93.5%) than the brandy aged in Limousin oak wood (DPPH scavenging: 45.7%) because of the greater gallic acid content (302.58 mg/L *vs.* 38.19 mg/L) [66].

3. *In Vitro* Cell-Based Methods

Chemical-based methods are cost effective, high-throughput and yield an index value that allows comparing and ranking different products. However, the chemical-based methods are limited because they could not reflect the antioxidant status at the cellular level such as lipophilicity and bioavailability [13, 28]. Biological systems are much more complex than the simple chemical mixtures employed. Thus, cell-based antioxidant assays have been increasingly used for accessing the antioxidant activity in recent years [67].

3.1. Inhibition of Cancer Cell Proliferation

Gallic acid helps reduce the risk of developing chronic diseases such as cancer, based partly on cellular studies establishing that gallic acid could inhibit the proliferation of cancer cells. Gallic acid might limit the proliferation of cancer cells or prevent a new tumor from increasing in its size [68]. It has been proven that gallic acid could inhibit the growth of cancer cells, including cancers of the esophagus, colon, breast, stomach, bladder, lung, prostate and cervix [69-73].

According to Jang *et al.* (2009), the cancer cells were inoculated into a 96-well plate [45]. After 24 h, gallic acid was added into the well and incubated for 72 h. The 3-(4,5-dimethylthiazol-2-yl)-2,5-diphenyl tetrazolium bromide (MTT) solution (5 mg/ml, 20 μL) was added and incubated for a further 4 h. The growth medium was removed and 150 μL of DMSO:ethanol (1:1) was added into each well and vortexed for 30 min. The absorbance of reaction mixture was read by an ELISA reader at 550 nm. The growth inhibition rate was calculated as follows:

$$Growth\ inhibition effect\ (\%) = \frac{A_1}{A_0} \times 100$$

where A_0 is the absorbance of the blank, a_1 is the absorbance of gallic acid.

It was shown that the inhibition activities of cancer cell proliferation inhibition activities by gallic acid at 10 ppm on G361 human melanoma, B16F10 mouse melanoma and A549 lung cancer were 51%, 23% and 3% respectively [45]. Chen *et al.* (2009) suggested that gallic acid had the potential to be developed into an anti-prostate cancer drug because it could block the growth of DU145 prostate cancer cells at G2/M phases by activating Chk1 and Chk2 and inhibiting Cdc25C and Cdc2 activities [74]. Furthermore, gallic acid has a synergistic effect with doxorubicin in suppressing the growth of prostate cancer cells. Gallic acid (0.275 mM) purified from *Phyllanthus emblica* L. fruit had apparent capacity of inhibiting the survival of MCF-7 human cancer cell line (IC_{50}: 64.88 μg/mL) [75]. Similarly, the extract of whole cowpeas (*Vigna unguiculata*) seeds at a concentration of approximately 100 mg gallic acid equivalent/L (the gallic acid content was 27.09 μg/g DW) inhibited the proliferation of MCF-7 cancer cells [76]. The effect of 0.01 mM gallic acid induced the same level on the inhibition of gastric adenocarcinoma cell metastasis as 4.0 mM protocatechuic acid [6].

3.2. Protective Effects of Normal Cell Death

Antioxidant activity of gallic acid can also be evaluated in different normal cell culture assay for the protective effects of external stimuli and carcinogenesis. Hydrogen peroxide (H_2O_2) has emerged as a pivotal molecule for determining the fate of cells [77]. Cell lines based model system, using cells exogenously exposed to H_2O_2 was used to evaluate the protective effect of gallic acid [78]. Gallic acid has antioxidative and protective effects against the cytotoxicity and clastogenesis induced by H_2O_2 [79]. The ability of reducing H_2O_2 toxicity was observed when cells were treated with gallic acid during and after exposure to H_2O_2 [80].

According to Li *et al.* (2013), the H9c2 cells were seeded in 96-well microstate plates at a density of $7x10^3$-$8x10^3$ cells per well [81]. The plate was incubated at 37°C for 24 h before further treatments. Oxidative stress was induced by treating with freshly prepared H_2O_2. Gallic acid was performed in the above described 96-well plates with pre-seeded cells and the treated cells were washed three times with PBS, and then added graded concentrations of H_2O_2 and incubated at 37 C for 3 h before staining. Cell viability was evaluated by MTT or sulforhodamine dye binding method [82].

Gallic acid was found to protect cell death in various normal cell lines and pretreatment with gallic acid led to a significant revival of the oxidative status. When the B14 Chinese hamster cells exposed to 1 μM gallic acid in the presence of 100 μM H_2O_2 for 1 h, the antioxidant activity of the cells was almost 2 times higher than the control [83]. Ozlem *et al.* (2012) suggested that gallic acid was a powerful antioxidant and exerted anti-apoptotic effects because of the protective effect of gallic acid against H_2O_2-induced oxidative stress and apoptotic cell death in HeLa cells, a model for human epithelial cells [84]. It was also shown that gallic acid (10 and 20 mg/kg) isolated from *Peltiphyllum peltatum* had protective effects against sodium fluoride-induced hepatotoxicity and oxidative stress [85].

3.3. Cellular Antioxidant Assay

Cellular antioxidant assay (CAA) has been developed to measure the antioxidant activity of an absorbed compound inside a living cell [28, 86]. CAA considers cellular uptake, distribution and efficiency of protection against peroxyl radicals under physiological conditions [87]. Oxidation of DCFH to DCF has been used as an indicator of oxidative stress. ABAP is used to generate peroxyl radicals to initiate oxidation and the final value was measured by a fluorescence microplate reader [88].

According to Wolfe *et al.* (2007), the HepG2 cells were seeded in black microfiber 96-well plates for 24 h [28]. The growth medium was removed and 25 μM DCFH-DA were added after the cells were treated with gallic acid for 1 h [62]. 600 μM ABAP (in HBSS) was added after washing the cells with PBS for three times. The formation of fluorescent dichlorofluorescein DCF (2′,7′-dichlorofluorescein), due to the oxidation of DCFH in the presence of various reactive oxygen species was measured every 5 min (excitation wavelength 485 nm, emission wavelength 538 nm) at 37°C [89]. The area under the curve for fluorescence versus time was integrated to calculate the CAA value and the IC_{50} values were converted to CAA values, expressed as μmol of quercetin per 100 μmol of gallic acid.

$$CAA\ unit = \left(100 - \frac{\int SA}{\int CA}\right) \times 100$$

where ∫SA is the integrated area under the sample fluorescence versus time curve and ∫CA is the integrated area from the control curve.

It was reported that 100 μM gallic acid was equivalent to approximately 450 μM caffeic acid, 230 μM catechin, 100 μM ascorbic acid, 25 μM myricetin and 9 μM quercetin in the cellular antioxidant activity [28]. The concentration of gallic acid to inhibit DCFH oxidation by 50% (IC_{50}) was 2.6 μM, which was much higher than α-tocopherol (119 μM) [90].

4. *In Vivo* Methods

Taking into account the complexity involved in the *in vivo* action of gallic acid, animal and human experiments are necessary. Some *in vivo* studies have shown promising health effects of gallic acid. For all *in vivo* methods, gallic acid is usually administered to the animals (mice, rats, *etc.*) at a definite dosage regimen [91-93]. After a specified period of time, the animals are usually sacrificed and the blood or tissues are used for the antioxdiant assays [94, 95]. Most of the activities could be estimated using commercial kits [96]. Gallic acid has been proven to have protective effect over oxidative damage *in vivo* mainly due to it redox property, which allows gallic acid to act as a reducing agent, hydrogen donator and singlet oxygen quenchers [97-99].

4.1. Glutathione Peroxidase (GSH-Px) Estimation

Glutathione and glutathione-related enzymes are involved in the metabolism as well as detoxification of cytotoxic and carcinogenic compounds [100]. GSH-Px was reported to catalyze the reaction of hydroperoxides with reduced glutathione to form glutathione disulfide and the reduction product of hydroperoxide [101]. Low activity of GSH-Px is one of the early consequences of a disturbance of the prooxidant/antioxidant balance [102, 103].

According to the method of Paglia and Valentine (1967), 50 μL of NADPH reagent (5 mM NADPH, 42 mM GSH, 10 units/mL of GSH-reductase in 1.25 mL of distilled water) were added to 890 μL of GSH-Px buffer (50 mM Tris HCl, pH 8.0, 0.5 mM EDTA) [104]. Then, 50 μL of serum sample and 10 μL of 30 mM tert-butyl hydroperoxide solution were added to the mixture. The reaction mixture was incubated at 37 C and the absorbance at 340 nm was determined [105]. Enzyme activity of GSH-Px was expressed in terms of mg of proteins [106].

It was reported that 50 mg/kg of gallic acid showed significant increase in hepatic GSH-Px activity (0.32 U/g) when compared with those administered with CCl_4 only (0.21 U/mg) [105]. Padma *et al.* (2011) found that gallic acid could significantly prevent the lindane induced alterations in kidney and liver tissues with a significant increase in antioxidant levels [107]. The GSH-Px activities of 50 mg/kg gallic acid and 100 mg/kg Lindane co-treatment group were 1.17 and 1.16 times higher than those of 100 mg/kg Lindane treatment group in the kidney and liver tissues of rats [107]. GSH-Px activities in the liver of mice were increased significantly by *Trachelospermum jasminoides* (Apocynaceae) extract (2.0 g/kg BW, the gallic acid content was 0.206 mg/g DW) at the third hour after carrageenan injection [108].

4.2. Reduced Glutathione (GSH) Estimation

GSH is an intra-cellular reductant and plays major role in catalysis, metabolism and transport [109]. GSH synthesis can be altered by the administration of glutathione precursors or GSH peroxidase and GSH reductase activities [110]. Glutathione plays an important role in the kidney and takes part in a transport system involved in the reabsorption of amino acids.

The method illustrated by Ellman (1959) can be used for determination of antioxidant activity [111]. The tissue homogenate is taken and added with equal volume of 20% trichloroacetic acid (TCA) containing 1 mM EDTA to precipitate the tissue proteins. The mixture is allowed to stand for 5 min prior to centrifugation for 10 min at 2000 rpm. The supernatant (200 μL) is then transferred to a new set of test tubes and added with 1.8 mL of the Ellman's reagent (0.1 mM 5,5 0-dithiobis-2-nitrobenzoic acid prepared in 0.3 M phosphate buffer with 1% of sodium citrate solution). Then all the test tubes are made up to the volume of 2 mL. After completion of the total reaction, solutions are measured at 412 nm against blank.

It was reported that the GSH activities of 50 mg/kg gallic acid and 100 mg/kg Lindane co-treatment group were 41.12 and 44.39 nmol/g in the kidney and liver tissues of rats, which were much higher than the 100 mg/kg Lindane treatment group (35.04 and 37.96 nmol/g) [107]. Co-treatment with CCl_4 and 200 mg/kg gallic acid for 24 h, the GSH value was increased significantly (6.3 μmol/g *vs* 3.7 μmol/g) in albino rats [112]. Moreover, 100 μM gallic acid could prevent cumene hydroperoxide (120 μM) induced GSH depletion for 3 h (GSH activity: 37.8 μM *vs* 10 μM) [113]. Gallic acid isolated from *Peltiphyllum peltatum* could inhibit sodium fluoride (NaF)-induced hepatotoxicity and oxidative stress [85]. Pretreatment with gallic acid at 20 mg/kg demonstrated a significant increase in GSH activity (41%) and leaded to a significant revival of the oxidative status.

4.3. Superoxide Dismutase (SOD) Method

Superoxide anion, a weak oxidant, it ultimately produces powerful and dangerous hydroxyl radicals as well as singlet oxygen, both of which contribute to oxidative stress [114, 115]. Several spectrophotometric SOD assays with different superoxide detecting dyes, such as nitroblue tetrazolium, adrenaline and sulphanilamide, have been reported [24].

The SOD activity can be measured as described by Spitz and Oberley (1989) [116]. Xanthine and xanthine oxidase are used to generate superoxide anion radical. First, 50 μL of the diluted sample and 1.7 mL of the mixed substrate (25 μmol/L 2-(4-iodophenyl)-3-(4-nitrophenol)-5-phenyltetrazolium chloride (INT) and 50 μmol/L xanthine) were added to 250 μL of xanthin oxidase. SOD activity is measured by the degree of reaction inhibition. The absorbance of samples was measured at a wavelength of 505 nm. The SOD activity was expressed as units (U) of SOD activity/mg protein [117].

It was shown that 25 μg/mL and 50 μg/mL gallic acid could increase the SOD activities (789 and 608 U/mg) of AAPH-induced oxidative hemolysis (242 U/mg) significantly [64]. Padma *et al.* (2011) also found that the SOD activities of 50 mg/kg gallic acid and 100 mg/kg Lindane co-treatment group were 1.39 and 1.23 times higher than those of 100 mg/kg Lindane treatment group in the kidney and liver tissues of rats [107]. Administration of CCl_4 induced a significant decrease in the SOD content in liver (80 U/mg) and kidney (75 U/mg) when compare to the control (150 U/mg) [118]. However, the extract of *Moringa oleifera* leave was found to inhibit the toxicity produced by CCl_4 administration and the SOD activity at 100 mg/kg was comparable to 50 mg/kg Vitamin E, which was caused by a high content of gallic acid (534.4 μg/g) [118].

4.4. Catalase (CAT)

Catalase exerts a dual function, decomposition of H_2O_2 to give H_2O and O_2, and oxidation of H donors [119]. In the ultraviolet range H_2O_2 shows a continual increase in absorption with decreasing wavelength [120]. One unit of catalase could be defined as the amount of enzyme required to decompose 1 μM of H_2O_2 in 1 min.

According to Aebi's method (1984), tissues were homogenized with 0.05 M phosphate buffer at pH 7.0 and centrifuged at 4 C for 15 min [121]. The supernatants were removed and kept at -20 C. In brief, 100 μL of the diluted sample and 1 mL of 59 mmol/L H_2O_2 (dissolved in 50 mmol/L potassium phosphate buffer, pH 7.0) were added to 1.9 mL of deionized water. CAT activity was measured spectrophotometrically at a wavelength of 570 nm and the results expressed as μmol of H_2O_2 consumed/mg protein or units (U) of CAT activity/mg protein.

Tung *et al.* (2009) evaluated the hepatoprotective effects of gallic acid against CCl_4-induced hepatotoxicity in rats and found that 50 mg/kg of gallic acid showed a significant increase in hepatic CAT activity (79 U/g) when compared with those administered with CCl_4 only (55 U/g) ($p < 0.05$) [105]. The CAT activities in the blood of the aging model induced by injection of D-galactose into 9-month-old male senescence accelerated mice declined significantly [94]. However, when these mice were treated by gallic acid purified from rose flowers for 30 days, it not only reinstated the activities of CAT in the blood, but also significantly reduced malondialdehyde (MDA) content in the liver, brain and kidneys (27.1%, 13.9% and 2.1%, respectively).

4.5. Lipid Peroxidation (LPO) Assay

LPO is an indicator of oxidative stress, which may cause peroxidative tissue damage in inflammation, aging, cancer and toxicity of xenobiotics [122]. Malondialdehyde (MDA) is formed during oxidative degeneration as a product of free oxygen radicals, which is one of the end products in the lipid peroxidation process and has been accepted as an indicator of lipid peroxi-dation [123, 124].

According to the method described by Ohkawa *et al.* (1979), lipid peroxidation was assessed indirectly by measuring the amounts of MDA produced primarily [125]. The tissues are homogenized in 0.1 M buffer pH 7.4 [124]. 0.2 mL of homogenate was mixed with 0.2 mL of 8.1% (w/v) sodium dodecyl sulphate solution, 1.5 mL of a 0.8% (w/v) solution of TBA and the final volume was adjusted to 4.0 mL with distilled water, and incubated at 95°C for 1 h, before cooling in an ice bath for 10 min prior to addition of 150 μL trichloroacetic acid (20%) [126]. After centrifugation at 13,000 rpm for 10 min, the absorbance of the supernatant fraction was measured at 532 nm and results were expressed as nmol of MDA formed/mg protein.

Acute administration of 1.5 mL/kg CCl_4 to rats could induce an increase in lipid peroxidation (1.6 nmol MDA/mg protein). It was shown that co-treatments with CCl_4 and 200 mg/kg (0.58 nmol MDA/mg protein) or 100 mg/kg gallic acid (0.82 nmol MDA/mg protein) in albino rats could protect the liver injury significantly [112]. It was also found that 200 mg/kg gallic acid was the most effective against CCl_4 induced liver and kidney damage [112]. In another experiment, gallic acid were intraperitoneally administrated for 1 week prior to sodium fluoride intoxica- tion and the results showed that pretreatment with 20 mg/kg gallic acid exhibited significant reduction in the thiobarbituric acid reactive substances level (37.06 nmol MDA/g) than VE (42.02 nmol MDA/g) in brain tissue [127]. Pre-treatment of green pods of *Acacia nilotica* (the gallic acid content was 105 mg/g DW) at 75 and 150 mg/kg body weight for 6 days caused a significant decrease in the level of MDA content in liver, lungs, kidneys and blood when compared to CCl_4-intoxicated rats [128].

Conclusion

There is sufficient data showing that gallic acid is a strong antioxidant. Antioxidant activity of gallic acid has been assessed in many ways. As per this review there are 9 *in vitro* methods and 5 *in vivo* methods that are being used for the evaluation of antioxidant activity of gallic acid and the extracts of food/plants rich in gallic acid. It was prepared based on literature search. DPPH method was found as the mostly used *in vitro* antioxidant assay while LPO was the mostly frequently used one for *in vivo* antioxidant activity evaluation. The combination of these methods available explains the measurement of antioxdiant activity of gallic acid cannot be evaluated by a simple test only. The food/plants rich in gallic acid also showed good antioxidant activity, which might be considered great contribution of gallic acid. This chapter provides an insight into the potential of using gallic acid to promote health and prevent diseases.

References

[1] Locatelli, C., Filippin-Monteiro, F. B. & Creczynski-Pasa, T. B. (2013). Alkyl esters of gallic acid as anticancer agents: a review. *European Journal of Medicinal Chemistry,* 60, 233-239.

[2] Chen, C.-Y., Lee, R.-J., Lee, V. S. Y., Dou, J., Preedy, V. R., Tzen, Jason t. C. & Lee, M.-R. (2013) Chapter 37: *Gallic Acid in Old Oolong Tea.* In Tea in Health and Disease Prevention. (Academic Press) pp. 447-456.

[3] Kroes, B. H., Van Den Berg, A. J. J., Quarles Van Ufford, H. C., Van Dijk, H. & Labadie, R. P. (1992). Anti-inflammatory activity of gallic acid. *Planta Medica,* 58, 499-504.

[4] Peng, C.-C., Hsieh, C.-L., Wang, H.-E., Chung, J.-Y., Chen, K.-C. & Peng, R. Y. (2012). Ferulic acid is nephrodamaging while gallic acid is renal protective in long term treatment of chronic kidney disease. *Clini- cal Nutrition,* 31, 405-414.

[5] Lu, Z., Nie, G., Belton, P. S., Tang, H. & Zhao, B. (2006). Structure–activity relationship analysis of antioxidant ability and neuroprotective effect of gallic acid derivatives. *Neurochemistry International,* 48, 263-274.

[6] Ho, H.-H., Chang, C.-S., Ho, W.-C., Liao, S.-Y., Wu, C.-H. & Wang, C.-J. (2010). Anti-metastasis effects of gallic acid on gastric cancer cells involves inhibition of NF-κB activity and downregulation of PI3K/AKT/small GTPase signals. *Food and Chemical Toxicology,* 48, 2508-2516.

[7] Priscilla, D. H. & Prince, P. S. M. (2009). Cardioprotective effect of gallic acid on cardiac troponin-T, cardiac marker enzymes, lipid peroxidation products and antioxidants in experimentally induced myocardial infarction in Wistar rats. *Chemico-Biological Interactions,* 179, 118-124.

[8] Punithavathi, V. R., Prince, P. S. M., Kumar, R. & Selvakumari, J. (2011). Antihyperglycaemic, antilipid peroxidative and antioxidant effects of gallic acid on streptozotocin induced diabetic Wistar rats. *European Journal of Pharmacology,* 650, 465-471.

[9] Abdelwahed, A., Bouhlel, I., Skandrani, I., Valenti, K., Kadri, M., Guiraud, P., Steiman, R., Mariotte, A.-M., Ghedira, K., Laporte, F., Dijoux-Franca, M.-G. & Chekir-Ghedira, L. (2007). Study of antimuta- genic and antioxidant activities of Gallic acid and 1,2,3,4,6-pentagalloylglucose from Pistacia lentiscus: Confirmation by microarray expression profiling. *Chemico-Biological Interactions,* 165, 1-13.

[10] Appeldoorn, C., Bonnefoy, A., Lutters, B., Daenens, K., Van Berkel, T., Hoylaerts, M. & Biessen, E. (2004). M.444 Gallic acid is a potent antagonist of P-selectin: Implications for the French paradox. *Athero-sclerosis Supplements,* 5, 102-103.

[11] Werner, R. A., Rossmann, A., Schwarz, C., Bacher, A., Schmidt, H.-L. & Eisenreich, W. (2004). Biosynthesis of gallic acid in Rhus typhina: discrimination between alternative pathways from natural oxygen isotope abundance. *Phytochemistry,* 65, 2809-2813.

[12] Tafulo, P. a. R., Queirós, R. B., Delerue-Matos, C. M. & Sales, M. G. F. (2010). Control and comparison of the antioxidant capacity of beers. *Food Research International,* 43, 1702-1709.

[13] López-Alarcón, C. & Denicola, A. (2013). Evaluating the antioxidant capacity of natural products: A review on chemical and cellular-based assays. *Analytica Chimica Acta,* 763, 1-10.

[14] Ghiselli, A., Serafini, M., Natella, F. & Scaccini, C. (2000). Total antioxidant capacity as a tool to assess redox status: critical view and experimental data. *Free Radical Biology and Medicine,* 29, 1106-1114.

[15] Bartosz, G. (2003) *Total antioxidant capacity.* In Advances in Clinical Chemistry. (Eds.) Herbert E. Spiegel, G. N. & Kwang-Jen, H.(Elsevier) pp. 219-292.

[16] Fardet, A., Rock, E. & Rémésy, C. (2008). Is the *in vitro* antioxidant potential of whole-grain cereals and cereal products well reflected *in vivo*? *Journal of Cereal Science,* 48, 258-276.

[17] Liu, J., Jia, L., Kan, J. & Jin, C.-H. (2013). *In vitro* and *in vivo* antioxi-dant activity of ethanolic extract of white button mushroom (*Agaricus bisporus*). *Food and Chemical Toxicology,* 51, 310-316.

[18] Reis, F. S., Martins, A., Barros, L. & Ferreira, I. C. F. R. (2012). Antioxidant properties and phenolic profile of the most widely appreciated cultivated mushrooms: A comparative study between *in vivo* and *in vitro* samples. *Food and Chemical Toxicology,* 50, 1201-1207.

[19] Wang, S., Meckling, K. A., Marcone, M. F., Kakuda, Y. & Tsao, R. (2011). Synergistic, Additive, and Antagonistic Effects of Food Mixtures on Total Antioxidant Capacities. *Journal of Agricultural and Food Chemistry,* 59, 960-968.

[20] Huang, P.-J., Hseu, Y.-C., Lee, M.-S., Senthil Kumar, K. J., Wu, C.-R., Hsu, L.-S., Liao, J.-W., Cheng, I. S., Kuo, Y.-T., Huang, S.-Y. & Yang, H.-L. (2012). *In vitro* and

in vivo activity of gallic acid and Toona sinensis leaf extracts against HL-60 human premyelocytic leukemia. *Food and Chemical Toxicology,* 50, 3489-3497.

[21] Prior, R. L. & Cao, G. (1999). *In vivo* total antioxidant capacity: comparison of different analytical methods. *Free Radical Biology and Medicine,* 27, 1173-1181.

[22] Niki, E. (2010). Assessment of antioxidant capacity *in vitro* and *in vivo*. *Free Radical Biology and Medicine,* 49, 503-515.

[23] Alam, M. N., Bristi, N. J. & Rafiquzzaman, M. Review on *in vivo* and *in vitro* methods evaluation of antioxidant activity. *Saudi Pharmaceutical Journal.* Available online 15 June 2012. DOI: 10.1016/j.bbr.2011.03. 031.

[24] Vives-Bauza, C., Starkov, A. & Garcia-Arumi, E. (2007) *Measurements of the Antioxidant Enzyme Activities of Superoxide Dismutase,* Catalase, *and Glutathione Peroxidase.* In Methods in Cell Biology. (Eds.) Liza, A. P. & Eric, A. S.(Academic Press) pp. 379-393.

[25] Prior, R. L., Wu, X. & Schaich, K. (2005). Standardized Methods for the Determination of Antioxidant Capacity and Phenolics in Foods and Dietary Supplements. *Journal of Agricultural and Food Chemistry,* 53, 4290-4302.

[26] Nenadis, N. & Tsimidou, M. Z. (2012). Contribution of DFT computed molecular descriptors in the study of radical scavenging activity trend of natural hydroxybenzaldehydes and corresponding acids. *Food Research International,* 48, 538-543.

[27] Takebayashi, J., Iwahashi, N., Ishimi, Y. & Tai, A. (2012). Development of a simple 96-well plate method for evaluation of antioxidant activity based on the oxidative haemolysis inhibition assay (OxHLIA). *Food Chemistry,* 134, 606-610.

[28] Wolfe, K. L. & Liu, R. H. (2007). Cellular antioxidant activity (CAA) assay for assessing antioxidants, foods, and dietary supplements. *Journal of Agricultural and Food Chemistry,* 55, 8896-8907.

[29] Huang, D., Ou, B. & Prior, R. L. (2005). The chemistry behind antioxidant capacity assays. *Journal of Agricultural and Food* Chemis- try, 53, 1841-1856.

[30] Ghiselli, A., Serafini, M., Maiani, G., Azzini, E. & Ferro-Luzzi, A. (1995). A fluorescence-based method for measuring total plasma antioxidant capability. *Free Radical Biology and Medicine,* 18, 29-36.

[31] Siger, A., Czubinski, J., Kachlicki, P., Dwiecki, K., Lampart-Szczapa, E. & Nogala-Kalucka, M. (2012). Antioxidant activity and phenolic content in three lupin species. *Journal of Food Composition and Analysis,* 25, 190-197.

[32] Awika, J. M., Rooney, L. W., Wu, X., Prior, R. L. & Cisneros-Zevallos, L. (2003). Screening methods to measure antioxidant activity of sorghum (*Sorghum bicolour*) and sorghum products. *Journal of Agricul- tural and Food Chemistry,* 51, 6657-6662.

[33] Ou, B., Hampsch-Woodill, M. & Prior, R. L. (2001). Development and validation of an improved oxygen radical absorbance capacity assay using fluorescein as the fluorescent probe. *Journal of Agricultural and Food Chemistry,* 49, 4619-4626.

[34] André, C. M., Oufir, M., Hoffmann, L., Hausman, J.-F., Rogez, H., Larondelle, Y. & Evers, D. (2009). Influence of environment and genotype on polyphenol compounds and *in vitro* antioxidant capacity of native Andean potatoes (*Solanum tuberosum* L.). *Journal of Food Com- position and Analysis,* 22, 517-524.

[35] Zulueta, A., Esteve, M. J. & Frígola, A. (2009). ORAC and TEAC assays comparison to measure the antioxidant capacity of food products. *Food Chemistry,* 114, 310-316.

[36] Rangkadilok, N., Sitthimonchai, S., Worasuttayangkurn, L., Mahidol, C., Ruchirawat, M. & Satayavivad, J. (2007). Evaluation of free radical scavenging and antityrosinase activities of standardized longan fruit extract. *Food and Chemical Toxicology,* 45, 328-336.

[37] Li, H., Deng, Z., Liu, R., Loewen, S. & Tsao, R. (2012). Ultra-performance liquid chromatographic separation of geometric isomers of carotenoids and antioxidant activities of 20 tomato cultivars and breed- ing lines. *Food Chemistry,* 132, 508-517.

[38] Wang, M., Tsao, R., Zhang, S., Dong, Z., Yang, R., Gong, J. & Pei, Y. (2006). Antioxidant activity, mutagenicity/anti-mutagenicity, and clastogenicity/anti-clastogenicity of lutein from marigold flowers. *Food and Chemical Toxicology,* 44, 1522-1529.

[39] Kranl, K., Schlesier, K., Bitsch, R., Hermann, H., Rohe, M. & Böhm, V. (2005). Comparing antioxidative food additives and secondary plant products: Use of different assays. *Food Chemistry,* 93, 171-175.

[40] Wu, T., Mccallum, J. L., Wang, S., Liu, R., Zhu, H. & Tsao, R (2012). Evaluation of antioxidant activities and chemical characterisation of staghorn sumac fruit (*Rhus hirta* L.). *Food Chemistry,* 138, 1333-1340.

[41] Berker, K. I., Güçlü, K., Tor, İ. & Apak, R. (2007). Comparative evaluation of Fe(III) reducing power-based antioxidant capacity assays in the presence of phenanthroline, batho-phenanthroline, tripyridyltria- zine (FRAP), and ferricyanide reagents. *Talanta,* 72, 1157-1165.

[42] Gliszczyńska-Świgło, A. (2006). Antioxidant activity of water soluble vitamins in the TEAC (trolox equivalent antioxidant capacity) and the FRAP (ferric reducing antioxidant power) assays. *Food Chemistry,* 96, 131-136.

[43] Khanizadeh, S., Tsao, R., Rekika, D., Yang, R., Charles, M. T. & Vasantha Rupasinghe, H. P. (2008). Polyphenol composition and total antioxidant capacity of selected apple genotypes for processing. *Journal of Food Composition and Analysis,* 21, 396-401.

[44] Alvarez-Suarez, J. M., Tulipani, S., Díaz, D., Estevez, Y., Romandini, S., Giampieri, F., Damiani, E., Astolfi, P., Bompadre, S. & Battino, M. (2010). Antioxidant and antimicrobial capacity of several monofloral Cuban honeys and their correlation with colour, polyphenol content and other chemical compounds. *Food and Chemical Toxicology,* 48, 2490-2499.

[45] Jang, A., Lee, N.-Y., Lee, B.-D., Kim, T. H., Son, J. H., An, B. J. & Jo, C. (2009). Biological functions of a synthetic compound, octadeca-9,12-dienyl-3,4,5-hydroxybenzoate, from gallic acid–linoleic acid ester. *Food Chemistry,* 112, 369-373.

[46] Zhang, R., Zeng, Q., Deng, Y., Zhang, M., Wei, Z., Zhang, Y. & Tang, X. (2013). Phenolic profiles and antioxidant activity of litchi pulp of different cultivars cultivated in Southern China. *Food Chemistry,* 136, 1169-1176.

[47] Giri, L., Dhyani, P., Rawat, S., Bhatt, I. D., Nandi, S. K., Rawal, R. S. & Pande, V. (2012). *In vitro* production of phenolic compounds and antioxidant activity in callus suspension cultures of Habenaria edge- worthii: A rare Himalayan medicinal orchid. *Industrial Crops and Products,* 39, 1-6.

[48] Amić, D. & Lučić, B. (2010). Reliability of bond dissociation enthalpy calculated by the PM6 method and experimental TEAC values in antiradical QSAR of flavonoids. *Bioorganic & Medicinal Chemistry*, 18, 28-35.

[49] Zalibera, M., Staško, A., Šlebodová, A., Jančovičová, V., Čermáková, T. & Brezová, V. (2008). Antioxidant and radical-scavenging activities of Slovak honeys – An electron paramagnetic resonance study. *Food Chemistry,* 110, 512-521.

[50] Apak, R., Güçlü, K., Demirata, B., Ozyürek, M., Celik, S. E., Bektaşoğlu, B., Berker, K. I. & Ozyurt, D. (2007). Comparative evalu- ation of various total antioxidant capacity assays applied to phenolic compounds with the CUPRAC assay. *Molecules,* 12, 1496-1547.

[51] Soong, Y.-Y. & Barlow, P. J. (2006). Quantification of gallic acid and ellagic acid from longan (*Dimocarpus longan* Lour.) seed and mango (*Mangifera indica* L.) kernel and their effects on antioxidant activity. *Food Chemistry,* 97, 524-530.

[52] Fang, Z., Hu, Y., Liu, D., Chen, J. & Ye, X. (2008). Changes of phenolic acids and antioxidant activities during potherb mustard *(Brassica juncea,* Coss.) pickling. *Food Chemistry,* 108, 811-817.

[53] Breksa Iii, A. P., Takeoka, G. R., Hidalgo, M. B., Vilches, A., Vasse, J. & Ramming, D. W. (2010). Antioxidant activity and phenolic content of 16 raisin grape (*Vitis vinifera* L.) cultivars and selections. *Food Chemis- try,* 121, 740-745.

[54] López-Giraldo, L. J., Laguerre, M., Lecomte, J., Figueroa-Espinoza, M. C., Baréa, B., Weiss, J., Decker, E. A. & Villeneuve, P. (2009). Kinetic and stoichiometry of the reaction of chlorogenic acid and its alkyl esters against the DPPH radical. *Journal of Agricultural and Food Chemistry,* 57, 863-870.

[55] Galano, A., lvarez-Diduk, R., Ramírez-Silva, M. T., Alarcón- ngeles, G. & Rojas-Hernández, A. (2009). Role of the reacting free radicals on the antioxidant mechanism of curcumin. Chemical Physics, 363, 13-23.

[56] Chen, Z., Bertin, R. & Froldi, G. (2013). EC_{50} estimation of antioxidant activity in DPPH assay using several statistical programs. *Food Chemis- try,* 138, 414-420.

[57] Villaño, D., Fernández-Pachón, M. S., Moyá, M. L., Troncoso, A. M. & García-Parrilla, M. C. (2007). Radical scavenging ability of polypheno- lic compounds towards DPPH free radical. *Talanta,* 71, 230-235.

[58] Wu, T., Yan, J., Liu, R., Marcone, M. F., Aisa, H. A. & Tsao, R. (2012). Optimization of microwave-assisted extraction of 1 phenolics from potato and its downstream waste using orthogonal array design. *Food Chemistry,* 133, 1293-1298.

[59] Pasanphan, W. & Chirachanchai, S. (2008). Conjugation of gallic acid onto chitosan: An approach for green and water-based antioxidant. *Carbohydrate Polymers,* 72, 169-177.

[60] Pérez-Jiménez, J. & Saura-Calixto, F. (2006). Effect of solvent and certain food constituents on different antioxidant capacity assays. *Food Research International,* 39, 791-800.

[61] Yen, G.-C., Duh, P.-D. & Tsai, H.-L. (2002). Antioxidant and pro-oxi- dant properties of ascorbic acid and gallic acid. *Food Chemistry,* 79, 307-313.

[62] Cho, Y.-S., Kim, S.-K., Ahn, C.-B. & Je, J.-Y. (2011). Preparation, characterization, and antioxidant properties of gallic acid-grafted-chitosans. *Carbohydrate Polymers,* 83, 1617-1622.

[63] Sroka, Z. & Cisowski, W. (2003). Hydrogen peroxide scavenging, antioxidant and anti-radical activity of some phenolic acids. *Food and Chemical Toxicology,* 41, 753-758.

[64] Hseu, Y.-C., Chang, W.-H., Chen, C.-S., Liao, J.-W., Huang, C.-J., Lu, F.-J., Chia, Y.-C., Hsu, H.-K., Wu, J.-J. & Yang, H.-L. (2008). Antioxi- dant activities of Toona

Sinensis leaves extracts using different antioxi- dant models. *Food and Chemical Toxicology,* 46, 105-114.

[65] Rockenbach, I. I., Rodrigues, E., Gonzaga, L. V., Caliari, V., Genovese, M. I., Gonçalves, A. E. D. S. S. & Fett, R. (2011). Phenolic compounds content and antioxidant activity in pomace from selected red grapes (*Vitis vinifera* L. and *Vitis labrusca* L.) widely produced in Brazil. *Food Chemistry,* 127, 174-179.

[66] Canas, S., Casanova, V. & Pedro Belchior, A. (2008). Antioxidant activity and phenolic content of Portuguese wine aged brandies. *Journal of Food Composition and Analysis,* 21, 626-633.

[67] Wolfe, K. L., Kang, X., He, X., Dong, M., Zhang, Q. & Liu, R. H. (2008). Cellular Antioxidant Activity of Common Fruits. *Journal of Agricultural and Food Chemistry,* 56, 8418-8426.

[68] Loo, G. (2003). Redox-sensitive mechanisms of phytochemical-mediated inhibition of cancer cell proliferation (review). *The Journal of Nutritional Biochemistry,* 14, 64-73.

[69] Faried, A., Kurnia, D., Faried, L. S., Usman, N., Miyazaki, T., Kato, H. & Kuwano, H. (2007). Anticancer effects of gallic acid isolated from Indonesian herbal medicine, Phaleria macrocarpa (Scheff.) Boerl, on human cancer cell lines. *International Journal of Oncology,* 30, 605-613.

[70] García-Rivera, D., Delgado, R., Bougarne, N., Haegeman, G. & Vanden Berghe, W. (2011). Gallic acid indanone and mangiferin xanthone are strong determinants of immunosuppressive anti-tumour effects of *Mangifera indica* L. bark in MDA-MB231 breast cancer cells. *Cancer Letters,* 305, 21-31.

[71] Liao, C.-L., Lai, K.-C., Huang, A.-C., Yang, J.-S., Lin, J.-J., Wu, S.-H., Gibson Wood, W., Lin, J.-G. & Chung, J.-G. (2012). Gallic acid inhibits migration and invasion in human osteosarcoma U-2 OS cells through suppressing the matrix metalloproteinase-2/-9, protein kinase B (PKB) and PKC signaling pathways. *Food and Chemical Toxicology,* 50, 1734-1740.

[72] Ho, H.-H., Chang, C.-S., Ho, W.-C., Liao, S.-Y., Lin, W.-L. & Wang, C.-J. (2013). Gallic acid inhibits gastric cancer cells metastasis and invasive growth via increased expression of RhoB, downregulation of AKT/small GTPase signals and inhibition of NF-κB activity. *Toxicology and Applied Pharmacology,* 266, 76-85.

[73] Chandramohan Reddy, T., Bharat Reddy, D., Aparna, A., Arunasree, K. M., Gupta, G., Achari, C., Reddy, G. V., Lakshmipathi, V., Subramanyam, A. & Reddanna, P. (2012). Anti-leukemic effects of gallic acid on human leukemia K562 cells: Downregulation of COX-2, inhibition of BCR/ABL kinase and NF-κB inactivation. *Toxicology in Vitro,* 26, 396-405.

[74] Chen, H.-M., Wu, Y.-C., Chia, Y.-C., Chang, F.-R., Hsu, H.-K., Hsieh, Y.-C., Chen, C.-C. & Yuan, S.-S. (2009). Gallic acid, a major component of Toona sinensis leaf extracts, contains a ROS-mediated anti-cancer activity in human prostate cancer cells. *Cancer Letters,* 286, 161-171.

[75] Luo, W., Zhao, M., Yang, B., Ren, J., Shen, G. & Rao, G. (2011). Antioxidant and antiproliferative capacities of phenolics purified from *Phyllanthus emblica* L. fruit. *Food Chemistry,* 126, 277-282.

[76] Gutiérrez-Uribe, J. A., Romo-Lopez, I. & Serna-Saldívar, S. O. (2011). Phenolic composition and mammary cancer cell inhibition of extracts of whole cowpeas (*Vigna unguiculata*) and its anatomical parts. *Journal of Functional Foods,* 3, 290-297.

[77] Kobayashi, H., Oikawa, S., Hirakawa, K. & Kawanishi, S. (2004). Metal-mediated oxidative damage to cellular and isolated DNA by gallic acid, a metabolite of antioxidant propyl gallate. *Mutation Research /Genetic Toxicology and Environmental Mutagenesis,* 558, 111-120.

[78] Khanduja, K. L., Avti, P. K., Kumar, S., Mittal, N., Sohi, K. K. & Pathak, C. M. (2006). Anti-apoptotic activity of caffeic acid, ellagic acid and ferulic acid in normal human peripheral blood mononuclear cells: A Bcl-2 independent mechanism. *Biochimica et Biophysica Acta (BBA) - General Subjects,* 1760, 283-289.

[79] Yang, H.-L., Chang, W.-H., Chia, Y.-C., Huang, C.-J., Lu, F.-J., Hsu, H.-K. & Hseu, Y.-C. (2006). Toona sinensis extracts induces apoptosis via reactive oxygen species in human premyelocytic leukemia cells. *Food and Chemical Toxicology,* 44, 1978-1988.

[80] Tayama, S. & Nakagawa, Y. (2001). Cytogenetic effects of propyl gallate in CHO-K1 cells. *Mutation Research/Genetic Toxicology and Environmental Mutagenesis,* 498, 117-127.

[81] Li, H., Deng, Z., Liu, R., Loewen, S. & Tsao, R. (2013). Carotenoid compositions of coloured tomato cultivars and contribution to antioxidant activities and protection against H_2O_2-induced cell death in H9c2. *Food Chemistry,* 136, 878-888.

[82] Andersson, B. S., Vidal, R. F., Sundberg, M., Rajs, J. & Sotonyi, P. (1996). Hydroperoxide-induced nucleotide degradation and adenosine formation in isolated rat cardiomyocytes. *Toxicology,* 106, 39-48.

[83] Labieniec, M. & Gabryelak, T. (2007). Antioxidative and oxidative changes in the digestive gland cells of freshwater mussels Unio tumidus caused by selected phenolic compounds in the presence of H_2O_2 or Cu^{2+} ions. *Toxicology in Vitro*, 21, 146-156.

[84] Erol-Dayi, O., Arda, N. & Erdem, G. (2012). Protective effects of olive oil phenolics and gallic acid on hydrogen peroxide-induced apoptosis. *European Journal of Nutrition,* 51, 955-960.

[85] Nabavi, S. F., Nabavi, S. M., Habtemariam, S., Moghaddam, A. H., Sureda, A., Jafari, M. & Latifi, A. M. (2013). Hepatoprotective effect of gallic acid isolated from Peltiphyllum peltatum against sodium fluoride-induced oxidative stress. *Industrial Crops and Products,* 44, 50-55.

[86] Sessa, M., Tsao, R., Liu, R., Ferrari, G. & Donsì, F. (2011). Evaluation of the stability and antioxidant activity of nanoencapsulated resveratrol during *in vitro* digestion. *Journal of Agricultural and Food Chemistry,* 59, 12352-12360.

[87] Prasetyo, E. N., Kudanga, T., Steiner, W., Murkovic, M., Wonisch, W., Nyanhongo, G. S. & Guebitz, G. M. (2010). Cellular and plasma anti- oxidant activity assay using tetramethoxy azobismethylene quinone. *Free Radical Biology and Medicine,* 49, 1205-1211.

[88] Roy, M. K., Juneja, L. R., Isobe, S. & Tsushida, T. (2009). Steam pro- cessed broccoli (Brassica oleracea) has higher antioxidant activity in chemical and cellular assay systems. *Food Chemistry,* 114, 263-269.

[89] Ngo, D.-H., Qian, Z.-J., Vo, T.-S., Ryu, B., Ngo, D.-N. & Kim, S.-K. (2011). Antioxidant activity of gallate-chitooligosaccharides in mouse macrophage RAW264.7 cells. *Carbohydrate Polymers,* 84, 1282-1288.

[90] Girard-Lalancette, K., Pichette, A. & Legault, J. (2009). Sensitive cell-based assay using DCFH oxidation for the determination of pro- and antioxidant properties of

compounds and mixtures: Analysis of fruit and vegetable juices. *Food Chemistry,* 115, 720-726.

[91] Ferk, F., Chakraborty, A., Jäger, W., Kundi, M., Bichler, J., Mišík, M., Wagner, K.-H., Grasl-Kraupp, B., Sagmeister, S., Haidinger, G., Hoelzl, C., Nersesyan, A., Dušinská, M., Simić, T. & Knasmüller, S. (2011). Potent protection of gallic acid against DNA oxidation: Results of human and animal experiments. *Mutation Research/Fundamental and Molecular Mechanisms of Mutagenesis,* 715, 61-71.

[92] Shinno, E., Shimoji, M., Imaizumi, N., Kinoshita, S., Sunakawa, H. & Aniya, Y. (2005). Activation of rat liver microsomal glutathione S-transferase by gallic acid. *Life Sciences,* 78, 99-106.

[93] Rajalakshmi, K., Devaraj, H. & Niranjali Devaraj, S. (2001). Assess- ment of the no-observed-adverse-effect level (NOAEL) of gallic acid in mice. *Food and Chemical Toxicology,* 39, 919-922.

[94] Li, L., Ng, T. B., Gao, W., Li, W., Fu, M., Niu, S. M., Zhao, L., Chen, R. R. & Liu, F. (2005). Antioxidant activity of gallic acid from rose flowers in senescence accelerated mice. *Life Sciences,* 77, 230-240.

[95] Hsiang, C.-Y., Hseu, Y.-C., Chang, Y.-C., Kumar, K. J. S., Ho, T.-Y. & Yang, H.-L. (2013). Toona sinensis and its major bioactive compound gallic acid inhibit LPS-induced inflammation in nuclear factor-κB transgenic mice as evaluated by *in vivo* bioluminescence imaging. *Food* Chemistry, 136, 426-434.

[96] Senevirathne, M., Jeon, Y.-J., Kim, Y.-T., Park, P.-J., Jung, W.-K., Ahn, C.-B. & Je, J.-Y. (2012). Prevention of oxidative stress in Chang liver cells by gallic acid-grafted-chitosans. *Carbohydrate Polymers,* 87, 876-880.

[97] Reckziegel, P., Dias, V. T., Benvegnú, D., Boufleur, N., Silva Barcelos, R. C., Segat, H. J., Pase, C. S., dos Santos, C. M. M., Flores, M. M. & Bürger, M. E. (2011). Locomotor damage and brain oxidative stress induced by lead exposure are attenuated by gallic acid treatment. *Toxicology Letters,* 203, 74-81.

[98] Gil-Longo, J. & González-Vázquez, C. (2010). Vascular pro-oxidant effects secondary to the autoxidation of gallic acid in rat aorta. *The Journal of Nutritional Biochemistry,* 21, 304-309.

[99] Jang, A., Srinivasan, P., Lee, N. Y., Song, H. P., Lee, J. W., Lee, M. & Jo, C. (2008). Comparison of hypolipidemic activity of synthetic gallic acid–linoleic acid ester with mixture of gallic acid and linoleic acid, gallic acid, and linoleic acid on high-fat diet induced obesity in C57BL/6 Cr Slc mice. *Chemico-Biological Interactions,* 174, 109-117.

[100] Knapen, M. F. C. M., Zusterzeel, P. L. M., Peters, W. H. M. & Steegers, E. a. P. (1999). Glutathione and glutathione-related enzymes in reproduction: A review. *European Journal of Obstetrics & Gynecology and Reproductive Biology,* 82, 171-184.

[101] Toribio, F., Martínez-Lara, E., Pascual, P. & López-Barea, J. (1996). Methods for purification of glutathione peroxidase and related enzymes. *Journal of Chromatography B: Biomedical Sciences and Applications,* 684, 77-97.

[102] Brigelius-Flohé, R. & Maiorino, M. (2012). *Glutathione peroxidases.* Biochimica et Biophysica Acta (BBA) - General Subjects, Available online 29 November 2012. DOI: 10.1016/j.bbr.2011.03.031.

[103] Penninckx, M. (2000). A short review on the role of glutathione in the response of yeasts to nutritional, environmental, and oxidative stresses. *Enzyme and Microbial Technology,* 26, 737-742.

[104] Paglia, D. E. & Valentine, W. N. (1967). Studies on the quantitative and qualitative characterization of erythrocyte glutathione peroxidase. The *Journal of Laboratory and Clinical Medicine,* 70, 158-169.

[105] Tung, Y.-T., Wu, J.-H., Huang, C.-C., Peng, H.-C., Chen, Y.-L., Yang, S.-C. & Chang, S.-T. (2009). Protective effect of Acacia confusa bark extract and its active compound gallic acid against carbon tetrachloride-induced chronic liver injury in rats. *Food and Chemical Toxicology,* 47, 1385-1392.

[106] Kim, H.-G., Yoo, S.-R., Park, H.-J., Lee, N.-H., Shin, J.-W., Sathyanath, R., Cho, J.-H. & Son, C.-G. (2011). Antioxidant effects of Panax ginseng C.A. Meyer in healthy subjects: A randomized, placebo-controlled clinical trial. *Food and Chemical Toxicology,* 49, 2229-2235.

[107] Vijaya Padma, V., Sowmya, P., Arun Felix, T., Baskaran, R. & Poornima, P. (2011). Protective effect of gallic acid against lindane induced toxicity in experimental rats. *Food and Chemical Toxicology,* 49, 991-998.

[108] Sheu, M.-J., Chou, P.-Y., Cheng, H.-C., Wu, C.-H., Huang, G.-J., Wang, B.-S., Chen, J.-S., Chien, Y.-C. & Huang, M.-H. (2009). Analge sic and anti-inflammatory activities of a water extract of Trachelosper- mum jasminoides (Apocynaceae). *Journal of Ethno pharmacology,* 126, 332-338.

[109] Pastore, A., Federici, G., Bertini, E. & Piemonte, F. (2003). Analysis of glutathione: implication in redox and detoxification. *Clinica Chimica Acta,* 333, 19-39.

[110] Roth, E., Oehler, R., Manhart, N., Exner, R., Wessner, B., Strasser, E. & Spittler, A. (2002). Regulative potential of glutamine—relation to gluta- thione metabolism. *Nutrition,* 18, 217-221.

[111] Ellman, G. L. (1959). Tissue sulfhydryl groups. *Archives of Biochemis- try and Biophysics,* 82, 70-77.

[112] Jadon, A., Bhadauria, M. & Shukla, S. (2007). Protective effect of Terminalia belerica Roxb. and gallic acid against carbon tetrachloride induced damage in albino rats. *Journal of Ethnopharmacology,* 109, 214-218.

[113] Pourahmad, J., Eskandari, M. R., Shakibaei, R. & Kamalinejad, M. (2010). A search for hepatoprotective activity of aqueous extract of Rhus coriaria L. against oxidative stress cytotoxicity. *Food and Chemical Toxicology,* 48, 854-858.

[114] Loprasert, S., Vattanaviboon, P., Praituan, W., Chamnongpol, S. & Mongkolsuk, S. (1996). Regulation of the oxidative stress protective enzymes, catalase and superoxide dismutase in Xanthomonas: A review. *Gene,* 179, 33-37.

[115] Donnelly, J. K., Mclellan, K. M., Walker, J. L. & Robinson, D. S. (1989). Superoxide dismutases in foods. A review. *Food Chemistry,* 33, 243-270.

[116] Spitz, D. R. & Oberley, L. W. (1989). An assay for superoxide dismutase activity in mammalian tissue homogenates. *Analytical Bio- chemistry,* 179, 8-18.

[117] Park, C. K., Lee, J. H., Cheong, H. T., Yang, B. K. & Kim, C. I. (1997). Effect of superoxide dismutase(SOD) on pronucleus formation of porcine oocytes fertilized *in vitro*. *Theriogenology,* 48, 1137-1146.

[118] Verma, A. R., Vijayakumar, M., Mathela, C. S. & Rao, C. V. (2009). *In vitro* and *in vivo* antioxidant properties of different fractions of Moringa oleifera leaves. *Food and Chemical Toxicology,* 47, 2196-2201.

[119] Chattopadhyay, K., Mondal, S., Chattopadhyay, B. & Ghosh, S. (2010). Ameliorative effect of sesame lignans on nicotine toxicity in rats. *Food and Chemical Toxicology,* 48, 3215-3220.

[120] Sepici-Dincel, A., Açıkgöz, Ş., evik, C., Sengelen, M. & Yeşilada, E. (2007). Effects of *in vivo* antioxidant enzyme activities of myrtle oil in normoglycaemic and alloxan diabetic rabbits. *Journal of Ethnopharma-* cology, 110, 498-503.

[121] Aebi, H. (1984). Catalase *in vitro. Methods Enzymol,* 105, 121-128.

[122] Mansouri, M. T., Farbood, Y., Sameri, M. J., Sarkaki, A., Naghizadeh, B. & Rafeirad, M. (2013). Neuroprotective effects of oral gallic acid against oxidative stress induced by 6-hydroxydopamine in rats. *Food Chemistry,* 138, 1028-1033.

[123] Orhan, N., Aslan, M., Orhan, D. D., Ergun, F. & Yeşilada, E. (2006). *In-vivo* assessment of antidiabetic and antioxidant activities of grapevine leaves (*Vitis vinifera*) in diabetic rats. *Journal of Ethnopharmacology,* 108, 280-286.

[124] Wahsha, M., Bini, C., Fontana, S., Wahsha, A. & Zilioli, D. (2012). Toxicity assessment of contaminated soils from a mining area in Northeast Italy by using lipid peroxidation assay. *Journal of Geochemi- cal Exploration,* 113, 112-117.

[125] Ohkawa, H., Ohishi, N. & Yagi, K. (1979). Assay for lipid peroxides in animal tissues by thiobarbituric acid reaction. *Analytical Biochemistry,* 95, 351-358.

[126] Verma, A. R., Vijayakumar, M., Rao, C. V. & Mathela, C. S. (2010). *In vitro* and *in vivo* antioxidant properties and DNA damage protective activity of green fruit of Ficus glomerata. *Food and Chemical Toxicolo- gy,* 48, 704-709.

[127] Nabavi, S. F., Habtemariam, S., Jafari, M., Sureda, A. & Nabavi, S. M. (2012). Protective Role of Gallic Acid on Sodium Fluoride Induced Oxidative Stress in Rat Brain. *Bulletin of Environmental Contamination and Toxicology,* 89, 73-77.

[128] Singh, B. N., Singh, B. R., Singh, R. L., Prakash, D., Sarma, B. K. & Singh, H. B. (2009). Antioxidant and anti-quorum sensing activities of green pod of Acacia nilotica L. *Food and Chemical Toxicology,* 47, 778-786.

In: Handbook on Gallic Acid
Editors: M. A. Thompson and P. B. Collins
ISBN: 978-1-62618-921-8

Chapter 5

GALLIC ACID: APPLICATIONS, ANALYSIS AND ELECTROCHEMICAL CHARACTERISATION

E. Skeva and S. Girousi*
Analytical Chemistry Laboratory, School of Chemistry,
Aristotle University of Thessaloniki, Thessaloniki, Greece

ABSTRACT

Electrochemical DNA biosensors are promising tools for fast, inexpensive and simple in vitro analysis for the determination of free radicals and antioxidants. Reactive Oxygen Spieces (ROS) and especially hydroxyl radicals ($^{\bullet}OH$), are known to damage DNA by oxidation of the bases, which results in their destruction, release or attack of the deoxyribose moieties by strand breaking. These lesions are related to the aging process, cancer, Alzheimer's disease and atherosclerosis. In living systems, hydroxyl radicals are generated from the metal (ferrous or cupric ions) ion-dependent breakdown of hydrogen peroxide by Fenton' s reaction, which is an important mediator of oxidative damage in vivo. Antioxidants such as phenolic acids and plant extracts containing high amounts of them, act as free radical terminators and reduce the effect of the oxidative damage on dsDNA.

In the present work, an overview of natural occurrences, antioxidant properties and health implications is being thoroughly considered.

Keywords: Gallic acid, natural occurrence, antioxidant properties, analytical methods, health implications

INTRODUCTION

The last decade, the scientific community has been trying intensively to understand and explain the connection between ROS, oxidative stress, DNA oxidative damage, various disorders and diseases and antioxidants, which present remarkable effects on the prevention

* Email: girousi@chem.auth.gr.

or even treatment of those harmful phenomena. Among the causes of serious chronic health problems, free radical and reactive oxygen species (ROS) proved to play a crucial role. Superoxide anion ($O_2^{\bullet-}$), hydroxyl radical ($^{\bullet}OH$) and peroxy radical ($ROO^{\bullet}$) have been implicated as mediators of DNA oxidative damage and many degenerative diseases and malfunctions such as aging, atherosclerosis, myocardial infarction, cancer, inflammations, autoimmune disorders, diabetes, Parkinson's and Alzheimer's disease and cataract formation [1,2].

Fundamental cellular and molecular structures such as membranes, proteins, enzymes and nucleic acids (DNA) present great sensitivity towards free radicals. The main lesions caused by their attack are lipid peroxidation, protein modification, enzyme deactivation and mutagenic alterations of DNA. In living systems, the majority of hydroxyl radicals is generated according to Fenton's reaction, which involves the reduction of hydrogen peroxide by transition metal ions. The produced hydroxyl radicals are extremely reactive and capable of oxidizing a wide variety of molecules. Although both Fe(II) and Cu(I) can initiate Fenton's reaction, the abundance of iron in biological systems increases the possibilities of its participation to the oxidation process. This procedure is of great importance since it is accused to be the main mediator of oxidative damage in vivo and the earliest means of ROS generation by nature [3,4]. During the application of Fenton's reaction on biological systems, especially DNA, the problem of site specific damages arise since iron is able to interact with DNA by unwinding the double helix. To the contrary, humans and aerobic organisms are protected from free radicals either through an endogenous defense mechanism that includes enzymes that scavenge the free radicals (SOD) and repair the oxidative damages on DNA (glycosylase), antioxidants (glutathione, $CoQ_{10,}$ ascorbic acid) and metal binding proteins (ferritin, ceruloplasmin) or dietary antioxidants (polyphenols, tocopherols, flavonoids) [5].

As antioxidant may be defined any substance, present at low concentrations compared to the oxidizable substrate, that delays or inhibits the oxidation of that substrate [6]. Antioxidants are divided into two general groups based on the protection mechanism [7]. Phenolic antioxidants or polyphenols comprise the majority of antioxidants that humans find in a variety of foodstuff such as fruits, herbs, vegetables, tea and coffee. Natural antioxidants have demonstrated beneficial effects in maintenance of health, management of age related diseases, ameliorating the harmful effects of toxic agents both chemical and physical [8]. Plant polyphenols are well known to show biological activity, such as antimutagenicity, anticarcinogenicity and antioxidant activity [9]. Phenolic compounds or polyphenols constitute one of the most numerous and ubiquitously distributed group of plant secondary metabolites, with more than 8000 phenolic structures currently known [10].

Natural Occurrence of Gallic Acid

Phenolic acids are ubiquitous antioxidants accounting for approximately one third of the phenolic compounds in our diet [11]. Two classes of phenolic acids can be distinguished depending on their structure: derivatives of benzoic acid and derivatives of cinnamic acid. They consist of benzene as a basis bond to a carboxylic group (benzoic acids) or to propenoic acid (cinnamic acids) [12,13].

Gallic acid, an organic acid, known as 3,4,5-trihydroxybenzoic acid ($C_6H_2(OH)_3COOH$), is an endogenous plant phenol (Figure 1), widely found throughout the plant kingdom. Gallic acid is the most potent antioxidant among the simple phenolic acids [10] and has both antioxidant and diverse biological effects [14]. It exists either as a free molecule, dimer, trimer and tetramer (ellagic acid, tergallic acid, gallagic acid) or as part of esterified tannins (particularly in black tea) [15]. Pure gallic acid is a colorless crystalline organic powder, while salts and esters of gallic acid are termed gallates [16]. The gallate moiety is crucial of its antioxidant and biological behaviour [17].

High gallic acid contents can be found in blackberries, raspberries, blueberries, strawberries, currants, green and black tea, clove buds, potato skin, grapes, wines (especially red wine), gallnuts, sumac, witch hazel, hops, oak bark, mangos, carob, apricots, apples, coffee and in several herbal remedies and medical plants used in traditional medicine [14,15,18-26]. Green tea is the most abundant source of tea phenolics, mostly simple hydroxybenzoic acids such as gallic acid and propyl gallate and its catechin derivatives (epigallocatechin-3-gallate, epigallocatechin, epicatechin-3-gallate and epicatechin) [16].

Gallic acid derivatives are present either as methylated gallic acids (syringic acid), or galloyl conjugates of catechin derivatives (flanvan-3-ols) and polygalloyl esters of glucose (quinic acid and glycerol) (Figure 2). Thus, they represent a large family of plant secondary polyphenolic metabolites present in many natural products such as tea. The latter two categories of polyphenols are also known as vegetable tannins, name attributed to their ability to transform animal skins into leather by interacting with collagen proteins [28].

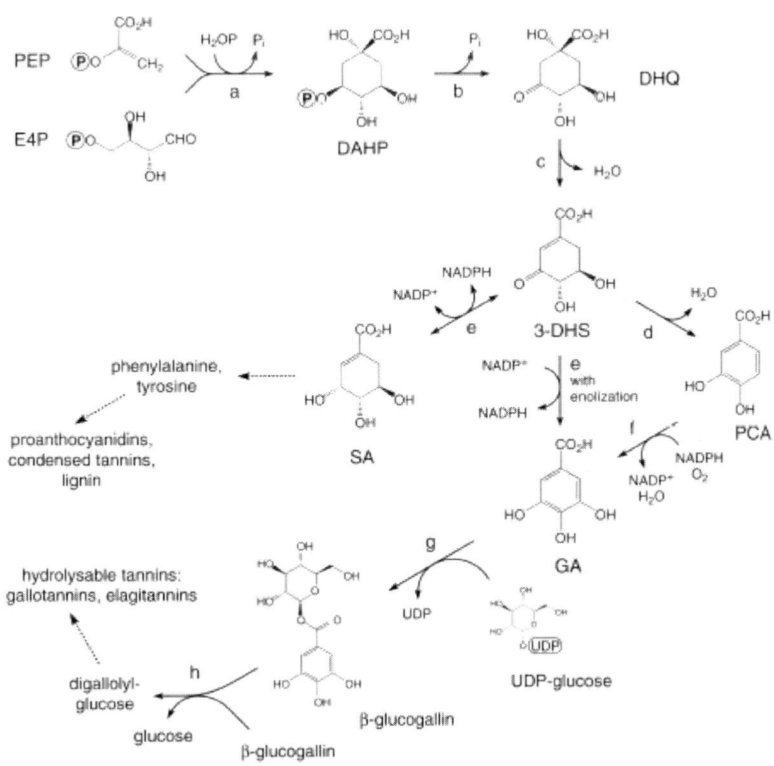

Figure 1. Synthesis of gallic and shikimic acids in plants, adapted from [27].

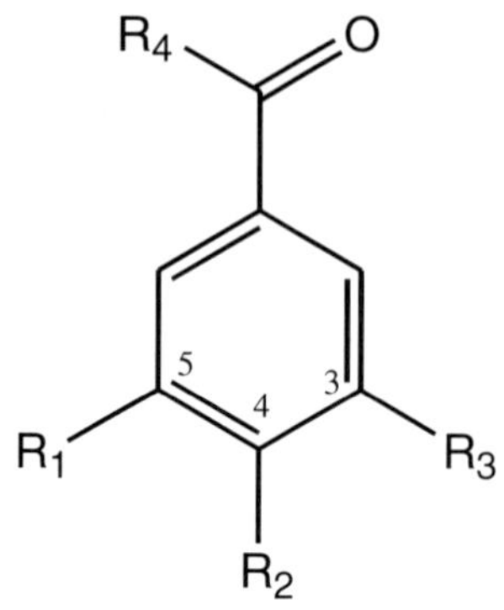

Gallic acid $R_1 = R_2 = R_3 = R_4 = OH$
5-Hydroxy-3,4-dimethoxybenzoic acid $R_2 = R_3 = OCH_3$ and $R_1 = R_4 = OH$
Syringic acid $R_1 = R_3 = OCH_3$ and $R_2 = R_4 = OH$
3,4,5-Trimethoxybenzoic acid $R_1 = R_2 = R_3 = OCH_3$ and $R_4 = OH$
Methyl gallate $R_1 = R_2 = R_3 = OH$ and $R_4 = CH_3$
n-Propyl gallate $R_1 = R_2 = R_3 = OH$ and $R_4 = CH_2CH_2CH_3$
n-Octyl gallate $R_1 = R_2 = R_3 = OH$ and $R_4 = (CH_2)_7CH_3$
n-Dodecyl gallate $R_1 = R_2 = R_3 = OH$ and $R_4 = (CH_2)_{11}CH_3$

Figure 2. Gallic acid and its esters.

Many gallic acid esters are used used as antioxidant additives in both food and pharmaceutical industry like E-310 (propyl gallate) and E-311 (octyl gallate) [29]. GA derivatives have also been found in many phytomedicines with a number of biological and pharmacological activities, including scavenging on free radicals, inducing apoptosis of cancer cells, inhibiting squalene epoxidase interfering the signal pathways involving Ca^{2+} and oxygen free radicals [30]. Moreover, gallic acid derivatives are known to cause apoptosis in tumour cell lines and to inhibit lymphocyte proliferation. It is also a fact that these compounds are excellent inhibitors of protein tyrosinase kinases (PTKs) [31].

Synthetic galloyl esters showed to be potent and selective enzyme inhibitors, as well as synergistic protectors against peroxyl radical damage in membranes. They were also found to inhibit cytokine-induced nuclear translocation of NF-kappa B, as well as expression of leukocyte adhesion molecules in vascular endothelium cells [31]. (−)-EGCG decreases the expression of proapototic genes and contributes to the integrity of the mitochondrial membrane and inhibits tumour growth by inhibiting VEGF induction in human colon carcinoma cells [32,33].

The recognised cytotoxic activity of phenolic derivatives depends on their rate of incorporation into cells, directly related to their lipophilicity, as much as to their antioxidant activity. Both these properties are affected by the presence of the ring substituent hydroxyl groups and by the length of the ester moiety, in polyhydroxylated phenolic esters [10,31].

USES OF GALLIC ACID

Gallic acid has many applications in chemical research and industry. First of all, it is commonly used as reference standard for the evaluation of the antioxidant activity of various compounds or plant extracts like in FC method [8,34,35]. Gallic acid and its ester (-)-epi-gallocatechin-3-gallate (EGCG) are also used as food preservatives in processed food, cosmetics and food packing materials to prevent lipid peroxidation and spoilage [9,20]. Apart from that, it is extensively used in tanning, ink dyes, as well as in the manufacturing of paper [16].

In recent years, a new application with great perpectives, which takes advantage of gallic acid's high antioxidant properties in order to improve the antioxidant activity and the biological properties of other compouds, gains importance [36]. The majority of natural antioxidants have limited therapeutic success because of their poor distribution and the various difficulties to attain the target sites [37,38]. Additionally, it is not always appropriate to directly apply gallic acid and other phenolic acids in foods and pharmaceutical products as antioxidant agents because undesirable reactions and chemical alterations may occur. In that concept, many reports proposed antioxidant polymer conjugates (macromolecular systems) that combine the advantages of both components and retain the excellent biological activities of antioxidant molecules and show higher stability and slower degradation rate than the antioxidant molecules [39]. To synthesize that kind of antioxidant–polymer conjugates chitosan, a novel, abundant natural polysaccharide which is consisted of primarily of 2-amino-2-deoxy-glucopyranose units linked by β-(1-4) linkage is appropriate [40,41]. Gallic acid-g-chitosans might also act beneficial toward diseases releated to oxidative damage and galloyl-cinnamic acid hybrid antioxidants have brain permeable properties [11,42]. Moreover, gallic acid was successfully incorporated into zein ultra-fine fibres in order to develop an encapsulating technology for functional ingredient delivery using electrospinning, preserving its phenolic character and antioxidant activity [43]. Perceiving the necessity of designing new drugs, the synthesis, the antioxidant and cytotoxic activities of hydrid molecules containing indole and gallic hydrazone moieties have been investigated [44].

ELECTROCHEMICAL BEHAVIOUR OF GALLIC ACID

Electrochemical techniques provide useful information for the physicochemical behaviour of compounds, such as the redox potential and mainly the number of electrons exchanged during the redox reaction, while revealing valuable conclusions on the redox reactions of these molecules in organisms [45, 46].

Antioxidant properties are related with the redox characteristics and consequently, knowledge of the redox behavior is very crucial. Accordingly, it seems very important to clarify the electrochemical oxidation mechanisms [47].

Additionally, electrochemistry has been used for the estimation of antioxidant capacity of polyphenols in natural samples [45]. Further information is given at the section about the methods used to estimate the antioxidant ability of gallic acid.

Cyclic voltammetry (CV) is usually the first experiment performed in an electrochemical study of a compound. The effectiveness of this technique results from its capability for rapidly observing the redox behavior over a wide potential range [48].

In each case the top scan represents the oxidation of the polyphenolic generating a positive (anodic) current, peaking at a particular electrode potential; the lower the potential of oxidation, the more powerful the phenolic is as a reducing agent. On the reverse scan a negative (cathodic) peak is produced where the oxidised form of the phenolic can be reduced back to its original form [49].

Gallic acid constists of an easily oxidizable ortho-diphenol group, a third phenol group adjacent to the ortho-diphenol group and an acidic -COOH group [45,50]. It is generally known that gallic acid is oxidized to quinone via semiquinone form. Thus, a proton (H+) participates in redox step [11,51].

The mechanism of the electrochemical oxidation of gallic acid in aqueous phosphate buffer solutions at different pH at glassy-carbon electrode using cyclic, convolution–deconvolution sweep voltammetry, chronoamperometry and chronocoulometry, has been studied. Two irreversible diffusion-controlled cyclic voltammetric waves at the entire range of pH are observed.

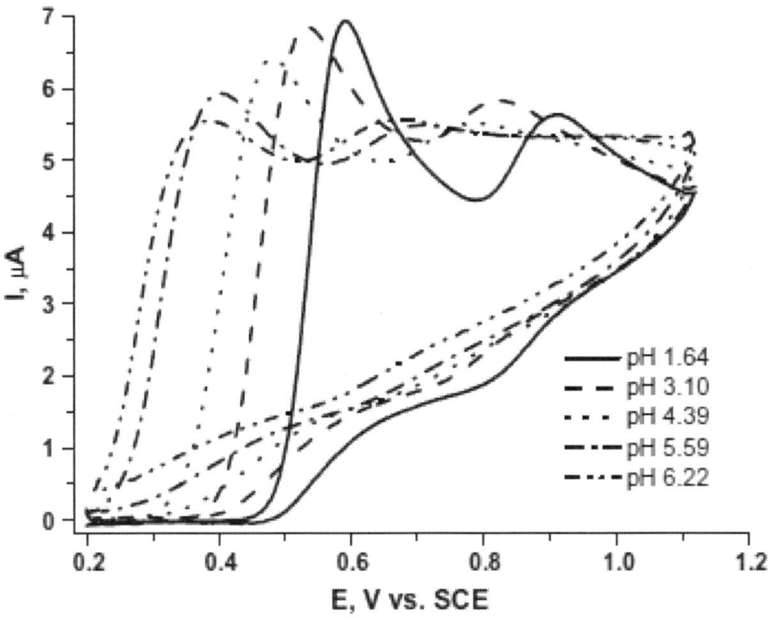

Figure 3. Cyclic voltammograms of gallic acid at different pH, adapted from [47].

The electrochemical oxidation mechanism was proposed to be an ECEC-first order mechanism in which the two electron transfer steps and the two chemical follow-up deprotonation reactions are irreversible [47].

Authors noticed that on increasing pH values (Figure 3), the oxidation potential of both peaks is moving to less positive values [45,47,52].

Specifically, the first wave corresponds to the irreversible oxidation of gallic acid to the semiquinone radical cation ($GA\cdot^{+}$) by irreversible electron transfer process. The formed radical cation loses a proton to form the semiquinone radical ($GA\cdot$). The one-electron oxidation product (GA·) is followed by a second irreversible electron transfer to the quinone cation (GA^+). The deprotonation of the quinone cation (GA^+) completes the two-electron process to give the quinone (GAO) which appears as an irreversible peak [47]. The oxidation mechanism of gallic acid is extensively described at Figure 4.

Gallic acid is irreversibly oxidised, in the absence of a return cathodic peak at glassy carbon electrode in phosphate buffer. Moreover, an increase in peak current for differrent solutions of gallic acid from 0.01 to 0.05 mM was observed, because of the limited electrode contamination compared to the catechin standards also studied [49].

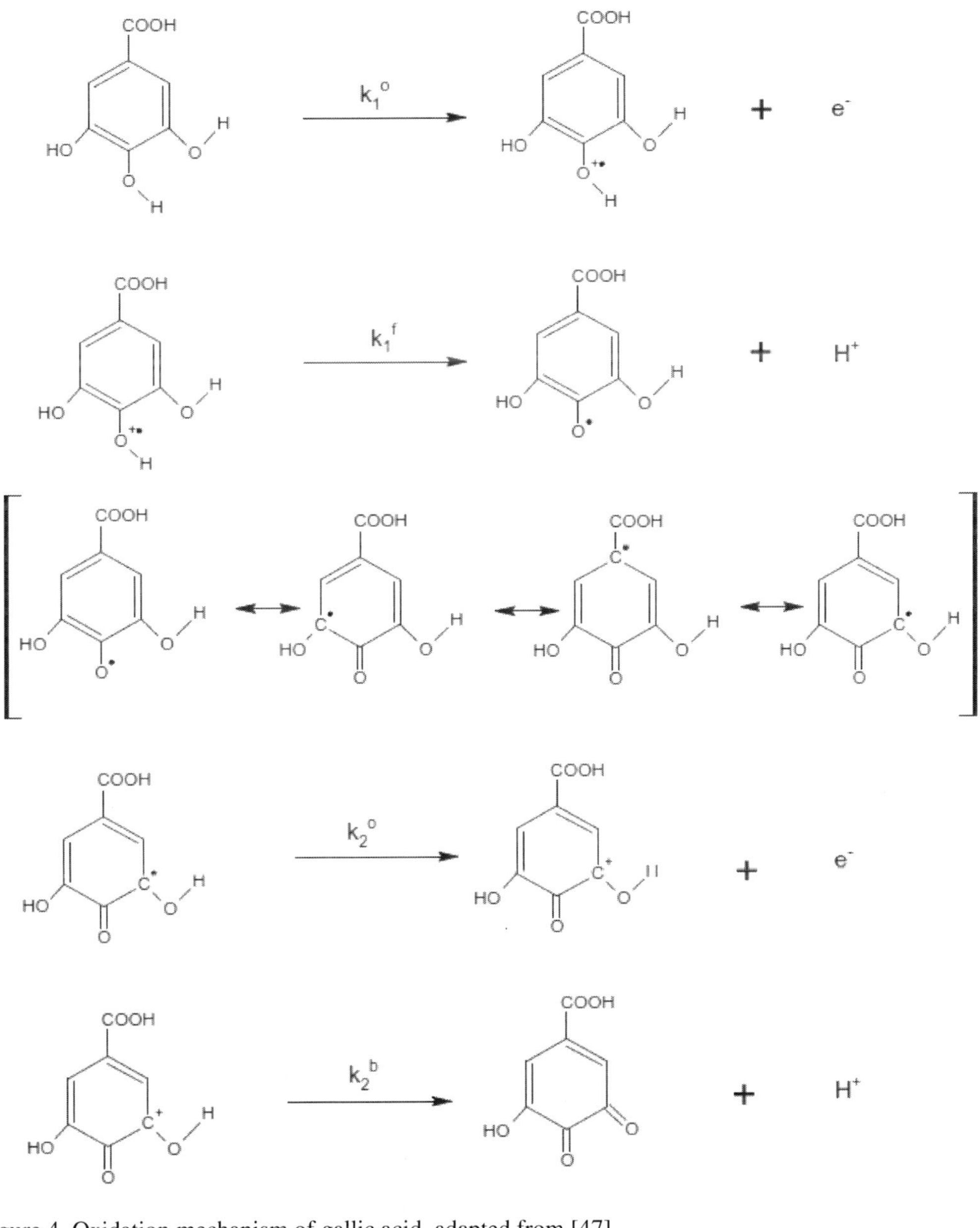

Figure 4. Oxidation mechanism of gallic acid, adapted from [47].

An other electrochemical study in model wine solutions showed that the irreversible peak appears at about +0.400 V, when the cathodic peak is missing, due to either further chemical actions that the oxidation product may under go or to its inability to be reduced at the glassy carbon electrode.

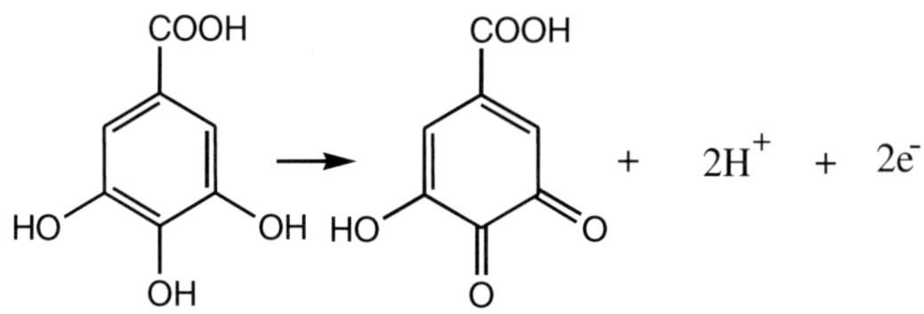

Figure 5. Chemical structure of gallic acid and oxidation mechanism at GCE.

A second oxidation process was observed at about +0.800 V, which is attributed to the third phenol group adjacent to the *ortho*diphenol previously oxidized [50], when at carbon paste electrode CV voltammograms reveal the two semi-reversible anodic peaks at +0.320 V and +0.690 V for gallic acid in acetate buffer solution pH=4.8 [48]. A pair of redox peaks also appeared at a bare GCE in acetate buffer solution using cyclic voltammetry. The anodic peak, at +0.329 V and the cathodic one, at +0.211 V, correspond to the oxidation of phenolic hydroxyl group of gallic acid (Figure 5) [46].

Cyclic voltammetric experiments also proved the irreversible behaviour of gallic acid and ethyl gallate at the glassy carbon electrode and the influence of an additional phenolic group on the stabilization of the semiquinone intermediate. The extra hydroxyl group could promote the stabilization of the radical produced by oxidation, expressed by the lower redox potentials attained. The oxidation mechanism involves an overall two-electron process: one-electron transfer to the semiquinone followed by a second electron transfer to give the *ortho*-quinone.

Differential pulse voltammograms of gallic acid, ethyl gallate and synthesised galloyl-cinnamic hybrid revealed two well defined anodic waves at physiological pH. The peaks at +0.117 V, +0.552 V and +0.057 V, +0.541V are connected to the oxidation of the pyrogallol unit. The galloyl-cinnamic hybrid is oxidized to quinone via semiquinone form as caffeic and gallic acid [11].

Also in square-wave voltammograms of gallic acid, two anodic peaks at about 0.450 V and 0.800 V were observed. These peaks are attributed to the oxidation of the galloyl group and the third -OH group of the galloyl moiety respectively. The electrooxidation of acidic -COOH group occurs at 2 V and does not appear at the voltammogram [45].

Antioxidant Properties

Antioxidant properties of gallic acid sententiously arise from their high reactivity as hydrogen or electron donors (chain-breaking function) and their potential to terminate Fenton reaction.

Gallic acid has been characterized as chain-breaking antioxidant and radical scavenger, by donating the phenolic hydrogen atom or electron and delocalizing/stabilizing the resulting phenoxyl radical (Figure 6) within the structure in order to give a stable compound and an antioxidant-derived radical [17,53,54].

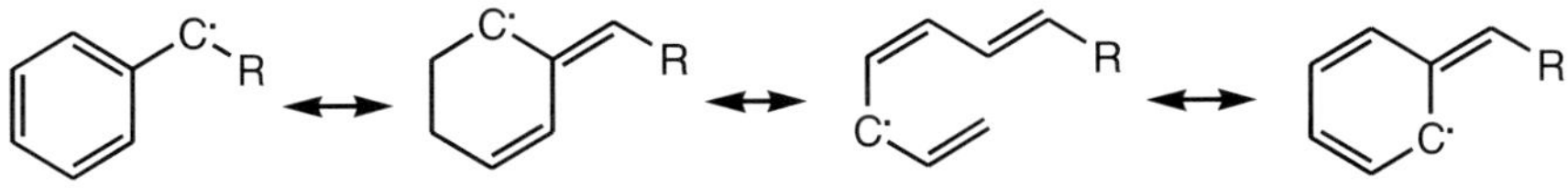

Figure 6. Schematic representation of the resonance stabilization of a radical in the benzylic position.

The actual free radical scavenging ability of antioxidants is generally designated by the rate of radical scavenging, the number of radical molecules scavenged, the fate of the antioxidant-derived radical, the interaction with other antioxidants, the absorption, the distribution, the retention and the metabolism of the antioxidants [55,56].

Two are the main mechanisms used to describe the radical scavenging activity of phenolic antioxidants such as gallic acid: the direct hydrogen-transfer process (HAT) (Equation 1) and the proton-coupled electron-transfer process (SET) (Equation 2).

$$R\cdot + ArOH \rightarrow RH + ArO\cdot \qquad \text{(Equation 1)}$$

$$R\cdot + ArOH \rightarrow R^{-} + ArOH\cdot^{-} \rightarrow RH + ArO\cdot \qquad \text{(Equation 2)}$$

HAT processes take place in nonpolar solvents and are characterized by the O–H bond dissociation enthalpy (BDE). The low O–H BDE and the stability of the generated radical are the criteria of effective antioxidants by HAT process [57]. The main factors that affect the O–H BDE values are the arrangment of functional groups on the aromatic ring, the number and configuration of H-donating hydroxyl groups, the presence of –OH group in ortho position to a phenol, the intra-molecular hydrogen bonds that can exist in pyrogallol moieties and lead to extra stabilization and the inter-molecular hydrogen bonds between these functional groups with polar protic solvents (H_2O, CH_3OH, CH_3CH_2OH) [53,58-61]. On the contrast, SET processes occur in polar solvents, determined by the ionization potential (IP). Other scientists support that the most relevant parameters for the free radical scavenging activity of antioxidants include the dissociation energy of the phenolic O–H bond (OH-BDE), the energy-eigenvalue of the highest occupied molecular orbital (EHOMO) and ionization potential (IP) of the antioxidant [30]. Experimental results are in agreement with the theoretical basis. In particular, phenolic acids are considered efficient hydrogen donors. Gallic acid, with 3-hydroxyl groups, was characterised by the highest antioxidant activity [10], although, some steric restrictions are involved among their hydroxyl groups [62]. These three aromatic phenoxyl groups are prone to oxidation with the formation of hydrogen peroxide and two distinct semiquinones [17]. Studies based on the DPPH method, showed that the galloyl moiety and ortho-dihydroxyl are related to radical scavenging, especially the numbers and the positions of the hydroxyl (–OH) groups [25]. Gallic acid reduced around seven DPPH· radicals but falls in the slow kinetic category [8]. Using Fenton reaction and dsDNA biosensor researchers proved the antioxidant ability of gallic acid and when compared to other phenolic compounds such as caffeic acid and Trolox, with two and one hydroxyl group respectively, was found to be the most efficient towards oxidative damage [48].

To conclude, not only the number of phenolic hydroxyl groups but also their position, the presence of other functional groups in the molecule such as double bonds and their conjugation to hydroxyl groups and ketonic groups affect the radical scavenging [8].

The metal chelation ability is named as the secondary antioxidant activity. Phenolic acids decelerate or intercept electron transfer from metals to hydrogen peroxide that initiates the hydroxyl radical generation. Chelating agents may inactivate metal ions and potentially inhibit the metal-dependent processes [63].

Gallic acid's ability to reduce metals like Fe(III) and Cu(II) and therefore terminate the OH· generation is attributed to its structure, since the aromatic ring and the hydroxyl group at position 3, can easily scavenge electron deficient free radicals by donating an electron and itself become oxidized into a quinonoid structure [14,64].

Gallic acid not only reduces Fe(III) but also chelates Fe(II) (Firure 7) [16,65,66].

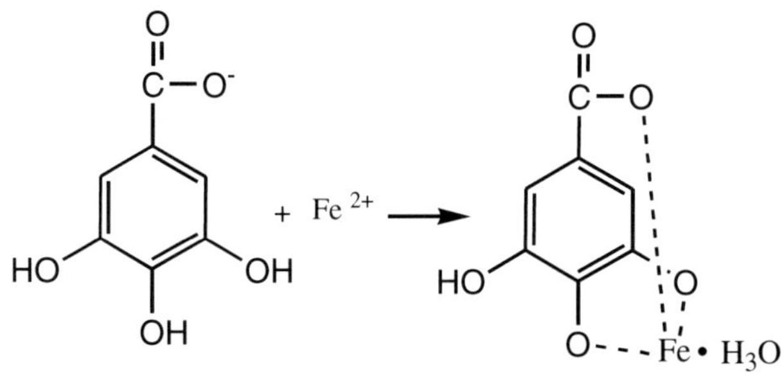

Figure 7. Fe^{2+} chelation by gallic acid.

METHODS

The study of antioxidants is very important and depends on the sample and the purpose of analysis. For example, scientists in the medical and biological field focus on the free radical scavenging capacity of the sample, choosing various methods, when food industry is interested in their quantitative determination. There are many publications dealing with the determination of phenolic compounds and phenolic acids in fruits, herbs, olive oil, cosmetic products using spectrophotometry, fluorescence, HPLC and GC-MS [48]. Studying the literature not only conventional methods to estimate the antioxidant capacity, such as Folin-Ciocalteu, DPPH, and FC, but also new perspectives using electrochemical techniques and biosensors, can be found. Both of them are described, following the previous observation, at the whole section. To conclude, the complexity and diverse range of research topics investigated have led to the development of a multitude of tests, but unfortunately not a universal and validated one to provide reliable measurements [67]. Thus, one of the major challenges is to know which method is best suited for the particular application and sample. It is recommended that the capacity of antioxidant compounds and their mixtures for antioxidation should be assessed from their effect on the levels of plasma lipid peroxidation in vitro and biomarkers of oxidative stress in vivo [56].

Major conventional antioxidant capacity assays can be generally divided into two groups: (i) hydrogen atom transfer (HAT) reaction based assays and (ii) single electron transfer (SET) reaction based assays. HAT-based methods estimate the ability of an antioxidant to quench free radicals by hydrogen donation and monitor competitive reaction kinetics such as ORAC, TRAP, TOSC, croton or β carotene bleaching, Low-Density Lipoprotein (LDL). HAT reactions are quick, solvent and pH independent. On the other hand, SET-based methods detect the ability of antioxidants to transfer one electron in order to reduce any compound, like metals, carbonyls, and radicals. The major representatives of this subcategory are TEAC, FRAP, DPPH-based assay, CUPRAC and total phenolics assay (TPC) by Folin- Ciocalteu (FC) [67,68]. SET reactions are slow, pH dependent, the antioxidant capacity calculations are based on the decreased concentration of the product and the reactivity in these methods is influenced by deprotonation and IP of the reactive functional group [69]. Details about these methods have been described elsewhere; in this article some principles and characteristics will be briefly discussed [67-71].

ORAC (Oxygen radical absorbance capacity assay). This method monitors antioxidant inhibition of peroxyl radical and reflects the radical chain breaking antioxidant activity by H atom transfer. It is based on the reduction of the signal of fluorencent probes after the rection with peroxyl radicals. Antioxidants essentially disable these radicals. Antioxidant capacity and ORAC values are expressed as Trolox equivalents.

TRAP (Total radical-trapping antioxidant parameter assay). This method uses a luminescence spectrometer to measure the fluorescence decay of R-phycoerythrin during a controlled peroxidation reaction. The principles are similar to those of ORAC. Typically, oxidation of the probe is quantified either optically or by fluorescence. TRAP values are usually expressed as reaction time of the sample compared to corresponding times for the standard Trolox or gallic acid.

TOSC (Total Oxidant Scavenging Capacity). This method gives the ability to evaluate the antioxidant capacity of several compounds in presence of various biologically generated radical sources. It is based on the reaction between three potent oxidants, hydroxyl radicals peroxyl radical and peroxynitrite and α-keto-γ-methiolbutyric acid (KMBA), which is oxidized to ethylene. The measurements are based on the reduction of the produced ethylene after the addition of antioxidants. Linear dose-response curves for antioxidants can be generated from kinetics of the reaction.

Croton or β-Carotene Bleaching. Carotenoids bleach via autoxidation, oxidation induced by light, heat or peroxyl radicals. Antioxidants that donate hydrogen atoms to quench radicals prevent this decolorization.

FRAP (Ferric ion reducing antioxidant parameter assay). The FRAP method quantifies the ability of antioxidants to reduce ferric iron. It is based on the reduction of the complex of ferric iron and 2,3,5-triphenyl-1,3,4- triaza-2-azoniacyclopenta-1,4-diene chloride (TPTZ) to the ferrous form at low pH and detection at 593 nm, using a diode-array spectrophotometer. The reaction is suitable for compounds with redox potentials lower than +0.7 V, redox potential of Fe^{3+}-TPTZ, rather helpful tool to estimate the ability to maintain redox status in cells or tissues. It has been used for the determination of the antioxidant capacity of botanicals and polyphenols, because the reducing power is related to hydroxylation and conjugation in their molecules. However, FRAP cannot detect compounds that act by radical quenching like thiols and proteins.

CUPRAC (Copper (II) reduction capacity). This assay is based on the reduction of Cu(II) to Cu(I) by antioxidants (reducing agents) in a sample, similar to FRAP. This method is more selective and has more applications than FRAP since copper, free or complexed, has lower redox potential than iron and less interferences than FRAP.

TEAC (Trolox equivalent antioxidant capacity assay). The TEAC assay is based on the scavenging ability of antioxidants towards the stable radical $ABTS\cdot^{+}$. Similar to ORAC, diode-array spectrophotometry is used to measure the loss of color when an antioxidant is added to the blue-green chromophore $ABTS\cdot^{+}$ (2,2'-azino-bis(3-ethylbenzthiazoline-6-sulfonic acid)). Results are expressed comperatively to the standard Trolox. Due to its simplicity, to the ability of its application on both hydrophilic and lipophilic substrates over a wide pH range, it has been used in many samples for studying the antioxidant capacity. However, the $ABTS\cdot^{+}$ radical is not found in mammalian organisms.

DPPH (2,2-Diphenyl-1-picrylhydrazyl assay). This assay measures by spectrophotometer or electron spin resonance (EPR) the ability of antioxidants to reduce 2,2-diphenylpicrylhydrazyl (DPPH), which bears a deep purple color, is commercially available but not commonly found in biological systems. The results can be evaluated by the decrease of absorbance. The DPPH assay is mainly based on a SET reaction. However, the results taken are not always accurate since DPPH is also decolorized by reducing agents.

Folin-Ciocalteu (F-C) or Total Phenolics Assay. There is always the controversy if only phenols are being detected in total antioxidant capacity assays. This method estimates the oxidation of phenols and polyphenolic antioxidants by a molybdotungstate reagent that yields a colored product at 745-750 nm and is usually applied on natural products, wines and herbs. The absorbance is proportional to the number of aromatic phenolic groups and is expressed as gallic acid equivalents. The method is simple and convenient but lacks specificity. Not only polyphenols, but also nonphenolic organic substances and some inorganic substances react with the FC reagent to give overestimated phenolic concentrations.

Table 1. Summary of methods used for the evaluation of the antioxidant capacity of gallic acid and its derivatives

	Real sample	Antioxidant	Antioxidant assay	Reference
1	-	gallic acid	mushroom tyrosinase inhibition GSH/GSSG in b16 cells	[82]
2	-	gallic acid	2,2′-azinobis-(3-ethyl-benzothiazoline-6-sulfonic acid) radical assay	[18]
3	Phytochemicals	gallic acid	DPPH / TRP	[63]
4	Pistacia lentiscus	gallic acid 1,2,3,4,6-pentagalloylglucose	DPPH / lipid peroxidation inhibition	[54]
5	Rose flowers	gallic acid	Animal models / CAT/ GPx activity	[83]
6	-	gallic acid	DPPH / Fe^{2+} chelating / inhibition of DNA oxidative damage	[9]
7	-	gallic acid	AAPH / ABTS / DPPH / 2'-deoxyguanosine assay	[53]
8	Human and animal intervention trial	gallic acid	ABTS / inhibition of DNA oxidative damage / SOD / GPx activity	[84]
9	-	gallic acid	HRP-H_2O_2 / oxidative stress biomarkers	[85]

	Real sample	Antioxidant	Antioxidant assay	Reference
10	Terminalia chebula Retz	gallic acid 1,3,6-tri-*O*-galloyl-β-D-glucopyranose	DPPH	[25]
11	-	gallic acid	DPPH	[8]
	Real sample	Antioxidant	Antioxidant assay	Reference
12	-	gallic acid	DPPH / FRAP	[14]
13	Mango fruit	gallic acid	DPPH / FRAP	[62]
14	-	gallic acid (−)-epigallocatechin (−)-epicatechin-3-gallate (−)-epigallocatechin-3-gallate	TEAC / FRAP / HOCL scavenging Deoxyribose assay	[10]
15	-	gallic acid gallic acid-grafted-chitosans	DPPH / H_2O_2 scavenging	[42]
16	-	gallic acid methyl gallate n-propyl gallate n-octyl gallate n-dodecyl gallate	DPPH	[30]
17	-	zein–gallic acid system	DPPH	[43]
18	-	gallic acid gallic acid-quaternized chitozan	DPPH / Hydroxyl and superoxide radical scavenging ability assay	[86]
19	-	gallic acid gallic acid-indole derivatives	DPPH	[44]
20	Leea indica (plant)	gallic acid	DPPH	[26]
21	Herb extracts	gallic acid	hydroxyl radical scavenging-dsDNA biosensor	[48]
22	Lemon flavoured water samples	gallic acid	hydroxyl radical scavenging-DNA biosensor / DPPH	[71]
23	-	galloyl-cinnamic hybrids	ABTS / DPPH	[11]
24	-	gallic acid	hydroxyl radical scavenging- DNA biosensor	[72]
25	Red wines	gallic acid polyphenols	ABTS / DPPH / DPV / FC	[87]
26	-	gallic acid	DPPH / CV	[88]
27	Apples	gallic acid	vitamin C equivalent antioxidant capacity (VCEAC) / ABTS / DPPH	[89]
28	Tea extracts	gallic acid	hydroxyl radical scavenging- DNA biosensor	[90]
29	Black raspberry blackberry red raspberry strawberry purple grape	gallic acid	DPPH / FRAP / ABTS	[91]
30	Herbs	gallic acid	CUPRAC / deoxyribose assay test	[92]
31	Flavours / Flavoured waters	gallic acid	superoxide radical scavenging-DNA biosensor	[93]
32	-	gallic acid epigallocatechin epigallocatechin gallate	CUPRAC	[94]
33	-	gallic acid	vitamin C equivalent antioxidant capacity (VCEAC)	[95]

Electrochemistry is nowadays an important tool for the evaluation of antioxidant capacity as it has the same principles as those exhibited by antioxidants in vitro systems. Electrochemical techniques allow both the simultaneous estimation of the total antioxidants in the real sample as well as the evaluation of their antioxidant activity [67]. The main electrochemical approaches used have been cyclic voltammetry and differential pulse voltammetry, which have been proposed as a useful tool in characterization of polyphenols in different food samples, such as teas , wines and orange juices [45].

Additionally, biosensors are attractive alternative tools for the determination of free radicals and antioxidants, characterized by selectivity, low cost, easy automatization and simplicity. Thus, the development of biosensors for the determination of the polyphenolic fraction and phenolic acids is a promising challenge. They can be defined as analytical tools or systems consisting of an immobilized biological material in contact with a suitable transducer. Biological recognition elements can be organisms, tissues, enzymes, antibodies and nucleic acids. Biosensors for the assessment of the antioxidant capacity are characterized by the in vitro production of ROS and the electrochemical detection, such as cytochrome c (cyt c), superoxide dismutase (SOD) and DNA biosensors [72]. Especially DNA electrochemical biosensors receive considerable attention since monitoring the DNA integrity is the only way to assess at cellural level the antioxidant capacity and literature provides a large variety of DNA biosensors [73-81].

A wide variety of the above methods have been used in order to estimate the antioxidant capacity of numerous compounds, extracts and real samples. Some of them are summarized in Table 1.

HEALTH IMPLICATIONS

Gallic acid is commonly used in the pharmaceutical industry as many in vivo and in vitro studies proved gallic acid's beneficial diverse properties. Their effects on health, though, have been more particularly attributed to their antioxidant properties [12].

Gallic acid presents considerable antioxidant behaviour in emulsion or lipid systems and protects human cells against oxidative damage [96]. It is more effective than several water-soluble antioxidants like ascorbic acid and almost as effective as Trolox [9]. Moreover, it shows cytotoxicity against cancer cells [97-99], it seems to have antifungal and antiviral properties [100,101], it can be used to treat albuminuria and diabetes [102,103], it is used as remote astringent in internal hemorrhage, to treat psoriasis and external hemorrhoids [16]. It can also inhibit apoptosis [104, 105], NO production in LPS-activated macrophages [25], prevent NSAID-induced gastric mucosal damage and accelerate the healing process of damaged gastric mucosa [14]. Scientific researches have also shown that scavenging antioxidants like gallic acid act towards melanogenesis and reduce hyperpigmentation via scavenging the ROS generated in cells from UV or MSH exposure [25,82,106].

When animals were treated with high doses of gallic acid, protection of liver against cytotoxic effects and prevention of radiation-induced lipid peroxidation (LP) have been reported [84]. Gallic acid prevents the production of proinflammatory cytokines, which play an important role in the development of gastroduodenal ulcer [107] and protects human hepatocytes from hydrogen peroxide induced oxidative stress. Other studies support that gallic acid may inhibit microsomal activation of AFB1 to electrophilic metabolite and have antimutagenic function [54].

However, to fully understand the antioxidant and health effects of phenolic acids, their bioavailability should also be examined. The physiological significance of dietary antioxidants depends on their mechanism of absorption and biotransformation. This leads to an assiduous and complete study to fully understand and estimate the beneficial antioxidant behaviour of these compounds. Findings suggest that consumption of common foodstuffs rich in gallic acid (tea, wine, fruits, and nuts) contributes substantially to the maintenance of its redox status and may improve the health status of individuals under oxidative stress [84]. It has been estimated that the daily intake of gallic acid in Southern Germany is 18mg/person/day, when in the USA is about 1g [30,84], when in UK and Ireland the major sources of gallic acid is the daily consumption of 3.4 cups of tea and red wine respectively [15]. Indeed tea is an important source of gallic acid as tea leaves may contain up to 4.5g/kg fresh wt [108]. On the other hand, consumers should also take into account toxicological studies and the ''no-observed-adverse-effect- level'' (NOAEL) of gallic acid, which is at least 120 mg/kg/day for rats and 1000 mg/kg/day gallates for mice [30, 109].

Several studies have shown that phenolic acids are extensively metabolized in vivo, mainly during transfer across the small intestine, by colonic microflora and in the liver, resulting in significant alteration in their redox potentials. After undergoing phase I deglycosylation, the phenolic aglycones are converted to glucuronides, sulphates and *o*-methylated derivatives during phase II metabolism [10]. Nevertheless, the total bioavailability of phenolic acids should be measured, which includes the direct absorption and metabolism of phenolic acids from food consumption and the bioavailability as a result of the cleavage on the main skeleton ring of flavonoids by the gut microflora [12].

Gallic acid is very well absorbed in humans, rats, rabbits and chickens [110]. Specifically, free and glucuronidated forms of gallic acid and its main metabolite 4-O-methylgallic (4OMGA) have been found in human blood plasma after ingestion of gallic acid-rich foods [111]. In vivo studies concluded that metabolic transformations occur in three ways: methylation (unconjugated and conjugated 4-O-methylgallic acid, 2-O-methylgallic acid), decarboxylation (unconjugated and conjugated pyrogallol, 4-O-methylpyrogallol) and dehydroxylation (resorcinol) [30,112].

In humans, the main metabolites of gallic acid found to be 4-O-methylated and O-glucuronidated derivatives regardless of the initial consumed form such as purified gallic acid, red wine and tea [111,113]. In some cases, the urinary excretion reached up to 39.6% of the ingested dose [12]. In other studies, after 50mg of gallic acid administration dose in volunteers, gallic acid and 4-O-methylgallic acid appeared rapidly in plasma and urine. The concentration levels of both compounds after deconjugation were tenfold higher and the urine excretion counted for 55.6% of the dose [15].

4-O-methylgallic acid (4OMGA) is also a biomarker for tea consumption and tea-derived polyphenol metabolism [114]. After black tea consumption, non-esterified gallic acid, rather than its esters, considered to the origin of urinary 4OMGA, as the majority of gallic acid present in black tea is non-esterified.

To the contrary, green tea contains non-esterified gallic acid. Non-esterified gallic acid is absorbed in the small intestine, followed by transportation to the liver, O-methylation and glucuronidation. Gallic acid may also be released from larger flavonoids, suggestion that justifies the urinary 4OMGA excretion after green tea consumption [115]. Additionally,

4OMGA has been used as an indicator of the intake of a variety of wines. In plasma, the major form of gallic acid was the 4-*O*-methylated one, but also the 3-*O*-methyl derivative appeared [113].

Many animal studies have also delineated the gastric absorption efficiency of gallic acid [116]. Especially in rats, concentrations of phenolic acids in plasma 5 minutes after administration increased at the following rank: gallic acid, chlorogenic acid < caffeic acid < p-coumaric acid, ferulic acid [117]. Other scientists suggest that gallic acid's rapid absorption involves the upper part of the gut, the stomach, the small intestine, or both. Indeed few quantities of gallic acid were absorbed from the stomach after an in situ gastric infusion [12,117].

In vitro studies performed to clarify the metabolism and bioavailability of n-propyl, n-octyl and n-dodecyl gallates, results shown that all test substances were extensively metabolised by the homogenate of the intestinal mucosa, while the microflora showed high metabolic activity, especially to n-propyl gallate.

A rather interesting finding is that when gallic acid or depsidic galloyl-gallic acid was intraperitoneally administrated to rats, decrease in the plasma triglycerides levels was observed [15].

Conclusion

Polyphenolic antioxidants from dietary sources are frequently a topic of interest due to widespread scientific agreement that they may help lower the incidence of certain cancers, cardiovascular and neurodegenerative diseases, and DNA damage and even may have antiaging properties. On the other hand, questions still remain as to whether some antioxidants could be potentially harmful to health, because an increase in glycation-mediated protein damage (carbonyl stress) has been reported in some cases. Although a diet rich in the polyphenolic forms of antioxidants does seem to offer hope in delaying the onset of age-related disorders, it is still too early to define their exact clinical benefit for treating age-related disease. This chapter critically examines gallic acid's antioxidant and health effects, its natural sources and bioavailiability with the hope that understanding may lead to better health and novel treatment approaches for various diseases.

References

[1] Tsao, R.; Deng, Z. Separation procedures for naturally occurring antioxidant phytochemicals. *J. Chromatogr. B* 2004, *812*, 85-99.

[2] Prior, R. L.; Hoang, H.; Gu, L.; Wu, X.; Bacchiocca, M.; Howard, L.; Hampsch-Woodill, M.; Huang, D.; Ou, B.; Jacob, R. Assays for Hydrophilic and Lipophilic Antioxidant Capacity (oxygen radical absorbance capacity (ORACFL)) of Plasma and Other Biological and Food Samples. *J. Agric. Food Chem.* 2003, *51*, 3273-3279.

[3] Koppenol, W. H.; The Haber-Weiss cycle –70 years later, *Redox Rep.*2001, *6*, 229-234.

[4] Winterbourn, C. C. Toxicity of iron and hydrogen peroxide: the Fenton reaction, *Toxicol. Lett.* 1995, *82/83*, 969-974.

[5] Evans, P.; Halliwell, B. Micronutrients: oxidant/ antioxidant status, *Brit. J. Nutr.* 2001, *85*, S67-S74.

[6] Halliwell, B. Antioxidant characterization: Methodology and mechanism, *Biochem. Pharmacol.* 1995, *49*, 1341-1348.

[7] Antolovich, M.; Prenzler, P. D.; Patsalides, E.; Mcdonald, S.; Robards, K. Methods for testing antioxidant activity. *Analyst* 2002, *127*, 183-198.

[8] Mishra, K.; Ojha, H.; Chaudhury, N.K. Estimation of antiradical properties of antioxidants using DPPH assay: A critical review and results. *Food Chemistry* 2012, *130*, 1036–1043.

[9] Yen, G.C.; Duh, P.D.; Tsai, H.L. Antioxidant and pro-oxidant properties of ascorbic acid and gallic acid. *Food Chemistry* 2002, *79*, 307–313.

[10] Soobrattee, M. A.; Neergheen, V. S.; Luximon-Rammaa, A.; Aruomab,O. I.; Bahorun, T. Phenolics as potential antioxidant therapeutic agents: Mechanism and actions. *Mutation Research* 2005, *579*, 200 213.

[11] Teixeira, S.; Silva, T.; Benfeito, S.; Gaspar, A.; Garrido, E. M.; Garrido, J.; Borges, F. Exploring nature profits: development of novel and potent lipophilic antioxidants based on galloyl-cinnamic hybrids. *European Journal of Medicinal Chemistry* 2013, in press.

[12] Lafay, S.; Gil-Izquierdo, A. Bioavailability of phenolic acids. *Phytochem. Rev.* 2008, *7*, 301–311.

[13] Robbins, R. J. Phenolic Acids in Foods: An Overview of Analytical Methodology *J. Agric. Food Chem.*, 2003, 51 (10), 2866-2887.

[14] Pal, C.; Bindu, S.; Dey, S.; Alam, A.; Goyal, M.; Iqbal, S. M.; Maity, P.; Adhikari, S. S.; Bandyopadhyay, U. Gallic acid prevents nonsteroidal anti-inflammatory drug-induced gastropathy in rat by blocking oxidative stress and apoptosis. *Free Radical Biology and Medicine* 2010, *49*, 258–267.

[15] Tomás-Barberán, F. A.; Clifford, M. N. Dietary hydroxybenzoic acid derivatives-nature, occurrence and dietary burden. *J. Sci. Food Agric.* 2000, *80*, 1024-1032.

[16] Fazary, A. E.; Taha, M. A.; Ju, Y.H. Iron complexation studies of gallic acid. *J. Chem. Eng. Data* 2009, *54, 35-42.*

[17] Eslami, A. C.; Pasanphan, W.; Wagner, B. A.; Buettner, G R. Free radicals produced by the oxidation of gallic acid: An electron paramagnetic resonance study. *Chemistry Central Journal* 2010, *4*, 15-19.

[18] Punithavathi, V. R.; Prince, P. S. M.; Kumar, R.; Selvakumari, J. Antihyperglycaemic, antilipid peroxidative and antioxidant effects of gallic acid on streptozotocin induced diabetic Wistar rats. *European Journal of Pharmacology* 2011, *650*, 465–471.

[19] Barreto, J. C; Trevisan, M. T.; Hull, W. E.; Erben, G.; de Brito, E. S; Pfundstein, B.; Würtele, G.; Spiegelhalder, B.; Owen, R. W.; Characterization and quantitation of polyphenolic compounds in bark, kernel, leaves, and peel of mango (Mangifera indica L.). *J. Agric. Food. Chem.* 2008, *56*, 5599–5610.

[20] Shi, J.; Yu, J.; Pohorly, J. E.; Kakuda, Y. Polyphenolics in grape seeds-biochemistry and functionality. *J. Med. Food* 2003, *6*, 291–299.

[21] Landrault, N.; Poucheret, P.; Ravel, P.; Gasc, F.; Cros, G.; Teissedre, P. L. Antioxidant capacities and phenolics levels of French wines from different varieties and vintages, *J. Agric. Food Chem.* 2001, *49*, 3341–3348.

[22] Zuo, Y.; Chen, H.; Deng, Y. Simultaneous determination of catechins, caffeine and gallic acids in green, Oolong, black and pu-erh teas using HPLC with a photodiode array detector. *Talanta* 2002, *57*, 307–316.

[23] Joubert, E.; Gelderblom, W. C.; Louw, A.; de Beer, D. South African herbal teas: aspalathus linearis, Cyclopia spp. and athrixia phylicoides – A review. *J. Ethnopharmacol.* 2008, *119*, 376–412.

[24] Monteiro, M.; Farah, A.; Perrone, D.; Trugo, L. C.; Donangelo, C. Chlorogenic acid compounds from coffee are differentially absorbed and metabolized in humans. *J. Nutr.* 2007, *137*, 2196–2201.

[25] Manosroi, A.; Jantrawut, P.; Akazawa, H.; Akihisa, T.; Manosroi, J. Biological activities of phenolic compounds isolated from galls of Terminalia chebula Retz, (Combretaceae). *Natural Product Research* 2010, *24*, 1915-1926.

[26] Srinivasan, G. V.; Ranjith, C.; Vijayan, K. K., Identification of chemical compounds from the leaves of Leea indica. *Acta Pharm.* 2008, *58*, 207–214.

[27] Muir, R. M.; Ibáñez, A. M.; Uratsu, S. L.; Ingham, E. S.; Leslie, C. A.; McGranahan, G. H.; Batra, N.; Goyal, S.; Joseph, J.; Jemmis, E. D.; Dandekar, A. M. Mechanism of gallic acid biosynthesis in bacteria (*Escherichia coli*) and walnut (*Juglans regia*). *Plant Mol. Biol.* 2011, *75*, 555–565.

[28] Tang, H. R.; Covington, A. D.; Hancock, R. A.; Synthesis and spectroscopic characterisation of polygalloyl esters of polyols: models for gallotannins. *J. Soc. Leather Chem. Tech.* 2003b, *87*, 179–188.

[29] Klein, E.; Weber, N. In vitro test for the effectiveness of antioxidants as inhibitors of thiyl radical-induced reactions with unsaturated fatty acids. *J. Agric. Food Chem.* 2001, *49*, 1224–1227.

[30] Lu, Z.; Nie, G.; Belton, P. S.; Tang, H.; Zhao, B. Structure–activity relationship analysis of antioxidant ability and neuroprotective effect of gallic acid derivatives. *Neurochemistry International* 2006, *48*, 263–274.

[31] Fiuza, S. M.; Gomes, C.; Teixeira, L. J.; Girao da Cruz, M. T.; Cordeiro, M. N. D. S.; Milhazes, N.; Borges, F.; Marques, M. P. M. Phenolic acid derivatives with potential anticancer properties— a structure–activity relationship study. Part 1: Methyl, propyl and octyl esters of caffeic and gallic acids. *Bioorganic and Medicinal Chemistry*, 2004, *12(13)*, 3581-3589.

[32] Levites, Y.; Amit, T.; Mandel, S.; Youdim, M. B. H. Neuroprotection and neurorescue against amyloid beta toxicity and PKCdependent release of non-amyloidogenic soluble precursor protein by green tea polyphenol (−)-epigallocatechin 3-gallate. *FASEB J.* 2003, *17*, 952–954.

[33] Jung, Y. D.; Kim, M. S.; Shin, B. A.; Chay, K. O.; Ahn, B. W.; Liu, W.; Bucana, C. D.; Gallick, G. E.; Ellis, L. M. EGCG, a major component of green tea, inhibits tumour growth by inhibiting VEGF induction in human colon carcinoma cells. *British Journal of Cancer* 2001, *84(6)*, 844–850.

[34] Nenadis, N.; Lazaridou, O.; Tsimidou, M. Z. Use of reference compounds in antioxidant activity assessment. *J. Agric. Food Chem.* 2007, *55*, 5452–5460.

[35] Erkan, N.; Ayranci, G.; Ayranci, E. Antioxidant activities of rosemary (Rosmarinus Officinalis L.) extract, blackseed (Nigella sativa L.) essential oil, carnosic acid, rosmarinic acid and sesamol. *Food Chemistry* 2008, *110*, 76–82.

[36] Pasanphan, W.; Buettner, G. R.; Chirachanchai, S. Chitosan gallate as a novel potential polysaccharide antioxidant: An EPR study. *Carbohydrate Polymers* 2010 *345,* 132–140.

[37] Gao, S.; Hu, M. Bioavailability challenges associated with development of anti-cancer phenolics. *Mini Rev. Med. Chem.* 2010, *10*, 550-567.

[38] Crozier, A.; Jaganath, I. B.; Clifford, M. N. Dietary phenolics: chemistry, bioavailability and effects on health. *Nat. Prod. Reports* 2009, *26*, 1001-1043.

[39] Arrua, D.; Strumia, M. C.; Nazareno, M. A. Immobilization of Caffeic Acid on a Polypropylene Film: Synthesis and Antioxidant Properties. *Journal of Agricultural and Food Chemistry* 2009, *58*, 9228–9234.

[40] Bordenave, N.; Grelier, S.; Coma, V. Advances on Selective C-6 Oxidation of Chitosan by TEMPO. *Biomacromolecules* 2008, *9*, 2377–2382.

[41] Ngo, D. H.; Qian, Z. J.; Vo, T. S.; Ryu, B.; Ngo, D. N.; Kim, S. K. Antioxidant activity of gallate-chitooligosaccharides in mouse macrophage RAW264.7 cells. *Carbohydrate Polymers* 2011, *84*, 1282–1288.

[42] Cho, Y. S.; Kim, S. K.; Ahn, C. B.; Je, J. Y. Preparation, characterization, and antioxidant properties of gallic acid-grafted-chitosans. *Carbohydrate Polymers* 2011, *83,* 1617–1622.

[43] Neo, Y. P.; Ray, S.; Jin, J.; Gizdavic-Nikolaidis, M.; Nieuwoudt, M. K.; Liu, D.; Quek, S. Y. Encapsulation of food grade antioxidant in natural biopolymer by electrospinning technique: A physicochemical study based on zein–gallic acid system. *Food Chemistry* 2013, *136*, 1013–1021.

[44] Khaledi, H.; Alhadi, A. A.; Yehye, W. A.; Ali, H. M.; Abdulla, M. A.; Hassandarvish, P. Antioxidant, cytotoxic activities, and structure-activity relationship of gallic acid-based indole derivatives. *Arch. Pharm. Chem. Life Sci.* 2011, *344*, 703–709.

[45] Novak, I.; Seruga, M.; Komorsky-Lovric, S. Electrochemical Characterization of Epigallocatechin Gallate Using Square-Wave Voltammetry. *Electroanalysis* 2009, *21,* 1019 – 1025.

[46] Yang, Z.; Zhang, D.; Long, H.; Liu, Y. Electrochemical behavior of gallic acid interaction with DNA and detection of damage to DNA. *J. Electr. Chem.* 2008, *624*, 91–96.

[47] Abdel-Hamid, R.; Newair, E. F. Electrochemical behavior of antioxidants: I. Mechanistic study on electrochemical oxidation of gallic acid in aqueous solutions at glassy-carbon electrode. *Journal of Electroanalytical Chemistry* 2011, *657*, 107–112.

[48] Skeva, E.; Girousi. S. A study of the antioxidative behavior of phenolic acids, in aqueous herb extracts, using a dsDNA biosensor. *Cent. Eur. J. Chem.* 2012, *10*, 1280-1289.

[49] Kilmartin, P. A.; Hsu, C. F. Characterisation of polyphenols in green, oolong, and black teas, and in coffee, using cyclic voltammetry. *Food Chemistry* 2003, *82*, 501–512.

[50] Kilmartin, P. A.; Zou, H.; Waterhouse, A. L. A Cyclic Voltammetry Method Suitable for Characterizing Antioxidant Properties of Wine and Wine Phenolics. *J. Agric. Food Chem.*, 2001, *49* (4), 1957-1965.

[51] Gunckel, S.; Santander, P.; Cordano, G.; Ferreira, J.; Munoz, S.; Nunez-Vergara, L. J.; Squella, J. A. Antioxidant activity of gallates: an electrochemical study in aqueous media. *Chem. Biol. Interact.* 1998, *114*, 45-59.

[52] Zare, H. R.; Golabi, S. M. Caffeic acid modified glassy carbon electrode for electrocatalytic oxidation of reduced nicotinamide adenine dinucleotide (NADH) *J. Solid State Electrochem.* 2000, *4*, 87.

[53] Siquet, C.; Pavia-Martins, F.; Lima, J.L; Reis, S.; Borges, F. Antioxidant profile of dihydroxy- and trihydroxyohenolicacids – A structure-activity relationship study. *Free Radical Research* 2006, *40*, 433–442.

[54] Abdelwahed, A.; Bouhlel, I.; Skandrani, I.; Valenti, K.; Kadri, M.; Guiraud, P.; Steiman, R.; Mariotte, A. M.; Ghedira, K.; Laporte, F.; Dijoux-Franca, M.G.; Chekir-Ghedira, L. Study of antimutagenic and antioxidant activities of Gallic acid and 1,2,3,4,6-pentagalloylglucose from Pistacia lentiscus Confirmation by microarray expression profiling. *Chemico-Biological Interactions* 2007, *165*, 1–13.

[55] Niki, E.; Noguchi, N. Evaluation of antioxidant capacity. What capacity is being measured by which method? IUBMB *Life* 2000, *50*, 323–329.

[56] Niki, E. Assessment of Antioxidant Capacity in vitro and in vivo *Free Radical Biology and Medicine* 2010, *49*, 503–515.

[57] Zhang, H. Y.; Sun, Y. M.; Wang, X. L. Substituent effects on O–H bond dissociation enthalpies and ionization potentials of catechols: A DFT study and its implications in the rational design of phenolic antioxidants and elucidation of structure– activity relationships for flavonoid antioxidants. *Chem. Eur. J.* 2003, *9*, 502–508.

[58] Singh, N.; O'Malley, P. J.; Popelier, L. A. Mechanistic aspects of hydrogen abstraction for phenolic antioxidants. Electronic structure and topological electron density analysis. *Phys. Chem.* 2005, *7*, 614–619.

[59] Heo, H.; Kim, Y.; Chung, D.; Kim, D. Antioxidant capacities of individual and combined phenolics in a model system. *Food Chemistry* 2007, *104*, 87–92.

[60] Cheng, Z.; Ren, J.; Yan, G.; Li, Y.; Chang, W.; Chen, Z. Quantitative elucidation of the molecular mechanisms of hydroxyl radical quenching reactivity of phenolic compounds. *Bioorg. Chem.* 2003, *31*, 149–162.

[61] Leopoldini, M.; Marino, T.; Nino, R.; Toscano, M. Antioxidant properties of phenolic compounds: H-atom versus electron transfer mechanism. *J. Phys. Chem. A.* 2004, *108*, 4916–492.

[62] Palafox-Carlos, H.; Yahia, E. M.;González-Aguilar, G. A. Identification and quantification of major phenolic compounds from mango (Mangifera indica, cv. Ataulfo) fruit by HPLC–DAD–MS/MS-ESI and their individual contribution to the antioxidant activity during ripening. *Food Chemistry* 2012, *135*, 105–111.

[63] Benherlal, P. S.; Arumughan, C. Studies on modulation of DNA integrity in Fenton's system by phytochemicals. *Mutation Research* 2008, *648*, 1–8.

[64] Firuzi, O.; Mladenka, P.; Petrucci, R.; Marrosu, G.; Saso, L. Hypochlorite scavenging activity of flavonoids. *J. Pharm. Pharmacol.* 2004, *56*, 801–807.

[65] Andjelković, M.; Van Camp, J.; De Meulenaer, B.; Depaemelaere, G.; Socaciu, C.; Verloo, M.; Verhe, R. Iron-chelation properties of phenolic acids bearing catechol and galloyl groups. *Food Chemistry* 2006, *98*, 123–31.

[66] Khokhar, S.; Owusu Apenten, R. K. Iron binding characteristics of phenolic compounds: some tentative structure–activity relations. *Food Chemistry* 2003, *81*, 133–140.

[67] Blasco, A. J.; Crevillen, A. G.; Gonzalez, M. C.; Escarpa, A. Direct Electrochemical Sensing and Detection of Natural Antioxidants and Antioxidant Capacity in Vitro Systems. *Electroanalysis* 2007, *19*, 2275 – 2286.

[68] Huang, D.; Ou, B.; Prior, R. L. The chemistry behind antioxidant capacity assays. *J. Agric. Food Chem.* 2005, *53*, 1841-1856.

[69] Prior, L. R.; Wu, X.; Schaich, K. Standardized Methods for the Determination of Antioxidant Capacity and Phenolics in Foods and Dietary Supplements *J. Agric. Food Chem.* 2005, *53*, 4290-4302.

[70] Dai, J.; Mumper, R. J. Plant Phenolics: Extraction, Analysis and Their Antioxidant and Anticancer Properties. *Molecules* 2010, *15*, 7313-7352.

[71] Mello, L. D.; Kubota, L. T.; Biosensors as a tool for the antioxidant status evaluation. *Talanta* 2007, *72*, 335–348.

[72] Barroso, M. F.; de-los-Santos-Alvarez, N.; Delerue Matos, C.; Oliveira, M. B. P. P.; Towards a reliable technology for antioxidant capacity and oxidative damage evaluation: Electrochemical (bio)sensors. *Biosens. Bioelectron* 2011, *30*, 1-12.

[73] Carocho, M.; Ferreira, I. C. F. R. A review on antioxidants, prooxidants and related controversy: Natural and synthetic compounds, screening and analysis methodologies and future perspectives. *Food and Chemical Toxicology* 2013, *51*, 15–25.

[74] Krishnaiah, D.; Sarbatly, R.; Nithyanandam, R. A review of the antioxidant potential of medicinal plant species. *Food and bioproducts processing* 2011, *89*, 217–233.

[75] Prieto-Simon, B.; Cortina, M.; Campas, M.; Calas-Blanchard, C. Electrochemical biosensors as a tool for antioxidant capacity assessment, *Sens. Act. B* 2008, *129*, 459-466.

[76] Barroso, M. F.; de-los-Santos-Alvarez, N.; Lobo-Castanon, M. J.; Miranda-Ordieres, A. J.; Deleruc-Matos, C.; Oliveira, M. B. P. P.; Tunon-Blanco, P. DNA-based biosensor for the electrocatalytic determination of antioxidant capacity in beverages. *Biosens. Bioelectron* 2011, *26*, 2396–2401.

[77] Liu, J.; Roussel, C.; Lagger, G.; Tacchini, P.; Girault, H. H. Antioxidant sensors based on DNA-modified electrodes. *Anal. Chem.* 2005, *77*, 7687–7694.

[78] Mello, L. D.; Hernandez, S.; Marrazza, G.; Mascini, M.; Kubota, L. T. Investigations of the antioxidant properties of plant extracts using a DNA-electrochemical biosensor. *Biosens. Bioelectron.* 2006, *21*, 1374–1382.

[79] Heilerova, L.; Buckova, M.; Tarapcik, P.; Silhar, S.; Labuda, J.; Comparison of antioxidative activity data for aqueous extracts of lemon balm (Melissa officinalis L.), oregano (Origanum vulgare L.), thyme (Thymus vulgaris L.), and agrimony agrimony (Agrimonia eupatoria L.) obtained by conventional methods and the DNA-based biosensor. *Czech J. Food Sci.* 2003, *21*, 77–84.

[80] Rusling, J. F.; Sensors for toxicity of chemicals and oxidative stress based on electrochemical catalytic DNA oxidation, *Biosens. Bioelectron.* 2004, *20*, 1022–1028.

[81] Qian, P.; Ai, S.; Yin, H.; Li, J. Evaluation of DNA damage and antioxidant capacity of sericin by a DNA electrochemical biosensor based on dendrimer-encapsulated Au-Pd/chitosan composite, *Microchim. Acta* 2010, *168*, 347–354.

[82] Kim, Y. J. Antimelanogenic and Antioxidant Properties of Gallic Acid. *Biol. Pharm. Bull.* 2007, *30(6),* 1052—1055.

[83] Li, L.; Ng, T.B.; Gao, W.; Li, W.; Fu, M.; Niu, S.; Zhao, M.; Chen, L.; Liu, F. Antioxidant activity of gallic acid from rose flowers in senescence accelerated mice. *Life Sciences* 2005, *77*, 230-240.

[84] Ferk, F.; Chakraborty, A.; Jager, W.; Kundi, M.; Bichler, J.; Misik, M.; Wagner, K.H.; Grasl-Kraupp, B.; Sagmeister, S.; Haidinger, G.; Hoelzl, C.; Nersesyan, A.; Dusinska, M.; Simic, T.,Knasmuller, S. Potent protection of gallic acid against DNA oxidation: Results of human and animal experiments. *Mutation Research - Fundamental and Molecular Mechanisms of Mutagenesis* 2011, *715*, 61-71.

[85] Serrano, J.; Cipak, A.; Boada, J.; Gonzalo, H.; Cacabelos, D.; Cassanye, A.; Pamplona, R., Zarkovic, N.; Portero-Otin, M. Double-edged sword behaviour of gallic acid and its interaction with peroxidases in human microvascular endothelial cell culture (HMEC-1). Antioxidant and pro-oxidant effects. *Acta Biochimica Polonica* 2010, *57*, 193-198.

[86] Ren, J.; Li, Q.; Dong, F.; Feng, Y.; Guo, Z. Phenolic antioxidants-functionalized quaternized chitosan: Synthesis and antioxidant properties. *International Journal of Biological Macromolecules* 2013, *53*, 77– 81.

[87] Šeruga, M.; Novak, I.; Jakobek, L. Determination of polyphenols content and antioxidant activity of some red wines by differential pulse voltammetry, HPLC and spectrophotometric methods. *Food Chemistry* 2011, *124*, 1208–1216.

[88] Arteaga, J. F.; Ruiz-Montoya, M.; Palma, A.; Alonso-Garrido, G.; Pintado, S.; Rodríguez-Mellado, J. M. Comparison of the Simple Cyclic Voltammetry (CV) and DPPH Assays for the Determination of Antioxidant Capacity of Active Principles. *Molecules* 2012, *17*, 5126-5138.

[89] Kim, O. K.; Lee, K. W.; Lee, H. J.; Lee, C. Y. Vitamin C Equivalent Antioxidant Capacity (VCEAC) of Phenolic Phytochemicals. *J. Agric. Food Chem.* 2002, *50*, 3713-3717.

[90] Šimková, D.; Beinrohr, E.; Labuda, J.; Flow-through electrochemical system with the DNA-based biosensor for the evaluation of deep DNA damage by chemicals and effect of antioxidants. *Acta Chimica Slovaca* 2009, *2,* 129 – 138.

[91] Ozgen, M.; Reese, R. N.; Tulio Jr, A. Z.; Scheerens, J. C.; Miller, A. R. Modified 2,2-azino-bis-3 ethylbenzothiazoline-6-sulfonic acid (ABTS) method to measure antioxidant capacity of selected small fruits and comparison to ferric reducing antioxidant power (FRAP) and 2,2'-diphenyl-1-picrylhydrazyl (DPPH) methods. *J. Agric. Food Chem.* 2006, *54*, 1151-1157.

[92] Ozyürek, M.; Bektaşoğlu, B.; Güçlü, K.; Apak, R. Hydroxyl radical scavenging assay of phenolics and flavonoids with a modified cupric reducing antioxidant capacity (CUPRAC) method using catalase for hydrogen peroxide degradation. *Anal. Chim. Acta* 2008, *616*, 196-206.

[93] Barroso, M. F.; Delerue-Matos, C.; Oliveira, M. B. P. P. Electrochemical DNA-sensor for evaluation of total antioxidant capacity of flavours and flavoured waters using superoxide radical damage. *Biosensors and Bioelectronics* 2011, *26*, 3748–3754.

[94] Apak, R.; Güçlü, K.; Ozyürek, M.; Bektas Oğlu, B.; Bener, M. Cupric ion reducing antioxidant capacity assay for food antioxidants: vitamins, polyphenolics, and flavonoids in food extracts. *Methods Mol. Biol.* 2008, *477*, 163-193.

[95] Kim, D. O.; Lee, C. Y. Comprehensive study on vitamin C equivalent antioxidant capacity (VCEAC) of various polyphenolics in scavenging a free radical and its structural relationship. *Crit. Rev. Food Sci. Nutr.* 2004, *44*, 253-273.

[96] Jittawan, K.; Sirithon, S. Phenolic contents and antioxidant activities of bitter gourd (Momordica charantia L.) leaf, stem and fruit fraction extracts in vitro. *Food Chem.* 2008, *110*, 881–890.
[97] Elvira, G. M.; Chandra, S.; MarcoVinicio, R. M.; Wenyi, W. Catalytic inhibition of human DNA topoisomerase by phenolic compounds in Ardisia compressa extracts and their effect on human colon cancer cells. *Food Chem. Toxicol.* 2006, *44*, 1191–1203.
[98] Saxena, H. O.; Faridi, U.; Srivastava, S.; Kumar, J.; Darokar, M.; Luqman, S.; Chanotiya, C. S.; Krishna, V.; Arvind, Negi, A. S.; Khanuja, S. P. S. Gallic acid-based indanone derivatives as anticancer agents. *Bioorganic and Medicinal Chemistry Letters* 2008, *18*, 3914–3918.
[99] Raina, K.; Rajamanickam, S.; Deep, G.; Singh, M.; Agarwal, R.; Agarwal, C. Chemopreventive effects of oral gallic acid feeding on tumor growth and progression in TRAMP mice. *Molecular Cancer Therapeutics* 2008, *7*(5), 1258-1267.
[100] Misao, U.; Hisashi, Y.; Yukiko, K.; Masanori, H.; Tomihiko, H.; Koyama, A. Antiviral effect of octyl gallate against DNA and RNA viruses. *AntiViral Res.* 2007, *73*, 85–91 A.
[101] Flausino, O.; Dufau, L.; O. Regasini, L.; S. Petronio, M.; H.S. Silva, D.; Rose, T.; S. Bolzani, V.; Reboud-Ravaux, M. Alkyl Hydroxybenzoic Acid Derivatives that Inhibit HIV-1 Protease Dimerization *Current Medicinal Chemistry* 2012, *19*, 4534-4540.
[102] Chiu-Lan, H.; Yuh-Charn, L.; Gow-Chin, Y.; Hui-Yin, C. Preventive effects of guava (Psidium guajava L.) leaves and its active compounds against R-dicarbonyl compounds-induced blood coagulation. *Food Chem.* 2007, *103*, 528–535.
[103] Punithavathi, V. R.; Prince, P. S. M.; Kumar, R.; Selvakumari, J. Antihyperglycaemic, antilipid peroxidative and antioxidant effects of gallic acid on streptozotocin induced diabetic Wistar rats *European Journal of Pharmacology* 2011, *650*, 465–471.
[104] Sameermahmood, Z.; Raji, L.; Saravanan, T.; Vaidya, A.; Mohan, V.; Balasubramanyam, M. Gallic acid protects RINm5F b-cells from glucolipotoxicity by its antiapoptotic and insulin-secretagogue actions. *Phytother. Res.* 2010, *24*, S83–S94.
[105] Ji, B. C.; Hsu, W. H.; Yang, J. S.; Hsia, T. C.; Lu, C. C.; Chiang, J. H.; Yang, J. L.; Lin, C. H.; Lin, J. J.; Suen, L. J. W.; Wood, W. G.; Chung, J. G. Gallic acid induces apoptosis via caspase-3 and mitochondrion-dependent pathways in vitro and suppresses lung xenograft tumor growth in vivo. *J. Agric. Food Chem.* 2009, *57*(16), 7596–7604.
[106] Yasui, H.; Sakurai, H. Age-dependent generation of reactive oxygen species in the skin of live hairless rats exposed to UVA light. *Experimental Dermatology* 2003, *12*, 655–661.
[107] Jainu, M.; Shyamala Devi, C. S. Attenuation of neutrophil infiltration and proinflammatory cytokines by Cissus quadrangularis: a possible prevention against gastric ulcerogenesis. *J. Herb. Pharmacother* 2005, *5*, 33–42.
[108] Manach, C.; Scalbert, A.; Morand, C.; Rémésy, C.; Jimenez, L. Polyphenols: food sources and bioavailability. *Am. J. Clin. Nutr. 2004, 79, 727-747.*
[109] Niho, N.; Shibutani, M.; Tamura, T.; Toyoda, K.; Uneyama, C.; Takahashi, N.; Hirose, M. Subchronic toxicity study of gallic acid by oral administration in F344 rats. *Food Chem. Toxicol.* 2001, *39*, 1063–1070.
[110] Manach, C.; Williamson, G.; Morand, C.; Scalbert, A.; Remesy, C.; Bioavailability and bioefficacy of polyphenols in humans. I. Review of 97 bioavailability studies. *Am. J. Clin. Nutr.* 2005, *81*, S230–S242.

[111] Shahrzad, S.; Aoyagi, K.; Winter, A.; Koyama, A.; Bitsch, I. Pharmacokinetics of gallic acid and its relative bioavailability from tea in healthy humans. *J. Nutr.* 2001, *131*, 1207–10].

[112] Murkovic, M. Phenolic compounds. In *Encyclopedia of Food Science and Nutrition*, Caballero, B.; Trugo, L. C.; Finglas, P. Eds.; Academic Press: New York, 2003, *7*, 4507-4514.

[113] Cartron, E.; Fouret, G.; Carbonneau, M. A.; Lauret, C.; Michel, F.; Monnier, L.; Descomps, B.; Leger, C. L. Red-wine beneficial long-term effect on lipids but not on antioxidant characteristics in plasma in a study comparing three types of wine–description of two O-methylated derivatives of gallic acid in humans. *Free Radic. Res.* 2003, *37*, 1021–1035.

[114] Hodgson, J. M.; Morton, L. W.; Puddey, I. B.; Beilin, L. J.; Croft, K. D. (Gallic acid metabolites are markers of black tea intake in humans. *J. Agric. Food Chem.* 2000, *48*, 2276–2280.

[115] Hodgson, J. M.; Chan, S. Y.; Puddey, I. B.; Devine, A.; Wattanapenpaiboon, N.; Wahlqvist, M. L.; Lukito, W.; Burke, V.; Ward, N. C.; Prince, R. L.; Kevin, D. Croft Phenolic acid metabolites as biomarkers for tea- and coffee-derived polyphenol exposure in human subjects *Brit. Jour. Nutr.* 2004, *91*, 301–305.

[116] Konishi, Y.; Hitomi, Y.; Yoshioka, E. Intestinal absorption of p-coumaric and gallic acids in rats after oral administration. *J. Agric. Food Chem.* 2004, *52*, 2527–2532.

[117] Konishi, Y.; Zhao, Z.; Shimizu, M. Phenolic acids are absorbed from the rat stomach with different absorption rates. *J. Agric. Food Chem.* 2006, *54*, 7539–7543.

In: Handbook on Gallic Acid
Editors: M. A. Thompson and P. B. Collins
ISBN: 978-1-62618-921-8

Chapter 6

THE ANTI-OXIDATIVE AND ANTI-INFLAMMATORY ROLES OF GALLIC ACID ON TRANSCRIPTIONAL REGULATION

***Nando Dulal Das*[*], *Amitabh Das*[*] *and Young Gyu Chai*[†]**
Department of Molecular and Life Sciences, Hanyang University,
Ansan, Korea

ABSTRACT

Gallic acid, a polyhydroxy phenolic compound, is found in various natural products such as tea leave, gallnuts, grapes, green tea, strawberries, lemons, pineapples and in red and white wines. Phenolic compounds are known to donate a proton from their hydroxyl (O–H) bond through hemolytic cleavage and form a stable phenoxy radical. Compounds possessing more than one phenolic hydroxyl group are further shown to generate an even more stabilized phenoxyl radical. As a result, gallic acid is regarded as an excellent antioxidant with a high free radical (reactive oxygen species or reactive nitrogen species-origin) scavenging effect. Among other properties, the anti-inflammatory, anti-bacterial, and antitumor properties are crucial. In this chapter, we mainly focus on the anti-oxidative and anti-inflammatory roles of gallic acid on transcriptional regulation of various cellular regulators. We will describe how gallic acid affects the suppression of anti-oxidative and anti-inflammatory regulators through the controlled regulation of key transcription factors. Since gallic acid and its esters can inhibit the TNFα-induced nuclear translocation of nuclear factor kappa-B (NF-κB) by way of a mechanism independent of inhibitors of kappa-B (IκB) degradation, we also deeply correlate the effect of gallic acid on the NF-κB-mediated transcriptional regulation of various early and late inflammatory mediators involved in related disease pathologies. In addition, we will focus on gallic acid's role on the epigenetic level since it exhibits potent p300/CBP-mediated anti-histone acetyl transferase (HAT) activity that inhibits the p65 acetylation-dependent NF-κB activation and production of inflammatory markers. Moreover, this review will show various gallic acid-mediated molecular targets, which affect signaling pathways within

---

[*] These authors contributed equally to this work.
[†] Corresponding Author, Young Gyu Chai (Ph.D.) Department of Molecular and Life Sciences, Hanyang University, 1271 Sa 3-dong, Ansan, Gyeonggi-do, 426-791, Korea. Tel: 82-31-400-5513, Fax: 82-31-436-8173, Email: ygchai@hanyang.ac.kr.

cells in different ways involving genetic or epigenetic mechanisms, and engage functional proteins in the formation of inhibiting complexes of strong affinities, ultimately perturbing the expression of oxidative and inflammatory molecules.

Keywords: Gallic acid, anti-inflammation, anti-oxidation, NF-κB, epigenetics, transcriptional regulation

Introduction

Gallic acid (GA) and its structurally related compounds are found in various natural products such as tea leaves, gallnuts, grapes, green tea, strawberries, lemons, pineapples and in red and white wines as its free or conjugated form with tannins. It is abundantly obtained by the solvent extraction of either nutgalls (*Quercus infectoria*, *Quercus stenophylla*) or Japanese gall (*Rhus javanica*) (Niho et al., 2001, p. 1,063). GA and its related compound are found in wine (Blanco et al., 1998, p. 327) and present as catechins in both green and black teas derived from the *Camellia sinensis* plant (Arce et al., 1998, p. 113; Constable et al., 1996, p. 189). Green tea contains polyphenols, consisting of flavanols, flavandiols, flavanoids, and phenolic acids, which make up 30% of dry weight materials in green tea (Graham, 1992, p. 334). A large proportion of the polyphenols contained in green tea are flavanols, commonly referred to as catechins.

The major catechins in green tea are (-) epigallocatechin-3-gallate (EGCG), (-) epigallocatechin, (-) epigallocatechin-3-gallate, (-) epicatechin, (+) gallocatechin, and (+) catechin (Graham, 1992, p. 334; Yang and Wang, 1993, p. 1,038). Among these polyphenols, ECGC is the most abundant one and is believed to be responsible for many of the beneficial effects of tea (Hour et al., 1999, p. 569; Kada et al., 1985, p. 127). In tea, GA may be present in free and esterified forms. Most of the GA found in black tea is derived originally from two flavanoids, epicatechin gallate and EGCG, which make up about 15% of the dry weight in the green tea leaf.

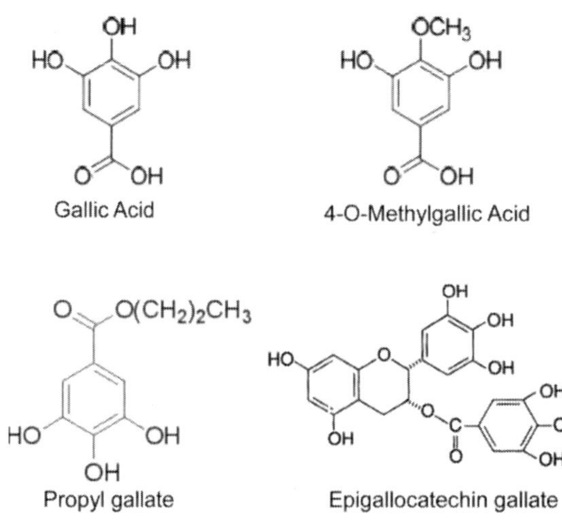

Figure 1. Structure of gallic acid and related compounds.

Approximately 35% of the mass of both epicatechin gallate and EGCG is GA, present as gallate esters (Wiseman et al., 1997, p. 705). Studies utilizing these compounds have found them to possess many potential therapeutic properties including antioxidant, anti-cancer (Chuang et al., 2010, p. 2,943) and anti-inflammatory activities (Kroes et al., 1992, p. 499).

As polyhydroxy phenolic compounds, GA and its related forms (Figure 1) are found in various natural products. Polyphenols are abundant micronutrients in our diet and are widespread constituents of fruits; vegetables; dry legumes; and beverages such as tea, coffee, or wine. Various molecules having a polyphenol structure have been identified in edible plants, and these molecules are classified into different groups as a function of the number of their phenol rings: the phenolic acids, flavonoids, stilbenes, and lignans (Manach et al., 2004, p. 727). In recent years, there has been a growing interest, supported by a large number of experimental and epidemiological studies (Dai and Mumper, 2010, p. 7,313; Middleton et al., 2000, p. 673; Scapagnini et al., 2011, p. 192), related to the beneficial effects of some phenolic substances against malignancies, cardiovascular diseases, and neurodegenerative diseases. Two classes of phenolic acids can be distinguished: derivatives of benzoic acid and derivatives of cinnamic acid. Hydroxyl-benzoic acids consist of GA, protocatechuic acid, and p-hydroxybenzoic acid. Among natural antioxidants, polyphenols successfully scavenge free radicals via their OH-groups (Hirano et al., 2001, p. 357). There is a high correlation between the content of phenolic substances and the total antioxidant activity of various plant extracts (Stratil et al., 2006, p. 607). Currently, GA (GA, 3, 4, 5-trihydroxybenzoic acid) and its derivatives are regularly applied in pharmaceutical and food industries (Lu et al., 2006, p. 263).

GA derivatives are exploited in phytomedicine, for example, as free radical scavengers, inducing the apoptosis of cancer cells (Saeki et al., 2000, p. 1,391) and interfering with the signal pathways associated with Ca^{2+} and oxygen radicals (Sohi et al., 2003, p. 221). GA and its derivatives also have a wide range of uses. Esters of GA, such as propyl gallate, octylgallate, lauryl gallate and dodecyl gallate, are commonly used as food additives and in cosmetics as antioxidants (van der Heijden et al., 1986, p. 1,067). Propyl gallate is also widely employed as an antioxidant in the pharmaceutical industry and is useful for preventing the deterioration and rancidity of fats and oils (van der Heijden et al., 1986, p. 1,067). GA has been investigated as a potential anticancer agent in various human cancer cell lines such as TE-2 (esophageal cancer), MKN-28 (gastric cancer), HT-29 and Colo201 (colon cancer), MCF-7 (breast cancer) and CaSki (cervix cancer) (Faried et al., 2007, p. 605). Additionally, previous studies have shown that GA-induced apoptosis in cancer cells is associated with oxidative stresses derived from reactive oxygen species (ROS), mitochondrial dysfunction and an increase in the intracellular Ca^{2+} level (Inoue et al., 2000, p. 1,153; Serrano et al., 1998, p. 350). However, the mechanism by which GA analogues induce apoptosis in some cell lines is not yet completely understood; the latter probably involving the paradoxical generation of ROS, which interferes with the homeostatic redox balance of the cell (Isuzugawa et al., 2001, p. 249). Furthermore, these compounds are excellent inhibitors of protein tyrosine kinases (PTKs) (Palacios et al., 2001, p. 527).

In another aspect, the amounts of GA in plant tissues may vary according to the plant species; however, it is recognized that external stimuli, such as UV radiation, microbial infections and chemical stressors such as glyphosate, may considerably alter the synthesis of these compounds (Daniel et al., 1999, p. 109). The complexity of determining health effects from the consumption of these food stuffs is therefore potentially augmented.

In this chapter, we mainly review the anti-oxidative and anti-inflammatory roles of GA, its esters, and gallic acid derivatives on the transcriptional regulation of various cellular regulators. We also focus on the effect of gallic acid on major transcription factors mediating the transcriptional regulation of various early and late inflammatory or oxidative mediators that are involved in related disease pathologies. In addition, the epigenetic role of GA for controlled transcription will also be discussed.

ANTI-INFLAMMATORY ROLE OF GALLIC ACID

Inflammation involves a complex web of intercellular cytokine signals. Activated monocytes and/or macrophages release a variety of inflammatory cytokines or mediators, such as tumor necrosis factor-α (TNF-α), interleukin- 1β (IL 1β), interleukin-6 (IL-6), reactive oxygen species (ROS), prostaglandin E2 (PGE2), and nitric oxide (NO), under physiological responses. These inflammatory cytokines and mediators regulate the functional activity of immune cells, and uncontrolled expressions are involved in the pathogenesis of many inflammatory diseases such as septic shock, hemorrhagic shock, multiple sclerosis, rheumatoid arthritis, ulcerative colitis, atherosclerosis, and cancer (Alfon et al., 1999, p. 325; Graeber and Streit, 2010, p. 89). It has been shown that inhibition in the production and biological activities of iNOS, COX-2, TNF-α, and IL-1β by selective inhibitors, neutralizing antibodies, or gene targeting resulted in a significant amelioration in the development of inflammation-mediated diseases such as rheumatoid arthritis, cardiovascular disease, and septic shock (Kagari et al., 2002, p. 1,459; Makarov, 2000, p. 441; Ohta et al., 2005, p. 180). Therefore, drugs that suppress the expression of these inflammation-associated genes have therapeutic potential for preventing inflammatory reactions and diseases.

There has been some evidence supporting GA's anti-inflammatory and pro-apoptotic activities (Kang et al., 2009, p. 760; Na et al., 2006, p. 1,597). GA treatment attenuates TNF-α and IL-6 expressions from human monocytes (Kuppan et al., 2010, p. 229). In addition, GA also inhibits mast cell-derived inflammation by suppressing pro-inflammatory cytokine expressions (Kim et al., 2006, p. 123). Recent reports showed that the expression levels of several important pro-inflammatory mediators from rheumatoid arthritis fibroblast-like synoviocytes were significantly suppressed by GA treatment (Yoon et al., 2012). We should now focus on how GA acts as an inflammatory antagonist. The answer lies in GA's role against key mediators of inflammation such as NF-κB, mitogen-activated protein kinases (MAPK), activator protein 1 (AP-1) etc.

NF-κB has a seminal role in immunity because NF-κB regulates the expression of many inflammatory genes encoding iNOS, COX-2, TNF-α, IL-1β, and IL-6 (Makarov, 2000, p. 441), which are involved in the pathogenesis of inflammatory diseases. Because of its critical role in inflammatory gene expression, NF-κB is a current target for treating various inflammatory diseases (Renard and Raes, 1999, p. 341). NF-κB, which is a heterodimer of a 50- and a 65-kDa subunit, resides as an inactive cytosolic protein due to its interaction with inhibitory proteins of the IκB family in most cell types, including blood monocytes. NF-κB activation is induced by a variety of stimuli, such as mitogens, cytokines, LPS, viruses, and UV light, which lead to the phosphorylation and the degradation of the IκB proteins (Mandrekar et al., 1999, p. 1,781). Most anti-inflammatory drugs have been shown to inhibit inflammatory cytokine expression by inhibiting the NF-κB activation pathway (Castrillo et al., 2001, p. 15,854). It has been reported that n-propyl gallate exhibits anti-inflammatory activity through the NF-κB/IκB-α pathway (Jung et al., 2011, p. 352). In addition, prodelphinidin B-4 3′-O-gallate and other tea polyphenols consisting of gallocatechin and epigallocatechin-3-gallate were identified to be involved in the inhibition of COX-2 and iNOS via the down-regulation of the TGF-β- activated kinase (TAK1)/NF-κB pathway in LPS-activated RAW264.7 macrophage cells (Hou et al., 2007a, p. 742). Furthermore, 4-O-methylgallic acid, a major metabolite of GA, inhibited the expression of iNOS, COX-2, TNF-α, and IL-1β as well as the production of NO and PGE2 in LPS-stimulated macrophages and endotoxemic animals by blocking NF-κB activation through the inhibition of redox-sensitive IκB kinase activity (Na et al., 2006, p. 1,597). These studies suggest that the GA-mediated NF-κB pathway contributes to the anti-inflammatory effect and the prevention of human diseases through the suppression of inflammatory gene expression (Keifer et al., 2001, p. 22,382). Thus, as a NF-κB inhibitor, gallic acid could be a potential therapeutic drug in clinical applications to regulate the uncontrolled immune responses in inflammatory diseases.

We also found that *Terminalia chebula* (TC) extract component GA can inhibit IκBα degradation and phosphorylation, which leads to the prevention of NF-κB translocation and ultimately suppresses NF-κB activity and downregulates two key inflammatory genes, IL-8 and MCP-1, in high-level NF-κB-transfected Jurkat cells (Das et al., 2011, p. 927). In another study, we again demonstrated that the TC extract component GA significantly inhibited NF-κB activity and affected the proteomic profile of Jurkat cells. Consequently, TC extract component GA treatment mainly suppressed the expression of β-tubulin, RCHY1, IGF1R, and HSP70, which may negatively affect lymphoma cells. Furthermore, Ingenuity Pathways Analysis identified top-scoring proteins that are involved in molecular networks of immunological and inflammatory diseases (Das et al., 2012, p. 651). It has been reported that HSP70 has been shown to be a potent activator of the innate immune system, capable of inducing the secretion of pro-inflammatory cytokines (Asea et al., 2000, p. 435) via a NF-κB pathway (Chase et al., 2007, p. 6,818). In addition, positive feedback regulation of HSP70 induction exists in innate immune cells and induced pro-inflammatory cytokine secretion via Toll-like receptor 4 (TLR4) (Lee et al., 2013, p. 88). Therefore, suppressing the expression of HSP70 by TC extract component GA could be an effective approach in treating inflammatory diseases.

Vascular inflammation is a primary event in the pathogenesis of many human diseases, including atherosclerosis, hypertension, restenosis, and septic shock (Schiffrin, 2002, p. 115; Shah, 2003, p. 2,175). There is *in vivo* evidence of increased expressions of the intercellular adhesion molecules, endothelial adhesion molecule-1 (ICAM-1) and vascular cell adhesion molecule-1 (VCAM-1) in inflammatory animal models and in human atherosclerotic plaques (Cybulsky and Gimbrone, 1991, p. 788). In most resting cells, NF-κB is sequestered in the cytoplasm in an inactive form associated with IκB. Upon stimulation of endothelial cells by inflammatory cytokines, such as TNF-α, IκB becomes phosphorylated and proteolytically degraded; permitting NF-κB to translocate into the nucleus where NF-κB binds to κB enhancer elements of inflammatory target genes, including ICAM-1 and VCAM-1, to induce their transcription (Beg et al., 1993, p. 3,301; Chen et al., 1995, p. 3,538). The latter evidence suggests that a selective suppression of the NF-κB signaling pathway prevents various inflammatory diseases including inflammatory atherosclerosis. A variety of natural substances, such as hematein and gallates, have been reported to possess anti-atherogenic properties (Choi et al., 2003, p. 287; Murase et al., 1999, p. 1,412). These compounds have been shown to inhibit the cytokine-induced expression of ICAM-1 and VCAM-1 in endothelial cells, probably by inhibiting NF-κB activation. Furthermore, 4-O-methylgallic acid prominently reduces the expression of ICAM-1 and VCAM-1 and also the adhesion of monocytes to TNF-α-treated human umbilical vein endothelial cells by inhibiting the redox-sensitive DNA-binding activity of NF-κB. These results indicate that these compounds have the potential to treat human vascular inflammatory diseases including atherosclerosis.

It is reported that grape seed extract containing GA, catechin, epigallocatechin gallate, and proanthocyanidin is shown to prevent amyloid beta (Aβ) deposition and attenuate inflammation in the brain of a mouse with Alzheimer disease (Wang et al., 2009, p. 3). GA also possesses potent anti-histone acetyltransferase (HAT) activity and inhibits RelA acetylation by directly inhibiting the activity of HAT enzymes, which finally leads to the downregulation of NF-κB function via diverse anti-inflammatory signals (Choi et al., 2009b, p. 2,011). Additionally, GA can efficiently block neuronal cell death by downregulating the expression of cytokines and the *in vivo* expression level of NF-κB acetylation (Kim et al., 2011, p. 1,798).

Mast cells, which are constituents of virtually all organs and tissue, are important mediators of allergic reactions. Immediate hypersensitivity (anaphylaxis) is mediated by histamine release in response to the antigen cross-linking of immunoglobulinE (IgE). Mast cell activation causes the process of degranulation, resulting in the release of mediators, such as histamine and an array of inflammatory cytokines (Church and Levi-Schaffer, 1997, p. 155). Among the inflammatory substances released from mast cells, histamine remains the best-characterized and most potent vasoactive mediator implicated in the acute phase of immediate hypersensitivity (Petersen et al., 1996, p. 672). Therefore, histamines are one of the targets for agents to suppress their expressions and/or activities, which could arrest the inflammation. GA efficiently suppressed histamine releases and proinflammatory cytokine production in mast cells (Kim et al., 2006, p. 123). Additionally, GA decreases the histamine release from rat basophilic leukemia cells and suppresses pro-inflammatory cytokine production in murine peritoneal macrophages (Kwon et al., 2004, p. 436; Matsuo et al., 1997, p. 58).

Matrix metalloproteinases (MMPs), zinc-dependent proteolytic enzymes, play an important role in the outcome of an inflammatory reaction, angiogenesis, and tissue remodeling, and cause the release of extracellular matrix-bound growth factors and cytokines that regulate many of these processes (Mott and Werb, 2004, p. 558). MMP-9, which is the 92-kDa gelatinase B expressed molecule in various cell lines, such as keratinocytes, osteoclasts, eosinophils, neutrophils, and macrophages, are increased and activated in many kinds of inflammatory and malignant diseases (Sorsa et al., 2006, p. 306). Therefore, MMPs are one of the targets for agents to suppress their expression and/or activities and could arrest the inflammation. Inhibition of MMP-9 activity by n-propyl gallate supports its anti-inflammatory activities (Jung et al., 2011, p. 352). However, how n-propyl gallate inhibits MMP-9 activity remains to be elusive. In addition, epigallocatechin-3-gallate has been found to inhibit the expression and secretion of MMP-9 in macrophage-differentiated HL-60 myeloid leukemia cells (Annabi et al., 2007, p. 1,277).

A mitogen-activated protein kinases (MAPKs) cascade is one of the important signaling pathways in immune responses (Arbabi and Maier, 2002, p. 74). The expression of TNF-α and IL-6 is also regulated by MAPKs. The exact signaling pathways among three types of

MAPKs, such as p38, ERK, and JNK, are still unclear; however, p38 MAPK is thought to play an important role in the regulation of inflammatory responses. Activation of p38 MAPK is essential for the expression of the pro-inflammatory cytokines (Manthey et al., 1998, p. 409). Interestingly, the idea is supported that GA has the inhibitory activity on p38 MAPK activation and downstream TNF-α and IL-6 production (Kim et al., 2006, p. 123). Although GA acts as a potent inhibitor of NF-κB and its derivative, epigallocatechin-3-gallate has been shown to suppress NF-κB activation and phosphorylation of p38 MAPK and JNK1/2 in human astrocytoma U373MG cells (Kim et al., 2007, p. 587). Moreover, prodelphinidin B-2 3,3′di-O-gallate, a green tea proanthocyanidin, results in the inhibition of COX-2 via the suppression of activated-MAPKs, including JNK, ERK, and p38 MAPK, and the subsequent blockage on the MAPK-mediated activation of NF-κB, and AP-1 and CCAAT/enhancer-binding protein (C/EBP) δ (Hou et al., 2007b, p. 67). In addition, GA-related polyphenol compounds, epigallocatechin-3-gallate and the aflavin-3, 3'–digallate also inhibit the activation of MAPKs (Chung et al., 2001, p. 2,022).

The transcription factor AP-1 has a pivotal role in a variety of physiological processes such as growth regulation, apoptosis, cell transformation, inflammation, innate immune response and tumor metastasis (Bohmann et al., 1987, p. 1,386; Eferl and Wagner, 2003, p. 859). It is being reported for the first time that the ethyl gallate attenuate LPS induced ICAM-1 and VCAM-1 through the modulation of AP-1 activity (Mehla et al., 2011, p. 1,345) where cell adhesion molecules, such as ICAM-1 and VCAM-1, were involved in the immune response by allowing leukocyte endothelial interaction and their subsequent migration at the site of insult (Luscinskas and Gimbrone, 1996, p. 413; Meager, 1999, p. 27). This suggests that the potential inhibition of AP-1 activity by GA or its ester could be relevant in the prevention of a number of diseases where AP-1 activation is involved in disease pathogenesis. Furthermore, it can also be useful for conditions where both AP-1 and NF-κB act coordinately (Naugler and Karin, 2008, p. 19; Shaulian and Karin, 2002, p. 131) in inflammatory diseases. In this context, by using a combinatorial approach, GA along with other NF-κB inhibitors may provide improved therapeutic efficacy for inflammatory-related diseases like atherosclerosis, asthma, acute lung injury, etc.

Anti-Oxidative Role of Gallic Acid

Free radicals in the human body can cause well recognized, harmful effects and are being continuously investigated. Free radical species viz. hydroxyl, nitric oxide (NO), superoxide work in an intricate way in biological systems and their overproduction deleteriously affects the membrane lipids, cellular proteins and enzymes, ultimately resulting in oxidative stress. This oxidative stress causes cell death and eventually leads to inflammatory disorders, cancer, diabetes, etc. A pathway leading to the activation of transcription factor NF-κB, a regulator of inflammation, is also under ROS-mediated control. Therefore, antioxidant compounds could have anti-inflammatory activities (Bubici et al., 2006, p. 6,731). Synthetic compounds viz. butylatedhydroxytoluene (BHT) and butylatedhydroxyanisole (BHA) are well known antioxidants, but are also carcinogenic (Ito et al., 1983, p. 343). Natural products from the plant sources are relatively safe and have immense potential to display over whelming biological activities which depend upon their nature, structure and interactions with other molecules in the assortment (Djeridane et al., 2006, p. 719).

Antioxidants preventing free-radical reactions attract intense scientific and economic interest in human health, (Aruoma et al., 1997, p. 389) food, and polymer industries (Balasundram et al., 2005, p. 319). Antioxidants, at low concentrations, delay or prevent molecular deterioration by adverse radical reactions and radical-related oxidation and protect the human body against damage by reactive oxygen species (Halliwell, 1996, p. 439). GA has been demonstrated to have a broad spectrum of biological activities, including antimicrobial, anti-inflammatory, and anti-oxidative activities.

Additionally, previous studies have shown that GA-induced apoptosis in cancer cells in association with oxidative stresses derived from reactive oxygen species (ROS), mitochondrial dysfunction and an increase in intracellular Ca^{2+} levels (Inoue et al., 2000, p.1,153; Serrano et al., 1998, p. 49). GA exhibiting antiapoptotic potential in normal human lymphocytes is well established (Li et al., 2005, p. 230); however, as a strong natural antioxidant scavenging ROS, GA's role is well reported in several studies (Abdelwahed et al., 2007, p.1; Lu et al., 2006, p. 263).

Reactive oxygen species (ROS) play a direct role in the pathogenesis of different disease conditions including inflammation, burns and neurodegenerative diseases (Basha and Madhusudhan, 2010, p. 1,017). It is well known that fluoride accumulation in tissues leads to an oxidative stress condition through increasing ROS production and suppressing endogenous antioxidant enzyme systems (Ranjan et al., 2009, p. 900). Recently, it has been suggested that GA ameliorated sodium fluoride induced an oxidative state in a rat brain (Nabavi et al., 2013, p. 261). Neuroprotective activities of GA against 6-hydrodopamine auto-oxidation- induced apoptosis in human SH-SY5Y cells (Lu et al., 2006, p. 263) and amyloid beta protein (25–35)-induced toxicity in cultured rat cortical neurons (Ban et al., 2008, p. 149) have been previously reported. Treatment with GA leads to an increase in glutathione (GSH) levels and a reduction in oxidized glutathione (GSSG). It is well known that the GSH/GSSG ratio enhancement by gallic acid is enhanced with the addition of the hydroxyl groups (Kim, 2007, p. 587). The presence of a hydroxyl group at a para position to the COOH group is responsible for the antioxidant activity of GA derivatives and may explain the protective mechanism against sodium fluoride-induced oxidative stress.

The accumulation of advanced glycation end products (AGEs), which are the product of the non-enzymatic glycation of proteins, plays a major role in stiffening the cardiac tissues and eventually leads to fibrosis (Ahmed, 2005, p. 67). The formation and accumulation of AGEs have been widely distributed for their key role in promoting vascular dysfunction and other complications. These AGEs impair cellular functions by binding to its receptor for AGEs (RAGE) (Forbes et al., 2008, p. 1,446) through the activation of NF-κB and MAPKs (Bierhaus et al., 2005, p. 876). The activation of RAGE induces an oxidative stress and a series of inflammatory process such as the up-regulation of cytokines and various adhesion molecules (Schleicher and Friess, 2007, p. 17). Dietary plants with abundant antioxidant phenolics have the potential to function as powerful anti-glycation agents in blocking the AGEs-induced toxicities (Farrar et al., 2007, p. 193). In this context, GA also receives much attention because of its potent free radical scavenging and anti-oxidant activities. Consistent with this concept, recently, GA has exhibited a protective role against AGEs-induced cardiovascular complications, probably through its free radical scavenging activity (Umadevi et al., 2012, p. 304).

Glutahione (GSH) is mainly involved on the synthesis of important macromolecules by conferring protection against reactive oxygen species. GSH contributes to the maintenance of the antioxidant activities of other antioxidant enzymes, such as glutathione peroxidase as well as vitamins C and E (Johansen et al., 2005, p. 5). In this regard, an increased concentration of superoxide dismutase, catalase, and GSH contents on the GA pretreatment implicates an augmentation in antioxidant capacity and reduced per oxidation in the membrane lipids on the GA pretreatment in cardiovascular complications (Umadevi et al., 2012, p. 304). Similar results were observed in cells induced with high glucose content with a concomitant depleted level of GSH (Niedowicz and Daleke, 2005, p. 289) and the restoration of a redox status in cells on the GA pretreatment in cultured human THP-1 monocytes (Kuppan et al., 2010, p. 229).

It is reported that GA induces the expression of several antioxidant enzymes, such as thioredoxin-1 (TXN), thioredoxin reductase 1 (TXNRD1), antioxidant enzyme 372 (AOE372), glutathione synthetase (GSS); and DNA repair molecules such as Ligase 4 (LIG4), DNA polymerase delta subunit 2 (POLD2), methylpurine DNA N-glycosylase (MPG), growth arrest and DNA-damage-inducible alpha (GADD45A), proliferating-cell nuclear antigen (PCNA), replication protein A 32 kDa subunit (RPA2), DNA damage-inducible transcript 3 protein (DDIT3), hemeoxygenase 2 (HMOX2), xero-dermapigmentosum, complementation group A (XPA), thymine-DNA glycosylase (TDG), excision repair cross-complementation group 1 (ERCC1), general transcription factor IIH, polypeptide 1 (GTF2H1). GA also induces the repression of glutathione peroxidase 1 (GPX1), selenoprotein W 1 (SEPW1), DNA polymerase delta subunit 1 (POLD1), and SHC-transforming protein 1 (SHC1) gene expressions using microarray expression profiling (Abdelwahed et al., 2007, p. 1). Previous studies investigated that mice lacking SHC display a prolonged life span, reduced production of intracellular oxidants, and an increased resistance to oxidative stress-induced apoptosis (Migliaccio et al., 1999, p. 309).

Accordingly, the majority of studies using SHC $^{-/-}$ mice have defined the pathophysiological role of SHC in cardiovascular disease where ROS represents a substantial triggering component (Camici et al., 2007, p. 5,217; Cosentino et al., 2008, p. 622). Other authors have also reported that another protein, such as TXN, is a highly conserved protein and has been recognized as a critical protection system against oxidative stress (Forstermann, 2008, p. 338). Such results suggest that pretreatment with GA exhibits a protective role against oxidative stress-induced apoptosis as well as physiological complications.

Suppressive Role of Gallic Acid from an Epigenetic Perspective

Epigenetic refers to any heritable influence (in the progeny of cells or individuals) on the chromosome or gene function that is not accompanied by a change in DNA sequence, constituting an important mechanism by which dietary components can selectively activate or inactivate gene expression (Davis and Ross, 2007, p. 88). Epigenetic mechanisms include changes in DNA methylation, histone modifications, and altered microRNA (miRNA) expression (Winter et al., 2009, p. 228; Yoo and Jones, 2006, p. 37). Changes to the structure of chromatin influence gene expression by either inactivating genes, which occurs when the chromatin is closed (heterochromatin), or by activating genes when the chromatin is open (euchromatin) (Rodenhiser and Mann, 2006, p. 341). The nucleosome, which is the fundamental repeating unit of chromatin, is composed of DNA wrapped around a histone octamer, formed by four histone partners, an H3-H4 tetramer, and two H2A-H2B-dimers. Each successive nucleosomal coreis separated by a DNA linker associated with a single molecule of histone H1. Chromatin modifications usually occur at the amino acids of the N-terminal tails of histones and either facilitate or hinder the association of DNA repair proteins and transcription factors with chromatin. These core histones undergo a wide range of post-translational modifications, including: acetylation, controlled by histone acetyltransferases (HATs) associated with gene expression (Zhang and Dent, 2005, p. 1,137); deacetylation, controlled by histone deacetylases (HDACs) associated with gene inactivation; and methylation, phosphorylation, ubiquitination, sumoylation, ADP-ribosylation, and possibly biotinylation (Davis and Ross, 2007, p. 88).

Among various epi-modifiers, GA acts as a potent HAT inhibitor. HATs, enzymes that acetylate conserved lysine aminoacids on histone proteins by transferring an acetyl group from acetyl CoA to form e-N-acetyl lysine, are other important targets for dietary components. HATs include atleast 25 members and are organized into 4 families based on primary structure homology (Lee and Workman, 2007, p. 284). Several studies have recently reported that GA is a potent HAT inhibitor. GA, an organic acid found in gallnuts, sumac, witch hazel, tea leaves, oak bark, and other plants, can uncompetitively inhibit p300/CBP-dependent HAT activities, increase the level of cytosolic IκBα, prevent lipopolysaccharide (LPS)–induced p65 translocation to the nucleus, and suppress LPS-induced NF-κB activation in A549 lung cancer cells (Choi et al., 2009b, p. 2,011).

Furthermore, they showed that gallic acid inhibits the acetylation of p65 and LPS-induced serum levels of interleukin-6 *in vivo*. The same group also examined the biological effect of GA as a HAT inhibitor (HATi) on Aβ-induced neuroinflammation and neuronal cell death, which were triggered by activated microglia. Furthermore, GA-pretreated mice showed a restoration of altered behavior and Aβ-induced memory impairment. They also found that GA inhibited NF-κB-mediated cytokine production in the brain by blocking RelA acetylation (Kim et al., 2011, p. 1,798). EGCG, the major polyphenol found in green tea, is reported to have an anti-NF-κB transactivation activity in a broad range of human malignancies such as colon cancer, lung cancer, breast cancer, and in chronic inflammation (Doss et al., 2005, p. 259; Yang et al., 2005, p. 135). EGCG has been shown to inhibit the production of nitric oxide synthase (NOS2) by blocking the NF-κB signal transduction pathway (Chan et al., 1997, p. 1281). EGCG is also known to suppress NF-κB activation and phosphorylation of the p38 mitogen-activated protein kinase and c-Jun NH2-terminal kinase (Kim et al., 2007, p. 587).

Furthermore, EGCG reduces the binding of p300 to the promoter region of the IL-6 gene with an increased recruitment of HDAC3, which highlights the importance of the balance between HATs and histone deacetylases in the NF-κB–mediated inflammatory signaling pathway (Choi et al., 2009a, p. 583).

DNA methylation, another epigenetic mechanism is a heritable modification of the DNA structure that does not alter the specific sequence of base pairs responsible for encoding the genome but that can directly inhibit gene expression (Das and Singal, 2004, p. 4,632). A DNA hypermethylation of CpG islands is usually associated with a silencing of the expression of genes in contrast to a loss of methylation, which often leads to gene reactivation. Abnormal patterns of methylation DNA may ultimately lead to genetic instability and cancer development through the epigenetic inactivation of certain critical cancer-related genes by promoter hypermethylation. These altered genes include tumor suppressor genes, such as the cell cycle checkpoint genes, $p21^{WAF1/CIP1}$ and $p16^{INK4a}$, and growth regulatory genes, such as the RAS association domain family 1A (*RASSF1A*) and retinoic acid receptor β (*RARβ*) (Baylin and Ohm, 2006, p. 107). DNA methylation is regulated by DNA methyltransferases (DNMT1, DNMT3a, and DNMT3b) in the presence of S-adenosyl-methionine, which serves as a methyl donor for the methylation of cytosine residues at the C-5 position to yield 5-methylcytosine (Herman and Baylin, 2003, p. 2042).

Although the role of GA on DNA methylation is yet to be disclosed, GA-related compounds, such as EGCG, exhibit a potential demethylating agent. The generation of S-adenosyl-S-homocysteine, a potent inhibitor of DNA methylation, is one of the mechanisms for the demethylating properties of this compound. EGCG can form hydrogen bonds with different residues in the catalytic pocket of DNMT and thus act as a direct inhibitor of DNMT1 (Fang et al., 2003, p. 7,563). Other groups reported that cancer inhibition generated from dietary EGCG is associated with gene reactivation through demethylation in the promoters of methylation-silenced genes such as $p16^{INK4a}$, *MGMT*, *hMLH1*, *GSTP1*, *WIF-1*, *RECK*, and *RARβ*.

The effects of dietary EGCG on DNMTs appear to be associated with direct inhibition from interactions with the catalytic site of the DNMT1 molecule, and may also indirectly influence the methylation status through metabolic effects associated with energy metabolism (Gao et al., 2009, p. 2,025; Kato et al., 2008, p. 647). Therefore, a reversal of the hypermethylation-induced inactivation of key tumor suppression genes by dietary DNMT inhibitors, such as GA or EGCG, could be an effective approach for cancer-inflammation prevention and therapy.

MicroRNAs are an evolutionary conserved class of ~22 endogenous nucleotide noncoding RNAs which are involved in post-transcriptional gene repression (Taganov et al., 2006, p. 12,481). MicroRNAs are being investigated for their role as post-transcriptional regulators of pro-inflammatory genes. MicroRNAs suppress protein synthesis by inhibiting the translation of proteins from mRNA or by promoting the degradation of mRNA, thereby silencing the gene expression. Certain miRNAs were discovered to be involved in various physiological and pathological processes such as vascular inflammation (Gantier et al., 2007, p. 458). Previous studies have demonstrated that miR-126 inhibited the expression of VCAM-1; therefore, decreasing miR-126 in endothelial cells increases TNF-α-stimulated VCAM-1 expression and enhances leukocyte adherence to endothelial cells (Urbich et al., 2008, p. 581). Recently, it has been demonstrated that polyphenolic red wine extract GA increased miR-126 and consequently decreased the mRNA expression of LPS-induced inflammatory mediators NF-κB, ICAM-1, VCAM-1 and platelet endothelial cell adhesion molecules (PECAM-1), indicating the potential role of miR-126 in the anti-inflammatory properties in human colon-derived CCD-18Co myofibroblast cells (Angel-Morales et al., 2012, p. 745). In addition, EGCG has also been found to modulate miRNA expression in human hepatocellular carcinoma HepG2 cells. A microarray analysis of the HepG2 cell line after EGCG treatment has revealed that EGCG modified the expression of 61 miRNAs. Furthermore, miR-16, which is one of the miRNAs up-regulated by EGCG treatment, is confirmed to mediate the EGCG induction of apoptosis in human hepatocellular carcinoma cells by targeting Bcl-2 (Tsang and Kwok, 2010, p. 12,481).

Conclusion

Currently, plant polyphenols GA and related compounds, the regular consumption of which have been claimed to be beneficial for human health, have achieved a greater recognition not only for anti-tumor, anti-inflammatory, anti-oxidative properties, but also for their presence and vast abundance in fruits, seeds, derived food stuffs and beverages. Here, we mainly review the anti-oxidative and anti-inflammatory roles of gallic acid and focus on key transcription factors, such as NF-κB, MAPK, and AP-1, for the suppression of major anti-oxidative and anti-inflammatory regulators. The effect of gallic acid on NF-κB, MAPK, and AP-1-mediated transcriptional regulation of various inflammatory mediators involved in related disease pathologies has been discussed in this review. The epigenetic role of gallic acid has also been investigated: it exhibits potent p300/CBP-mediated anti-histone acetyl transferase (HAT) activity that inhibits the p65 acetylation-dependent NF-κB activation and production of inflammatory markers; a gallic acid-related compound, EGCG, exhibits a potential demethylating agent; finally, the affect of GA on mir-126 expression resulted in a decreased expression of various inflammatory mediators such as NF-κB, ICAM-1, VCAM-1, and PECAM-1. Further research is necessary to determine whether, and to what extent, GA plays an epigenetic role as an anti-inflammatory and anti-oxidative molecule.

Moreover, our study suggests various gallic acid-mediated molecular targets (Figure 2), which affect signaling pathways within cells in different ways involving genetic or epigenetic mechanisms.

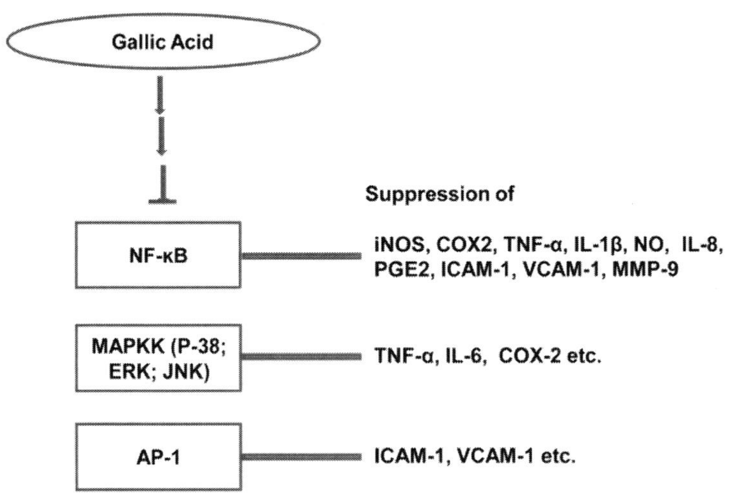

Figure 2. Possible site of action and targets of gallic acid.

The mechanisms underlying the anti-oxidative and anti-inflammatory effects of gallic acid are multifactorial, and it is likely that multiple molecular mechanisms, rather than a single receptor or molecular target, are involved. It is reasonable to suggest that gallic acid initially binds to one or more of the target proteins, which may be transmembrane receptors, kinases or other enzymes. These actions may inhibit key signaling and metabolic pathways that are essential for the development of inflammation and oxidative stress. Knowledge gained from the studies on the biological properties and activities of gallic acid reviewed in this article will be useful in the design of prospective studies and in the selection of target biomarkers for intervention trials against inflammation and oxidative stress.

ACKNOWLEDGMENTS

This work was supported by the National Research Foundation of Korea Grant funded by the Korean Government (No. 2012-0009212).

REFERENCES

Abdelwahed, A., Bouhlel, I., Skandrani, I., Valenti, K., Kadri, M., Guiraud, P., Steiman, R., Mariotte, A.M., Ghedira, K., and Laporte, F. (2007). Study of antimutagenic and antioxidant activities of gallic acid and 1,2,3,4,6-pentagalloylglucose from Pistacia lentiscus. *Chemico-biological interactions,* 165; 1-13.

Ahmed, N. (2005). Advanced glycation endproducts--role in pathology of diabetic complications. *Diabetes research and clinical practice,* 67; 3-21.

Alfon, J., Guasch, J.F., Berrozpe, M., and Badimon, L. (1999). Nitric oxide synthase II (NOS II) gene expression correlates with atherosclerotic intimal thickening. Preventive effects of HMG-CoA reductase inhibitors. *Atherosclerosis,* 145; 325-331.

Angel-Morales, G., Noratto, G., and Mertens-Talcott, S. (2012). Red wine polyphenolics reduce the expression of inflammation markers in human colon-derived CCD-18Co myofibroblast cells: potential role of microRNA-126. *Food and function,* 3; 745-752.

Annabi, B., Currie, J.C., Moghrabi, A., and Beliveau, R. (2007). Inhibition of HuR and MMP-9 expression in macrophage-differentiated HL-60 myeloid leukemia cells by green tea polyphenol EGCg. *Leukemia research,* 31; 1277-1284.

Arbabi, S., and Maier, R.V. (2002). Mitogen-activated protein kinases. *Critical care medicine,* 30; S74-S79.

Arce, L., Rios, A., and Valcarcel, M. (1998). Determination of anti-carcinogenic polyphenols present in green tea using capillary electrophoresis coupled to a flow injection system. *Journal of chromatography,* A 827; 113-120.

Aruoma, O.I., Whiteman, M., England, T.G., and Halliwell, B. (1997). Antioxidant action of ergothioneine: assessment of its ability to scavenge peroxynitrite. *Biochemical and biophysical research communications,* 23; 389-391.

Asea, A., Kraeft, S.K., Kurt-Jones, E.A., Stevenson, M.A., Chen, L.B., Finberg, R.W., Koo, G.C., and Calderwood, S.K. (2000). HSP70 stimulates cytokine production through a CD14-dependant pathway, demonstrating its dual role as a chaperone and cytokine. *Nature medicine,* 6; 435-442.

Balasundram, N., Ai, T.Y., Sambanthamurthi, R., Sundram, K., and Samman, S. (2005). Antioxidant properties of palm fruit extracts. *Asia Pacific journal of clinical nutrition,* 14; 319-324.

Ban, J.Y., Nguyen, H.T., Lee, H.J., Cho, S.O., Ju, H.S., Kim, J.Y., Bae, K., Song, K.S., and Seong, Y.H. (2008). Neuroprotective properties of gallic acid from Sanguisorbae radix on amyloid beta protein (25--35)-induced toxicity in cultured rat cortical neurons. *Biological and pharmaceutical bulletin,* 31; 149-153.

Basha, P.M., and Madhusudhan, N. (2010). Pre and post natal exposure of fluoride induced oxidative macromolecular alterations in developing central nervous system of rat and amelioration by antioxidants. *Neurochemical research,* 35; 1017-1028.

Baylin, S.B., and Ohm, J.E. (2006). Epigenetic gene silencing in cancer - a mechanism for early oncogenic pathway addiction? *Nature reviews Cancer,* 6; 107-116.

Beg, A.A., Finco, T.S., Nantermet, P.V., and Baldwin, A.S., Jr. (1993). Tumor necrosis factor and interleukin-1 lead to phosphorylation and loss of I kappa B alpha: a mechanism for NF-kappa B activation. *Molecular and cellular biology,* 13; 3301-3310.

Bierhaus, A., Humpert, P.M., Morcos, M., Wendt, T., Chavakis, T., Arnold, B., Stern, D.M., and Nawroth, P.P. (2005). Understanding RAGE, the receptor for advanced glycation end products. *J Mol Med (Berl),* 83; 876-886.

Blanco, V.Z., Auw, J.M., Sims, C.A., and O'Keefe, S.F. (1998). Effect of processing on phenolics of wines. *Advances in experimental medicine and biology,* 434; 327-340.

Bohmann, D., Bos, T.J., Admon, A., Nishimura, T., Vogt, P.K., and Tjian, R. (1987). Human proto-oncogene c-jun encodes a DNA binding protein with structural and functional properties of transcription factor AP-1. *Science,* 238; 1386-1392.

Bubici, C., Papa, S., Dean, K., and Franzoso, G. (2006). Mutual cross-talk between reactive oxygen species and nuclear factor-kappa B: molecular basis and biological significance. *Oncogene,* 25; 6731-6748.

Camici, G.G., Schiavoni, M., Francia, P., Bachschmid, M., Martin-Padura, I., Hersberger, M., Tanner, F.C., Pelicci, P., Volpe, M., Anversa, P., *et al.* (2007). Genetic deletion of p66(Shc) adaptor protein prevents hyperglycemia-induced endothelial dysfunction and oxidative stress. Proceedings of the National Academy of Sciences of the United States of America, 104; 5217-5222.

Castrillo, A., de Las Heras, B., Hortelano, S., Rodriguez, B., Villar, A., and Bosca, L. (2001). Inhibition of the nuclear factor kappa B (NF-kappa B) pathway by tetracyclic kaurene diterpenes in macrophages. Specific effects on NF-kappa B-inducing kinase activity and on the coordinate activation of ERK and p38 MAPK. *The Journal of biological chemistry,* 276; 15854-15860.

Chan, M.M., Fong, D., Ho, C.T., and Huang, H.I. (1997). Inhibition of inducible nitric oxide synthase gene expression and enzyme activity by epigallocatechin gallate, a natural product from green tea. *Biochemical pharmacology,* 54; 1281-1286.

Chase, M.A., Wheeler, D.S., Lierl, K.M., Hughes, V.S., Wong, H.R., and Page, K. (2007). Hsp72 induces inflammation and regulates cytokine production in airway epithelium through a TLR4- and NF-kappaB-dependent mechanism. *J Immunol,* 179; 6318-6324.

Chen, C.C., Rosenbloom, C.L., Anderson, D.C., and Manning, A.M. (1995). Selective inhibition of E-selectin, vascular cell adhesion molecule-1, and intercellular adhesion molecule-1 expression by inhibitors of I kappa B-alpha phosphorylation. *J Immunol,* 155; 3538-3545.

Choi, J.H., Jeong, T.S., Kim, D.Y., Kim, Y.M., Na, H.J., Nam, K.H., Lee, S.B., Kim, H.C., Oh, S.R., and Choi, Y.K. (2003). Hematein inhibits atherosclerosis by inhibition of reactive oxygen generation and NF-kappaB-dependent inflammatory mediators in hyperlipidemic mice. *Journal of cardiovascular pharmacology,* 42; 287-295.

Choi, K.C., Jung, M.G., Lee, Y.H., Yoon, J.C., Kwon, S.H., Kang, H.B., Kim, M.J., Cha, J.H., Kim, Y.J., and Jun, W.J. (2009a). Epigallocatechin-3-gallate, a histone acetyltransferase inhibitor, inhibits EBV-induced B lymphocyte transformation via suppression of RelA acetylation. *Cancer research,* 69; 583-592.

Choi, K.C., Lee, Y.H., Jung, M.G., Kwon, S.H., Kim, M.J., Jun, W.J., Lee, J., Lee, J.M., and Yoon, H.G. (2009b). Gallic acid suppresses lipopolysaccharide-induced nuclear factor-kappaB signaling by preventing RelA acetylation in A549 lung cancer cells. *Molecular cancer research,* MCR7; 2011-2021.

Chuang, C.Y., Liu, H.C., Wu, L.C., Chen, C.Y., Chang, J.T., and Hsu, S.L. (2010). Gallic acid induces apoptosis of lung fibroblasts via a reactive oxygen species-dependent ataxia telangiectasia mutated-p53 activation pathway. *Journal of agricultural and food chemistry,* 58; 2943-2951.

Chung, J.Y., Park, J.O., Phyu, H., Dong, Z., and Yang, C.S. (2001). Mechanisms of inhibition of the Ras-MAP kinase signaling pathway in 30.7b Ras 12 cells by tea polyphenols (-)-epigallocatechin-3-gallate and theaflavin-3,3'-digallate. *FASEB journal,* 15; 2022-2024.

Church, M.K., and Levi-Schaffer, F. (1997). The human mast cell. *The Journal of allergy and clinical immunology,* 99; 155-160.

Constable, A., Varga, N., Richoz, J., and Stadler, R.H. (1996). Antimutagenicity and catechin content of soluble instant teas. *Mutagenesis,* 11; 189-194.

Cosentino, F., Francia, P., Camici, G.G., Pelicci, P.G., Luscher, T.F., and Volpe, M. (2008). Final common molecular pathways of aging and cardiovascular disease: role of the p66Shc protein. *Arteriosclerosis, thrombosis, and vascular biology,* 28; 622-628.

Cybulsky, M.I., and Gimbrone, M.A., Jr. (1991). Endothelial expression of a mononuclear leukocyte adhesion molecule during atherogenesis. *Science,* 251; 788-791.

Dai, J., and Mumper, R.J. (2010). Plant phenolics: extraction, analysis and their antioxidant and anticancer properties. *Molecules,* 15; 7313-7352.

Daniel, O., Meier, M.S., Schlatter, J., and Frischknecht, P. (1999). Selected phenolic compounds in cultivated plants: ecologic functions, health implications, and modulation by pesticides. *Environmental health perspectives,* 107 Suppl *1*; 109-114.

Das, N.D., Jung, K.H., Park, J.H., Choi, M.R., Lee, H.T., Kim, M.S., Lee, S.R., and Chai, Y.G. (2012). Proteomic analysis of Terminalia chebula extract-dependent changes in human lymphoblastic T cell protein expression. *Journal of medicinal food,* 15; 651-657.

Das, N.D., Jung, K.H., Park, J.H., Mondol, M.A., Shin, H.J., Lee, H.S., Park, K.S., Choi, M.R., Kim, K.S., Kim, M.S., Lee S.R., and Chai, Y.G. (2011). Terminalia chebula extract acts as a potential NF-kappaB inhibitor in human lymphoblastic T cells. *Phytotherapy research,* 25; 927-934.

Das, P.M., and Singal, R. (2004). DNA methylation and cancer. *Journal of clinical oncology,* 22; 4632-4642.

Davis, C.D., and Ross, S.A. (2007). Dietary components impact histone modifications and cancer risk. *Nutrition reviews,* 65; 88-94.

Djeridane, A., Yousfi, M., Nadjemi, B., Maamri, S., Djireb, F., and Stocker, P. (2006). Phenolic extracts from various Algerian plants as strong inhibitors of porcine liver carboxylesterase. *Journal of enzyme inhibition and medicinal chemistry,* 21; 719-726.

Doss, M.X., Potta, S.P., Hescheler, J., and Sachinidis, A. (2005). Trapping of growth factors by catechins: a possible therapeutical target for prevention of proliferative diseases. *The Journal of nutritional biochemistry,* 16; 259-266.

Eferl, R., and Wagner, E.F. (2003). AP-1: a double-edged sword in tumorigenesis. *Nature reviews Cancer,* 3; 859-868.

Fang, M.Z., Wang, Y., Ai, N., Hou, Z., Sun, Y., Lu, H., Welsh, W., and Yang, C.S. (2003). Tea polyphenol (-)-epigallocatechin-3-gallate inhibits DNA methyltransferase and reactivates methylation-silenced genes in cancer cell lines. *Cancer research,* 63; 7563-7570.

Faried, A., Kurnia, D., Faried, L.S., Usman, N., Miyazaki, T., Kato, H., and Kuwano, H. (2007). Anticancer effects of gallic acid isolated from Indonesian herbal medicine, Phaleria macrocarpa (Scheff.) Boerl, on human cancer cell lines. *International journal of oncology,* 30; 605-613.

Farrar, J.L., Hartle, D.K., Hargrove, J.L., and Greenspan, P. (2007). Inhibition of protein glycation by skins and seeds of the muscadine grape. *Biofactors,* 30; 193-200.

Forbes, J.M., Coughlan, M.T., and Cooper, M.E. (2008). Oxidative stress as a major culprit in kidney disease in diabetes. *Diabetes,* 57; 1446-1454.

Forstermann, U. (2008). Oxidative stress in vascular disease: causes, defense mechanisms and potential therapies. *Nature clinical practice Cardiovascular medicine,* 5; 338-349.

Gantier, M.P., Sadler, A.J., and Williams, B.R. (2007). Fine-tuning of the innate immune response by microRNAs. *Immunology and cell biology,* 85; 458-462.

Gao, Z., Xu, Z., Hung, M.S., Lin, Y.C., Wang, T., Gong, M., Zhi, X., Jablon, D.M., and You, L. (2009). Promoter demethylation of WIF-1 by epigallocatechin-3-gallate in lung cancer cells. *Anticancer research,* 29; 2025-2030.

Graeber, M.B., and Streit, W.J. (2010). Microglia: biology and pathology. *Acta neuropathologica,* 119; 89-105.

Graham, H.N. (1992). Green tea composition, consumption, and polyphenol chemistry. *Preventive medicine,* 21; 334-350.

Halliwell, B. (1996). Vitamin C: antioxidant or pro-oxidant in vivo? *Free radical research,* 25; 439-454.

Herman, J.G., and Baylin, S.B. (2003). Gene silencing in cancer in association with promoter hypermethylation. *The New England journal of medicine,* 349; 2042-2054.

Hirano, R., Sasamoto, W., Matsumoto, A., Itakura, H., Igarashi, O., and Kondo, K. (2001). Antioxidant ability of various flavonoids against DPPH radicals and LDL oxidation. *Journal of nutritional science and vitaminology,* 47; 357-362.

Hou, D.X., Luo, D., Tanigawa, S., Hashimoto, F., Uto, T., Masuzaki, S., Fujii, M., and Sakata, Y. (2007a). Prodelphinidin B-4 3'-O-gallate, a tea polyphenol, is involved in the inhibition of COX-2 and iNOS via the downregulation of TAK1-NF-kappaB pathway. *Biochemical pharmacology,* 74; 742-751.

Hou, D.X., Masuzaki, S., Hashimoto, F., Uto, T., Tanigawa, S., Fujii, M., and Sakata, Y. (2007b). Green tea proanthocyanidins inhibit cyclooxygenase-2 expression in LPS-activated mouse macrophages: molecular mechanisms and structure-activity relationship. *Archives of biochemistry and biophysics,* 460; 67-74.

Hour, T.C., Liang, Y.C., Chu, I.S., and Lin, J.K. (1999). Inhibition of eleven mutagens by various tea extracts, (-)epigallocatechin-3-gallate, gallic acid and caffeine. *Food and chemical toxicology,* 37; 569-579.

Inoue, M., Sakaguchi, N., Isuzugawa, K., Tani, H., and Ogihara, Y. (2000). Role of reactive oxygen species in gallic acid-induced apoptosis. *Biological and pharmaceutical bulletin,* 23; 1153-1157.

Isuzugawa, K., Ogihara, Y., and Inoue, M. (2001). Different generation of inhibitors against gallic acid-induced apoptosis produces different sensitivity to gallic acid. *Biological and pharmaceutical bulletin,* 24; 249-253.

Ito, N., Fukushima, S., Hagiwara, A., Shibata, M., and Ogiso, T. (1983). Carcinogenicity of butylated hydroxyanisole in F344 rats. *Journal of the National Cancer Institute,* 70; 343-352.

Johansen, J.S., Harris, A.K., Rychly, D.J., and Ergul, A. (2005). Oxidative stress and the use of antioxidants in diabetes: linking basic science to clinical practice. *Cardiovascular diabetology,* 4; 5.

Jung, H.J., Kim, S.J., Jeon, W.K., Kim, B.C., Ahn, K., Kim, K., Kim, Y.M., Park, E.H., and Lim, C.J. (2011). Anti-inflammatory activity of n-propyl gallate through down-regulation of NF-kappaB and JNK pathways. *Inflammation,* 34; 352-361.

Kada, T., Kaneko, K., Matsuzaki, S., Matsuzaki, T., and Hara, Y. (1985). Detection and chemical identification of natural bio-antimutagens. A case of the green tea factor. *Mutation research,* 150; 127-132.

Kagari, T., Doi, H., and Shimozato, T. (2002). The importance of IL-1 beta and TNF-alpha, and the noninvolvement of IL-6, in the development of monoclonal antibody-induced arthritis. *J. Immunol.,* 169; 1459-1466.

Kang, M.S., Jang, H.S., Oh, J.S., Yang, K.H., Choi, N.K., Lim, H.S., and Kim, S.M. (2009). Effects of methyl gallate and gallic acid on the production of inflammatory mediators interleukin-6 and interleukin-8 by oral epithelial cells stimulated with Fusobacterium nucleatum. *J. Microbiol.,* 47; 760-767.

Kato, K., Long, N.K., Makita, H., Toida, M., Yamashita, T., Hatakeyama, D., Hara, A., Mori, H., and Shibata, T. (2008). Effects of green tea polyphenol on methylation status of RECK gene and cancer cell invasion in oral squamous cell carcinoma cells. *British journal of cancer,* 99; 647-654.

Keifer, J.A., Guttridge, D.C., Ashburner, B.P., and Baldwin, A.S., Jr. (2001). Inhibition of NF-kappa B activity by thalidomide through suppression of IkappaB kinase activity. *The Journal of biological chemistry,* 276; 22382-22387.

Kim, M.J., Seong, A.R., Yoo, J.Y., Jin, C.H., Lee, Y.H., Kim, Y.J., Lee, J., Jun, W.J., and Yoon, H.G. (2011). Gallic acid, a histone acetyltransferase inhibitor, suppresses beta-amyloid neurotoxicity by inhibiting microglial-mediated neuroinflammation. *Molecular nutrition and food research,* 55; 1798-1808.

Kim, S.H., Jun, C.D., Suk, K., Choi, B.J., Lim, H., Park, S., Lee, S.H., Shin, H.Y., Kim, D.K., and Shin, T.Y. (2006). Gallic acid inhibits histamine release and pro-inflammatory cytokine production in mast cells. *Toxicological sciences,*91; 123-131.

Kim, S.J., Jeong, H.J., Lee, K.M., Myung, N.Y., An, N.H., Yang, W.M., Park, S.K., Lee, H.J., Hong, S.H., and Kim, H.M. (2007). Epigallocatechin-3-gallate suppresses NF-kappaB activation and phosphorylation of p38 MAPK and JNK in human astrocytoma U373MG cells. *The Journal of nutritional biochemistry,* 18; 587-596.

Kim, Y.J. (2007). Antimelanogenic and antioxidant properties of gallic acid. *Biological and pharmaceutical bulletin,* 30; 1052-1055.

Kroes, B.H., van den Berg, A.J., Quarles van Ufford, H.C., van Dijk, H., and Labadie, R.P. (1992). Anti-inflammatory activity of gallic acid. *Planta medica,* 58; 499-504.

Kuppan, G., Balasubramanyam, J., Monickaraj, F., Srinivasan, G., Mohan, V., and Balasubramanyam, M. (2010). Transcriptional regulation of cytokines and oxidative stress by gallic acid in human THP-1 monocytes. *Cytokine,* 49; 229-234.

Kwon, K.H., Murakami, A., and Ohigashi, H. (2004). Suppressive effects of natural and synthetic agents on dextran sulfate sodium-induced interleukin-1beta release from murine peritoneal macrophages. *Bioscience, biotechnology, and biochemistry,* 68; 436-439.

Lee, K.H., Jeong, J., and Yoo, C.G. (2013). Positive feedback regulation of heat shock protein 70 (Hsp70) is mediated through Toll-like receptor 4-PI3K/Akt-glycogen synthase kinase-3beta pathway. *Experimental cell research,* 319; 88-95.

Lee, K.K., and Workman, J.L. (2007). Histone acetyltransferase complexes: one size doesn't fit all. *Nature reviews Molecular cell biology,* 8; 284-295.

Li, L., Ng, T.B., Gao, W., Li, W., Fu, M., Niu, S.M., Zhao, L., Chen, R.R., and Liu, F. (2005). Antioxidant activity of gallic acid from rose flowers in senescence accelerated mice. *Life sciences,* 77; 230-240.

Lu, Z., Nie, G., Belton, P.S., Tang, H., and Zhao, B. (2006). Structure-activity relationship analysis of antioxidant ability and neuroprotective effect of gallic acid derivatives. *Neurochemistry international,* 48; 263-274.

Luscinskas, F.W., and Gimbrone, M.A., Jr. (1996). Endothelial-dependent mechanisms in chronic inflammatory leukocyte recruitment. *Annual review of medicine,* 47; 413-421.

Makarov, S.S. (2000). NF-kappaB as a therapeutic target in chronic inflammation: recent advances. *Molecular medicine today,* 6; 441-448.

Manach, C., Scalbert, A., Morand, C., Remesy, C., and Jimenez, L. (2004). Polyphenols: food sources and bioavailability. *The American journal of clinical nutrition,* 79; 727-747.

Mandrekar, P., Catalano, D., and Szabo, G. (1999). Inhibition of lipopolysaccharide-mediated NFkappaB activation by ethanol in human monocytes. *International immunology,* 11; 1781-1790.

Manthey, C.L., Wang, S.W., Kinney, S.D., and Yao, Z. (1998). SB202190, a selective inhibitor of p38 mitogen-activated protein kinase, is a powerful regulator of LPS-induced mRNAs in monocytes. *Journal of leukocyte biology,* 64; 409-417.

Matsuo, N., Yamada, K., Shoji, K., Mori, M., and Sugano, M. (1997). Effect of tea polyphenols on histamine release from rat basophilic leukemia (RBL-2H3) cells: the structure-inhibitory activity relationship. *Allergy,* 52; 58-64.

Meager, A. (1999). Cytokine regulation of cellular adhesion molecule expression in inflammation. *Cytokine and growth factor reviews,* 10; 27-39.

Mehla, K., Balwani, S., Kulshreshtha, A., Nandi, D., Jaisankar, P., and Ghosh, B. (2011). Ethyl gallate isolated from Pistacia integerrima Linn. inhibits cell adhesion molecules by blocking AP-1 transcription factor. *Journal of ethnopharmacology,* 137; 1345-1352.

Middleton, E., Jr., Kandaswami, C., and Theoharides, T.C. (2000). The effects of plant flavonoids on mammalian cells: implications for inflammation, heart disease, and cancer. *Pharmacological reviews,* 52; 673-751.

Migliaccio, E., Giorgio, M., Mele, S., Pelicci, G., Reboldi, P., Pandolfi, P.P., Lanfrancone, L., and Pelicci, P.G. (1999). The p66shc adaptor protein controls oxidative stress response and life span in mammals. *Nature,* 402; 309-313.

Mott, J.D., and Werb, Z. (2004). Regulation of matrix biology by matrix metalloproteinases. *Current opinion in cell biology,* 16; 558-564.

Murase, T., Kume, N., Hase, T., Shibuya, Y., Nishizawa, Y., Tokimitsu, I., and Kita, T. (1999). Gallates inhibit cytokine-induced nuclear translocation of NF-kappaB and expression of leukocyte adhesion molecules in vascular endothelial cells. *Arteriosclerosis, thrombosis, and vascular biology,* 19; 1412-1420.

Na, H.J., Lee, G., Oh, H.Y., Jeon, K.S., Kwon, H.J., Ha, K.S., Lee, H., Kwon, Y.G., and Kim, Y.M. (2006). 4-O-Methylgallic acid suppresses inflammation-associated gene expression by inhibition of redox-based NF-kappaB activation. *International immunopharmacology,* 6; 1597-1608.

Nabavi, S.F., Nabavi, S.M., Habtemariam, S., Moghaddam, A.H., Sureda, A., and Mirzaei, M. (2013). Neuroprotective Effects of Methyl-3-O-methyl gallate Against Sodium Fluoride-Induced Oxidative Stress in the Brain of Rats. *Cellular and molecular neurobiology,* 33; 261-267.

Naugler, W.E., and Karin, M. (2008). NF-kappaB and cancer-identifying targets and mechanisms. *Current opinion in genetics and development,* 18; 19-26.

Niedowicz, D.M., and Daleke, D.L. (2005). The role of oxidative stress in diabetic complications. *Cell biochemistry and biophysics,* 43; 289-330.

Niho, N., Shibutani, M., Tamura, T., Toyoda, K., Uneyama, C., Takahashi, N., and Hirose, M. (2001). Subchronic toxicity study of gallic acid by oral administration in F344 rats. *Food and chemical toxicology,* 39; 1063-1070.

Ohta, H., Wada, H., Niwa, T., Kirii, H., Iwamoto, N., Fujii, H., Saito, K., Sekikawa, K., and Seishima, M. (2005). Disruption of tumor necrosis factor-alpha gene diminishes the development of atherosclerosis in ApoE-deficient mice. *Atherosclerosis,* 180; 11-17.

Palacios, C., Cespon, C., Martin De La Vega, C., Roy, G., Serrano, A., Salinas, M., and Gonzalez-Porque, P. (2001). Lauryl gallate inhibits the activity of protein tyrosine kinase c-Src purified from human platelets. *Journal of enzyme inhibition,* 16, 527-533.

Petersen, L.J., Mosbech, H., and Skov, P.S. (1996). Allergen-induced histamine release in intact human skin in vivo assessed by skin microdialysis technique: characterization of factors influencing histamine releasability. *The Journal of allergy and clinical immunology,* 97; 672-679.

Ranjan, R., Swarup, D., Patra, R.C., and Chandra, V. (2009). Tamarindus indica L. and Moringa oleifera M. extract administration ameliorates fluoride toxicity in rabbits. *Indian journal of experimental biology,* 47; 900-905.

Renard, P., and Raes, M. (1999). The proinflammatory transcription factor NFkappaB: a potential target for novel therapeutical strategies. *Cell biology and toxicology,* 15; 341-344.

Rodenhiser, D., and Mann, M. (2006). Epigenetics and human disease: translating basic biology into clinical applications. *CMAJ : Canadian Medical Association journal = journal de l'Association medicale canadienne,* 174; 341-348.

Saeki, K., Yuo, A., Isemura, M., Abe, I., Seki, T., and Noguchi, H. (2000). Apoptosis-inducing activity of lipid derivatives of gallic acid. *Biological and pharmaceutical bulletin,* 23; 1391-1394.

Scapagnini, G., Vasto, S., Abraham, N.G., Caruso, C., Zella, D., and Fabio, G. (2011). Modulation of Nrf2/ARE pathway by food polyphenols: a nutritional neuroprotective strategy for cognitive and neurodegenerative disorders. *Molecular neurobiology,* 44; 192-201.

Schiffrin, E.L. (2002). Beyond blood pressure: the endothelium and atherosclerosis progression. *American journal of hypertension,* 15; 115S-122S.

Schleicher, E., and Friess, U. (2007). Oxidative stress, AGE, and atherosclerosis. *Kidney international Supplement,* S17-26.

Serrano, A., Palacios, C., Roy, G., Cespon, C., Villar, M.L., Nocito, M., and Gonzalez-Porque, P. (1998). Derivatives of gallic acid induce apoptosis in tumoral cell lines and inhibit lymphocyte proliferation. *Archives of biochemistry and biophysics,* 350; 49-54.

Shah, P.K. (2003). Inflammation, neointimal hyperplasia, and restenosis: as the leukocytes roll, the arteries thicken. *Circulation,* 107; 2175-2177.

Shaulian, E., and Karin, M. (2002). AP-1 as a regulator of cell life and death. *Nature cell biology,* 4; E131-136.

Sohi, K.K., Mittal, N., Hundal, M.K., and Khanduja, K.L. (2003). Gallic acid, an antioxidant, exhibits antiapoptotic potential in normal human lymphocytes: A Bcl-2 independent mechanism. *Journal of nutritional science and vitaminology,* 49; 221-227.

Sorsa, T., Tjaderhane, L., Konttinen, Y.T., Lauhio, A., Salo, T., Lee, H.M., Golub, L.M., Brown, D.L., and Mantyla, P. (2006). Matrix metalloproteinases: contribution to pathogenesis, diagnosis and treatment of periodontal inflammation. *Annals of medicine,* 38; 306-321.

Stratil, P., Klejdus, B., and Kuban, V. (2006). Determination of total content of phenolic compounds and their antioxidant activity in vegetables--evaluation of spectrophotometric methods. *Journal of agricultural and food chemistry,* 54; 607-616.

Taganov, K.D., Boldin, M.P., Chang, K.J., and Baltimore, D. (2006). NF-kappaB-dependent induction of microRNA miR-146, an inhibitor targeted to signaling proteins of innate immune responses. *Proceedings of the National Academy of Sciences of the United States of America,* 103; 12481-12486.

Tsang, W.P., and Kwok, T.T. (2010). Epigallocatechin gallate up-regulation of miR-16 and induction of apoptosis in human cancer cells. *The Journal of nutritional biochemistry,* 21; 140-146.

Umadevi, S., Gopi, V., Simna, S.P., Parthasarathy, A., Yousuf, S.M., and Elangovan, V. (2012). Studies on the cardio protective role of gallic acid against AGE-induced cell proliferation and oxidative stress in H9C2 (2-1) cells. *Cardiovascular toxicology,* 12; 304-311.

Urbich, C., Kuehbacher, A., and Dimmeler, S. (2008). Role of microRNAs in vascular diseases, inflammation, and angiogenesis. *Cardiovascular research, 79*; 581-588.

van der Heijden, C.A., Janssen, P.J., and Strik, J.J. (1986). Toxicology of gallates: a review and evaluation. *Food and chemical toxicology,*24; 1067-1070.

Wang, Y.J., Thomas, P., Zhong, J.H., Bi, F.F., Kosaraju, S., Pollard, A., Fenech, M., and Zhou, X.F. (2009). Consumption of grape seed extract prevents amyloid-beta deposition and attenuates inflammation in brain of an Alzheimer's disease mouse. *Neurotoxicity research,* 15; 3-14.

Winter, J., Jung, S., Keller, S., Gregory, R.I., and Diederichs, S. (2009). Many roads to maturity: microRNA biogenesis pathways and their regulation. *Nature cell biology,* 11; 228-234.

Wiseman, S.A., Balentine, D.A., and Frei, B. (1997). Antioxidants in tea. *Critical reviews in food science and nutrition,* 37; 705-718.

Yang, C.S., Liao, J., Yang, G.Y., and Lu, G. (2005). Inhibition of lung tumorigenesis by tea. *Experimental lung research,* 31; 135-144.

Yang, C.S., and Wang, Z.Y. (1993). Tea and cancer. *Journal of the National Cancer Institute,* 85; 1038-1049.

Yoo, C.B., and Jones, P.A. (2006). Epigenetic therapy of cancer: past, present and future. *Nature reviews Drug discovery,* 5; 37-50.

Yoon, C.H., Chung, S.J., Lee, S.W., Park, Y.B., Lee, S.K., and Park, M.C. (2012). Gallic acid, a natural polyphenolic acid, induces apoptosis and inhibits proinflammatory gene expressions in rheumatoid arthritis fibroblast-like synoviocytes. Joint, bone, spine : revue du rhumatisme.

Zhang, K., and Dent, S.Y. (2005). Histone modifying enzymes and cancer: going beyond histones. *Journal of cellular biochemistry,* 96; 1137-1148.

In: Handbook on Gallic Acid
Editors: M. A. Thompson and P. B. Collins
ISBN: 978-1-62618-921-8

Chapter 7

GALLIC ACID EXTRACTION AND ITS APPLICATION TO PREVENTION AND TREATMENT OF CANCER

Mayela Govea-Salas,[1] Marily González-Castillo,[1] Cristóbal N. Aguilar,[1] Raúl Rodríguez-Herrera,[1] Alejandro Zugasti-Cruz,[1] Sonia Y. Silva-Belmares[1] and Jesús A. Morlett-Chávez[1,2,*]

[1]Department of Food Research, School of Chemistry, Universidad Autónoma de Coahuila, Saltillo, Coahuila, México
[2]Laboratory of Clinical Analysis and Molecular Diagnostics, School of Chemistry, Universidad Autónoma de Coahuila, Saltillo, Coahuila, México

ABSTRACT

Gallic Acid (GA) is one of the most important compounds of the hydrolysable polyphenols group; and can be found in a variety of plant tissues. Extraction of this compound has been performed using optimized conditions such as temperature and solvent type used that ensure the best yields without affecting its properties. Polyphenols are associated with color and flavor of fruits and vegetables; however, the nutritional value of this compound is that it may promote a wide variety of biological activities. Also, it has been reported that Gallic acid participates in regulation of damage by inflammatory diseases, microbial infections, cardiovascular conditions as well as antitumor and anticancer protection. These effects occur because it is a potent antioxidant, and can delay, inhibit, or prevent oxidation of oxidizable materials by scavenging free radicals and decreasing oxidative stress. For this reason several investigations are aimed at the evaluation of phenolic compounds like gallic acid for prevention or treatment of degenerative diseases such as cancer. Results have

* Corresponding author: Jesús A. Morlett-Chávez. Department of Food Research, School of Chemistry. Laboratory of Clinical Analysis and Molecular Diagnostics, School of Chemistry, Universidad Autónoma de Coahuila. C.P. 25282, Saltillo, Coahuila, México. Ph. +52 (844) 4-15-95-34, 4-15-57-52. E-mail: antoniomorlett@uadec.edu.mx.

demonstrated that polyphenols affect some mechanisms of tumor developed in various stages, including initiation, promotion and progression of cancer, which triggers the metastasis process. The objective of the present chapter is to present a review about extraction process of gallic acid as well as the steps and factors implicates in this extraction, properties of this phenolic acid and its applications for cancer treatment as an alternative for the control of the one disease of high incidence, worldwide.

INTRODUCTION

There is a huge variety of plants that are considered source of therapeutic agents due to their properties which may generate a health benefit. Use of natural products, especially plants, to cure and/or treat diseases comes from thousands of years ago, per example; there are records that Hippocrates used about 400 different species of plants for medicinal purposes (Sarker *et al.*, 2006). In Latin-American especially in Mexico, herbal medicine goes back beyond pre-Columbian times, representing a natural and inexpensive alternative for health care, having an important role in traditional medicine (FHEUM, 2001). According to data from the World Health Organization (WHO), about 75% of people use medicinal plant extracts for primary health care worldwide.

Therefore, interest and research level in natural products as source of new medicines and industrial products have increased, so about 40% of modern drugs in use have been developed from natural products. Since these natural products have a wide chemical structures diversity, mainly small molecules (<2000 Da), which are able to be absorbed and metabolized quickly in the human body, called secondary metabolites, its interactions give positive biological effects resulting in health benefits (Sarker *et al.*, 2006; Heber, 2007). Some authors reported that about 39% of drugs approved from 1983 to 1994 were natural products or their derivatives, including a 60-80% antibacterial and anticancer agents (Cragg and Newman, 2001; Cragg *et al.*, 2007). On the other hand, about 60% of the tested drugs in clinical trials were used because the increase in the rate of chronic degenerative diseases, these drugs were completely natural, including some from marine flora (Kaminski *et al.*, 2010; Sithranga-Boopathy and Kathiresan, 2010). This progress of science and technology have brought to further knowledge about the relationship between natural products and health, rediscovering the usefulness of medicinal herbs as dietary supplements of botanical origin, being potentially important as preventive and medicinal purposes around the world.

ANTIOXIDANTS

Some herbs contain pharmacologically active compounds that may act individually as health promoters (Gurib-Fakim, 2006). One of the most important substances extracted from some plants are antioxidants. These compounds are used to retard or prevent interaction with oxidation by free radicals, ceding electrons to transform molecule into a nontoxic free radical. Furthermore, importance of antioxidants lie on reduction of the quantity of harmful molecules known as free radicals, which are produced by physiological processes such as respiration, metabolism, food or exercise. In addition, these molecules are induced by external factors such as pollution, radiation, chemicals, food additives, pesticides, drugs, etc.

FREE RADICALS

Free radicals are unstable structures that in its chemical structure possess an unpaired electron, which makes them highly reactive. Its half-life range goes from a few nanoseconds to more reactive compounds, to hours for little more stable radicals. Furthermore, such structures produce chain reactions which result in oxidation of macromolecules, in order to be chemically stable (Aristidis *et al.*, 2012). Free radicals are divided into four main categories based on their central atom, reactive oxygen species (ROS), reactive nitrogen species (RNS), reactive sulfur species (RSS), and reactive chlorine species (RCS), producing oxygen, nitrogen, sulfur and chlorine derivates, respectively (Halliwell, 2008).

ROS are the main category of reactive species produced from oxygen molecules (O_2) by partial chemical reductions. The products of these reductions are: superoxide anion (O^{-2}), hydrogen peroxide (HOO^-) and hydroxyl radical (HO^-) (Hazra, 2010). Being the superoxide anion (O^{-2}) one of the most abundant reactive species generated by O_2 due to electron reduction. When O_2 is reduced by two or three electrons, hydrogen peroxide and hydroxyl radical are produced, respectively (Simic 1988). Currently, it is known that ROS may initiate peroxidation of phospholipids in cell membranes, causing damage due to lipoprotein oxidation by an oxidative chain reaction, generating oxidative stress.

The importance of research on therapy against free radicals in human body is due to that they contribute to more than one hundred human disorders, among which can be mentioned atherosclerosis, arthritis, ischemia, central nervous system injuries, gastritis, cancer, AIDS, etc. (Pourmorad *et al*, 2006). By this reason, antioxidant defense systems have co-evolved with aerobic metabolism to counteract oxidative damage from ROS. Therefore, there has been a resurgence of interest in the therapeutic potential of medicinal herbal antioxidants to reduce the damage induced by free radicals in tissues, as well as the aging process, being flavonoids and other polyphenolic compounds those who have received the most attention (Finkel and Holbrook, 2000; Hazra *et al.*, 2010). Recent research has shown that antioxidant properties of plants could be correlated with oxidative stress defense, various diseases and the aging process. Being flavonoids and other derived polyphenolic compounds, which have received the most attention because of their biological properties (Hazra *et al.*, 2010).

POLYPHENOLS

Polyphenols are chemical substances that contain more than one phenol group per molecule and are one of the most numerous and widely distributed natural products in the vegetal kingdom. Currently, more than 8,000 phenolic structures are known, of which over 4,000 flavonoids have been identified (Cheynier, 2005). Although polyphenols are chemically characterized as compounds with specific structural characteristics, this group of natural products is very diverse and contains several sub-groups. These was due to secondary metabolites found in various plants and are also present in some foods such as fruits, vegetables, whole grains, and in beverages such as tea, chocolate and wine (Bravo, 1998). Because of their diversity and variety, polyphenols have been classified of different ways. Polyphenols are classified by their source of origin, their biological function and chemical structure. Also, most of plant polyphenols exist as glycosides with different sugar units and acylated sugars in different positions of the skeletons of polyphenol structure (Chandrasekara and Shahidi, 2010). Many polyphenols of plants have been studied for their chemopreventive and chemotherapeutic properties against human cancer, including green tea polyphenols, genistein (found in soybean), apigenin (celery, parsley), luteolin (broccoli), quercetin (onions), kaempferol (broccoli, grapefruit), turmeric (curcuma), etc. (Riboli and Norat, 2003).

Epidemiological studies suggest that high intake of polyphenols is associated with a reduced risk of diseases, including cardiovascular disease (CVD), some cancers and neurodegenerative diseases (Kuriyama *et al.*, 2006). Because they can alter lipid metabolism, inhibiting oxidation of low density lipoprotein (LDL), reducing atherosclerotic lesion formation, inhibiting platelet aggregation, improving endothelial function and reducing blood pressure (Zern *et al.*, 2005; Jeong *et al.*, 2005). Furthermore, it was shown that these compounds exert beneficial effects in reversing cognitive neurodegeneration (Hubbard *et al.*, 2006), and also generating a variety of anticancer effects, including the ability to induce apoptosis in tumor cells (Mantena *et al.*, 2006), inhibiting proliferation of cancer cells (Fini *et al.*, 2008), preventing angiogenesis and tumor cell invasion (Corona *et al.*, 2009).

The main polyphenolic compounds that have the last biological properties are divided into two families, the condensed tannins (flavonoids, catechins, anthocyanidins), which are oligomers and polymers of flavonoid units linked by carbon-carbon bonds, these compounds are not susceptible to the formation of bonds after hydrolysis, and hydrolysable tannins (gallic acid and glucose) which are molecules with a polyol (generally D-glucose) as a central core, these tannins are present in small amounts in plants (Hernández *et al.,* 1999; Mueller-Harvey, 2001).

Extraction of Phenolic Compounds

The principal sources of polyphenols are fruits and wild plants; however, the polyphenols extraction method depends on physical and chemical characteristics of raw material. In the case of wines and juice requires minimal manipulation, whereas from whole fruits require a sequence of steps which implies more time. Soft fruits as berries and grapes are crushed or pressed followed by grinding to generate a powder of a defined particle size (Häkkinen *et al.*, 1998).

Fruit shells can be used for extraction after dehydration and grinding. In addition, agro-industrial residues are other alternative for extraction of polyphenolic compounds. Large amounts of these materials, including seeds, peels, husks, whole pomace, among others, are generated every year in form of wastes. This plant material is poorly valorized or discarded, however, in recent years there are reports on use of these residues for phenolic compounds recovery (Martins *et al.*, 2011).

Extraction Processes

Extraction techniques of polyphenols need to take into account the location of these compounds in plant tissues. More phenolic acid derivatives present in the plant matrix are stored in vacuoles and are generally extracted using alcoholic or organic solvents.

Yield of chemical extraction depends of the type of solvents, polarities, extraction time and temperature, sample-to-solvent ratio as well as on the chemical composition and physical characteristics of samples (Dai and Mumper, 2010).

The common solvents for phenolic compounds extraction are water, methanol, ethanol, acetone and ethyl acetate. In particular, methanol has been generally found to be more efficient in extraction of lower molecular weight polyphenols while the higher molecular weight flavanols are better extracted with aqueous acetone. Ethanol is another good solvent for polyphenol extraction and is safe for human consumption (Shi *et al.*, 2005).

In some cases, physical methods are employed to enhance molecular interaction, per example: vortex followed by centrifuge, sonication, mechanical stirring, and continuous rotary extraction. However, conventional extraction methods such as maceration and soxhlet extraction have shown low efficiency and potential environmental pollution. By this reason, in recent years, a number of methods have been developed such as microwave, ultrasound-assisted extractions, and techniques based on use of compressed fluids as extracting agents as supercritical fluid extraction (SFE) and pressurized fluid extraction (PFE) (Robbins, 2003).

During polyphenols extraction commonly are used moderate temperatures and short times. Temperature is one of the most important factors affecting antioxidant activity.

Generally, heating causes an acceleration of initiation reactions, and hence a decrease in the activity of antioxidants (Pokorný, 1986). Increase of temperature cans promote higher analyte solubility by get bigger solubility and mass transfer rate. In addition, viscosity and surface tension of solvents decrease at higher temperature, which help to solvents, reach the sample, improving extraction rate. On the other hand, long extraction times and high temperature increase chance of oxidation of phenolics which decrease extraction yield (Robards, 2003). There are some reports mentioned that lower sample particle size increased polyphenols concentration and reduced extraction time because distance between solute and solvent is shorter and expands extraction area (Kossah *et al.*, 2010; Wang *et al.*, 2011).

Quantification

The methods used for the analysis of phenolic compounds are usually classified as either those measuring total polyphenols content, or those quantifying a specific group or class of phenolic compounds. Quantification of these compounds is influenced by the chemical nature of analyte, as well as assay method, selection of standards and presence of interfering substances (Naczk and Shahidi, 2006).

Different methods have been reported, some of them are Folin-Ciocalteu method, permanganate titration, colorimetric with iron salts, and ultraviolet absorbance, however, Folin-Ciocalteu method has been found to be more used than other methods because its easy reproducibility (Singleton *et al.*, 1999). This assay relies on electrons transference in an alkaline medium from phenolic compounds to phosphomolybdic/phosphotungstic acid complexes to form blue coloration that is determined spectroscopically at approximately 760 nm. Gallic acid is widely used as the comparison standard and values are usually compared as milligram of gallic acid equivalent (GAE) per kilogram or liter of extract (Singleton and Rossi, 1965).

However, other oxidation substrates present in the extract sample can interfere in total phenolic measurement, acting as inhibitors or increasing final concentration (Singleton *et al.*, 1999).

Hydrolysable tannins can be quantified by a number of approaches including the potassium iodate, rhodanine and sodium nitrite methods. Potassium iodate method is most widely used.

It is based on the reaction of methyl gallate, formed upon methanolysis of hydrolysable tannins in presence of strong acids, with potassium iodate to produce a red chromophore with a maximum absorbance between 500 nm and 550 nm (Hartzfeld *et al.*, 2002). Rhodanine method can be used for estimation of gallotannins specifically and is based on determination of gallic acid in a sample subject to acid hydrolysis under conditions that must be anaerobic to avoid oxidation of the product (Inoue and Hagerman, 1988).

Although spectrophotometric assays provide simple and fast screening methods to quantify different classes of phenolic compounds in crude plant samples, that are methods more specific, sensitive and accurate for this class of determination such as HPLC analysis (Anttonen and Karjalainen, 2006; Harris *et al.*, 2007).

The variety of supports and mobile phases in HPLC method enables analysis of a wide range of compounds and specifically is the most popular and reliable technique for analysis of phenolic compounds and its degradation products (Mertz *et al.*, 2007; Downey and Rochfort, 2008).

Purification

Plant crude extracts usually contain large amounts of carbohydrates and/or lipid material which may be part of the extract polyphenols concentration. To concentrate and obtain polyphenols-rich fractions, are performed a sequential extraction or liquid-liquid partitioning and/or solid phase extraction (SPE) based on polarity and acidity of samples. Elimination of lipid material can be achieved by washing the crude extract with non-polar solvents. This process is based on distribution of solutes between two immiscible phases in which the compound and matrix have a different solubility. The solvents more employers are hexane, dichloromethane and chloroform (Ramirez-Coronel *et al.*, 2004; Neergheen., *et al.*, 2006; Zhang *et al.*, 2008).

To remove polar non-phenolic compounds such as sugars and organic acids, a SPE process is usually carried out. Solid-phase extraction is used to fractionate as well as to remove unwanted components from the sample. C18 cartridges have been the most widely used in phenolic compound separation by eluting with solvents of varying pH; larger phenolics (flavanol aglycones, flavanol-3-ols and ellagic acid) and sugars are separated from the smaller phenolic components (Rommel and Wrolstad, 1993).

Other sorbents such as Amberlite XAD-2 (Llorach *et al.*, 2003), XAD-7 (Moreno *et al.*, 2005), XAD-16 (Ascacio-Valdés *et al.*, 2010) and Oasis HLB (Jeffery *et al.*, 2008) have also successfully been used to purify phenolic compounds in crude extracts or wine samples.

These types of polymeric resins are highly porous, whose internal surface can be adsorbed and desorbed a wide variety of compounds. Ethanol is a solvent widely used for elution of phenolic compounds due to affinity with them.

GALLIC ACID: EXTRACTION AND BIOLOGICAL PROPERTIES

Gallic acid (GA), also known as 3, 4, 5-trihydroxybenzoic (C_6H_2 $(OH)^3$ COOH) is a monomeric phenolic acid, solid and colorless. It has a molecular weight 170.12 g / mol. It was discovered in 1787 by Schcele in cold aqueous oak galls, which today remains as one of the major sources of this acid. It comes as free molecule or as one of the chemical components of tannins (Hemingway *et al.*, 1999). Gallic acid exists can be found in plant material as free acid, esters, catechin derivatives and hydrolysable tannins. It is present in relatively high concentrations in a number of vegetal samples and in industrial wastes from where it could be extracted (Cláudio *et al.*, 2012).

Extraction of Gallic Acid

GA is obtained directly from different plants, fruits and foods like strawberry, banana, grape, lemon, red wine, nuts, tea leaves, apple skin, among others. Is also obtained by hydrolysis of tannic acid ($C_{76}H_{52}O_{46}$), which is a polymer formed by monomers of gallic acid, glucose and some derivatives of alkyl esters (methyl gallate, ethyl gallate, propyl gallate, cetyl gallate and lauryl gallate), (Seth and Chand, 2000). GA extraction from natural sources has been performed with polar solvents as water, ethanol and methanol or combinations of them (Rakesh *et al.*, 2009).

The most used mixture is water-ethanol; Pawar and Surana (2010) optimized the hydroalcoholic GA extraction from *Caesalpinia decapetala* which is a wild medicinal plant of Sub-Himalayan region and is a rich source of tannins. A composition 70:30 ethanol: water was the best combination of solvents used; also temperature (65-70 °C) and time was determinate (48 h) as optimum extraction parameters.

As mentioned above temperature factor in phenolic compounds extraction may cause changes on its chemical composition and its biological activities.

Per example, Réblová in 2012 evaluated the effect of temperature on antioxidant activity of different phenolic acids including gallic acid. This type of compounds may be easily oxidizable, and showed a slower decrease in antioxidant activity with the increasing of temperature in comparison with less oxidizable compounds as ferulic, sinapic and vanillic acids which presented a fast decrease of this biological property.

Different studies have been performed in order to evaluate solvent effect on GA extraction. Karvat *et al.*, (2012) analyzed the effect of 12 different solvents (water, dimethylsulfoxide, ethylene glycol, acetonitrile, dimethylformamide, methanol, acetone, acetic acid, ethyl acetate, chloroform, diethyl ether and toluene) on GA extraction from *Emblica officinalis* Gaertn, which is a plant used in Indian traditional medicine. These authors demonstrated that extraction of this compound is possible using polar and non polar solvents; however, the highest concentration of GA was observed with ethyl acetate, a medium polarity solvent. This information indicates that amount of extracted phenolics depending of sample chemical origin and its interaction with the solvent used. In addition, different GA extraction methods have been evaluated. Amirah *et al.*, (2012) compared three techniques for recovery of GA from *Jatropha curcas*. In this study Shoxhlet, Ultrasonic-Assisted (UAE) and Microwave-Assisted (MAE) extraction methods were performed. The highest recovery of GA was observed with the MAE. In this extraction process, microwave energy is used to facilitate partition of analytes from sample matrix into the solvent. The main advantage of this technique is the shorter extraction time and solvent volume as compared to conventional extraction techniques (Eskilsson and Bjorklund, 2000). There are also reports of employment of MAE for extraction of small-molecule phenolic compounds such as gallic and ellagic acids (Li *et al.*, 2004).

Other important process for GA obtaining is biotransformation of tannic acid to gallic acid by enzymes. Tannase is the principal enzyme acting in the oxidation process. This enzyme catalyzes the hydrolysis of ester and break linkages of gallotannins to release gallic acid and glucose. In the pathways for degradation of tannins, gallic acid is converted into pyrogallol by the action of decarboxylase (Mingshu *et al.*, 2006). It can also be obtained by chemical methods, using acid or base in a sugar, a polyalcohol and a phenol carboxylic acid. GA recovery is accomplished by adding ethyl acetate and acidified extract thus obtained gallic acid crystals. (Aguilar and Gutiérrez-Sánchez, 2001).

Karamac *et al.* (2006) measured gallic acid content in ethanolic extracts of green teas, bearberry leaves, hazelnuts, evening primrose and grape seeds. All samples were treated with tannase in an effort to liberate gallic acid from the phenolic esters and hydrolysable tannins. Results indicated that plant material treated with tannase has great potential for preparation of gallic acid-rich fractions which can be used in functional foods.

GA can be obtained by microbial action. Some bacteria such as *Achromobacter* sp., *Bacillus* sp., *Corynebacterium* sp., *Klebsiella* sp. and *Citrobacter* sp. may degrade gallotannins (Deschamps, 1985; Kumar et al. 1999). Deschamps *et al.*, in 1980 reported degradation of gallotannins by aerobic bacteria, and isolated different bacterial strains belonging to the genera *Bacillus*, *Staphylococcus* and *Klebsiella* by the enrichment culture technique, using gallotannins as a sole carbon source. They found that a few bacteria such as those from the genera *Bacillus* can produce an extracellular tannase to degrade gallotannins to gallic acid.

On the other hand, different fungal strains belonging to the genera *Aspergillus*, *Fusarium*, *Penicillium*, *Sporotrichum*, *Rhizoctonia*, *Cylindrocarpon* and *Trichoderma* can degrade tannins, particularly gallotannins for GA production (Mahadevan and Muthukumar 1980; Makkar *et al.*, 1994). Particularly *Aspergillus* and *Penicillium* have been reported to grow at high concentration of gallotannins as a sole carbon source (Scalbert, 1991). However, several factors affecting microbial growth and enzyme activities for tannin degradation have been reported among them: nitrogen sources, pH, temperature, oxygen diffusion and combined parameters. For example *A. niger* could tolerate to 5% (w/v) gallotannins at pH 3.5 and 28 °C, with agitation speed of 150 rpm and urea of 10% (w/v) according with those results reported by Ruiz-Aguilar *et al.*, (2004).

Meyer in 2005 affirmed that particle size of mashed samples can be a main factor to increase enzyme action and extraction efficiency of phenolic compounds using the enzyme-assisted extractions. In accordance with this, Lokeswari (2012) worked on optimization of parameters for maximum tannase production from Cashew husk.

In this study was found that microbial synthesis of enzymes in SSF (Solid State Fermentation) are affected by factors like particle size of substrate, water content, relative humidity, type and size of inoculum and temperature, principally. Besides, these factors also affected GA production.

The variation of several factors in extraction process of phenolic compounds included GA affected its physical and chemical characteristics at different levels depending of raw material and therefore the biological properties are modified.

Many biological activities in plants have been attributed at the polyphenols presence in this and so that the recent investigations are aimed at the search of its application in pharmaceutical, food and biotechnological industries.

Properties and Applications of Gallic Acid

GA has many applications in different areas such as a) pharmaceutical manufacturing of broad-spectrum antibiotics such trimethoprim and b) industrial processing treatment of skin, leather and animal hair and is also used as a photographic developer and for different types of inks (Lekha-Lonsane, 1997).

Similarly, GA is used as oils and fats antioxidant and beverage additive in some types of drinks and foods avoiding oxidation (Hocman, 1988). It has also been reported that GA is capable of regulating some biological activities as anti-inflammatory, antibacterial, antiviral, anticancer and cardiovascular protection.

These processes may occur because it is a potent antioxidant, with higher levels even than some vitamins. However, GA can also promote cellular oxidative stress because it has pro-oxidant activity, in some cases causing cytotoxicity (Kim *et al.*, 2006).

On the other hand, Inoue *et al.*, (2000) mentioned that generation of ROS produced by gallic acid induces *in vitro* apoptosis in cultured cancer cells. For this reason has been attributed to GA anticancer properties.

ANTICANCER THERAPY

Cancer is considered a set of uncontrolled cellular changes caused by a number of selective genetic changes in the main tumor suppressor genes and oncogenes (Link *et al.*, 2010). Furthermore, it is known that exist an alteration in metabolism and cell growth control, favoring formation of tumors and metastatic processes (Plass, 2002).

Although it is known that a small proportion of tumors can be inherited, it is believed that most cancers are result of accumulating changes throughout life due to exposure to some endogenous factors such as nutrients, infections, physical activity, social behavior and other environmental factors (Dolinoy *et al.*, 2007).

Even when cancer initiation and progression is driven by genetic alterations acquired as epigenetic disorder of some genes expression, which play a very important role in the development of disease (Dolinoy *et al.*, 2007).

For this reason, diet and epigenetic alterations affected by environment, play a crucial role in progression of human cancer (Herceg, 2007). Due to this, some authors mention that lifestyle is an essential component for prevention of cancer.

According to the American Cancer Society, there were 1, 479, 350 new cases of cancer registered with 562.340 deaths in 2009 (Jemal *et al.*, 2009). However, comparison of diagnoses and mortality from cancer in 1971, in which the National Cancer Act was signed into federal law in the US, and 2010 indicate that while the number of cancer diagnoses increases, mortality rate decreased in those years. This explains the increase in cancer survivors, more than 3 million in 1971 to almost 12 million in 2007 (Altekruse *et al.*, 2010), of which an estimated 328.652 are survivors of childhood cancer. This was due to new therapies that have evolved over years (Mariotto *et al.*, 2009). However, despite the existing treatment methods, many researchers still seek to find a less invasive healing strategy to this recurrent disease. Interestingly, many epidemiological and experimental studies have demonstrated protective effects of dietary factors in some human cancers (Tsao, 2010).

Some compounds derived from plants, such those polyphenol compounds mentioned above, are useful tools in treatment of cancer and other pathologies, because it has been found that these compounds, such as gallic acid, lock mechanisms in various stages, including initiation, promotion and progression of cancer (Lu *et al.*, 2010).

These results have allowed strengthen chemoprevention, halting or delaying the process of carcinogenesis stages before malignancy development (Riboli and Norat, 2003). Some of the processes which allow cancer cells malignancy development are related with the protein family known as cyclins and cyclin dependent kinases (CDKs), because they are critical components for cell cycle regulation in eukaryotes.

For this reason, some researchers have shown that some polyphenols mainly from plants and green tea leaves, such as epigallocatechin gallate (EGCG) and gallic acid (GA), can affect and inhibit expression of these cell cycle regulators (Ahmad *et al.*, 1997; Qiao *et al.*, 2009). It has also been demonstrated that EGCG and AG induced cell cycle arrest and apoptosis in several cancer cell lines, including lung, colon, prostate and skin, generating a minimal effect on normal cells (Khan *et al.*, 2006).

Cancer and Gallic Acid

In the last decade, some authors mentioned that incidence of cancer is significantly lower in people whose diet consists mainly of fruits and vegetables, than people whose food consumption is mainly animal products. This has led researchers to search natural anticancer compounds obtained from plants or foods such as phytochemicals and polyphenols, which may be considering alternative treatments (Nair *et al.*, 2007).

Some studies with GA obtained from various foods and plants have shown that this compound has anti-cancer activity against several types of cancer interfering in some stages of tumor growth, suppressing angiogenesis and metastasis in some cases (Chi *et al.*, 2007). Albini *et al*, (2007) reported that GA is responsible, in part, of anti-angiogenic activity in some tumor lines such as lung, prostate, bladder, brain, kidney and leukemia. Also, Raina *et al.*, (2008) mentioned that GA may interfere at different stages of tumor development, for example, decreasing the ornithine decarboxylase response related to tumor promotion.

It also suppresses tumoral angiogenesis (Liu *et al.*, 2006) and inhibits metastasis of P815 liver cells (Kawada *et al.*, 2001).

Some studies have reported that GA has properties that inhibit *in vitro* growth and proliferation of various cancer cells by anti-angiogenic, apoptotic and oxidative processes (Lu *et al.*, 2010). Also, Ji, *et al.* (2009) reported that GA significantly decreases viability, proliferation and invasion of cancer cells inducing apoptosis in a dose dependent, through the p53 pathway, Bcl-2 and caspases family, these being a coded signaling p53 pathway. The p53 protein is the product of a tumor suppressor gene whose product can act as an inhibitor of cell transformation which is induced by DNA damage (Kastan *et al.*, 1991). While Bcl-2 are proteins regulating the pro and anti-apoptosis, acting protecting cells individually, altered mitochondrial integrity or inducing apoptosis (Larsen, 1994).

Other biological process which is related to the malignancy of cancer cells is apoptosis, which consists in some biochemical events that lead to a variety of morphological changes, including changes in cell membrane (Roy *et al.*, 2005).

This process is suppressed during development of cancer cells, facilitating the uncontrolled growth thereof. However, GA and EGCG have shown to inhibit expression of anti-apoptotic proteins (Bcl-2 and Bcl-XL), while promoting expression of pro-apoptotic proteins (Bax and Bak) (Qin *et al.*, 2007).

Furthermore, it has been reported that effectively inhibit cellular proliferation and induced apoptosis of chondrosarcoma cells CRL-7891 and SW1353, significantly reducing the levels of expression of Bcl-2 and rise significantly Bax levels (Chi *et al.*, 2009). Also, these natural compounds induced apoptosis in Hep2 cells in squamous cell larynx carcinoma through the caspase pathway, mediated by p53 (Tang *et al.*, 2010).

Moreover, GA induces *in vitro* death in DU145 and 22Rv1 cell prostate cancer in athymic mice, by a dose-dependent manner, inhibiting growth and proliferation of tumors, also promoting apoptosis.

It also exerts *in vivo* anti-angiogenic activity because reduces development of blood microvessels, which also contributes to the property of gallic acid against prostate cancer.

In contrast, for PWR-1E cells that are not cancerous, have no significant change, indicating that GA has a selective toxicity to cancer cells of prostate cancer (Kaur, 2006).

Later, Maurya *et al.*, (2010) mentioned that GA induces *in vitro* cell death in A549 human lung carcinoma by a dose-dependent manner, also apoptosis is induced by elevated ROS through stress oxidative favoring activation of caspase-3. Besides, Elango *et al.*, (2011) evaluated GA anticancer potential effect on mitochondrial respiration, cell-membrane integrity, apoptotic body formation and DNA fragmentation in cultured A549 cells. They observed increased level of reactive oxygen species, as well as activation of Caspase-9 in treated cells, and therefore induced apoptosis in A549 cells through intrinsic pathway. In 2010, Cai *et al*, showed that GA significantly reduces viability of U251n and U87 human cells glioma after 24 h of treatment at dose-dependent manner. In contrast, GA showed a reduction in toxicity of normal epithelial cells from mouse brain at the same concentrations, suggesting that GA has a selective cytotoxicity for glioma cells at dose-dependent manner.

Antitumor effects of GA on human stomach cancer (KATO III) cells and human colon adenocarcinoma (COLO 205) cells have been also investigated. Exposures of KATO III and COLO 205 cells to this compound led to both growth inhibition and induction of apoptosis. Morphological changes showing apoptotic bodies were observed in both treated cell lines (Yoshioka *et al.*, 2000).

Furthermore, more evidence for the hypothesis that antioxidants from foods can execute beneficial health effects, including acting as inducers of mechanisms related to antioxidant defense (Hubbard *et al.*, 2006), longevity (Ludwig *et al.*, 2004), cell maintenance and DNA repair was reported (Vauzour *et al.*, 2010).

GA is considered one of the most widely used antioxidants for alternative therapies due to its biological properties, particularly to cancer *in vitro* and *in vivo* in many biological models (Hodgson and Croft, 2006).

However, few clinical studies affirm that GA is fast metabolized and absorbed by organism and it is easy eliminated and so that it has a good bioavailability and can be administered by oral via (Shahrzad *et al.*, 2001; Al-Salih, 2010).

Conclusion

In the last 20 years, studies on the biological effects of GA have indicated an inverse correlation between consumption of fruits and vegetables with development of different cancers. While exist a variety of phenolic antioxidants sources, must have the knowledge about which of this are more desirables for each type of cancer and which mechanisms are implicated in this activity. However, it requires further investigations with different biological models to continue testing these epidemiological associations. Moreover, assays with human patients are necessary for evaluation of GA bioavailability as well as known its behavior against cancer and normal cells. Gallic acid might be a candidate drug for cancer treatment to overcome resistance to chemotherapeutic drugs, but taking into account secondary effects that this may cause and always trying to search the best way for that active compound reaches target cells.

References

Aguilar, C. N., Gutiérrez-Sánchez, G. 2001. Sources, properties, and potential uses of tannin acyl hydrolase (3.1.1.20). *Food Science and Technology International*. 7(5):373-382.

Ahmad, N., Feyes, D. K., Nieminen, A. L., Agarwal, R., Mukhtar, H. 1997. Green tea constituent epigallocatechin-3-gallate and induction of apoptosis and cell cycle arrest in human carcinoma cells. *J. Natl. Cancer Inst.;* 89(24):1881–6. [PubMed: 9414176].

Albini, A. 2007. Molecular pathways for cancer angio-prevention. *Clin. Cancer Res.*13,4320-4325.

Al-Salih, R. M. H. 2010. Clinical Experimental Evidence: Synergistic effect of gallic acid and tannic acid as antidiabetic and antioxidant agents. *Thi-Qar Medical Journal*, 4 :(4), 109-119.

Altekruse, S. F., Kosary, C. L., Krapcho, M. 2009. SEER Cancer Statistics Review, 1975-2007; based on November 2009 SEER data submission, posted to the SEER web site, 2010. http://seer.cancer.gov/ csr/1975_2007/.

Amirah, D. M., Reddy, P., Maksudur, R. K. 2012. Comparison of extraction techniques on extraction of gallic acid from stem bark of *Jatropha curcas*. *J. Applied Sci.*, 12 (11): 1106-1111.

Anttonen, M. J., Karjalainen, R. O. 2006. High-performance liquid chromatography analysis of black currant (*Ribes nigrum* L.) fruit phenolics grown either conventionally or organically. *J. Agric. Food Chem.*, 54, 7530-7538.

Aristidis, S., Veskoukis-Aristidis, M., Tsatsakis-Dimitrios, K. 2012. Dietary oxidative stress and antioxidant defense with an emphasis on plant extract administration. *Cell Stress and Chaperones* (2012) 17:11–21

Ascacio-Valdés, J. A., Aguilera-Carbó, A., Martínez-Hernández, J. L., Rodríguez-Herrera, R., Aguilar, C. N. 2010. *Euphoria antisyphilitica* residues as a new source of ellagic acid. *Chemical Papers*, 64 (4), 528–532.

Ashughyan, V. R., Marihart, S., Djavan, B. 2006. Chemopreventive trials in urologic cancer. *Rev. Urol.* 8(1):8–13. [PubMed: 16985555]

Bravo, L. 1998. Polyphenols: Chemistry, Dietary Sources, Metabolism, and Nutritional Significance. *Nutr. Rev.* 56, 317-333

Cai, Y., Chopp, M., Katakowski, M., Jiang, F., Jiang, H., Lu, Y., Shing-Shun, T. T., Wu, K., Zhen, X. 2010. Gallic acid suppresses cell viability, proliferation, invasion and angiogenesis in human glioma cells. *Eur. J. Pharmacol.* 641(2-3): 102–107. doi:10.1016/j.ejphar.2010.05.043.

Chandrasekara, A., Shahidi, F. 2010. Content of insoluble bound phenolics in millets and their contribution to antioxidant capacity. *J. Agric. Food Chem.* 58, 6706-6714.

Cheynier, V. 2002. Polyphenols in foods are more complex than often thought. *Am. J. Clin. Nutr.* 81, 223S-229S.

Chi, A., Norden, A. D., Wen, P. Y. 2007. Inhibition of angiogenesis and invasion in malignant gliomas. *Expert Rev. Anticancer Ther.* 7:1537–1560.

Chi, H. T., Vu, H. A., Iwasaki, R., Thao le, B., Hara, Y., Taguchi, T., Watanabe, T., Sato, Y. 2009. Green tea (−) epigalocatechin-3-gallate inhibits KIT activity and causes caspase-dependent cell death in gastrointestinal stromal tumor including imatinib-resistant cells. *Cancer Biol. Ther.* 8(20): 1934–9. [PubMed: 19770581].

Cláudio, A. F. M., Ferreira, A. M., Freire, C. S. R., Silvestre, A. J. D., Freire, M. G., Coutinho, J. A. P. 2012. Optimization of the gallic acid extraction using ionic-liquid-based aqueous two-phase systems. *Separation and Purification Technology*, 97 : 142–149.

Corona, G., Deiana, M., Incani, A., Vauzour, D., Dessi, M. A., Spencer, J. P. 2009. Hydroxytyrosol inhibits the proliferation of human colon adenocarcinoma cells through inhibition of ERK1/2 and cyclin D1. *Mol. Nutr. Food Res.* 53, 897-903.

Cragg, G. M., Newman, D. J. 2001. Natural product drug discovery in the next millennium. *Pharmaceutical Biology* 39:8–17.

Cragg, G. M., Newman, D. J. 2007. Medicinal for the millennia: the historical record. *Annals of the New York Academy of Sciences* 953:3–25. DOI 10.1007/s12192-011-0293-3

Dai, J., Mumper, R. (2010). Plant Phenolics: Extraction, analysis and their antioxidant and anticancer properties. *Molecules*, 15, 7313-7352.

Deschamps, A. M. 1985. Evaluation de la degradation de deux types de tannin condense par de bacteries isolees d'ecorees en decomposition. *Can. J. Microbiol.*, 31, 499–505.

Deschamps, A. M., Mohudeau, G., Conti M., Lebeault, J. M. 1980. Bacteria degrading tannic acid and related compounds. *J. Ferment. Technol.*, 58, 93–97.

Dolinoy, D. C., Huang, D., Jirtle, R. L. 2007. Maternal nutrient supplementation counteracts bisphenol an induced DNA hypomethylation in early development. *Proc. Natl. Acad. Sci. US.* 104 (32):13056–13061. [PubMed: 17670942].

Dolinoy, D. C., Weidman, J. R., Jirtle, R. L. 2007. Epigenetic gene regulation: linking early developmental environment to adult disease. *Reprod. Toxicol.* Apr.; 23(3):297–307. [Pub Med: 17046196]

Downey, M. O., Rochfort, S. 2008. Simultaneous separation by reversed-phase high-performance liquid chromatography and mass spectral identification of anthocyanins and flavonols in Shiraz grape skin. *J. Chromatogr.*, 1201, 43-47.

Elango, S. Balwas, R., Padma, V. 2011. Gallic acid isolated from pomegranate peel extract induces reactive oxygen species mediated apoptosis in A549 cell line. *Journal of Cancer Therapy*, 2, 638-645.

Eskilsson, C. S., Bjorklund, E. 2000. Analytical-scale microwave-assisted extraction. *J. Chromatogr.*, 902, 227-250.

Fini, L., Hotchkiss, E., Fogliano, V., Graziani, G., Romano, M., De Vol, E. B., Qin, H., Selgrad, M., Boland, C. R. Ricciardiello, L. 2008. Chemopreventive properties of pinoresinol-rich olive oil involve a selective activation of the ATM-p53 cascade in colon cancer cell lines. *Carcinogenesis* 29, 139-146.

Finkel, T., Holbrook, N. J. 2000. Oxidants, oxidative stress and the biology of ageing. *Nature* 2000, 408:239-247.

Gurib-Fakim, A. 2006. Medicinal plants: traditions of yesterday and drugs tomorrow, *Molecular Aspects of Medicine* 27: 1–93.

Häkkinen, S. H., Kärenlampi, S. O., Heinonen, I. M., Mykkänen, H. M., Törrönen A. 1998. Riitta HPLC method for screening of flavonoids and phenolic acids in berries. *J. Sci. Food Agric.* 77, 543-551.

Halliwell, B. 2008. Are polyphenols antioxidants or pro-oxidants? What do we learn from cell culture and in vivo studies? *Arch. Biochem. Biophys.* 476, 107-112.

Hallund, J., Bugel, S., Tholstrup, T., Ferrari, M., Talbot, D., Hall, W. L., Reimann, M., Williams, C. M., Wiinberg, N. 2006. Soybean isoflavone-enriched cereal bars affect markers of endothelial function in postmenopausal women. *Br. J. Nutr.* 95, 1120-1126.

Harris, C. S., Burt, A. J., Saleem, A., Le, P. M., Martineau, L. C., Haddad, P. S., Bennett, S. A., Arnason, J. T. 2007. A single HPLC-PAD-APCI/MS method for the quantitative comparison of phenolic compounds found in leaf, stem, root and fruit extracts of *Vaccinium angustifolium*. *Phytochem. Anal.*, 18, 161-169.

Hartzfeld, P. W., Forkner, R., Hunter, M. D., Hagerman, A. E. 2002. Determination of hydrolysable tannins (gallotannins and ellagitannins) after reaction with potassium iodate. *J. Agric. Food Chem.*, 50, 1785-1790.

Hazra, B., Sarkar, R., Biswas, S., Mandal, N. 2010. Comparative study of the antioxidant and reactive oxygen species scavenging properties in the extracts of the fruits of *Terminalia chebula, Terminalia belerica* and *Emblica officinalis*. Hazra et al. *BMC Complementary and Alternative Medicine* 2010, 10:20.

Heber, D. 2007. *PDR for Herbal Medicines*. 3a. ed. Thomson PDR, NJ, US, pp. 400-404

Hemingway, R. W., Gross, G. G., Yoshida, T. 1999. Plant Polyphenols: *Chemistry and Biology*, pp. 495–505. New York: Plenum Press.

Herceg, Z. 2007. Epigenetics and cancer: towards an evaluation of the impact of environmental and dietary factors. *Mutagenesis*. 22(2):91–103. [PubMed: 17284773]

Hernández, A., Prieto, M., González, E. A. 1999. Plantas que contienen polifenoles. Antioxidantes dentro del estilo de vida. *Rev. Cubana de Inv. Biomed.* 18 (1): 12 – 4.

Hocman, G. 1988. Chemoprevention of cancer: phenolic antioxidants (BHT, BHA). *Int. J. Biochem.* 20(7): 639-651.

Hodgson, J., Croft, K. 2006. Dietary flavonoids: Effects on endothelial function and blood pressure. *J. Sci. Food Agric.* 86, 2492-2498.

Hubbard, G. P., Wolffram, S., de Vos, R., Bovy, A., Gibbins, J. M., Lovegrove, J. A. 2006. Ingestion of onion soup high in quercetin inhibits platelet aggregation and essential components of the collagen-stimulated platelet activation pathway in man: A pilot study. *Br. J. Nutr.* 96, 482-488.

Hubbard, G. P., Wolffram, S., de Vos, R., Bovy, A., Gibbins, J. M., Lovegrove, J. A. 2006. Ingestion of onion soup high in quercetin inhibits platelet aggregation and essential components of the collagen-stimulated platelet activation pathway in man: A pilot study. *Br. J. Nutr.* 96, 482-488.

Inoue, K. H., Hagerman, A. E. 1988. Determination of gallotannin with rhodanine. *Anal. Biochem.*, 169, 363-369.

Inoue, M., Sakaguchi, N., Isuzugagua, K., Tani, H., Ogihara, Y.2000. Role of reactive oxygen species in gallic acid-induced apoptosis. *Biol. Pharm. Bull.;* 23:1153-1157.

Jeffery, D. W., Mercurio, M. D., Herderich, M. J., Hayasaka, Y., Smith, P. A. 2008. Rapid isolation of red wine polymeric polyphenols by solid-phase extraction. *J. Agric. Food Chem.*, 56, 2571-2580.

Jemal, A., Siegel, R., Ward, E., Hao, Y., Xu, J, Thun, M.J. 2009. Cancer Statistics, 2009. *CA Cancer J. Clin.* 59(4):225–249. [PubMed: 194743 85].

Jeong, Y. J., Choi, Y. J., Kwon, H. M., Kang, S. W., Park, H. S., Lee, M., Kang, Y. H. 2005. Differential inhibition of oxidized LDL-induced apoptosis in human endothelial cells treated with different flavonoids. *Br. J. Nutr.* 93, 581-591.

Ji, B. C., Hsu, W. H., Yang, J. S., Hsia, T. C., Lu, C. C., Chiang, J. H., Yang, J. L., Lin, C. H., Lin, J. J., Suen, L. J., Gibson, W. W., Chung, J. G. 2009. Gallic acid induces apoptosis via caspase-3 and mitochondrion-dependent pathways in vitro and suppresses lung xenograft tumor growth in vivo. *J. Agric. Food Chem.* 2009; 57: 7596–7604.

Karamac, M., Kosinska, A., Pegg, R. B. 2006. Content of Gallic acid in selected plant extracts. *Pol. J. Food Nutr. Sci.*, 15/56, 1, 55–58.

Karvat, M., Debnath, M., Sukla, V. J. 2012. Solvent effect on extraction of gallic acid from amalaki churna (*Emblica officinalis* Gaertn.) to reduce matrix effect using HPTLC and UV-Spectroscopy with 12 different nature solvents. *International Research Journal of Pharmacy*, 3 (6): 155-158.

Kastan, M. B., Onyekwere, O., Sidransky, D., Vogelstein, B., Craig, R. W. 1991. Participation of p53 protein in the cellular response to DNA damage. *Cancer Res.* 51, 6304 – 6311.

Kaur, M., Singh, R. P., Gu, M., Agarwal, R., Agarwal, C. 2006. Grape seed extract inhibits in vitro and in vivo growth of human colorectal carcinoma cells. *Clin. Cancer Res.*;12: 6194–6202.

Kawada, M., Ohno, Y., Ri, Y. 2001. Anti-tumor effect of gallic acid on LL-2 lung cancer cells transplanted in mice. *Anticancer Drugs*; 12: 847–852.

Khan, N., Afaq, F., Saleem, M., Ahmad, N., Mukhtar, H. 2006. Targeting multiple signaling pathways by green tea polyphenol (−)-epigallocatechin-3-gallate. *Cancer Res.* 66(5): 2500–5. [PubMed: 165105 63]

Kim, S. H., Jun, C. D., Suk, K., et al. 2006. Gallic acid inhibits histamine release and pro-inflammatory cytokine production in mast cells. *Toxicol. Sci.*; 91:123–131.

Kossah, R., Nsabimana, C., Zhang, H., Chen, W. 2010. Optimization of Extraction of Polyphenols from Syrian Sumac (*Rhus coriaria* L.) and Chinese Sumac (*Rhus typhina* L.) Fruits. *Research Journal of Phytochemistry,* pp. 8.

Kumar, R. A., Gunasekaran, P., Lakshmanan, M. 1999. Biodegradation of tannic acid by *Citrobacter freundii* isolated from a tannery effluent. *J. Basic Microbiol.*, 39, 161–168.

Kuriyama, S., Shimazu, T., Ohmori, K., Kikuchi, N., Nakaya, N., Nishino, Y., Tsubono, Y., Tsuji, I. 2006. Green tea consumption and mortality due to cardiovascular disease, cancer, and all causes in Japan: The Ohsaki study. *JAMA* 296, 1255-1265.

Larsen, C. J. 1994. The BCl-2 gene prototype of a gene family that controls programmed cell death (apoptosis). *Ann. Genet.* 37,121-134. In French.

Lekha, P., Lonsane, B. 1997. Production and application of tannin acyl hydrolase: state of the art. *Advances in Applied Microbiology* 44: 215-260.

Li, H., Chen, B., Nie, L., Yao, S. 2004. Solvent effects on focused microwave assisted extraction of polyphenolic acids from *Eucommia ulmodies*. *Phytochem. Anal.*, 15, 306-312.

Link, A., Balaguer, F., Goel, A. 2012. Cancer chemoprevention by dietary polyphenols: promising role for epigenetics. *Biochem. Pharmacol.*; 80(12): 1771–1792. doi:10.1016/j.bcp.2010.06.036.

Liu, Z., Schwimer, J., Liu, D., Lewis, J., Greenway, F. L., York, D. A., Woltering, E. A. 2006. Gallic acid is partially responsible for the antiangiogenic activities of *Rubus* leaf extract. *Phytother. Res.*; 20:806–813.

Llorach, R., Gil-Izquierdo, A., Ferreres, F., Tomas-Barberan, F. A. 2003. HPLC-DAD-MS/MS ESI characterization of unusual highly glycosylated acylated flavonoids from cauliflower (*Brassica oleracea* L. var. botrytis) agroindustrial byproducts. *J. Agric. Food Chem.*, 51, 3895-3899.

Lokeswari, N. 2012. Optimization of environmental parameters for maximum tannase production from cashew husk. *Int. J. Pharm.*, 2(2): 375-379.

Lu, Y., Jiang, F., Jiang, H., Wu, K., Zheng, X., Cai, Y., Katakowski, M., Chopp, M., Shing-Shun, T. T. 2010. Gallic acid suppresses cell viability, proliferation, invasion and angiogenesis in human glioma cells. *Eur. J. Pharmacol.* 641(2-3): 102–107. doi:10.1016/j.ejphar.2010.05.043.

Ludwig, A., Lorenz, M., Grimbo, N., Steinle, F., Meiners, S., Bartsch, C., Stangl, K., Baumann, G., Stangl, V. 2004. The tea flavonoid epigallocatechin-3-gallate reduces cytokine-induced VCAM-1 expression and monocyte adhesion to endothelial cells. *Biochem. Biophys. Res. Commun.* 316, 659-665.

Mahadevan, A., Muthukumar, G. 1980. Aquatic microbiology with reference to tannin degradations. *Microbiologia*, 72, 73–79.

Makkar, H. P. S., Singh, B., Kamar, D. N. 1994. Biodegradation of tannins in oak (Quercus *incana*) leaves by *Sporotrichum pulveulentum*. *Lett. Appl. Microbiol.*, 18, 39–41.

Mantena, S. K., Baliga, M. S., Katiyar, S. K. 2006. Grape seed proanthocyanidins induce apoptosis and inhibit metastasis of highly metastatic breast carcinoma cells. *Carcinogenesis* 2006, 27, 1682-1691.

Mariotto, A. B., Rowland, J. H., Yabroff, K. R., 2009. Long-term survivors of childhood cancers in the United States. *Cancer Epidemiol. Biomarkers Prev.* 18(4):1033-1040.

Martins, S., Mussatto, S., Martínez-Avila, G., Montañez-Saenz, J., Aguilar, C. N., Teixeira, J. A. (2011). Bioactive phenolic compounds: Production and extraction by solid-state fermentation. A review. *Biotechnology Advances,* 29, 365–373.

Maurya, D. K., Nandakumar, N., Asir-Devasagayam, T. P. 2011. Anticancer property of gallic acid in A549, a human lung adenocarcinoma cell line, and possible mechanisms *J. Clin. Biochem. Nutr.* 48(1): 85–90.

Mertz, C., Cheynier, V., Gunata, Z., Brat, P. 2007. Analysis of phenolic compounds in two blackberry species *(Rubus glaucus* and *Rubus adenotrichus*) by high-performance liquid chromatography with diode array detection and electrospray ion trap mass spectrometry. *J. Agric. Food Chem.*, 55, 8616-8624.

Meyer, A. S. 2005. Enzymatic upgrading of antioxidant phenolics in berry juices and press residues. *Fruit Process.*, 15, 382-387.

Mingshu, L., Kai, Y., Qiang, H., Dongying, J. 2006. Biodegradation of gallotannins and ellagitannins. *J. Basic Microbiol.*, 46 (1): 68–84.

Moreno, Y. S., Sanchez, G. S., Hernandez, D. R., Lobato, N. R. 2005. Characterization of anthocyanin extracts from maize kernels. *J. Chromatogr. Sci.,* 43, 483-487.

Mueller-Harvey, I. 2001. Analysis of hydrolysable tannins. *Animal Feed Science and Technology* 91. 3-20.

Naczk, M., Shahidi, F. 2006. Phenolics in cereals, fruits and vegetables: occurrence, extraction and analysis. *J. Pharm. Biomed. Anal.*, 41:1523-1542.

Nair, S., Li, W., Kong, A. N. T. 2007. Natural dietary anticancer chemopreventive compounds: redox mediated differential signaling mechanisms in cytoprotection of normal cells versus in cytotoxicity in tumor cells. *Acta Pharmacologica Sinica* 28 (4): 459-472.

Neergheen, V. S., Soobrattee, M. A., Bahorun, T., Aruoma, O. I. 2006. Characterization of the phenolic constituents in Mauritian endemic plants as determinants of their antioxidant activities *in vitro. J. Plant Physiol.*, 163, 787-799.

Pawar, C. R., Surana, S. J. 2010. Optimizing conditions for gallic acid extraction from *Caesalpinia decapetala* wood. *Pak. J. Pharm. Sci.*, 23 (4): 423-425.

Plass C. 2002. Oxygen species scavenging properties in the extracts of the fruits of *Terminalia chebula, Terminalia belerica* and *Emblica officinalis*. Cancer epigenomics. *Hum. Mol. Genet.* 11(20):2479–2488. [PubMed: 12351584]

Pokorný, J. 1986. Addition of antioxidants for food stabilization to control oxidative rancidity. *Czech Journal of Food Sciences*, 4: 299–307.

Pourmorad, F., Hosseinimehr, S. J., Shahabimajd, N. 2006. Antioxidant activity, phenol and flavonoid contents of some selected Iranian medicinal plants. *Afric. J. Biotech.* 11: 1142-1145.

Qiao, Y., Cao, J., Xie, L., Shi, X. 2009. Cell growth inhibition and gene expression regulation by (−)- epigallocatechin-3-gallate in human cervical cancer cells. *Arch. Pharm. Res.* 32 (9):1309–15. [PubMed: 1978 4588]

Qin, J., Xie, L. P., Zheng, X. Y., Wang, Y. B., Bai, Y., Shen, H. F., Li, L. C., Dahiya, R. 2007. A component of green tea, (−)-epigallocatechin-3-gallate, promotes apoptosis in T24 human bladder cancer cells via modulation of the PI3K/Akt pathway and Bcl-2 family proteins. *Biochem. Biophys. Res. Commun.* 354(4):852–7. [PubMed: 17266926]

Raina, K., Rajamanickam, S., Deep, G., Singh, M., Agarwal, R., Agarwal, C. 2008. Chemopreventive effects of oral gallic acid feeding on tumor growth and progression in TRAMP mice. *Mol. Cancer Ther.*; 7:1258–1267.

Rakesh, S. U., Salunkhe, V. R., Dhabale, P. N., Burade, K. B. 2009. HPTLC Method for Quantitative Determination of Gallic Acid in Hydroalcoholic Extract of Dried Flowers of *Nymphaea Stellata* Willd. *Asian J. Research Chem.* 2(2): 131-134.

Ramirez-Coronel, M. A., Marnet, N., Kolli, V. S., Roussos, S., Guyot, S., Augur, C. 2004. Characterization and estimation of proanthocyanidins and other phenolics in coffee pulp (*Coffea arabica*) by thiolysis-high-performance liquid chromatography. *J. Agric. Food Chem.*, 52, 1344-1349.

Réblová, Z. 2012. Effect of temperature on the antioxidant activity of phenolic acids. *Czech J. Food Sci.*, 30, (2): 171–177.

Riboli, E, Norat, T. 2003. Epidemiologic evidence of the protective effect of fruit and vegetables on cancer risk. *Am. J. Clin. Nutr.* 78(3 Suppl.): 559S–569S. [PubMed: 129369 50]

Robards, K. 2003. Strategies for the determination of bioactive phenols in plants, fruit and vegetables. *J. Chromatogr.* A, 1000, 657-691.

Robbins, R. J. 2003. Phenolic Acids in Foods: An overview of analytical methodology. *J. Agric. Food Chem.*, 51 (10), 2866-2887.

Rommel, A., Wrolstad, R. E. 1993. Influence of acid and base hydrolysis on the phenolic composition of red raspberry juice. *J. Agric. Food Chem.*, 41, 1237-1241.

Roy, A. M., Baliga, M. S., Katiyar, S. K. 2005. Epigallocatechin-3-gallate induces apoptosis in estrogen receptor-negative human breast carcinoma cells via modulation in protein expression of p53 and Bax and caspase-3 activation. *Mol. Cancer Ther.* 4(1):81–90. [Pub Med: 156573 56]

Ruiz-Aguilar, G. M. L., Rios-Leal, E., Tomasini-Campocosio, A., Veloz-Rodriguez, M., Barajas-Aceves, M. 2004. Effect of culture parameters on the degradation of a hydrolysable tannin extracted from cascalote by *Aspergillus niger*. *Bull. Environ. Contam. Toxicol.*, 73, 45–52.

Scalbert, A. 1991. Antimicrobial properties of tannins. *Phytochem.*, 30, 3875–3883.

Seth, M., Chand, S. 2000. Biosynthesis of tannase and hydrolysis of tannins to gallic acid by *Aspergillus awamori* – optimization of Process parameters. *Process Biochemistry*, 36, 39-44.

Shahrzad, S., Aoyagi, K., Winter, A., Koyama, A., Bitsch, I. 2001. Pharmacokinetics of Gallic Acid and Its Relative Bioavailability from Tea in Healthy Humans. *The journal of nutrition*, 1207-1210.

Shi, J., Nawaz, H., Pohorly, J., Mittal, G., Kakuda, Y., Jiang, Y. 2005. Extraction of polyphenolics from plant material for functional foods-engineering and technology. *Food Rev. Int.,* 21, 139-166.

Simic, M. G. *1988 Oxygen radicals in biology and medicine*. Plenum, New York

Singleton, V. L., Orthofer, R., Lamuela-Raventos, R. M. 1999. Analysis of total phenols and other oxidation substrates and antioxidants by means of Folin-Ciocalteu reagent. *Methods Enzymol.*, 299, 152-178.

Singleton, V. L., Rossi, J. A., Jr. 1965. Colorimetry of total phenolics with phosphomolybdic-phosphotungstic acid reagents. *Am. J. Enol. Vitic.*, 16, 144-158.

Sporn, M. B. 1976. Approaches to prevention of epithelial cancer during the preneoplastic period. *Cancer Res.* 36(7 PT 2):2699–702. [PubMed: 1277177]

Taitzoglou, I. A., Tsantarliotou, M., Zervos, I., Kouretas, D., Kokolis, N. A. 2001. Inhibition of human and ovine acrosomal enzymes by tannic acid in vitro. *Reproduction* 121, 131-137.

Tang, G. Q., Yan, T. Q., Guo, W., Ren, T. T., Peng, C. L., Zhao, H., Lu, X. C., Zhao, F. L., Han, X. 2010. Epigallocatechin-3-gallate induces apoptosis and suppresses proliferation by inhibiting the human Indian Hedgehog pathway in human chondrosarcoma cells. *J. Cancer Res. Clin. Oncol.*

Tsao, R. 2010. Chemistry and Biochemistry of Dietary Polyphenols. *Nutrients*, 2, 1231-1246; doi:10.3390/nu2121231

Vauzour, D., Rodriguez-Mateos, A., Corona, G., Oruna-Concha, M. J., Spencer, J. P. E. 2010. Polyphenols and human health: prevention of disease and mechanisms of action. *Nutrients* 1106-1131; doi:10.3390/ nu2111106

Wang, Z., Pan, Z., Ma, H., Atungulu, G. 2011. Extract of phenolics from pomegranate peels. *The Open Food Science Journal*, 5, 17-25.

Yoshioka, K., Kataoka, T., Hayashi, T., Hasegawa, M., Ishi, Y., Hibasami, H. 2000. Induction of apoptosis by gallic acid in human stomach cancer KATO III and colon adenocarcinoma COLO 205 cell lines. *Oncol. Rep.*, 7 (6):1221-3.

Zern, T. L., Wood, R. J., Greene, C., West, K. L., Liu, Y., Aggarwal, D., Shachter, N. S., Fernandez, M. L. 2005. Grape polyphenols exert a cardioprotective effect in pre- and postmenopausal women by lowering plasma lipids and reducing oxidative stress. *J. Nutr.* 135, 1911-1917.

Zhang, Y., Seeram, N. P., Lee, R., Feng, L., Heber, D. 2008. Isolation and identification of strawberry phenolics with antioxidant and human cancer cell antiproliferative properties. *J. Agric. Food Chem.*, 56, 670-675.

In: Handbook on Gallic Acid
Editors: M. A. Thompson and P. B. Collins
ISBN: 978-1-62618-921-8

Chapter 8

GALLIC ACID: OCCURRENCE IN PLANT FOODS AND EFFECTS OF AGRICULTURAL PRACTICES, VEGETATIVE STAGE AND PROCESSING

Antonietta Baiano*
Dipartimento di Scienze Agrarie, degli Alimenti e dell'Ambiente,
University of Foggia, Foggia, Italy

ABSTRACT

Gallic acid or trihydroxybenzoic acid belongs to the class of phenolic acids, has antioxidant and antimicrobial properties, and shows cytotoxicity against cancer cells. It occurs in different amounts in a great number of fruits and vegetables and, thus, in their food derivatives. Gallic acid can also arise from the breakdown of tannins. Berries supply gallic acid together with anthocyanins. A cup of hot chocolate contains higher amounts of this compound than beverages such as wine and tea. Other sources of gallic acid are walnuts, hops, apples, and flax seeds.

Agricultural practices, vegetative stage and processing parameters affect the phenolic content and profile of foods. Furthermore, processing operations including washing, cutting, size reduction, mechanical separations, and thermal treatments can modify the gallic acid content.

This review represents a survey of the gallic acid content of plants and plant-based foods and highlights the effects of agricultural practices, vegetative stage, and processing on the content of this phenolic acid.

INTRODUCTION

Gallic acid, a phenolic compound with primary antioxidant properties, is one of the most bio-active natural compounds and is found in about 70% of foods.

* a.baiano@unifg.it, Via Napoli, 25 - 71122 Foggia, Italy.

Together with its derivatives, it represents a large family of plant secondary polyphenolic metabolites. It occurs in plant materials in form of either methylated gallic acids (e.g., syringic acid) or galloyl conjugates of catechin derivatives, i.e., flanvan-3-ols, or polygalloyl esters of glucose, quinic acid or glycerol. [1].

Plants, bacteria, and fungi synthesize and accumulate gallic acid according to a pathway that is largely unknown. However, two potential pathways for gallic acid synthesis have been postulated: (1) direct oxidation of 3-dehydroshikimate and (2) dehydration of 3-dehydroshikimate to protocatechuic acid followed by hydroxylation of protocatechuic acid. According to retrobiosynthetic NMR studies with 13C-labeled glucose and oxygen isotope ratio mass spectrometry, gallic acid is synthesized entirely or predominantly by dehydrogenation of DHS [2-4]. Shikimate dehydrogenase, a shikimate pathway enzyme essential for aromatic amino acid synthesis, is also required for gallic acid production [5].

Agricultural wastes such as apple bagasse, green tea waste, mango seed kernel, olive mill, palm kernel cake, peat moss, and tamarind have gallotannin content. Tannase is an extracellular and inducible enzyme, which catalyses the hydrolysis of gallotannins releasing glucose and gallic acid [6] and is usually obtained from bacteria, some yeasts and filamentous fungi, mainly from the following species: *Aspergillus, Penicillium, Fusarium* and *Thichoderma* [7]. Among the above mentioned agricultural wastes, the production of gallic acid by *A. niger* AUMC 4301 was the highest for mango seed kernel [8].

Gallic acid is known as a potent antimicrobial compound. It was found to be effective against two human pathogens, the bacteria *Salmonella typhi* and *Staphylococcus aureus,* with the minimum inhibitory concentration (MIC) values equal to 2500 and 1250 μg/ml, respectively [9] but also against a plant pathogen (*Erwinia carotovora*), and a human pathogenic yeast (*Candida albicans*) [10].

According to another paper [11], gallic acid is able to inhibit the production of aflatoxin by moulds. The researchers found that *Aspergillus flavus* and *A. parasiticus,* which are the best known species capable of forming aflatoxins and able to infect walnuts, almonds, pistachios, peanuts, corn, and cotton, help free up the gallic acid in walnut tannin by producing the tannase enzyme.

Gallic acid could represent a pharmacologically important compound. Its effects on melanogenesis inhibition may be related to its ability to scavenge reactive oxygen species [12]. It also showed a significant inhibition of cell proliferation in a series of cancer cell lines and induces apoptosis in esophageal cancer cells but not in non-cancerous cells [13]. Some authors [14] studied the effects of gallic acid and its derivatives on drug metabolising enzymes in the light of their distribution and consequent ingestion, current uses and suggested therapeutic potential. Gallic acid is used as a remote astringent in cases of internal haemorrhage in uterine, pulmonary, and nephritic haemorrhages ad was found beneficial in treating albuminuria and diabetes [15]. According to other studies [1], the protective effects of gallic acid derivatives in cell systems under oxidative stress depend on both their antioxidant capacities and hydrophobicity. Gallic acid was also found to be beneficial for the treatment of myocardial damage associated with type-1 diabetes [16].

As antioxidants, the esters of gallic acid have a wide range of industrial uses, in food, cosmetics, and pharmaceutical industry. Due to the use of gallic acid esters as food additives to prevent food oxidations, the average U.S. intake of gallic acid from food is about 1 g/day. In food, drugs, and cosmetics, it prevents rancidity induced by lipid peroxidation and spoilage. Several ayurvedic herbs and formulations are already in use in treatments of different diseases and their gallic acid content ranges from 0.593 mg/g of *Acacia arabica* (Babool) to 2.234 mg/g of Chyavanprash [15]. Gallic acid and related compounds are effective against *Trypanosoma brucei brucei*, a protozoan with flagella species that causes African trypanosomiasis (or sleeping sickness) in humans and nagana in animals in Africa. Studies elucidated that the pyrogallol moiety could be responsible for this activity [17].

A study on the ability of gallic acid and other phenolic acids (ferulic, protocatechuic, sinapic) to protect naturally α-tocopherol during the heating of sunflower, rapeseed, olive, and soybean oils was performed by Réblová et al. [18]. Gallic acid was the only phenolic acid able to protect α-tocopherol in all the studied conditions.

Gallic acid is generally recognized as safe in the form of gallotannins by the Food and Drug Administration.

Toxicology studies in rodents show that the no observed adverse effect level is 120 mg/kg, a dose equivalent to 2.9 g in a 70–kg man [19]. Nevertheless, in order to solve possible toxicity implications of gallic acid and its esters as food antioxidant, they can be used in form of microcrystalline cellulose ester, whose ability to remove all kinds of free radicals is higher than or approximate equal to gallic acid and whose safety is higher than gallic acid and gallic acid propyl [20].

Phenolics are essential to life but, when metabolized incorrectly, they can cause major and minor physical, mental, and emotional disturbances in a large number of patients.

In these cases, instead of recommending a restrictive diet, neutralization of gallic acid is less traumatic.

Frequently, parents report a marked improvement in the school performances of their children and a normalization of hyperactivity and craving for sweets. Gallic acid has effects upon the muscular skeletal system, the lower back, and the chronic chest pain that is not of cardiac origin [21].

Gallic acid can be used in multifunctional packaging material. Chitosan is a good material for this type of packaging because of its biocompatibility, biodegradability, antibacterial properties, secondary antioxidant activity, film forming ability, and resistance to lipids.

Gallic acid can be grafted to chitosan to obtain a material that readily forms films usable to pack peanuts and other oxidisable foods, having good mechanical performance and stability as highlighted by a stable antioxidant activity over a 15 week storage period at 50°C [22].

Gallic acid can be used to modify the mechanical and barrier characteristics of edible biodegradable films made with vital wheat gluten, a by-products of wheat starch production. Gallic acid acts like a plasticizer and films including gallic acid are more elastic than the control [23]. For little addition of gallic acid, films have higher water vapour permeability values than the control due to the penetration of this low molecular substance into the protein network and the consequent disruption of the intermolecular interactions among peptide chains.

At concentrations higher than 2%, the water vapour permeability decreases respect to the control, as a consequence of the excessive presence of the gallic acid molecules that filled the free space in the film network.

In addition, gallic acid is employed as a source material for inks, paints and colour developers [14], and dye-making, leather, and chemical industries [8].

Occurrence of Gallic Acid in Plant Foods

Gallic acid and its structurally related compounds are found widely distributed in various plants and foods such as walnuts, tea leaves, bark, green tea, apple-peels, grapes, strawberries, pineapples, bananas, lemons, and in red and white wines.

Gallic acid was identified in mango (*Mangifera indica* L.) kernel by the use of reversed-phase high-performance liquid chromatography (RP-HPLC) coupled with photodiode array detection (DAD). The frieeze-dried mango kernel contains 185 mg of gallic acid per 100 g of seeds and the concentration increases with hydrolysis and heat treatment up to 838 mg/100 g possibly attributable to: the degradation of high molecular weight compounds, the releasing of free gallic acid, the formation of phenolic compounds during the heating process, and, the non-enzymatic interconversion between phenolic molecules [24].

According to De La Rosa et al. [25], gallic acid greatly occurrs in Mexican pecan (*Carya illinoinensis*). In particular, the free gallic acid contents ranges from 64.3 to 80.7 μg/g whereas the total gallic acid ranges from 189.0 to 274.5 μg/g.

Tsujita et al. [26] hypothesizes that the main polyphenols of chestnut astringent skin might be represented by oligomeric proanthocyanidins with gallic and ellagic acids.

Gallic acid is the main detectable components of water extractable polyphenols of carob pulp together with (-)-epigallocatichin, (+)-catichin, (-)-epicatichin gallate, and (-)-epigallocatichin gallate [27].

Honey can be defined as a highly concentrated solution of a complex mixture of sugars and contains small amounts of minerals, proteins, vitamins, organic acids, and phenolic compounds. A characterization study was performed by Kalil et al. [28] on Malaysian honeys: Tualang, multi-floral honey produced by *Apis dorsata*; Gelam, produced by similar bees collecting nectar and pollen from the plant *Melaleuca cajaputi Powell*; Borneo, produced by *Apis cerana* collecting the nectar mainly from *Acacia mangium*. Six phenolic acids (gallic, syringic, benzoic, trans-cinnamic, *p*-coumaric, and caffeic) and five flavonoids (catechin, kaempferol, naringenin, luteolin, and apigenin) were detected. Catechin occurred in higher amounts (from 11.32 to 36.77 mg/kg) than the other phenolic compounds. Gallic acid was detected in only two of the seven honey samples in concentrations of about 0.40 mg/kg. Michalkiewicz et al. [29] analysed samples of both linden and heather honey, founding 0.9 mg of gallic acid per kilogram in the first one and the absence of gallic acid from the second one.

Propolis is very popular as food supplementary material for human health. According to Gülçin et al. [30], gallic acid concentration is equal to 419.1 mg per kilogram of lyophilized aqueous extract of propolis.

Pear is one of the most consumed fruit and has also been used in Chinese folk medicine to treat respiratory disease. Five Chinese varieties of pears (*Shuijing, Fengshui, Xuehua, Ya*, and *Xiang*), analysed for their phenolic content, gave the following results, expressed as μg/g: gallic acid ranged from 5.23 to 10.72; catechin from 0.41 to 28.83; chlorogenic acid from 485.11 to 873.03; caffeic acid from 0 to 1.16; epicatechin from 6.73 to 131.49, and rutin from 0.92 to 104.64 [31].

Muscadine (*Vitis rotundifolia*) is a grape vine of the southeastern United States. Its fruits can be eaten fresh, or used in making wine, juice, and jelly. Noble and Carol are two muscadine varieties used for making a red and a white wine, respectively. The skins of their berries were found to contain 17.47 (Noble) and 36.8 (Carol) milligrams of gallic acid per 100 g of fresh weight whereas the content of the same phenolic acid in their seeds was 25.9 mg/100g. Catechin, epicatechin, and epicatechingallate in seeds were 47.9, 74.8, and 15.87 mg/100g, respectively [32].

Gallic acid is also contained in cherry wines in concentrations ranging from 0.0066 to 0.0836 g/L [33].

The gallic acid content was analysed in wines obtained by five American wine grape cultivars (Cynthiana, the French hybrids Villard blanc and Chambourcin, the *V. labrusca* Catawba and Noah) and two traditional European grapes (*V. vinifera* Merlot and Cabernet sauvignon) [34]. For the red wines, the concentration of the phenolic acid ranged from 7.69 (Cabernet sauvignon) to 8.35 (Cynthiana) mg/l. The rosè wine from Catawba showed a gallic acid concentration equal to 3.07 mg/l whereas this compounds was not detected in wines from the grapes Villard blanc and Chambourcin. According to a review of de Beer et al. [35], the gallic acid content ranges from 3.1 to 320 mg/l in red wines and from 2.8 to 11 mg/l in white wines.

Rice flour is obtained by dehulling and grinding rice grains. In a study of Irakli et al. [36], the gallic acid contents amounted to 2.11, 2.35, and 2.56 μg/g in non-pigmented, red, and black rice, respectively. In the same order, the concentrations of catechin were 2.60, 8.78, and 4.90 whereas the concentrations of epicatechin were 3.43, 4.28, and 14.14. Instead, the contents of gallic acid (μg/g) in rice bran were found to be equal to: 25.9 in normal rice, 39.0 in the red rice, and from 85.3 to 161.1 in the black one [37].

Black tea quality can be graded on four scales of quality. According to one of these scales, commonly used in Turkey, it is classified as high quality (grades from 1 to 3) and low-quality (grades from 4 to 7). The gallic acid content ranges from 96.8 to 116 mg/100 g of fresh weight but without significant differences among the various grades. Gallic acid-derivatives such as epigallocatechin and gallocatechin are present in higher amounts, ranging from 1038 to 1214 and from 467 to 650 mg/100 g, respectively. Nevertheless, the high and the low-quality black teas should no be distinguished on the basis of the nutritional and functional characteristics [38].

Berries are rich sources of phenolics. In particular, the gallic acid concentrations (as grams per kilogram of fresh weight) detected in the main berries are the following: blueberry, 3-7; black currant, 4-13; red currant, 3-8; gooseberry, 9-14; red raspberry, 15-59; strawberry, 1-44 [39]. Wada and Ou [40] analysed five types of caneberries (evergreen blackberries (*Rubus laciniatus*), marionberries (*Rubus ursinus*), boysenberries (*Rubus ursinus* × *idaeus*), red raspberries (*Rubus idaeus*), and black raspberries (*Rubus occidentalis*) for their phenolic contents and antioxidant activity. Gallic acid was present in evergreen blackberries, marionberries, and boysenberries at oncentrations of 0.02, 0.03, and 0.09 mg/g, respectively.

The cashew tree (*Anacardium occidentale* L.) is a hardy tree that produced fruits (commonly called cashew apples, often considered as by-products) and the most popular nuts. Nevertheless, the cashew apple is five times richer in vitamin C than orange juice and 10 times more than the pineapple juice [40] and its juice is considered as an excellent remedy for sore throat and chronic dysentery [41]. It is also rich in phenolic compounds. In particular, the gallic acid content of the juice deriving from red cashew apples ranges from 1.9 to 4.4 mg/l whereas those obtained from yellow fruits ranges from 13.3 to 15.4 mg/l [42]. Fig syrup, a Mediterranean fig derivative, has been proposed as a nutraceutical supplement. Its content of gallic acid was determined in 0.021 mg per gram of syrup [43].

Cocoa leaves are sources of antioxidants. In a study of Osman et al. [44], cocoa shoot, young cocoa leaves, and tea leaves were processed according to the green tea processing procedures and gave products rich in epicatechin, epigallocatechingallate, epigallocatechin, gallic acid, and epicatechin gallate. In particular, the gallic acid contents (as mg/100 mg sample) were: cocoa shoot, 0.58: young cocoa leaves, 0.39; green tea, 0.33.

Carob pods are used as a cocoa substitute in baking, cereal bars, chocolate, confectionery, ice creams and light products. Gallic acid was found to be the principal free phenolic acid in carob pods accounting from 17.96 to 32.92% of the total phenolics present [45]. According to Corsi, et al. [46] the phenolic fraction of carob pods amounts to 1.36 mg/kg and was dominated by gallic acid (88%) with relevant contributions by catechins, epigallocatechin and epicatechin gallate.

Chocolate preparation, made from the fruit of the cacao tree, is a popular food and is also used as a flavouring agent and an ingredient of beverages and confectionery. Its composition included polyphenol flavanolic compounds. Tokusoglu and Ünal [47] determined the phenolic composition of dark and milk chocolate of several brands. The gallic acid content was in the range 0.44-1.36 mg/g. Catechin, epicatechin, epigallocatechin, epigallocatechin gallate, and epicatechin gallate were detected in analogous concentrations.

Gallic acid is also contained in orange juice (3.33 mg/l) and orange wine (1.93 mg/l) although in concentrations lower than those of other phenolic acids [48]. Commercial grape and prune juices were found to contained 12.5 and 20.9 μg/g of gallic acid, respectively whereas the concentration of this acid was found to be below 5 μg/g in both apple and pomegranate juice [49].

Apple pomace is a by-products of the juice production and is a heterogeneous mixture consisting of skin, core, seed, calyx, stem and soft tissue. It is a rich source of phenolic compounds mainly deriving from skin and seeds. Apple pomace from Gala cultivar was found to contain gallic acid as free acid (0.065 mg/g of fresh weight), esterified acid (0.175 mg/g), and insoluble-bound acid (0.077 mg/g) [50].

Gallic acid is the main phenolic acid in cucumber with 10.10 mg/g [51].

Red chicory is a source of gallic acid, containing an average value of 14.56 mg/100g of fresh weight [52].

Kales are a poor source of gallic acid (free, ester, glycoside, ester-bound), with 13.8 ng per gram of fresh leaves and 10.9 ng per gram of dry seeds [53].

The gallic acid content of English walnut (*Juglans regia*) varies with cultivar and has been determined in the range 1.4-3.4% on dry weight. In the case of Black walnut (*Juglans nigra* "Thomas" and *Juglans hindsii* "Rawlins"), the concentration is about 1%, always on dry weight. Pistach walnut (*Pistacia vera*) contains 0.5 % of gallic acid whereas almonds (Prunus dulcis) content is below 0.1% [54].

Euphoria longana Lam. (longan) is a member of the Sapindaceae family. This family also includes litchi (*Litchi chinensis* L.), rambutan (*Nephelium lappaceum* L.), and horse chestnut (*Aesculus hippocastanum* L.). Longan fruit are extensively consumed in Thailand. A large variation wis observed in the contents of gallic acid among cultivars. The contents of gallic acid are higher in dried tissues (peel and seed) than fresh tissues. On dry weigh basis, fresh pulp contains higher levels of gallic acid than dried pulp. Gallic acid content is highest in dried seed (0.8-2.3 mg/g dry weigh), being 5-10 times higher than fresh seed [55].

Chenopodium album is an annual weed diffused in cultivated fields, due to the high protein content and a balanced amino acid spectrum. It has also been cultivated largely for centuries as a leafy vegetable as well as an important subsidiary grain crop for human and animal food-stuff [56]. Gallic acid was detected in fruits (from 0.39 to 0.58 mg/100 g) but not in leaves [57].

Vinegar can also be considered a source of gallic acid even if its consumption is rather limited. A common rosè vinegar can have about 6.7 mg of gallic acid per liter whereas a red vinegar can show concentrations of about 27.9 mg/l [58].

Onion (*Allium cepa* L.) has been reported to have several beneficial health effects. In a study on the phenolic composition and antioxidant activity of this product [59], it was found that the white and violet onions had gallic acid contents higher than the red and green ones and that the outer part is richer than middle and inner layers. In the case of the violet variety, the gallic acid contents were 322, 266, and 158 μg/g of dry weight for the outer, middle, and inner layers, respectively. The poorest variety, the green one, showed concentrations of 29.5 (outer), 21.6 (middle), and 9.3 (inner) μg/g.

The genus *Amaranthus* belongs to the family Amaranthaceae and includes a great number of species. In particular, *Amaranthus hypochondriacus*, *A. cruentus*, and *A. caudatus* are the essential grain species. Amaranth can supply grains and tasty leafy vegetables of high nutritional quality as a food and animal feed. The contents of gallic acid in *A. caudatus* leaves was found to be equal to 0.083 μg/g od dry weight [60].

Hop is an interesting source of gallic acid. The concentration of this phenolic acid ranges from 3.2 to 6.5 mg/100 g in Czech hop cultivars whereas the content of catechin is included between 15.6 and 42.3 mg/100 g and that of epicatechin is 4.0-9.9 mg/100 g [61]. Beer also contains gallic acid. Floridi et al. [62] found an average value of 0.6 mg/l on 23 different samples.

EFFECTS OF AGRICULTURAL PRACTICES, VEGETATIVE STAGE, AND PROCESSING ON GALLIC ACID CONTENT

Environmental factors have a major effect on phenolic content and profile. These factors may be pedoclimatic (soil type, sun exposure, rainfall) or agronomic (culture in greenhouses or fields, biological culture, hydroponic culture, fruit yield per tree, etc). In general, phenolic acid concentrations decrease during ripening. Storage may also affect the content of phenols that are easily oxidized. Oxidation reactions result in the formation of more or less polymerized substances, which lead to changes in colour and organoleptic characteristics. Such changes may be beneficial (as is the case with black tea) or harmful (browning of fruit) to consumer acceptability [62].

Gallic acid is one of the most important phenolic compounds in sweet bell peppers and, according to a recent study of Hallmann and Rembialkowska [63], its content is higher when an organic growing system is applied with respect to the conventional system (865.9 vs. 649.2 mg/kg dry weight). Authors explained that, in organic farming, the exclusion of the use of synthetic protection products activates natural mechanisms of plant defence against diseases and pests.

Safflower (*Carthamus tinctorius* L.) is a member of the Asteraceae family and is an important aromatic and medicinal plant. Its flower is the most interesting part of the plant and is used in fields such as food, pharmaceutics, and cosmetics. Phenolic composition changes quantitatively and qualitatively during the various stages of flower development (floral budding, floral formation, and full flowering), with gallic acid as the most abundant phenolic compound. The higher concentrations of this phenolic acid, expressed as μg/g of dry weight, were found at full flowering in the orange flowers (~102.57), at floral budding in the yellow flowers (~207.74), and at floral formation in the red flowers (~192.63) [64].

Methods of culinary preparation also have effects on the phenolic content of foods. For example, simple peeling of fruit and vegetables can eliminate a significant portion of phenols because these substances are often present in higher concentrations in the outer parts than in the inner parts. Cooking may also have a major effect, determining consistent loss of phenolics. Industrial food processing also decreases phenolic content. As fruit peeling, dehulling of legume seeds and decortication and bolting of cereals can result in a loss of some polyphenols. Grinding of plant tissues may lead to oxidative degradation of polyphenols which are then transformed into brown pigments [62].

Mango, one of the most important tropical fruits with banana, pineapple, papaya and avocado (65), has interesting vitamin C, β-carotene and vitamin E contents and moderate levels of phenolic compounds. Minimally processing can affect in a negative way these attractive characteristics. In a study on the possible losses of bioactive compounds that occur during storage in minimally processed mangoes [66], fresh-cut fruits simply immersed in water and designed as the control were compared to slice dipped for 3 min in a 1% calcium chloride solution as firmness stabilizer also containing 1% ascorbic acid and 1% citric acid as anti-browning agents. The initial value of gallic acid was 28 mg per 100 g of dry weight for both treated and control slices. This concentration was maintained during the first 6 and 10 days of storage for treated and control slices, respectively. Afterwards gallic acid increased until the end of the storage period, reaching values of 68 mg and 53 mg per 100 g of dry weight for treated and control slices, respectively.

In a study on the effects of abiotic postharvest stress on mango polyphenolic compounds throughout ripening, an abiotic postharvest stress was assessed using high temperature/short time thermal quarantine. The results showed that changes in free gallic acid and gallotannin concentrations due to ripening were minor, with no significant difference between means of fruit at days 0 and 24 [67]. These data suggest that these compounds are not synthesized during ripening.

In a study on the phenolic profile of eight types of tea [68], the gallic acid contents, expressed as mg/g, were in an increasing order: green teas (Shangai green 0.37, Meifoo green 0.74, Jasmine tea 1.13, and Hangzhou Lung Ching 1.84), Oolong teas (Fujian 1.42 and Jiangxi 1.67), black tea (Fujian black 2.06), and pu-erh tea (5.53). These data can be explained taking into account the different operations they were submitted to. In fact, green tea is obtained by drying and steaming the fresh leaves whereas oolong samples derives from fresh leaves subjected to a partial fermentation stage before drying. Both black and pu-erh teas unergo a full fermentation stage before drying and steaming, consisting in an oxidation (black samples) and in a microbic fermentation (pu-erh). The fermentation step increases the liberation of gallic acid from catechin gallate.

Karamac [69] analysed the content of gallic acid of selected plant before and after treatment with tannase. The average contents (mg/g of extract) of this phenolic acid before and after treatment were the following: green tea (*Camellia sinensis* L.) 4 and 91; bearberry leaves (*Arctostaphylos uva-ursi* L.) 2 and 42; hazelnut (*Corylus avellana*) 1 and 5; evening primrose (*Oenothera biennis*) 15 and 36; grape seeds (*Vitis vinifera* L.) 2 and 16. Green tea and bearberry leaves were extracted with 95% ethanol whereas the other matrices were extracted with 80% acetone after defatting. The high content of gallic acid liberated by tannase in these preparations was due to the presence of hydrolisable tannins.

Centella asiatica (L.) Urban is used in ayurvedic medicine for the treatment of leprosy, asthma, ulcer, etc. Is is also used as memory enhancer and, in Sri Lanka, as porridge for feeding pre-school children to avoid nutritional deficiencies. Ariffin et al. [70] compared the phenolic contents and antioxidant activities of unprocessed, partially fermented, and fully fermented *Centella asiatica* herbal teas. When the water extraction method was used, the average gallic acid contents (mg/g) were 2105, 1319, and 11 for unfermented, partially fermented, and fully fermented *Centella asiatica*, respectively. The reduction of the phenolic acid concentration with the fermentation process was due to the formation of degradation products.

Free and bound gallic acid and condensed tannins occur in hazelnuts (*Corylus avellana* L.). Alasalvar et al. [71] and Pelvan et al. [72] investigated the effects of roasting on those and other taste-active compounds. They found that the contents of gallic acid and condensed tannins considerably varied among unroasted hazelnuts varietes, ranging from 0.159 and 0.871 mg/100g (the first) and from 399 and 4056 mg/100g (the latter). Roasting caused losses of gallic acid and condensed tannins equal to 67 and 97.3%, respectively, as a consequence of the removal of the brown skin and also, in the case of the phenoli acid, of the thermal decomposition.

The effects of the addition of grape seed tannins on gallic acid of a red wine from the Estremadura region of Portugal was studied by Neves et al. [73]. Increasing amounts of grape seed tannins increased the gallic acid concentration in red wine. In fact, although the grape seed tannins did not contain gallic acid, they had a large amount of polymeric proanthocyanidins with a high percentage of galloylation. During wine-making and storage, galloylated proanthocyanidins release gallic acid in wine under the wine pH conditions.

Grape pomace, made of grape seeds (40 to 60% of dry weight) and skins, is a waste product generated during wine-making that can be used as food ingredient in the production of value-added products. In particular, grape pomace and seeds can be dried and ground into a flour that can be included in baking products such as cookies, cakes, and bread. Baking products are submitted to more or less severe heat treatments and thus the thermal stability of the grape seed flour phenolic compounds represents a critical factor. A study performed by Ross et al. [74] showed that, when submitted to heat treatments up to 180°C, the gallic acid content increased with heating time, starting from 61.99 mg/100g of dry weight (untreated grape seed flour) and reaching the maximum concentration (320.73) after 90 min. This behaviour could be explained by the thermal decomposition of catechin, gallocatechin, and epicatechin.

Defatted rice bran is the waste material deriving from rice milling and oil extraction. As in the case of grape pomace, it still contains phenolic substances than can be extracted with supercritical water. The content of phenolic acids is low but gallic acid is the one that occurs in high concentrations (about 0.1, 1.4, and 1.8 mg/g of bran at 125, 150, and 175°C, respectively) [75]. The thermal analysis showed that *p*-coumaric, caffeic, and ferulic acids started to decompose at about 170°C whereas gallic acid decomposition started at about 200°C.

Several phenolic acids, such as gallic acid, protocatechuic, p-hydroxybenzoic, vanillic acid, caffeic acid, syringic, *p*- and *o*-coumaric acid, ferulic acid, and cinnamic acid have been detected in extra-virgin olive oil but the sum of their concentrations is often lower than 1 mg per kg of oil [76]. The recovery of gallic acid changes with the crushing methods. According to Veillet et al. [77], its concentrations range from 1.32 (hammer crushing system) to 1.60 stone mill) mg per kg of oil. One exception is represented by the Tunisian virgin olive oils. In fact, an investigation concerning eight Tunisian virgin olive oil samples (Chemlali cv. from the locations of Sfax, Jerba, and Thibar; Chetoui cv. from Sfax and Thibar; Zarrazi cv. from Mareth and Chemmekh; Zalmati from the location of Zarzis) determined gallic acid contents ranging from 2.573 (Zalmati) to 9.226 (Zarrazi from Chemmekh) g per kg of oil [78]. Gallic acid was detected neither in full fat- nor in defatted olive cake [79].

Inglett et al. [80] studied the processing techniques for enhancing the bioactive and physical properties of corn bran for human health benefits. In particular, the influence of jet-cooking and alkaline treatments on corn bran fractions was studied. The results showed that the jet-cooking released more gallic acid if combined with an alkaline treatment (54.25 mg/g in the soluble fraction and 29.74 mg/g in the insoluble fraction) than when performed alone (6.85 and 3.5 mg/g, in the soluble and insoluble fraction, respectively). The starting corn brans contained only 4.64 mg of gallic acid per gram.

Gallic acid is also detectable in brandies as a consequence of the transfers occurred between the wood of the oak barrels and the liquid. According to the European Community regulations [81] the minimum storage time for brandy in an oak barrel is one year, or six months if the capacity of the barrel is less than a thousand litres. Giménez et al. [82] analysed 38 samples of commercial brandies produces in several European Countries and aged *a)* along the traditional or static system, *b)* along the dynamic or criaderas and soleras system. The concentration of the gallic acid in traditional brandies ranged from 0 to 12.62 mg/l whereas was between 0 and 17.43 mg/l in the case of brandies aged along the soleras system. The authors explained the differences between the two methods taking into account: the variety of oak used in the construction of barrels (French oak in the traditional system, American oak in the soleras one); the age and life of the barrels (in the soleras method, barrels which were previously used to age sherry show lower gallic acid contents); length of time the brandy is kept in the barrel.

Gallic acid incorporated into the spirits as a consequence of hydrolysis of tannins is related to the aging/maturation process and, together with other age-related compounds such as phenolic aldehydes (vanillin, syringaldehyde, coniferaldehyde and sinapaldehyde) and acids (vanillic acid and syringic acid) has significant bearing on the authenticity of the beverages. According to Ng et al. [83], the concentrations of gallic acid in cognac, expressed as milligrams per liter, were: VS (a blend of 2–4-year-old cognacs) 6.4; VSOP (with a very high percentage of 4–6-year-old cognacs) 6.7; XO (with a very high percentage of > 6-year-old cognacs) 11.8 Concerning scotch, the concentrations were: 2.4 (12 years), 7.0 (17 years), and 37.6 (30 years) mg/l. Regarding whisky, the concentrations of gallic acid were: 1.8 (6 years) and 2.4 (12 years) mg/l.

Cachaca is the exclusive name of the beverage produced by distillation of fermented sugarcane juice. The study performed by Vicente et al. [84] highlighted that the gallic acid content of non-aged cachaca was 1.07 mg/l. Cachaca was aged for 18 months in 20 l-casks of white oak (*Quercus* sp.), and of the Brazilian wood species white peroba (*Paratecoma peroba*), balsamo (*Myroxylon balsamum*), timborana (*Piptadenia* sp.) and jaboty (*Erisma uncinatum*). Cachaca with the highest concentration of gallic acid was that aged in balsam casks (49.39 mg/l) followed by that aged in timborana (8.95 mg/l). Cachacas aged in the other types of casks did not show significant increase of gallic acid concentration with respect to the non aged cachaca.

CONCLUSION

Many plants, plant-derived foods, and by-products are rich of gallic acid and other natural antioxidants. Within the same species, gallic acid occurs in a range of concentrations that depend on the varieties. Food processes should be projected in order to minimise the loss of bioactive compounds whereas agricultural wastes should not be considered only as disposal problems but as low-cost sources of healthy compounds, which can be extracted, concentrated, and used to prepare functional foods and nutraceuticals.

Gallic acid can be found in form of free acid or of esters and hydrolysable tannins. In the latter case, the extraction of the phenolic compound requests the application of the enzymatic treatments. The concentrations of gallic acid detected in plants, foods, and agricultural wastes are often lower than those of the other phenolic compounds, nevertheless its recovery is of economic interest due to the great number of applications it is involved in.

REFERENCES

[1] Lu, Z.; NIe, G.; Belton, P. S.; Tang, H.; Zhao, B. Structure–activity relationship analysis of antioxidant ability and neuroprotective effect of gallic acid derivatives. *Neurochem. Int.* 2006, 48, 263-274.

[2] Werner, I.; Bacher, A.; Eisenreich, W. Retrobiosynthetic NMR studies with C-13-labeled glucose—formation of gallic acid in plants and fungi. *J. Biol. Chem.* 1997, 272, 25474–25482.

[3] Werner, R. A.; Bacher, A.; Eisenreich, W. Analysis of gallic acid biosynthesis via quantitative prediction of isotope labeling patterns. In: Gross GG, Hemingway RW, Yoshida T, editors. Plant polyphenols 2: chemistry, biology, pharmacology, ecology. 1. NY: *Kluwer Academic/Plenum Publishers*; 1999. pp. 43–61.

[4] Werner, R. A.; Rossmann, A.; Schwarz C, Bacher A, Schmidt H-L, Eisenreich W. Biosynthesis of gallic acid in Rhus typhina: discrimination between alternative pathways from natural oxygen isotope abundance. *Phytochemistry*. 2004; 65:2809–2813.

[5] Muir, R. M.; Ibáñez, A. M.; Uratsu, S. L.; Ingham, E. S.; Leslie, C. A.; McGranahan, G. H.; Batra, N.; Goyal, S.; Joseph, J.; Jemmis, E. D.; Dandekar, A. M. Mechanism of gallic acid biosynthesis in bacteria (*Escherichia coli*) and (*Juglans regia*). *Plant Mol. Biol.* 2011, 75, 555-65.

[6] Barthomeuf, C.; Regerat, F.; Pourrat, H. Production, purification and characterization of a tannase from *Aspergillus niger* LCF 8. *J. Ferment. Bioeng.* 1994, 77, 320-323.

[7] Aguilar, C. N.; Rodríguez-Herrera, R.; Gutiérrez-Sánchez, G.; Augur, C.; Favela-Torres, E.; Prado-Barragán, L. A.; Ramírez-Coronel, A.; Contreras-Esquivel, J. C. Microbial tannases: advances and perpectives. *Appl. Microbiol. Biotechnol.* 2007, 76, 47-59.

[8] El-Fouly, M. Z;. El-Awamry, Z.; Shahi, A. A. M.; El-Bialy, H. A.; Naeem, E.; El-Saeed, G. E. Gallic acid formation from gallotannins-rich agricultural wastes using *Aspergillus niger* AUMC 4301 or its tannase enzyme. *Arab Journal of Nuclear Sciences and Applications* 2012, 45, 489-496.

[9] Chanwitheesuk, A.; Teerawutgulrag, A.; Kilburn, J. D.; Rakariyatham, N. Antimicrobial gallic acid from *Caesalpinia mimosoides* Lamk. *Food Chem.* 2007, 100, 1044–1048.

[10] Fogliani, B.; Raharivelomanana, P.; Bianchini J.–P.; Madjebi, S. B.; Hnawia, R. Bioactive ellagitannins from *Cunonia macrophylla*, an endemic Cunoniaceae from New Caledonia. *Phytochem.* 2005, 66, 241–247.

[11] Mahoney, N.E.; Molyneux, R. J.; Mcgranahan, G.; Mckenna, J.; Leslie, C. A.; Resistance of Tulare walnut (*Juglans regia* Cv. Tulare) to aflatoxigenesis. *J. Food Sci.* 2003, 68, 619-622.

[12] Jadon, A.; Bhadauria, M.; Shukla, S. Protective effect of *Terminalia belerica* Roxb. and gallic acid against carbon tetrachloride induced damage in albino rats. *J. Ethnopharmacol.* 2007, 109, 214–218.

[13] Faried, A.; Kurnia, D.; Faried, L. S.; Usman, N.; Miyazaki, T.; Kato, H.; Kuwano, H. Anticancer effects of gallic acid isolated from Indonesian herbal medicine, *Phaleria macrocarpa* (Scheff.) Boerl, on human cancer cell lines. *Int. J. Oncol.* 2007, 30, 605-13.

[14] Ow, Y. Y.; Stupans, I. Gallic acid and gallic acid derivatives: effects on drug metabolizing enzymes. *Curr. Drug Metab.* 2003, 4, 241-8.

[15] Borde, V. U.; Pangrikar, P. P.; Tekale S. U. Gallic acid in ayurvedic herbs and formulations. *Recent Res. Sci. Technol.* 2011, 3, 51-54.

[16] Patel, S. S.; Goyal, R. K. Cardioprotective effects of gallic acid in diabetes-induced myocardial dysfunction in rats. *Pharmacognosy Res.* 2011, 3, 239–245.

[17] Koide, T.; Nose, M.; Inoue, M.; Ogihara, Y.; Yabu, Y.; Ohta, N. Trypanocidal effects of gallic acid and related compounds. *Planta Med.* 1998, 64, 27-30.

[18] Réblová, Z.; Fišnar, J.; Tichovská, D.; Doležal, M.; Joudalová, K. Effect of temperature and oil composition on the ability of phenolic acids to protect naturally present α-tocopherol during the heating of plant oils. *Czech J. Food Sci.* 2012, 30, 351–357.

[19] Niho. N.; Shibutani, M.; Tamura, T.; Toyoda, K.; Uneyama, C.; Takahashi, N.; Hirose, M. Subchronic toxicity study of gallic acid by oral ad- ministration in F344 rats. *Food Chem. Toxicol.* 2001, 39, 1063–1070.

[20] Ying, Y.; Wang, X. synthesis and the characterizations of structure and function of gallic acid microcrystalline cellulose ester. *Advanced Materials Research* 2009, 79-82, 2059-2062.

[21] http://www.desbio.com/phenolics-intro.html, available on 5th November 2012.

[22] Schreiber, S. B. (2012). Chitosan-gallic acid films as multifunctional food packaging, Master's Thesis, University of Tennessee. http://trace. tennessee.edu/utk_gradthes/1204. sschrei2@utk.edu.

[23] Hager, A. S.; Vallone, K. J. R.; Arendt, E. Influente of gallic acid and tannic acid on the mechanical and barrier properties of wheat gluten films. *J. Agric. Food Chem.* 2012, 60, 6157-6163.

[24] Soong, Y. Y.; Barlow, P. J. Quantification of gallic acid and ellagic acid from longan (Dimocarpus longan Lour.) seed and mango (*Mangifera indica* L.) kernel and their effects on antioxidant activity. *Food Chem.* 2006, 97, 524-530.

[25] De La Rosa, L.; Alvarez-Parrilla, E.; Shahidi, F. Phenolic compounds and antioxidant activity of kernels and shell of Mexican pecan (*Carya illinoinensis*). *J. Agric. Food Chem.* 2011, 59, 152-162.

[26] Tsujita, T.; Yamada, M.; Takaku, T.; SHintani, T.; Teramoto, K.; Sato, T Purification and characterization of polyphenols fro chestnut astringent skin. *J. Agric. Food Chem.* 2011, 59, 8646-8654.

[27] Avallone, R.; Cosenza, F.; Farina, F.; Baraldi, C.; Baraldi, M. Extraction and purification from Ceratonia siliqua of compounds acting on central and peripheral benzodazepine. Fitoterapia 2002, 73, 390-396.

[28] Khalil, M. I.; Moniruzzaman, M.; Sulaiman, S. A.; Gan, S H. Phenolic acid coposition and antioxidant properties of Malaysian honeys. *J. Food Sci.* 2011, 76, C921-C928.

[29] Michalkiewicz, A.; Biesaga, M.; Pyrzynska, K. Solid-phase extraction procedure for determination of phenolic acids and some flavonols in honey. *J. Chromatogr. A* 2008, 1187, 18–24.

[30] Gülçin, I.; Bursal, E.; Sehitoglu, M. H.; Bilsel, M.; Go□ren, A. C. Polyphenol contents and antioxidant activity of lyophilized aqueous extract of propolis from Erzurum, Turkey. *Food Chem. Toxicol.* 2010, 48, 2227–2238.

[31] Li, X.; Gao, W. Y.; Huang, L. J.; Zhang, J. Y.; Guo, X. H. Antioxidant and antiinflammation capacities ofsome pear cultivars. *J. Food Sci.* 2011, 76, C985-C990.

[32] You, Q.; Chen, F.; Sharp, J. L.; Wang, X.; You, Y.; Zhang, C. High-performance liquid chromatography-mass spectrometry and evaporative light scattering detector to compare phenolic profiles of muscadine grapes. *J. Chromatogr. A* 2012, 1240, 96-103.

[33] Niu, Y.; Zhang, X.; Xiao, Z.; Song, S.; Jia, C.; Yu, H. Characterization of taste-active copounds of various cherry wines and their correlation with sensory attributes. *J. Chroatogr. B* 2012, 902, 55-60.

[34] Zhu, L.; Zhang, Y.; Deng, J.; Li, H.; Lu, J. Phenolic concentration and antioxidant properties of wines made from North Aerican grapes grown in China. *Molecules* 2012, 17, 3304-3323.

[35] De Beer, D.; Joubert, E.; Gelderblom, W. C. A.; Manley, M. Phenolic copounds: a review of their possible role as *in vivo* antioxidants of wine. *S. Afr. J Enol. Vitic.* 2002, 23, 48-61.

[36] Irakli, M. N.; Samanidou, V. F.; Biliaderis, C. G.; Papadoyannis, I. N. Simultaneous determination of phenolic acids and flavonoids in rice using solid-phase extraction and RP-HPLC with photodiode array detection. *J. Sep. Sci.* 2012, 35, 1603-1611.

[37] Laokuldilok, T.; Shoemaker, C.; Jongkaewwattana, S.; Tulyathan, V. Antioxidant and antioxidant activity of several pigmented rice brans. J *Agric. Food Chem.* 2011, 59, 193-199.

[38] Serpen, A.; Pelvan, E.; Alasalvar, C.; Mogol, B. A.; Yavuz, H. T.; Gökmen, V.; Özcan, N.; Özçelik, B. Nutritional and functional characteristics of seven grades of blacktea produced in Turkey. *J. Agric. Food Chem.* 2012, 60, 7682-7689.

[39] Schuster, B.; Herrmann, K. Hydroxybenzoic and hydroxycinnamic acid derivatives in soft fruits. *Phytochem.* 1985, 24, 2761–2764.

[40] Akinwale, T. O. Cashew apple juice: its use in fortifying the nutritional quality of some tropical fruits. *Eur. Food Res. Tech.* 2000, 211, 205-207.

[41] Morton J. Cashew Apple. In: Fruits of warm climates. Julia F. Morton, Miami. 1987, 239–240.

[42] Adou, M.; Kouassi, D. A.; Tetchi, F. A.; Amani N'guessan, G. Phenolic profile of cashew apple juice (*Anacardium occidentale* L.) from Yamoussoukro and Korhogo (Côte d'Ivoire). *J. Appl. Biosci.* 2012, 49, 3331– 3338.

[43] Puoci, F.; Iemma, F.; Spizzirri, U. G.; Restuccia, D.; Pezzi, V.; Sirianni, R.; Manganaro, L.; Curcio, M.; Parisi, O. I.; Cirillo, G.; Picci, N. Antioxidant activity of a mediterranean food product: "fig syrup". *Nutrients* 2011, 3, 317-329.

[44] Osman, H.; Nasarudin, R.; Lee, S. L. Extracts of cocoa (*Theobroma cacao* L.) leaves and their antioxidation potential. *Food Chemistry* 2004, 86, 41-46.

[45] Rakib, E. M.; Chicha, H.; Abouricha, S.; Alaoui, M.; Bouli, A. A.; Hansali, M.; Owen, R. W. Determination of phenolic composition of carob pods grown in different regions of Morocco. *Journal of Natural Products* 2010, 3, 134-140.

[46] Corsi, L.; Avallone, R.; Cosenza, F.; Farina, F.; Baraldi, C.; Baraldi, M. Antiproliferative effects of *Ceratonia Siliqua* L. on mouse hepatocellular carcinoma cell line. *Fitoterapia* 2002, 73, 674–684.

[47] Tokusoglu, O.; Ünal, M. K. Optimized method for simultaneous determination of catechin, gallic acid, and methylxanthine compounds in chocolate using RP-HPLC. *Eur. Food Res. Technol.* 2002, 215, 340–346.

[48] Kelebek, H.; Selli, S.; Canbas, A.; Cabaroglu, T. HPLC determination of organic acids, sugars, phenolic compositions and antioxidant capacity of orange juice and orange wine made from a Turkish cv. Kozan. *Microchem. J.* 2009, 91, 187-192.

[49] Amakura, Y.; Okada, M.; Tsuji, S.; Tonogai, Y. Determination of phenolic acids in fruit juices by isocratic column liquid chromatography. *J. Chromatogr. A* 2000, 891, 183–188.

[50] Soares, M. C.; Ribeiro, E. T.; Kuskoski, E. M.; Gonzaga, L. V.; Lima, A.; Mancini Filho, J.; Fett, R. Composition of phenolic acids content in apple (*Malus sp*) pomace. *Semina: Ciências Agrárias Londrina* 2008, 29, 339-348.

[51] Le Thi, H.; Phuong Lan, P. T.; Chin, D. V.; Kato-Noguchi, H. Allelopathic potential of cucumber (*Cucumis sativus*) on barnyardgrass (*Echinochloa crusgalli*). *Weed. Biol. Manag.* 2008, 8, 129–132.

[52] Rossetto, M.; Lante, A.; Vanzani, P.; Spettoli, P.; Scarpa, M.; Rigo, A. Red chicories as potent scavengers of highly reactive radicals: a study on their phenolic composition and peroxyl radical trapping. *J. Agric. Food Chem.* 2005, 53, 8169-8175.

[53] Ayaz, F. A.; Hayırlıoglu-Ayaz, S.; Alpay-Karaoglu, S.; Grúz, J.; Valentová, K.; Ulrichová, J.; Strnad, M. Phenolic acid contents of kale (Brassica oleraceae L. var. *acephala* DC.) extracts and their antioxidant and antibacterial activities. *Food Chem.* 2008, 107, 19–25.

[54] Molyneux, R. J.; Mahiney, N.; Kim, J. H.; Campbell, B. C. In Tree nuts: composition, phytochemicals and health effects; Shahidi, F. and Alasalvar, C.; Ed.; Health effectsand aflatoxigenic activity of phytochemicals in tree nuts; CRC Press: *Boca Raton*, FL 2008; Vol. 3, pp 95-107.

[55] Rangkadilok, N.; Worasuttayangkurn, L.; Bennett, R. N.; Satayavivad, J. Identification and quantification of polyphenolic compounds in longan (*Euphoria longana* Lam.) fruit. http://www. aseanfood.info/Articles/ 13005839.pdf.

[56] Bhargava, A.; Shukla, S.; Ohri, D. Genetic variability and heritability of selected traits during different cuttings of vegetable *Chenopodium*. *Indian J. Gen. Plant Breed* 2003, 63, 359–360.

[57] Hafeez Laghari, A.; Memon, S.; Nelofar, A.; Khan, K. M.; Yasmin, A. Determination of free phenolic acids and antioxidant activity of methanolic extracts obtained from fruits and leaves of *Chenopodium album*. *Food Chemistry* 2011, 126, 1850–1855.

[58] Lòpez, F.; Pescador, P.; Güell, C.; Morales, M. L.; García-Parrilla, M. C.; Troncoso, A. M. Industrial vinegar clarification by cross-flow microfiltration: effect on colour and polyphenol content. http://www. aseanfood.info/Articles/11013310.pdf.

[59] Prakash, D.; Singh, B. N.; Upadhyay, G. Antioxidant and free radical scavenging activities of phenols from onion (Allium cepa). http://www. aseanfood.info/articles/11018103.pdf.

[60] Paranthaman, R.; Praveen Kumar, P.; Kumaravel, S. GC-MS analysis of phytochemicals and simultaneous determination of flavonoids in *Amaranthus caudatus* (Sirukeerai) by RP-HPLC. *J. Anal. Bioanal. Techniques* 2012, 3, 147. doi:10.4172/2155-9872.1000147.

[61] Jelínek, L.; Sneberger, M.; Karabín, N.; Dostálek, P. Comparison of Czech hop cultivars based on their contents of secondary metabolites. Czech J Food Sci 2010, 28, 309–316.

[62] 62. Manach, C.; Scalbert, A.; Morand, C.; Rémésy, C.; Jiménez, L. Polyphenols: food sources and bioavailability. *Am. J. Clin. Nutr.* 2004, 79, 727-747.

[63] Hallmann, E.; Rembialkowska, E. Characterisation of antioxidant compounds in sweet bell pepper (*Capsicum annuun* L.) under organic and conventional growing systems. *J. Sci. Food Agric.* 2012, 92, 2409-2415.

[64] Salem, N.; Msaada, K.; Hamdaoui, G.; Limam, F.; Marzouk, B. Variation in phenolic composition and antioxidant activity during flower development of safflower (*Carthamus tinctorius* L.). *J. Agric. Food Chem.* 2011, 59, 4455-4463.

[65] FAO, Food and Agriculture Organization of the United Nations. Statistical Database. 2007. FAO/United Nations, Rome. http://faostat. fao.org/.

[66] Robles-Sánchez, R. M.; Rojas-Grau□b, M A.; Odriozola-Serrano, I.; González-Aguilara, G. A.; Martín-Belloso, O. Effect of minimal processing on bioactive compounds and antioxidant activity of fresh-cut 'Kent' mango (*Mangifera indica* L.). *Postharvest Biol. Technol.* 2009, 51, 384-390.

[67] Lounds-Singleton, A. J. Influence of thermal postharvest stress on mango (*Magnifera indica*) polyphenolics during ripening. Thesis presented to the Graduate School of the University of Florida in partial fulfillment of the requirements for the degree of Master of Science 2003. http://etd.fcla.edu /UF/UFE0002724/lounds_a.pdf.

[68] Zuo, Y.; Chen, H., Deng, Y. Simultaneous determination of catechins, caffeine and gallic acid in green, Oolong, black and pu-erh tea using HPLC with a photo diode array detector. *Talanta* 2002, 57, 307-316.

[69] Karamac, M.; Kosinska, A., Pegg, R. B. Content of gallic acid in selected plant extracts. *Pol. J. Food Nutr. Sci.* 2006, 15/56, 55-58.

[70] Ariffin, F.; Chew, S. H.; Bhupinder, K.; Karim, A. A. Antioxidant capacity and phenolic composition of fermented *Centella asiatica* herbal teas. *J. Sci. Food Agric.* 2011, 91, 2731-2739.
[71] Alasalvar, C.; Amarowicz, R. Effects o roasting on taste-active compounds of Turkish hazelnut varieties (*Corylus avellana* L.). *J. Agric. Food Chem.* 2010, 58, 8674-8679.
[72] Pelvan, E.; Alasalvar, C.; Uzman, S. Effects of roasting on the antioxidant status and phenolic proiles of commercial Turkish hazelnut varieties (*Corylus avellana* L.). *J. Agric. Food Chem.* 2012, 60, 1218-1223.
[73] Neves, A. C.; Spranger, M. I.; Zhao, Y.; Leandro, M. C.; Sun, B. Effect of addition of commercial grape seed tannins on phenolic composition, chroatic characterstics, and antioxidant activity of red wine. *J. Agric. Food Chem.* 2010, 58, 11775-11782.
[74] Ross, C. F.; Hoye, C.; Fernandez-Plotka, V. C. Influence of heating on the polyphenolic content and antioxidant activity of grape seed flour. *J. Food Sci.* 2011, 76, C884-C890.
[75] Fabian, C.; Thi, N. Y., T.; Kasim, N. S.; Ju, Y.H. Release of phenolic acids from defatted rice bran by supercritical water treatment. *J. Sci. Food Agric.* 2010, 90, 2576-2581.
[76] Bendini, A.; Cerretani, L.; Carrasco-Pancorbo, A.; Gómez-Caravaca, A. M.; Segura-Carretero, A.; Fernández-Gutiérrez, A. Phenolic molecules in virgin olive oils: A survey of their sensory properties, health effects, antioxidant activity and analytical methods—An overview of the last decade. *Molecules* 2007, 12, 1679–1719.
[77] Veillet, S.; Tomao, V.; Bornard, I.; Ruiz, K.; Chemat, F. Chemical changes in virgin olive oils as a function of crushing systems: Stone mill and hammer crusher. *C. R. Chim.* 2009, 12, 895–904.
[78] Smaoui, S.; Hlima, H. B.; Jarraya, R.; Kamoun, N. G.; Ellouze, R.; Damak, M. Physico-chemical characterization and analysis of various tunisian olive oil varieties: minor compounds and fatty acids. *J. Chem. Soc. Pak.* 2012, 34, 491-498.
[79] Aludatt, M. H.; Alli, I.; Ereifej, K.; Alhamad, M.; Al-Tawaha, A. R.; Rababah, T. Optimisation, characterisation and quantification of phenolic compounds in olive cake. *Food Chem.* 2010, 123, 117–122.
[80] Inglett, G. E.; Chen, D.; Berhow, M. Influence of Jet-Cooking Corn Bran on Its Antioxidant Activities, Phenolic Contents and Viscoelastic Properties. *Food Nutr. Sci.* 2011, 2, 521-529.
[81] Council Regulation (EEC) N° 1576/89 of 29 May 1989 laying down general rules on the definition, description and presentation of spirit drinks. *Official Journal of European Communities L* 160, 1-17.
[82] Giménez, R.; Villalón, m.; López, H.; Navarro, M.; Cabrera, C.; Olalla, M.; JQuesadaa, J. J.; López, M. C. Determination of gallic acid in commercial brandies using high performance liquid chromatography. *Cienc. Tecnol. Aliment.* 2000, 3, 13-20.
[83] Ng, L. K. Lafontaine, P.; Harnois, J. Gas chromatographic–mass spectrometric analysis of acids and phenols in distilled alcohol beverages. Application of anion-exchange disk extraction combined with in-vial elution and silylation. *J. Chromatogr. A* 2000, 873, 29-38.
[84] Vicente, C. D.; De Abreu, F. C.; Goulart, M. O. F.; De Vasconcelos, J. N. Phenolic constituents, furfuraldehyde and antioxidant capacity of sugar cane spirit aged in woods casks. *Am. J. Food Technol.* 2011, 6, 631-646.

In: Handbook on Gallic Acid
Editors: M. A. Thompson and P. B. Collins
ISBN: 978-1-62618-921-8

Chapter 9

GALLIC ACID: OCCURRENCES, ANTIOXIDANT ACTIVITY AND HEALTH IMPLICATIONS

Miriam Rejane Bonilla Lemos[1*] and Rui Carlos Zambiazi[2]
[1]Health Sciences Postgraduation Program, Health Sciences Faculty, Campus Universitário Darcy Ribeiro, Universidade de Brasília, Brasília, DF, Brazil
[2]Posgraduate Program of Science and Agro industrial Technology, Laboratory of Chromatography. Center of Chemical, Farmaceutical and Food Science, Universidade Federal de Pelotas (UFPel), RS, Brazil

ABSTRACT

Studies confirm that oxidative stress may be involved in the etiology of several chronic diseases, episodes genetic, metabolic and neurodegenerative processes such as cardiovascular, diabetes, cancer, Alzheimer's, among others, also working in the acceleration of aging and apoptosis. However it is known that consumption of fruit and vegetables have shown a positive correlation reducing the chances of developing pathologies.

The aim of this study is to relate the bioactive compounds contained in foods derived from plants, among these, the almond baru [Dipteryx alata Vog.], A fruit of savannah, the Cerrado and its relationship to antioxidant activity as a prevention of chronic degenerative diseases and their complications. Almonds were obtained in local trade Brasilia-DF, from three regions of the Cerrado (MT, MG and GO), which were previously selected and randomly assigned into two groups: raw almonds and almond raw film without film. Part of the kernels of these two groups was packaged in transparent polyethylene and stored at -80ºC.

The remainder of almonds with and without film was subjected to roasting, which was held in an oven without air circulation at 150°C/45 minutes. The raw almonds and toast were crushed and analyzed in triplicate.

Thus, knowing that the cellular components are not fully protected by endogenous mechanisms is well established that antioxidants obtained from nature, are indispensable

* Corresponding author: Health Sciences Postgraduation Program, Health Sciences Faculty; Campus Universitário Darcy Ribeiro, Universidade de Brasília, Brasília, DF, Brazil. P.O. Box 70910900. e-mail: rebonilla@gmail.com.

for the proper defense against oxidation and thus begin to play an important role in maintaining health, as These, and contain essential nutrients and micronutrients, contribute to the compounds present in the leaves, fruits, films, and seeds, from its chemical structure, promote action varied in the body. Certain phenolic substances such as flavonoids and phenolic acids, are considerably more potent antioxidant than vitamin C and vitamin E. Gallic acid in a group of phenolic compounds simpler and is inserted in the group of hydroxybenzoic acids or non-flavonoids, being commonly in grapes, strawberries, nuts, cereals and leguminous seeds that act as potential antioxidants with proven preventive action degenerative processes in cardiovascular, and activity against some mutagenic chemicals, striking chemopreventive action, antiviral and antibacterial.

Recent studies show that daily consumption of almond baru [Dipteryx alata Vog.], a seed of edible oilseed feature, reduces oxidative stress induced in rats.

It was confirmed these kernels, the presence of substances with a high content of bioactive compounds, wherein, of the eight phenolic compounds identified by liquid chromatography (HPLC), at concentrations ranging between 67.7 and 224.0 mg/100 acid, gallic was the predominant compound and found that a thermal process is not capable of reducing the content of phenol. Furthermore, the antioxidant activity involves a set of intrinsic factors that protects the body against harmful damage, being able to prevent and heal pre-existing diseases.

Keywords: Potential antioxidant, bioactive compounds, almond baru, gallic acid

1. Introduction

Studies in the context of food and orthomolecular medicine show that many compounds present in vegetables are able to prevent and even reduce the development of diseases related to oxidative stress. Consumption of fruit and vegetables shows a positive correlation in preventing several pathologies. The mechanism of these bioactive compounds is related to its antioxidant activity which prevents or even inhibits the outbreak of diseases caused by oxidative reactions [1, 2, 4].

The antioxidant activities of these compounds include the action as free radicals scavengers or by chelating metals [3, 5, 6,].

Gallic acid (3,4,5-trihydroxybenzoic) is a phenolic compound widely distributed in nature, especially in the husks and nuts, almonds [7, 8, 9], wine, tea, grapes and other plant derivatives.

Belonging to the group phenols simpler hydroxybenzoic acid derivatives such as p-hydroxybenzoic, protocatéico, vanillic and syringic, a phenolic compound is flavonoid that has undergone no substitutions at positions *Meta* and *Para* to its chemical structure, and its antioxidant activity associated with the position of the hydroxyl grouping and also with proximity-CO_2H group with the phenyl group [3, 4, 5]. Therefore, the closer the group is the phenyl group, the greater the antioxidant capacity of the hydroxyl group in the *Meta* position [1, 10, 11, 12].

This gives the antiradical activity gallic acid, ability to stabilize free radicals responsible for participating in cellular degenerative processes that cause diseases [6, 12, 13].

The stress condition is a matter of great magnitude involving the etiology of several chronic diseases, episodes genetic, metabolic and neurodegenerative disorders, including, cardiovascular, diabetes, cancer, Alzheimer's processes, among others.

The activity, even in cases of accelerated aging and apoptosis due to disruption of homeostasis, may be determinant of the decline in life expectancy of individuals of the same species.

Based on this understanding, the purpose of this chapter is to compare the triggering episodes of pathologies, which act as determinants of low immunity and the body's protective agents, which benefit the whole cell, in actions against diseases, with emphasis on gallic acid, a phenolic compound that is recognized with antiradical action and has in its structure, characteristics that make it stand out among the other phenolic compounds [2, 3, 6, 14].

2. THE EVOLUTION PRINCIPLES AND SPECIES ADAPTATION TO ADVERSITIES

From the historical point of view, humanity records the contraction of the humans by the nature and science domain and of the other and to keep balance in promoting life. This is how the relationship between lifestyle, diet and adaptability of human and animal species to the adversities shows up as the basis for the balance between internal (cells) and external (environment) medium.

In the study of epidemiology in public health this relationship has been considered as a factor in triggering and understanding diseases. Thus, it is possible to understand how the processes change in the incidence or prevalence of diseases, as well as the main causes of death in the light of this relationship during the evolution of species [13, 14, 70].

From this perspective, epidemiology engages with the disease process in human collectivities, seeking "specific measures to prevent *control or eradicate diseases and providing indicators which support the planning, management and evaluation of health programs*" [15].

From the aforementioned relationship of the imbalances is evidenced the onset of epidemiological interaction between agents, the process of individual or collective desease (Figure 1).

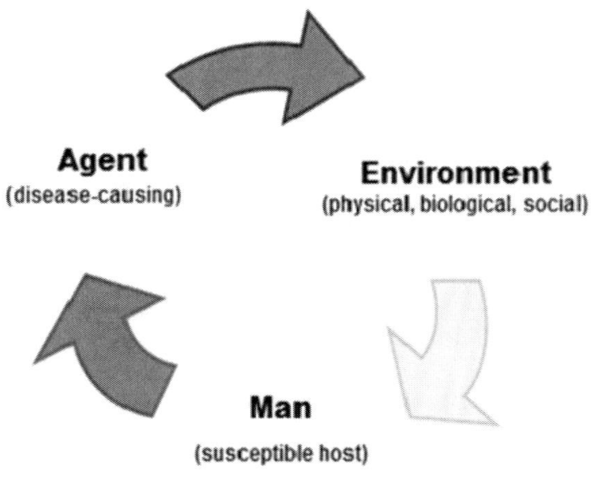

Figure 1. Diagram of interaction between epidemiological agents [15].

As can be seen by the diagram, once triggered the interaction alluded it will induce the epidemiological transition. It is evident therefore that there are evolutionary factors that are crucial for understanding the phenomena that staked the occurrence of chronic diseases and related free radicals (FR). The transition of diseases by which *Homo sapiens* lived shows important aspects that can be understood the origin of chronic diseases in contemporary society. These elements allow us to identify the factors predisposing to disease (stressors) and their preventive factors (antioxidants) [16].

From these factors, the evolutionary process and its effects on the today diseases become unquestionable. Initially, it was during the paleontolitic period that occurred contact of humans (hominids) with unknown phenomena, leading to stressful reactions. It became clear a lifestyle (oxidants) and a plant-based diet and its constituents (antioxidants). The gallactotannins, which Gallic acid is part of, are found in abundance in bark, roots, buds, seeds, leaves and fruits [13].

At that time the tribes of nomadic hominids were composed predominantly by hunters and pickers, which presenting in their evolutionary history, a low life expectancy since there was no guarantee of obtaining food energy and nutrients. The food was dependent on the nature, oscillating between periods with plenty of food and others with total shortages. Such instability often generated hunger, one of the elements of evidence that are favorable to oxidative stress [16].

Because they were nomadic, these individuals did not grow nor produce their food, and due to their physical activity because they wordhunters not reflected in these individuals' sedentary habits and the energy expenditure depended specifically by the abundance or scarcity of food. The diet, while plentiful, was based on hunting and fishing products, eggs, seafood and vegetables also obtained from nature. The grains and cereals, in this period, were not yet known and food preparation was very rudimentary or even crude, which favored the ingestion of dietary compounds in nature, with full utilization of the natural [16, 17].

The plant-based diet and its constituents (roots, shoots, stems, bark, films, fruits and leaves), were rich in bioactive compounds, including gallic acid.

From a physiological standpoint [18], the alternation of conditions for obtaining food, fluctuations in energy stocks and constant physical activity, caused changes in life expectancy and vulnerability to disease, marking a kind of natural selection on genes involved in the regulation oxidative metabolism that was decisive. The inaccuracy between long periods of hunger and energy expenditure to ensure survival could only be supported by those individuals with greater ability to store energy, spending it so restrained, thus preventing the extinction of the human species [18, 19, 13].

In the Neolithic period, subsequent to Paleontolític, the Agricultural Revolution took place, featuring the man domain over food production and the establishment of settled life, to the detriment of the nomadic lifestyle. The development of primary food production led to changes in the pattern of diseases still unknown by the hominids that once inhabited the land. The sedentary lifestyle and agricultural production led abruptly to the accumulation of surplus food, totally opposite to the previous period, driving major changes in the lives of human beings [16, 17, 18, 19, 13].

With the rapid increase in population settled sedentary lifestyle and changing habits in man. The investment in the domestication of animal species (favoring zoonoses) promoted the extensive breakdown of the ecological balance (due to the occupation of territories for monocultures, with insertion of exotic species), and promoted the emergence of social inequality with increased competitiveness for food and consequent maintenance of life [19, 20]. The proximity of water resources, the disposal of human waste and animal health without education projects have created a favorable environment for disease vectors, which were determinants individuals become susceptible to various diseases.

The adaptation of species to adversity was also verified with subsistence agriculture, increasing nutritional deficiencies that had major implications for the health of populations, due to a decrease in the variety of foods available to all individuals, due to the high population density and by reducing access to food and its nutrients, resulting in vulnerability to infectious diseases [17].

However, individuals overcame these adversities, maintaining the survival and reproductive patterns, gradually reaching an increase in life expectancy, while that from changing eating habits, there was an imbalance of energy metabolism, predisposing to NCD (non-communicable Chronic Diseases) in populations. The transition from a change of scenery between instability of dieters to an abundance of foods rich in energy and less need for energy expenditure certainly caused a significant metabolic impact and stressor.

Over the years and the advent of modernity, a new evolutionary period settled, the Industrial Revolution, which was the period where appeared inumerous cities around the factories, with the majority of them, without any planning, culminating in many social and environmental problems, that reflected in public health [21, 22, 23]. There was a lower incidence of disease and as a consequence, the decrease in infant mortality rates and fertility (birth control), factors that contributed to the increase in life expectancy, where individuals were living longer, causing changes in the epidemiological profile due to the incidence and prevalence of degenerative diseases characteristic of individuals with advanced age. This period is linked to scientific and technological development, enabling improved survival rates of human population, giving individuals greater control and prevention of infectious diseases [16].

So, finally settled the contemporary epidemiological, which happened to modernization and urbanization degree, both arising from industrialization, observing significant epidemiological changes in populations, among them, changes in dietary habits and life style changes, with the increasing incidence and prevalence of chronic diseases.

The consumption of foods rich in simple carbohydrates, easily absorbed, saturated fats, spices and seasonings to salt-based, among other elements, were introduced into the diet at the expense of those rich in fiber and vitamins and bioactive, favoring the higher incidence of chronic diseases, especially cancer and cardiovascular processes and their complications [17, 18].

This period was characterized by the resurgence of infectious diseases even more aggressive than those already known. Many of which are resistant to antibiotics (bacterial resistance caused by mutations) with high potential epidemiologic and pandemic risks with global trends in a population weakened in their immune defenses, determined by constant stress to which they are subjected on a daily basis due to endogenous and accelerators exogenous reactive oxidizing species [16, 23].

However, the advancement of science was not enough to prevent diseases such as Hemorrhagic Fever, Hantavirus infection and Human Immunodeficiency virus (HIV/AIDS), atypical pneumonias and chronic diseases evolve. The man of this period tried to adapt to overcome adversity although, as ever, more susceptible to the action of reactive species and stressful, man now changed markedly the balance between your lifestyle and diet [18, 20, 21].

The behavioral model of today expresses the contrast between the man of the past, with constant activity and sedentary gift. The latter is exposed to an exacerbation of reactive species and stress due to fewer hours of sleep in, the continuous stress, high calorie diet (rapid intestinal absorption and undiversified), smoking, alcohol, drugs and medicines indiscriminately. All this coupled with the constant attacks on the environment because an imbalance between internal (cellular) and external (environmental) medium associated with little or no dietary intake of antioxidant-rich compounds.

Facing this reality, man's struggle for survival and their constant search for food is historical. The man has since used its power plants that have become a source of prophylactic agents preventing and treating the disease [24].

3. The Oxidative Stress and Chronic Diseases

Understanding the role of food in maintaining health and curing diseases had the origins in ancient cultures that was registered by historical documents, such as in the Chinese scriptures, the Indian ayurveda (ancient Indian medicine) and by Greek with traditional knowledge of Hippocrates (at 434-460 a.C) that was considered the father of medicine. Hippocrates contextualized the relationship between the use of appropriate foods for health preservation and its therapeutic benefits [25, 24, 26] by the statement: "The diseases originate from nature and could be cured by it, when it strikes a balance between the environment, food intake and spirit". Since then, the search for compounds that control the action of stressors of the body has been steady. Hippocrates also highlighted the close relationship between food and its ability to control and prevent disease. In this context, one can infer that the Paleontolitic period the individuals were healthy because they were active and not sedentary and they eat mostly raw vegetables and their constituents: roots, bark, buds, fruit and fresh seeds, fresh. Therefore, it was preserved the antioxidant potential of these structures, due to not cooking, what guaranteed the properties against free radicals [16, 17].

Phenols in general are highly susceptible to enzymatic and no enzymatic oxidation because of the presence of double bonds. Recent research has shown the stability of gallic acid during heating which one shown to be thermo stable under conditions of heat exposure, being able to maintain its antioxidant capacity [27, 13].

These factors refer to a profound reflection on the intrinsic components present in fruits and vegetables, especially phenolic compounds content, with possible positive association between diets rich in fruits and vegetables and reduced risk of diseases related to the content of bioactive compounds present and potential antiradical with varied mechanisms of action.

Therefore, the use of plant extracts for therapeutic is already an inevitable reality. With the advances in research, this contribution will be even more in control and by the knowledge of the path physiology of chronic degenerative diseases [28, 29]. From this perspective it is postulated that oxidative stress as a causative agent of chronic diseases. The health state is characterized by the process of balance or homeostasis, which is a condition in which the body's internal environment remains within certain physiologic limits.

When the principle of homeostasis is interrupted, the stress is established, which is characterized by an imbalance in the body's internal environment, either by chemical, physical or biological means [32].

The fact is that there is evidence that the reactive species, determined by free radicals may be involved in the etiology of more than fifty types of degenerative diseases, including, cancer, diabetes, Alzheimer's, osteoporosis, Parkinson's disease, atherosclerosis, chronic inflammation, heart disease and rheumatoid arthritis. Oxidative stress has also been considered a triggering factor of premature aging, muscle degeneration caused by intense physical activity, anxiety, obesity and acute myocardial infarction, wich is responsible for high morbidity and mortality of the population [31, 32, 33]. The chronic diseases are responsible for 59% of deaths worldwide and the projection for 2020 is that these affects more than three quarters of all deaths that will occur worldwide and that 71% of these deaths will be caused by ischemic heart disease, 75% of deaths from brain blood vessels accident (CVA) and 70% of deaths from diabetes, especially those in developing countries [16, 31, 13]. The Oxidative Stress originates from reactive processes that manifest in the individual in the form of emotional stress, chemical, physical or infectious, altering the cell structure and therefore increasing the risk of diseases if unchecked by antioxidants. However, the generation of free radicals is not always harmful to the body, instead, is necessary in many biological processes such as cell signaling, inflammation, muscle contraction and immune system [34, 35].

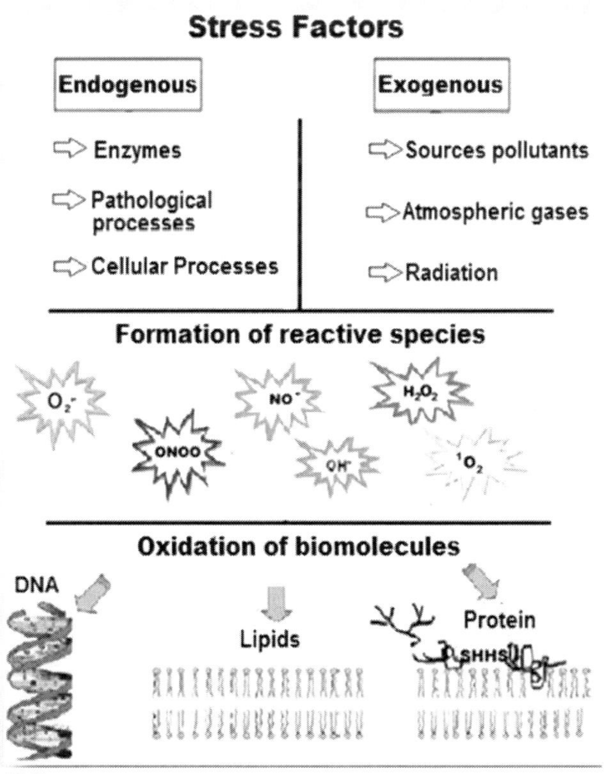

Source: Adapted from [36].

Figure 2. Diagram of cellular damage from the action of reactive species that cause stress.

When cells are attacked by some stressor, they are stimulated to produce more radicals, fighting the offending agent. However, the aggravating would be generated when the total levels of free radicals becomes greater than the defense capability, thus causing significant cellular damage. When there is an overproduction of reactive species (exacerbation) or inadequate production or absorption of antioxidants, this balance is disrupted, occurring damage to vital components (biological systems), such as in the DNA, proteins and lipids.

When the overproduction of reactive species is excessive and the antioxidant defense system insufficient, the broken balance will generate oxidative stress, which is the main causative or aggravating illnesses.

3.1. Systems of the Body's Defense

The continuous production of free radicals (FR) during the metabolic processes lead cells to develop a defense mechanism, which came to control and reduce the induction of cellular damage [16]. These defenses shape the evolution of the response to the inevitable existence of molecules in living conditions stressful aerobic, working at different levels in protecting the body by preventing the loss of cellular integrity.

These defense systems are classified as endogenous (enzymes) and (exogenous (non-enzymic) mechanisms. The non-enzymic refers to the external components, which are acquired through diet rich in bioactive compounds that are capable of protecting the body from oxidative damage [16, 36, 37].

3.1.1. Enzyme Defense System

The cellular components are not fully protected by endogenous mechanisms, and it is well established that antioxidants obtained from food are indispensable for the proper defense against oxidation and thus begin to play an important role in maintaining health [3. 38, 1].

The vegetables contains essential nutrients and micronutrients with a variety of chemical structure, which is dependent of the type and degree of maturity of the plant, as well as climatic conditions, soil management and cultivation system [38, 39].

The phytochemical terminology refers to a very large group of compounds (Figure 3), which are produced and accumulates in plants. Many of these compounds present biological activities, although some may exhibit toxic effects when ingested in high doses.

The interest in these substances comes from epidemiological studies which indicate that a diet rich in fruits and vegetables can bring health benefits, contributing as preventive agents for developing several chronic degenerative diseases which is reinforced by the positive correlation between total antioxidant activity and their content found in these experimental foods [3, 4, 38].

Exogenous antioxidants in general possess the ability to scavenge free radicals. Among these, there are groups that are considerably more potent than other antioxidants, which are phenolic compounds, presenting a diverse mechanism of action. The antioxidant activity depends on their chemical structure and may be determined by the action of the molecule as a reducing agent (rate of inactivation of free radical reactivity with other antioxidants and metal chelating potential).

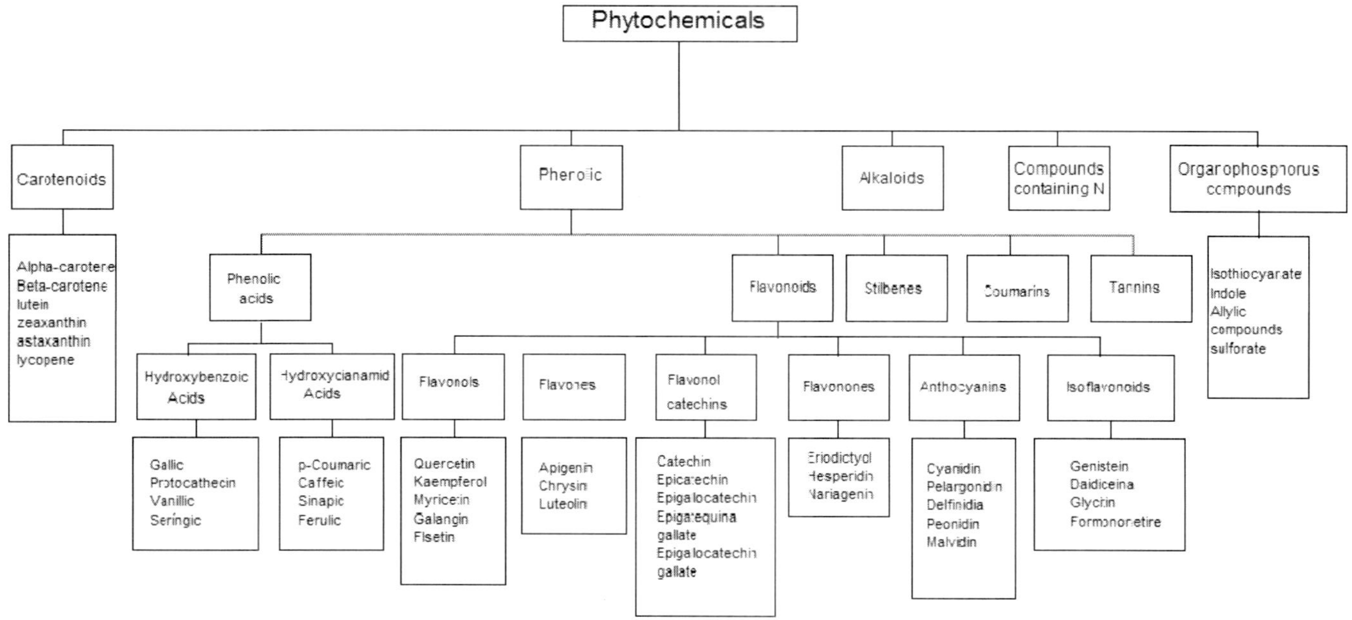

Source: Adapted from [40].

Figure 3. Classification and distribution of the main Phytochemicals in nature.

3.1.2. Phenolic Compounds

Phenolic compounds are considered important natural antioxidants and are widely distributed in nature [41, 43]. These compounds are products of secondary metabolism of plants, that are involved on environmental stress conditions, serving as active in the defense again pathogens, parasites and predators.

Among these, we highlight the phenolic acids (hydroxybenzoic acids) that are non-flavonoid compounds represented by p-hydroxybenzoic, protocateic, vanillic, gallic and syringic acids (figure 4), wchich are commonly present in grapes, strawberries, walnuts, almonds, cereals and seeds [41, 4, 40, 12].

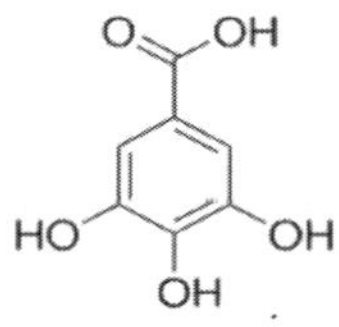

Figure 4. Structure Chemical gallic acid [41, 4, 40, 12].

3.1.2.1. Gallic Acid and its N- Alkyl esters

Gallic acid (3, 4, 5-trihydroxybenzoic) is widely distributed in nature as a major phenolic compounds present in black and green teas.

It is present in free form and as part of tannins. Among its pharmacological properties, can be cited the antiradical action, promoters prevention and protecting the body against various etiologies, acting as a potent anticancer, antimutagenic antifungal, antibacterial, antiviral, anti-inflammatory, trypanocidal, among others [41, 43, 44, 45].

Derivatives thereof, n-alkyl esters, are used in the food industry, as examples of propyl, octyl and dodecyl gallate. In the pharmaceutical industry, the decyl, undecyl, dodecyl and tetradecila gallate, are promising prophylactic and chemotherapeutic action in several pathologies, including as antiparasitic [46, 47, 13].

Studies have shown that octyl gallate has the ability to accelerate the killing of cells infected by herpes and also against two types of RNA viruses (vesicular stomatitis virus and polio virus).

Phenolic compounds with the galoil grouping also present in gallates, have positive evidenced action against HIV integrase-type 1. It has been shown in experiments that the gallic acid is able to inhibit the viral agent when bound to the CD4 receptor, interfering with symptoms and improving the framework of patients. Under the immunological point of view is likely that the biological activity of these compounds is conditioned to the size of the alkyl chain [56, 53, 47].

3.1.2.2. Tannins

Tannins are polyphenolic compounds whose concentration in plants depends on regional factors, collection time, climate and soil. They are present in the diet and some undesirable effects have been attributed to them, probably due to their capacity to bind to proteins inactivating them, and also with some essential dietary minerals, making them unavailable or partially available for absorption (antinutritional factors).

Nevertheless, these polyphenols exert important protective effects on degenerative diseases [48, 49, 40]. The tannins are high molecular weight polymers that are grouped according to their structural and chemical properties (Figure 5).

These compounds may be subdivided into two well-defined classes: hydrolysable tannins, which comprise polymers of ellagic acid and gallic acid, that are present in fruits, nuts, almonds, grapes, seeds, among others; and condensed tannins, which comprise the polymers of catechins, present in fruits, vegetables, nuts, seeds and flowers [41, 50, 51, 52, 48].

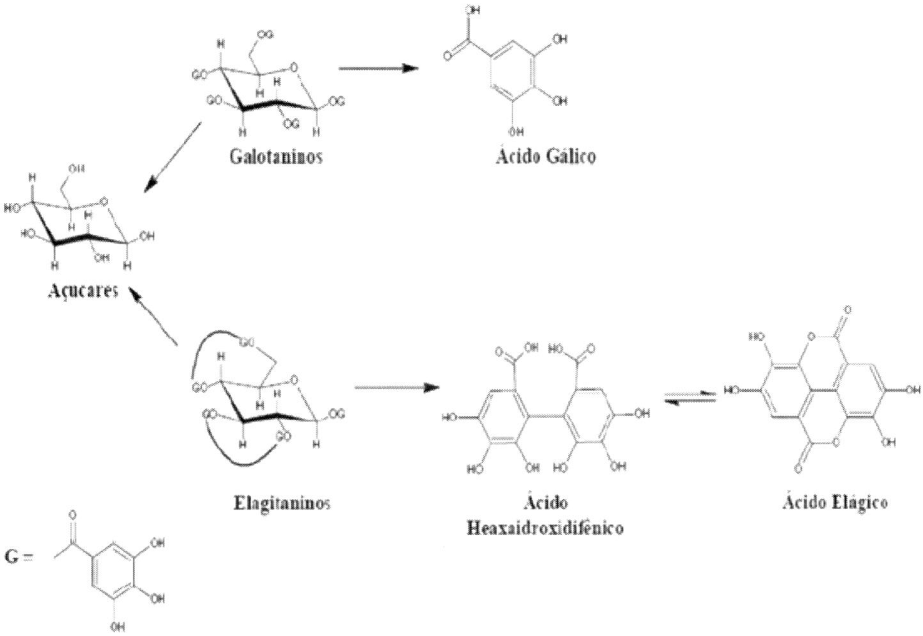

Figure 5. Chemical structure of tannins [48].

Hydrolysable Tannins

The hydrolysable tannins are the largest group of tannins known and can be found in foods such as hazelnuts, walnuts, almonds, pomegranates, persimmons peach, berries and wine. They are formed from gallic acid metabolism joining with sugars (mainly aglicana), represented in (Figure 6).

The tannins are subdivided into hydrolyzable gallotannins, which produce gallic acid and ellagitannins. Of these, the hydrolyzable esters of gallotannins are found abundantly in all parts of plants (roots, flowers, fruits, leaves, husks, skins and stem), that just as ellagitannins, are used as medicinal herbs in the manufacture of food and beverages. In the food industry, contribute to the astringent taste of food and drink, especially in reds, greens and fruit teas. In nature protect plants against herbivores and pathogens attack, and in the body they are potentially active against free radicals [52, 48, 50, 55].

Ellagic acid is a dimmer condensation of gallic acid, and like phenol possesses some properties of this compound. Studies with ellagic acid and ellagitannins demonstrated inhibitory properties against the replication of the HIV / AIDS virus and human herpes virus (HPV), and inhibitory activity similar to that of gallic acid. Therefore, together with gallic acid act as potent protective degenerative processes in cardiovascular desease, antimicrobial, antimutagenic and other therapeutic actions mentioned above [53, 54, 56].

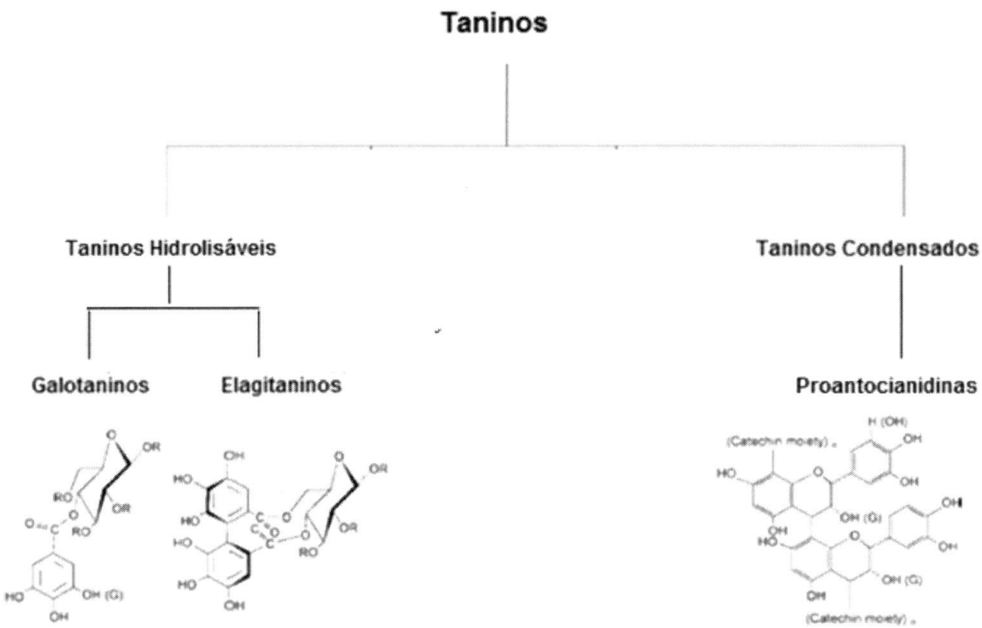

Figure 6. Structure of hydrolysable tannins [50].

In traditional medicine, certain plants rich in tannins are used as astringent the gut and also in the treatment of skin abrasions.In the treatment of burns, tissue proteins are precipitated when exposed to tannins and form a protective and slightly antiseptic coating, and may occur under tissue regeneration. Tannin-rich plants, particularly rich in gallic acid, are also employed in traditional medicine for treatment of several diseases, including diarrhea, hypertension, rheumatism, bleeding, wounds, burns, gastrointestinal disorders, urinary disorders and inflammatory processes in general [13, 46, 47].

Then, from a knowledge of how these radical species are able to act adversely on quality of life and the way in which the body is able to respond to these effects through antioxidant mechanisms, it can be understood because one of the major factors influencing the health status of individuals is the eating habit, and this can influence the present state of health and also determine the life expectancy [13].

4. Brazilian Flora With Medicinal Potential

Brazil has a immense and diversified flora still unexplored, that comprise a rich source of natural resources including native fruits with peculiar characteristics that have therapeutic properties, offering therefore, beyond their exotic flavors, compounds with medicinal properties [2].

However, the significant potential due to plant diversity, is also unknown to the majority of individuals. Therefore, by the intense devastation there is a risk of irreparable loss of promising materials, including a vast botanical material [56, 57, 55] that could contain substances that hold active ingredients against the many ills, including cancer, Parkinson's, diabetes, cardiopulmonary disorders and their consequences and even AIDS [55].

Six major biomes in Brasil are the richtest by the diversity: Amazon Atlantic Forest, Caatinga, Cerrado, Pantanal and Pampa, making Brazil, a country with a greater diversity of flora and fauna on the planet, with a unique diversity in the world [57].

4.1. The Cerrado Biome

The Cerrado biome occupais a large territory in the central Brazilian plateau, with borders in the Amazon, Caatinga to the northeast pantanal to southeast and the Atlantic ocean to east. It contains an area of 2.039.386km², equivalent to all of Western Europe, that extends to the federal District and for several Brazilian states, including: Minas Gerais, Goiás, Tocantins, Bahia, Maranhão, Mato Grosso, Mato Grosso do Sul and Piaui [57].

The Cerrado is evident in a large territory with a privileged geographical position, climate and diversity among native species [58]. This bioma is originated from the three main sources of the nation's river basins: Amazon, Paraná and São Francisco; therefore, havingan extremely strategic location, not only for the biodiversity, but also by the possibility of preservation of natural water resources [59, 9, 60, 61].

The typical vegetation of the Cerrado include trees and shrubs, such as Buriti, Paineira Angico, Jatoba, Ipe and Barueiro, that are especially resistant to dry weather. The dry season extends from April to September, with rain periods from October to March, with an average annual of 1,500 mm rainfall, and mild temperatures throughout the year, ranging between 22 °C to 27 °C [62, 57].

Many species of native flora of the Cerrado have economic potential (ornamental, edible, medicinal, textile, wood and others). The exploitation of genetic resources of plants, fruits and natural products targeted to medical activities, largely related to the extensive collection and extraction of wild material. Despite the considerable volume of exportation of various medicinal species in raw form or byproducts, very few of these rise to the level of being cultivated, even on a small scale.

The native fruits have a prominent place in the Cerrado ecosystem, because many of them are sold and consumed "in natura" or benefited by homemade fabrications in several products.

For the intrinsic characteristics and economic medicinal potential, stand out in this scenario some native fruits, including the guariroba (Syagrus oleracea), gabiroba (Camponesia *cambessedeana*), pequi (Caryocar *brasiliense*), cagaita (Eugenia *dysenterica*), mangaba (Hancornia *speciosa*), araticum (Annona *crassiflora*), de-inga-meter (Inga sp), cashew-of-field or cajuí (Anacardium *humile*), Curriola (Pouteria *ramiflora*) and baru (Dipteryx *alata* Vog.) [10, 14, 63, 62].

4.2. The Baru

Scientific name: *Dipteryx* alata Vog.

Family: *Leguminosae* (Fabaceae)

Division: Magnoliophyta (Angiospermae)

Class: Magnoliopdida (Dicotiledonae)

Order: Rosales

Popular names: baru, barujó chestnut rail, coco bean, tonka bean-the-big leaf, cumarurana, cumaru-true, tonka bean, purple, cumbaru, cumbary, emburena-mad, coco beans, meriparagé, dick –cumaru.

Occurrence: Cerrado, Mata Seca and Savana.

Distribution: Amazonas, Bahia, Distrito Federal, Goiás, Tocantins, Maranhão, Piauí, Mato Grosso, Mato Grosso do Sul, Minas Gerais, Sao Paulo

Flowering: November to May.

Native fruit of the savannah, the baru (Dipteryx alata Vog.) (Fig.7) is among the 10 most promising tree for cultivation in areas to be reclaimed, what can bring benefits to the conservation and preservation of the environment species, which is also threatened by the devastation that has been suffering the Cerrado [61].

Figura 7. Baru (*Dipteryx alata* Vog.) [64].

The Baru produces a drup fruit with intense fruiting, slightly flattened and brown, with a single seed or almond with ovoid shape, and with average weigh of 25g (with 30% of pulp, 65% of endocarp and 5% of seed. The size of these fruits can vary greatly between regions, depending on the soil conditions, water and plant genetics [10, 11].

According to the botanical classification, oilseeds are classified into two classes: real nuts and edible seeds. Walnuts are described as real dried fruit, that are thick and may contain spines covering the seeds (walnuts, pecans, brown-and-stop, cashews, pistachios, hazelnuts, macadamia nuts, and nuts in general), while the edible seeds despite having similar nuts characteristics have a different botanical classification. Standing out in this group, peanut and baru nuts [54, 64, 65, 68].

The baru nut is a seed that has edible film or coating in bright color, with shades that vary from brown to dark red [65, 66, 67] as shown in (Figure 8).

The baru nuts are a rich oil source, and also a rich composition of micro and macro nutrients such as galic acid (224,0 mg/100 g), tannins and phytic acid (1073.6 ± 114.9 mg/100g and 472.2 ± 12.5 mg/100 g, respectively), and zinc, copper, iron, phosphorus and magnesium (4.2 ± 0.4, 1.4 ± 0.1, 4.7 ± 0.3, 273.4 ± 8.8, 139.0 ± 6.0 mg/100 g, respectively), tocopherols, fiber and proteins [54, 66, 68, 69].

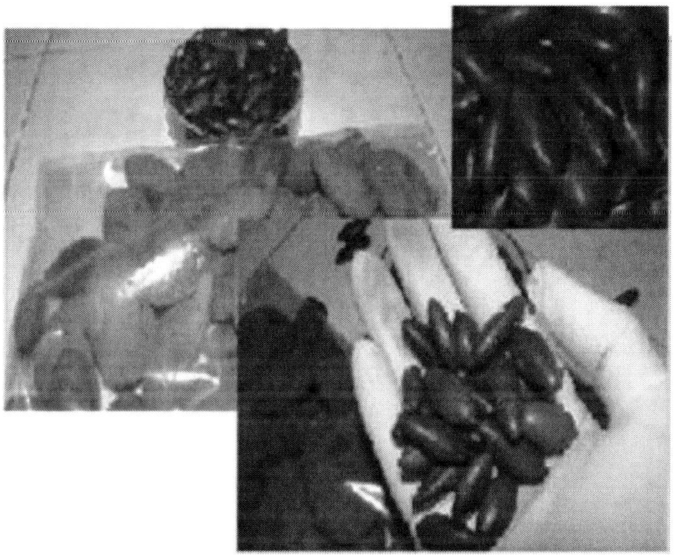

Source: author's personal archive.

Figure 8. Fruit and baru nuts (Dipteryx alata Vog.).

5. Stability and Antioxidant Activity of Baru Nuts

Recently, two studies were conducted at the University of Brasilia (UNB/Brasil). The first was a test performed in vivo [14] that evaluated the reduction of iron-induced oxidative stress in rats supplemented with baru nuts (Dipteryx alata Vog.). Mediated by the bioactive compounds present in the nuts it was observed an antiradical protection, infering the antioxidant potential and probable preventive action against chronic diseases and their complications [13, 14, 70].

The second test, *in vitro*, was developed in the Chromatography Laboratory, Department of Agroindustrial Science and Technology, Federal University of Pelotas (DCTA/UFPEL/Brasil), using the same baru nuts (Lemos et al., 2012), caught the attention by the high phenolic compounds content, including high proportion of galic acid. These compounds were related as responsible for the protective effect against stress induced in animals supplemented with baru nuts [13, 14, 70].

The baru nuts contain an antinutritional compound that must be inactivated by heat. Therefore, baru nuts were separated into two groups: whole raw nuts and peeled raw nuts that were submmited to roasting process at 150°C in a circulating air oven for a period of 45 minutes.

The typical chromatogram of phenolic compounds present in samples of baru nuts shows that the gallic acid stands out among the others and it was the major compound found in these almonds.

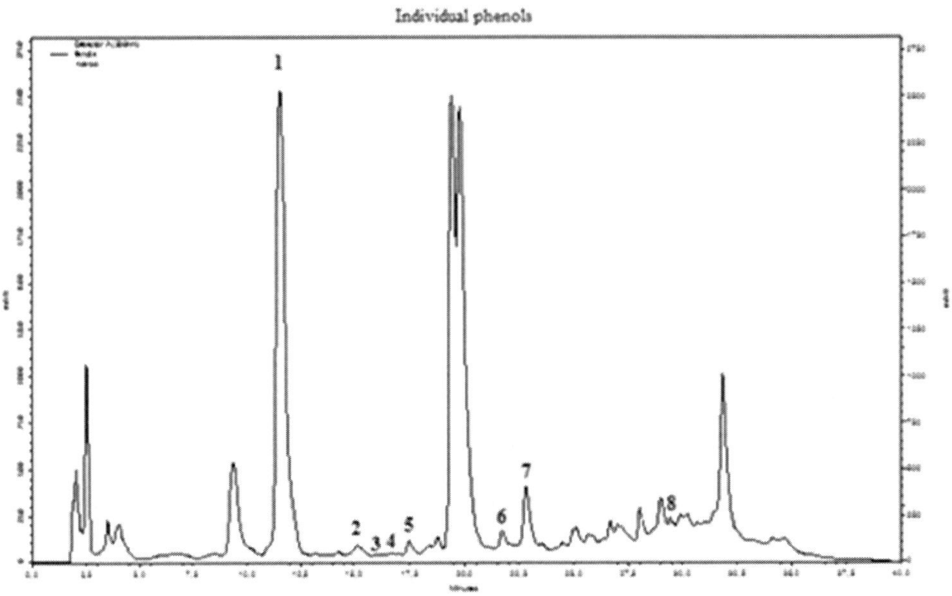

Source: author's personal archive.

Figure 10. Typical chromatogram of phenolic compounds in baru nuts obtained in the HPLC system (Shimadzu) with a reverse phase column and UV detector (280 nm). Mobile phase: gradient of acetic acid in water (1:99 v / v) and methanol with a flow 0.9 ml / min. Phenolic compounds: (1) Gallic acid; (2) Cathehin; (3) p-Hidroxybenzoic acid; (4) Caffeic acid; (5) Epicatechin; (6) p-Coumaric acid; (7) Ferulic acid; (8) Ellagic acid.

Among the quantified compounds, it was found that gallic acid was the phenolic compound with the highest stability when submmited to heat process (Figure 11). The roasting process (heat stress) resulted in reduction of catechin, ferulic acid and of epicatechin content, without reducing significantly the gallic acid content. The galic acid present a highly hydroxylated wich is chemically vulnerable to the redox reaction, which could result in losses when subjected to heat; however, it behaved differently than other phytochemicals analyzed, which showed br more thermolabile [70].

The observed differences in individual behavior of these compounds are justified by the intrinsic diversity of occurrences in biological materials, that may change the antioxidant capacity or loss of natural antioxidants or even to strengthen its antioxidant activity, or by formation of new compounds with antioxidative or pro –oxidative activity, or even not occur any kind of change.

The content of phenolic compounds is associated with the system of protection of plants against biotic and abiotic means, what justify the higher content of bioactive compounds in baru nuts, because the Brazilian Cerrado is a region where the climate is Tropical, presenting extreme weather conditions, times of extreme drought and temperature (with an annual average of 25 °C, reaching markings to 40°C).

The presence of the peel is another key factor and probably contributed to the protective effect against heat damage in baru nuts, compared with those without peel, which showed greater deterioration in their bioactive compounds. The gallotannins, formed from the metabolism of gallic acid, are found in abundance in husks and fruits, walnuts, hazelnuts, almonds and edible seeds, proving the protection of baru nuts, toasted and with peel when subjected to heat.

Thus, in this study it was possible to demonstrate that, despite the partial degradation of phenolic compounds during the roasting process, the baru nuts maintained its antioxidant activity, which was evidenced in vitro antioxidant activity and in animal model (*in vivo*). Its bioactive compounds, including gallic acid, acted as significant protective against stress, being a promising study for evaluation of these compounds in the control and prevention of various diseases.

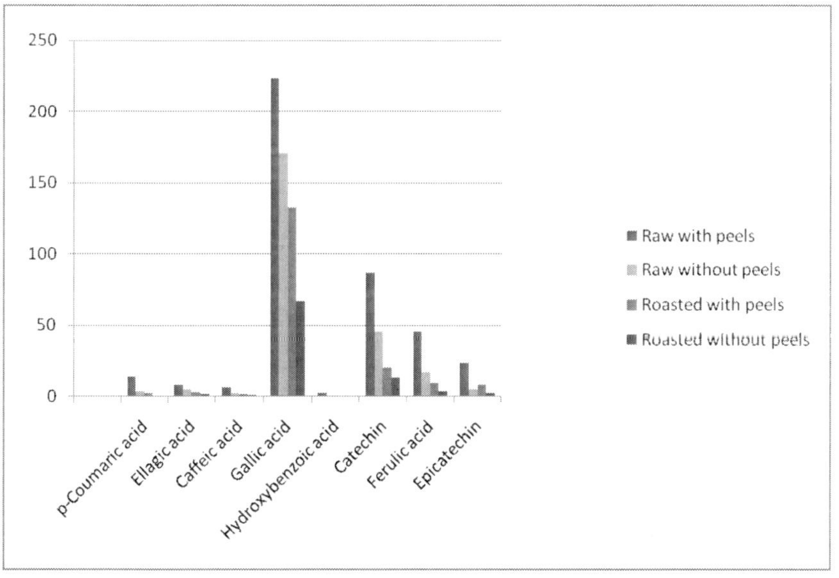

Source: author's personal archive.

Figure 11. Content of phenolic compounds in raw baru nuts (mg/100g), with and without peel, submitted or not to the roasting process.

References

[1] Cohen MV. Free radicals in ischemic and reperfusion myocardial injury: is this time for clinical trials? *Ann. Intern. Med.* 1989 111: p 918-31.

[2] Phung OJ, Makanji SS, White CM, Coleman CI. Almonds have a neutral effect on serum lipid profiles: a meta-analysis of randomized trials. *J. Am. Diet. Assoc.* 2009 May;109(5): p 865-73.

[3] Van Horn L, McCoin M, Kris-Etherton PM, Burke F, Carson JAS, Champagne CM, Karmally W, Sikand G. The Evidence for Dietary Prevention and Treatment of

Cardiovascular Disease. *Journal of the American Dietetic Association* 2008 108: p 287-331.

[4] Rochfort S., Panozzo J. Phytochemicals for health, the role of pulses. *Journal of Agricultural and Food Chemistry,* 2007 55: p 7981–7994.

[5] Liu RH, Liu J, Chen B. Apples Prevent Mammary Tumors in Rats. *J. Agric. Food Chem.* 2005, 53 (6): p 2341-43.

[6] Tiburg LB, Mattern T, Folts JD, Weisgerber UM, Katan MB..Tea flavonoids and cardiovascular disease: a review. *Crit. Rev. Food Sci. Nutr.* 1997 Dec;37(8):p 771-85.

[7] Klink, C. A. and Machado, R. B. A conservação do Cerrado brasileiro. Megadiversidade 2005 1(1): p 147-155.

[8] Sano, S. M., Ribeiro, J. F., Brito, M. A. Baru: biologia e uso. Planaltina, DF: Embrapa Cerrados, 2004. 52 p. (Embrapa Cerrados. Documentos, p 116.

[9] Lorenzi, H., Bacher L., Lacerda M., Sartori S. Frutas brasileiras e exóticas cultivadas de consumo *in natura.* São Paulo: Instituto Plantarum de Estudos da Flora, 2006.640p.

[10] Rice-Evans, C.; Nicolas, J.; Miller, J.; Paganga, G. *Structure- antioxidant activity relationships of flavonoids and phenolic acids. Free Radic. Biol. Med.,* v.20, p.933-956, 1996.

[11] Baranowski, J.D.; Nagel, C. W. *Isolation and identification of the hydroxycinnamic acid derivatives in white riesling wine. Am. J. Enol. Vitic*., v. 32, n. 1, p. 5-13, 1981.

[12] Kratz, Jadel Müller- Estudo do mecanismo da ação anti-herpética do ácido gálico e do galato de pentila / *Florianópolis,* 2007. 79 p.

[13] Lemos, M R B Caracterização e estabilidade dos compostos bioativos em amêndoas de baru (*Dipteryx alata* Vog.), submetidas a processo de torrefação-Tese de doutorado, 2012, Universidade de Brasilia-UnB.

[14] Siqueira, E M A, Marin, AM F, Arruda S F, Cunha M S B, Fustinoni A M, and Dourado, L. P. Consumption of baru seeds [Dipteryx alata Vog.], a Brazilian savanna nut, prevents iron-induced oxidative stress in rats. *Food Research International 2012, 45*,p 427– 433, 2012.

[15] Rouquayrol MZ, Almeida Filho N, organizadores. Epidemiologia and saúde. 6ª Ed. Rio de Janeiro: Editora Medsi; 2003. p 499-513.

[16] Gottlieb, MGV, Morassutti, AL, and Cruz, I B M Transição epidemiológica, estresse oxidativo e doenças crônicas não transmissíveis sob uma perspectiva evolutiva. Scientia Medica. Porto Alegre, 2011 21: 2, p 69-80.

[17] Eaton, S.B., Eaton, S.B III., and Konner, M.J. (1997). Paleolithic nutrition revisited: a twelve-year retrospective on its nature and implications. European *Journal of Clinical Nutrition,* 51, p 207-216.

[18] Armelagos GJ, Brown PJ, Turner B. Evolutionary, historical and political economic perspectives on health and disease. *Soc. Sci. Med.* 2005;61:p 755-65.

[19] Futuyma DJ Biologia evolutiva. 2a ed. Sociedade Brasileira de Genética. 57: p 1216–1220, 2003.

[20] Hoberg EP, Jones A, Rausch RL, Eom KS, Gardner SL. A phylogenetic hypothesis for species of the genus *Taenia* (Eucestoda*: taeniidae). J. Parasitol.* 2000;86:p 89–98.

[21] Hogan, DJ and Vieira, PF (org.). Dilemas Socioambientais e Desenvolvimento Sustentável. Campinas, Editora da Unicamp, 1992.

[22] Silverthorn DU. Fisiologia humana: uma abordagem integrada. 5ª ed. Porto Alegre, Artmed, 2010.

[23] Pignatti MG. Saúde e Ambiente: As doenças emergentes no Brasil. Revista Ambiente and Sociedade. VII. 1, 2004.

[24] Hanninen, O, and Sem, KC Nutritional Supplements and Functional Foods: Functional Significance and Global Regulations. *Nutraceutical and food Regulations*, Columbus, 2008 p 35-11.

[25] Bagchi, D. Preface. Nutraceutical and Functional Food Regulations in the United States and Around the World Elsevier professional, p. 462, 2008.

[26] Yunes, RA, and Cechinel F°, V: Breve análise histórica da Química de Plantas Medicinais: Sua importância na atual concepção de fármaco segundo os paradigmas Ocidental e Oriental: In: RA Yunes, JB Calixto, Plantas Medicinais sob a ótica da Química Medicinal Moderna. Argus, Chapecó-SC: 2001 p 523.

[27] Popkin, B. M: Global nutrition dynamic: the world is shifting rapidly toward a diet linked with non communicable diseases. American Journal of Clinical Nutrition, *Chapel Hill* 2006 84(2), p 298-289.

[28] Li H, Wang X, Li Y, Li P, Wang H: Polyphenolic compounds and antioxidant properties of selected China wines. *Food Chem.* 2009, 112:p 454-460.

[29] García, A., M.; Pascual, T. S.; Santos, B. C., and Rivas, G. J. C. Evaluation of the antioxidant properties of fruits. *Food Chemistry, Salamanca,* 2004 84 (3), p18.

[30] OMS Organização Mundial da Saúde. Classificação Estatística Internacional de Doenças e Problemas Relacionados à Saúde – 10° revisão. v. 1. São Paulo: Edusp; 2000.

[31] Barreiros, L B S, David, J M, and David, J P: Oxidative stress: relations between the formation of reactive species and the organism's defense. *Química Nova*, 2006 29, p 113-123.

[32] Halliwell B, Gutterdge JMC. Antioxidant defenses. In: Free radicals in biology and medicine. 3rd ed. Oxford: Clarenton Press; 1999. p105-245.

[33] Ross D, Moldeus PR. Antioxidant defense systems and oxidative stress. In. Vigo-Pelfrey C, ed. Membrane Lipid Oxidation, Vol. II Boca Raton: CRC,1991: p 151-170.

[34] Guaratini, T.; Medeiros, M. H. G., and Colepicolo, P. Antioxidantes na manutenção do equilíbrio redox cutâneo: uso e avaliação de sua eficácia. *Quím. Nova*, São Paulo, 2007:30; 1.

[35] Matsuo W; and Kaneco T. The chemistry of reactive oxygen species and related free radicals. In Radák Z, editor. Free Radicals in Exercise and Aging Champaign: Human Kinetics, 2001.

[36] Halliwell, B. Antioxidants and human diseases: A general introduction. *Nutrition Rev.*, 1997, 55: p S44-S49.

[37] Pompella, A. Biochemistry and histochemistry of oxidant stressand lipid peroxidation. *International Journal of Vitamin and Nutrition Research*, Bern, v.67, n.5, p.289-297, 1997.

[38] Sies, H. and Stahl, W. Vitamins E and C, β-carotene, and other carotenoids as antioxidants. *American Journal of Clinical Nutrition, Houston,* v.62, n.6, p.1315-132, 1995.

[39] Prior, Rl. Phytochemicals. In: Shils M.E.; Olson J.A.; Shike M.; Ross A.C. Modern nutrition in: health and disease. Philadelphia: Lippincott Willians and Wilkins, p.582-94, 2006.

[40] Liu, RH. Potential synergy of phytochemicals in cancer prevention: mechanism of action. *Journal of Nutrition*, v.134, n.12 Suppl, p.3479S-3485S, 2004.

[41] Vaccari, N. F. S.; Soccol; Marcilene, C. H., and Ide, G.M. Compostos fenólicos em vinhos e seus efeitos antioxidantes na prevenção de doenças. Revista de Ciências Agroveterinárias. Lages, v.8, n.1, p. 71-83, 2009.

[42] Hoffmann-Ribani, R. and Rodriguez-Amaya, D. Otimização de método para determinação de flavonóis e flavonas em frutas por cromatografia líquida de alta eficiência utilizando delineamento estatístico e análise de superfície de resposta. *Quim. Nova,* v.31, n. 6, p.1378-1384, 2008.

[43] Prior, Rl. Phytochemicals. In: Shils M.E.; OLSON J.A.; SHIKE M.; ROSS A.C. Modern nutrition in: health and disease. Philadelphia: Lippincott Willians and Wilkins, p.582-94, 2006.

[44] Degáspari, C. H., and Waszczynsky (2004). Propriedades antioxidantes de compostos fenólicos. Visão Acadêmica. Curitiba, v. 5, n. 1, p. 33 – 40, 2004.

[45] Balasundram, N., Sundram, K, and Samman, S.Phenolic Compounds In Plants and Agri-Industrial By-Products: Antioxidant activity, occurrence, and potential uses. *Food Chemistry,* 99 (1), p. 191-203, 2006.

[46] Figueiredo, Sérgio S. Efeito do ácido Gálico sobre a fibrinogênese hepática murina. Ribeirão Preto, 2012, 47p.

[47] Eger, Iriane-Potencial quimioterapêutico de diferentes derivados do ácido gálico para o tratamento da infecção pelo Trypanossoma cruzi (tese): Avaliação in vitro e in vivo, *Florianópolis,* 2010, 105p.

[48] Queiroz, C.R.A.A.; Morais, S.A.L. and Nascimento, E.F. Caracterização dos taninos da aroeira-preta (Myracrodruon urundeuva). *Rev. Árvore*, v. 26, p. 485-492, 2002.

[49] Bernays, E. A.; G. C. Driver; M. Bilgener. Herbivores and plant tannins. *Advances in Ecological Research.* v. 19, p.263-302, 1989.

[50] Khanbabaee K, and Van Ree T. Tannins: classification and definition. *Nat. Prod. Rep,* v.18, n., p. 641-649, 2001.

[51] Monteiro, J.; Albuquerque, M.; Araújo U. P.; Amorim, E. L., Cavalcanti; E. L. Taninos: uma abordagem Taninos: uma abordagem da química à ecologia. *Quím. Nova,* v.28, 2005.

[52] Mullen Et al.. Ellagitannins, flavonoids, and other phenolics in red raspberries and their contribution to antioxidant capacity and vasorelaxation properties. *J. Agric. Food Chem.* v. 50, p.5191-5196, 2002.

[53] Wang, S.Y, Maas, J.L., Payne, J.A, et al. Ellagic acid content in small fruits mayhaws, and other plants.Journal Small Fruit and Viticulture, *Baton Rouge*, v. 2, n. 4, p. 11-49, 1994.

[54] Freitas, J. B. and Naves, M. M. Composição química de nozes e sementes comestíveis e sua relação com a nutrição e saúde. Revista de Nutrição, *Campinas,* v.23, n.2, p.269-279, 2010.

[55] Lugo, A.E. Estimating Reductions in the Diversity of Tropical Forest Species. In: WILSON, E.O. and PETER, F.M. Biodiversity. Washinton. National Academy Press, p. 3-18, 1988.

[56] Saito. T S, Welzel A, Suyenaga E S, Bueno F. Method for fast determination of epigallocatechin gallate (egcg), epicatechin (ec), catechin (c) and caffeine (caf) in green tea using HPLC. *Cienc. Technol. Aliment.* v. 26, p. 394-398, 2006.

[57] EMBRAPA CERRADO: acesso em 24/02/2012. Disponivel em: http www. agencia.cnptia.embrapa.br/Agencia16/AG01/Abertura.html.

[58] Borlaug, N. Feeding a world of 10 billion people: the miracle ahead. In: R. Bailey (ed.). Global warming and other eco-myths. p. 29-60. Competitive Enterprise Institute, *Roseville,* EUA, 2002.

[59] Silva, F. A. M., Assad, E. D. and Evangelista, B. A. Caracterização climática do Bioma Cerrado. In *Cerrado: ecologia e flora* (S.M. Sano, S.Paulo: Almeida and J.F. Ribeiro, eds.). Embrapa Cerrados, Brasília, p. 61-88, 2008.

[60] Caramori, P.H.; Kathounian, C.A.; Morais, H.; Leal, A.C.; Gorreta, R, and Androcioli Filho, A. Arborização de cafezais e aspectos climatológicos. In: Matsumoto, S.N. (Org.). Arborização de cafezais no Brasil. Edições Uesb, *Vitória da Conquista*, p. 21-38, 2004.

[61] Myers, N., Mittermeier, R. A., Mittermeier, C. G., Fonseca, G. A. B. and Kent, J. Biodiversity hotspots for conservation priorities. *Nature,* 403, p.853-858, 2000.

[62] Roque, P. A. Colonização do Cerrado: savanas e celeiro do mundo. Ed. Premio. São Paulo. 2006.

[63] Aguiar, L. M. S.; Camargo, A. J. A. Cerrado: Ecologia e Caracterização. Planaltina-DF: Embrapa Cerrados. Brasília: DF. Embrapa Informação Tecnológica, 2004.

[64] Baru- Acesso em 24/02/2012. Disponível (site de consultas do Cerrado)htp://www.portalsaofrancisco.com.br/alfa/baru/baru-6.php.

[65] Freitas, J. B. Qualidade nutricional e valor protéico da amêndoa de baru em relação ao amendoim, castanha-de-caju e castanha-do-pará. Dissertação de mestrado: Programa de Pós-Graduação em Ciência e Tecnologia de Alimentos da Universidade Federal de Goiás-UFG. Goiás: 2009.

[66] Correa, et al. Determinações físicas em frutos e sementes de baru (Dipteryx alata Vog.), cajuzinho (Anacardium othonianum Rizz.) e pequi (Caryocar brasiliense Camb.), visando melhoramento genético. *Bio-science Journal,* v. 24, n. 4, p. 42-47, 2008.

[67] Sano S. M.; Brito M. A. De, and Ribeiro, J. F. *Baru.* In: *Frutas nativas da região Centro-Oeste do Brasil.* 1st Ed.; Vieira, R. F.; Costa, T. da S. A.; Silva, D. B. da; Ferreira, F. R.; Sano, S. M., (Ed.) Embrapa. *Recursos Genéticos e* Biotecnológicos. Brasília: (pp. 76-98). Brasil: Empresa Brasileira de Pesquisa Agropecuária do Brasil (Embrapa-CPAC), 2006.

[68] Marin, A. M., Arruda, S. F., and Siqueira, E. M. A. Minerals, phytic acid nd tannin contents of 18 fruits from the Brazilian savanna. International *Journal of Food Science,* 60, p.180–190, 2009.

[69] Takemoto, E.; Okada, I. A.; Garbelotti, M. L.; Tavares, M., and Aued-Pimentel, S. Composição química da semente e do óleo de baru (*Dipteryx alata* Vog.) nativo do Município de Pirenópolis, Estado de Goiás. Revista do Instituto Adolfo Lutz, v.*60*, p.113–117, 2001.

[70] Lemos, M R B; Siqueira, E M; Arruda, S F; Zambiazi, R C The effect of roasting on the phenolic compounds and antioxidant potential of baru nuts [Dipteryx alata Vog.]. *Food Research International*, vol. 48, No. 2. (October), pp. 592-597, 2012.

In: Handbook on Gallic Acid
Editors: M. A. Thompson and P. B. Collins
ISBN: 978-1-62618-921-8

Chapter 10

ANTIOXIDANT, ANTITUMORAL AND ANTI-INFLAMMATORY ACTIVITIES OF GALLIC ACID

Claudriana Locatelli[1], Fabíola Branco Filippin-Monteiro[2], Ariana Centa[2] and Tânia Beatriz Creczinsky-Pasa[2*]

[1]Curso de Farmácia, Universidade do Oeste de Santa Catarina, Videira, SC, Universidade do Alto Vale do Rio do Peixe, Caçador, SC, Brazil

[2]Departamento de Ciências Farmacêuticas, Universidade Federal de Santa Catarina, Florianópolis, SC, Brazil

ABSTRACT

Previous studies have demonstrated a variety of biological activity of gallic acid (GA) including antitumoral, antioxidant and antiinflammatory properties. Among these numerous data, *in vitro* assays showed that the GA and its derivatives are active against several types of tumor cells. Particularly, the studies have shown that the antitumor activity seems to be related to the induction of apoptosis involving different signalling pathways. Apoptosis induced by GA may be associated with oxidative stress derived from reactive oxygen species (ROS) and mitochondrial dysfunction. Interestingly, GA has both pro-oxidant and antioxidant properties. Intracellular ROS levels were increased in cancer cells treated with GA and their derivatives and these results suggested that GA plays a central role as a pro-oxidant in cancer cell models.

It also has been reported that the cell death promoted by GA in different cell lines may be related with glutathione (GSH) depletion. Since the intracellular GSH has a decisive effect on anticancer drug-induced apoptosis, those apoptotic effects were inversely related to GSH content. The reduction of GSH levels might be related to the capability of GA to decrease the expression, as well as the activity of γ-glutamyl cysteine synthase (γ-GCS).

Regarding tumorigenesis, it has been widely reported that GA is responsible for suppression of tumor angiogenesis. The mechanisms underlying this event might be

* Corresponding author: Tel: + 55 48 3721 2212; fax: + 55 48 3721 9542. E-mail: tania.pasa@ufsc.br; taniabcp@gmail.com.

related to alterations in the interaction among the stromal cell-derived factor-1 (SDF-1), an alpha-chemokine, and CXCR4, a G-protein-coupled receptor. These changes are involved in angiogenesis, upregulating key angiogenic factors including vascular endothelial growth factor (VEGF). Furthermore, GA suppressed ADAM17 expression, a transmembrane protease, which may be associated with the inhibition of invasiveness through the inactivation of the well-known cell-survival signalling pathways PI3K/Akt and Ras/MAPK. Additionally, GA might inhibit the activity of matrix metalloproteinases (MMP) involved in tumor angiogenesis and metastasis, such as MMP-2 and MMP-9. These MMP are directly involved in degradation of extracellular matrix and play pivotal roles in cancer cell migration and invasion.

Finally, GA induced caspase-3 and -8 expression, alteration in the Bcl-2/Bax ratio, and an inhibition in tyrosine phosphorylation by BCR/ALB kinase, as well as a down-regulation in COX-2 levels in leukemia cell lines. The GA was also able to inhibit COX-1 and COX-2 activities in these cell lines, suggesting an antiinflammatory activity of the compound. Increased COX-2 expression seems to be involved in the development of cancer by triggering cell division and inhibiting apoptosis. These studies suggest that GA induces cell death mediating multivariant mechanisms, being a good candidate to clinical studies for cancer treatment.

1. INTRODUCTION

Gallic acid (GA) or 3,4,5-trihydroxybenzoic acid (Figure 1), consisting of a tri-hydroxylated phenolic structure, is an intermediate of secondary plant metabolism in higher plants [1]. GA, may be obtained by acid hydrolysis of hydrolysable tannins, and, being a major component of this raw material, its biological properties and those of their analogs have been extensively investigated. In its acid function, GA allows the introduction of various substituents, enabling thereby obtaining numerous analogs esters with distinct pharmacological properties.

In 1945, Bucher [2] described the ability of GA in protecting the salmon oil rancidity. Thompson et al [3] reported the antimalarial activity of GA and other similar chemical structures. Mahadevam and Reddy [4], demonstrated the antifungal activity of GA, catechol, among others molecules. In the same year, Isoyama and colleagues demonstrated the antibacterial activity of synthetic derivatives of GA against *Bacillus subtilis, Staphylococcus aureus* and *Escherichia coli*. In 1970, Frolov and Mishenkova [5] described the anti-influenza activity of GA *in vivo* and *in vitro*, demonstrating a reduction in mortality of monkeys exposed to the virus by reducing the development of pneumonia. All these important findings allowed establishment of different and important activities for GA opening a new scenario for this important natural compound.

More recently, studies by Kawada et al [6] showed that GA was able to induce apoptosis in human lung cancer cell lines. Moreover, the association of GA with cisplatin showed an increase in apoptosis of these tumor cells when compared to cisplatin alone. These data suggested that the combination of GA with other anticancer drugs could be effective in lung cancer treatment trials [6]. These studies suggested one possible application of GA in malignant tumors with intrinsic or acquired drug resistance such as cisplatin [7].

Although, the toxicity of GA in tumor cells was expressive, some researchers have been reported that GA has not presented cytotoxicity against non-tumoral human lymphocytes at concentrations up to 4.17 mM [8], suggesting some selectivity for tumor cells. Additionally, *in vivo* studies demonstrated that no toxic effects were observed in rats treated with GA at 120 mg/kg/day [9]. Conversely, Locatelli et al. [10] showed that there was no clear evidence that tetradecyl gallate, one derivative of GA, produced toxicity when mice were treated each three days in a period of 28 days with 3.7mg/Kg weight body. These results showed that GA and alkyl esters derivatives could be a good alternative for cancer treatment, since they exhibited low toxicity and a relative good selectivity.

GA can induce programmed cell death in various tumor cell lines like in leukemia, lung, colon and stomach cancer and exhibit excellent antioxidant activities. Given the above, this chapter aims to review antioxidant, antitumoral and anti-inflammatory activities of GA and derivatives.

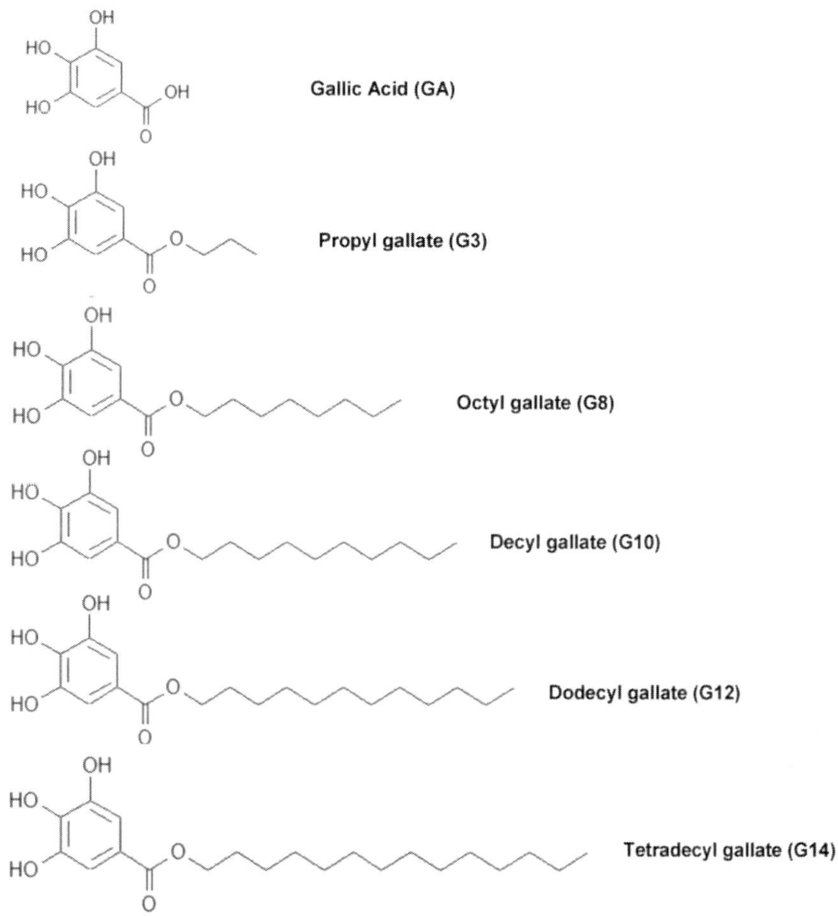

Figure 1. Chemical structure of GA and its derivatives.

2. Antioxidant Activity of Gallic Acid

Many disorders have been associated with reactive oxygen substances (ROS) exposure due to their ability to produce oxidative damages to DNA, proteins, and lipid membranes among other macromolecules. Despite the fact that almost all organisms are prepared against these injuries with antioxidant defenses and repair systems, an imbalance between ROS generation and antioxidant systems may occur, leading to tissue damages [11]. The impact of ROS on tissues are devastating regarding oxidative stress-induced cell death [12]. This scenario may contribute to tissue injuries leading to important consequences to the organs as well as the whole organism.

Gallic acid has been reported to exert both pro-oxidant and antioxidant activities [13]. Interestingly, a phenolic compound like GA was recently associated with cell death caused by oxidative stress derived from ROS and mitochondrial dysfunction [14]. The intracellular ROS generation was assessed using a specific probe and GA increased ROS after only 30 minutes of exposure in lung cancer cells [15]. These phenomena were also reported in two well-established lung carcinoma cell lines A549 and Calu-6 [15]. The levels of a well-known ROS superoxide anion (O_2•-) were decreased in the presence of GA from 100 μM to 400 μM. However, concentrations up to 50 μM slightly decreased the generation of O_2•- [15]. The authors suggested that GA generally plays a role as a pro-oxidant but its effect is dependent on the concentration used in the assays, period of incubation, and cell types.

Despite the abundant data regarding ROS-induced cell death, especially in tumoral cells, GA is well-known as a natural antioxidant involved in both inhibitory and scavenging actions of ROS [16]. Conversely, some studies have been reported that GA exerts DNA protective properties [17, 18]. However, concentrations and period of incubations as well as the means of assessment of these properties vary substantially. According to Ferk [19], the doses discussed in such reports vary in 2–3 orders of magnitude which exemplify the difficulties in establishing the gap between a pro-oxidant and an antioxidant agent.

Some investigations raised the hypothesis that GA and its derivatives are able to prevent the lipid peroxidation occurring in cell membranes. Some authors evaluated the importance of GA regarding its molecular properties in preventing lipid peroxidation in liposome systems [20, 21]. The hydrophobicity and steric effect in scavenging free radicals were important information to point out especially in the field of structure–activity relationships of GA and its derivatives [22].

Antitumoral Activity of Gallic Acid and Its Derivatives

Previous studies have demonstrated a variety of biological activity of GA including antitumoral effects. Among these numerous data, *in vitro* assays showed that GA and its derivatives exerted activity against several types of tumor cells, including leukemia, melanoma, lung cancer and breast cancer cell lines. Particularly, the studies have shown that the antitumor activity seems to be related to the induction of apoptosis involving different signalling pathways. Apoptosis induced by GA may be associated with oxidative stress derived from ROS, mitochondrial dysfunction and an increase in intracellular Ca^{2+} levels. Interestingly, GA has both pro-oxidant and antioxidant properties.

GA analogues such as cholesteryl gallate [23] and alkyl ester of GA [24] led also to apoptosis in tumor cells [25]. Apoptosis induced by GA and its derivatives is Ca^{2+} dependent which was shown in HL-60 leukemia cells; however, other mechanisms seem to be involved in this pathway of death such as the inhibition of superoxide dismutase during apoptosis [26, 27].

Although the mechanism by which GA induces cell death seems to be distinct among various cell types, the production of ROS and the increase of intracellular Ca^{2+} concentration were required as common signals [26]. It was also shown that GA-sensitive cells produced small amounts of catalase, in contrast to the insensitive cells, which produced large amounts of catalase and released them into the medium. This may be explained by the cell death mechanism induced by GA, which involves the generation of hydrogen peroxide [27]. This induction of apoptotic cell death throughout the increased of ROS production by mitochondria, leading to pore formation and subsequent cytochrome *c* release, has been described as an effect induced by several natural compounds. Jacobi et al. [28] and Kobayashi et al. [29] proposed that the damage caused by propyl gallate oxidizing the DNA was associated with ROS production. The researchers showed that propyl gallate was converted to GA by a cytoplasmic esterase releasing H_2O_2, $O_2{\bullet}^-$ and the semiquinone reactive species, which were responsible for the DNA damage and subsequently apoptosis induction.

The hypothesis that the apoptotic cell death in a murine melanoma cell line is a consequence of the pro-oxidative action of GA and derivatives is supported by their ability to activate NF-kB, an ubiquitous transcription factor very well characterized responsible for several cell processes including oxidative stress [30]. Although, the NF-kB activation might promote the transcription of both anti- and pro-apoptotic proteins, it was reported that its activation occurs in pro-oxidant conditions [31]. In this context, it is possible to infer that the NF-kB induction by esters of GA demonstrated in many studies would be directly related to the induction of ROS generation by these compounds, conversely with the fact that GA derivatives decrease the reduced/oxidized glutathione ratio [30, 32]. Changes in the GSH levels and in the redox state of mitochondria were associated with oxidative stress induced by various oxidizing agents. Additionally, it has been shown evidences that, at least in particular conditions, the release into the cytosol of lysosomal constituents may initiate the events related to apoptosis. The triggering of the apoptotic cascade may involve early release of lysosomal constituents leading to an increase in mitochondrial oxidant production, additional lysosomal rupture followed by mitochondrial cytochrome *c* release. The observed loss of mitochondrial membrane potential reflects a mitochondrial dysfunction, which was accompanied by an elevation of expression of the proapoptotic protein Bax as well as, by an inhibition of the antiapoptotic protein Bcl-2. Both events seem to be due to a pro-oxidative effect of GA derivatives in B16F10 mouse melanoma cells [33].

Considering the chemical properties of the alkyl gallate molecules, the length of the lipophilic alkyl side chain has a highly important influence on the binding to mitochondrial inner membrane of these compounds [34] and the three adjacent phenolic hydroxyl groups are responsible for the cytotoxicity [35]. It is possible to consider that one of these hydroxyl groups is an electron-donating moiety. An alkyl gallate derived free radical intermediary may be formed as result of enzymatic oxidation into the Complex I, probably by the protein-bound fast-relaxing semiquinone anion radical, which is observable only in the presence of proton-transmembrane electrochemical potential [36]. This radical could undergo dismutation to another radical intermediate, forming the original compound and the o-quinone species. Then, the o-quinone species could nucleophilically attack proteins of the Complex I, inhibiting the electron flow. This concept is supported by the observation that the semi-quinone form of nordihydroguaiaretic acid might interact with thiols groups modifying proteins [37]. In addition, o-quinones might alter proteins attacking directly the ε-amino groups, resulting in an amino-quinone adduct; consequently the quinonic groups are reduced [38].

Serrano et al [24] reported that the triphenolic compounds, and especially derivatives of GA, are excellent inhibitors of partially purified protein tyrosine kinases (PTKs) from human spleen, exhibiting an inhibitory potency, superior to that of widely used PTK inhibitors such as genistein or tyrphostins. This fact may explain the difficulties found for the interpretation of the results obtained with these compounds when used in different systems. GA ester derivatives, such as octyl and dodecyl gallates, showed an inhibitory potency on PTK activity, which results in a 50–250 times greater apoptosis induction than that of their precursor GA for various human cell lines, indicating a selectivity for fast-growing cells.

Gallic acid also has potent inhibitory effects on AGS cell migration, a human gastric adenocarcinoma, by suppressing the expression of matrix metalloproteinases (MMP) -2/-9 and cytoskeletal F-actin. The anti-migratory effect of GA may involve the inhibition of NF-kB activity and multiple proteins related to metastasis and cytoskeletal reorganization signal pathways, including Ras, Cdc42, Rac1, RhoA, RhoB, PI3K, and p38 [39]. The suppression of ADAM metallopeptidase domain 17 (ADAM17), also called tumor necrosis factor-a-converting enzyme (TACE), and the down-regulation of PI3K/Akt and ERK/MAPK signaling pathways may contribute to a GA-induced decrease of U87 cell invasiveness [40, 41].

Many studies of tumor invasion and metastasis have supported that the degradation of the extracellular MMP play a central role in invasiveness. Two of these enzymes, MMP-2 and MMP-9 correlate with the processes of tumor cell invasion and metastasis in human cancer [42]. Moreover, MMP gene expression is chiefly regulated by transcriptional factors (for example, NF-kB and AP-1) via the PI3K/AKT pathway [43]. NF-kB is maintained in the cytoplasm through interactions with an inhibitor of NF-kB (IkB). Upon dissociation, NF-kB moves into the nucleus and promotes cancer cell proliferation, angiogenesis, and metastasis. Several studies have documented that the quantity of IkB could control NF-kB nuclear translocation and consequently influence the expression of MMP-2 in several types of human cells [44, 45]. This finding suggested that the increase in cytoplasmic IkB could facilitate its bound to NF-kB and then inhibits the NF-kB activity. Akt, in turn, when activated might directly bind to IkB kinase (IKK) unblocking it, and finally trigger the IkB destruction [46]. The protein levels of PI3K, AKT-1 and P-AKT also decreased in a dose-dependent manner when treated with GA. These results indicate that GA might increase IkB binding to NF-kB then suppress the PI3K/AKT pathway subsequently avoiding metastasis of AGS cell. These findings highlighted the therapeutic potential of GA for controlling tumor metastasis through its inhibitory effect on the motility of AGS cells. A possible mechanism of the inhibitory effects of GA on AGS cells might be partly through the Ras/PI3K/AKT signaling pathway. In addition, the increased levels of cytoplasmic IkB, which exert inhibitory effects on transcriptional factor NF-kB decreasing MMP-2 and MMP-9 activities increasing antimetastatic effects [47].

The relationship between GA and the inhibition of cell migration and metastasis can be exemplified by some specific enzymes related to cell invasion. It has been shown that GA is able to suppress the expression of ADAM17, p-Akt, and p-Erk. In addition to the downregulation of its expression, the activity of ADAM17 was also significantly reduced by GA. ADAMs are best known as ectodomain sheddases, and their domains function as metalloproteases. The ADAM family belongs to one of the Zn-dependent metalloproteinases [48, 49]. ADAM17 is an important member of the ADAM family and it is involved in proteolysis of collagen IV of the ECM and the release from the cell surface of several integrins, suggesting that ADAM17 influences the invasive activity of different cells [50].

Regarding proteins involved in cell survival, GA significantly down-regulated the phosphorylation of members of both PI3K/Akt and Ras/MAPK signal transduction pathways, which have been implicated in cell proliferation, invasion and survival. These findings suggest that suppression of ADAM17 by GA might be responsible for decreased invasiveness through the downregulation of PI3K/Akt and Ras/MAPK pathways [40].

Cooperatively, GA has multiple anti-metastatic activities and has the potential to be developed in an anti-metastatic agent for prostate cancer. The possible signal pathways for GA to inhibit migration and invasion in PC-3 (human prostate cancer) cells may be through down-regulation of PKC, p38, ERK, JNK and PI3K/AKT signaling pathways and NF-κB resulting in inhibition of MMP-2 and MMP-9.

GA has also been reported to exert antitumoral activity against colon cancer cells. Bernhaus et al [51] showed that digallolyresveratrol (DIG), a novel synthetic compound structurally related to both resveratrol and GA, inhibited the growth of HT-29 human colon cancer cells in concentrations comparable with those needed for resveratrol or GA. Resveratrol and GA are both known to inhibit the enzyme ribonucleotide reductase (RR) which catalyzes the conversion of ribonucleotides to deoxyribonucleotides. Since the conversion to dNTPs is a rate-limiting step for the *de novo* DNA synthesis in cells, RR inhibited cell proliferation. Hence RR is an excellent target for cancer therapy and it is significantly up-regulated in tumor cells to suit the need of rapidly proliferating malignant cells for dNTPs [52]. Inhibition of RR prevented DNA repair, caused cell cycle arrest and eventually led to apoptosis. Bernhaus et al [51] demonstrated that DIG also inhibited the *in situ* RR activity dose-dependently and led to an alteration of the dNTP pool balance ultimately blocking the DNA synthesis in HT-29 colon cancer cells. Apoptosis caused by DIG could also be observed in HT-29 human colon carcinoma cells. In a morphologic analysis employing a double staining method, it could be demonstrated that apoptosis was mainly induced by DIG whereas necrosis barely occurred. DIG caused an arrest of HT-29 cells in G0–G1 cell cycle phase and led to a depletion of S and G2-M phase cells contrary to the effect of resveratrol in HT-29 cells where a block of the G2 phase was observed [55]. Conversely, previous studies demonstrated that in HL-60 human promyelocytic leukemia cells, both resveratrol and GA caused a significant depletion of dNTP pools [53, 54].

Inhibition of RR and cyclooxygenases (COXs) in human HL-60 promyelocytic leukemia cells was reported [54]. Increased COX-2 expression appears to be involved in the development of cancer by promoting cell division and apoptosis inhibition [56, 57]. The inhibition of COX-2 activity by traditional NSAIDs blocks these activities and thus might account for their anti-carcinogenic activity.

Another important effect of GA derivatives is the ability to inhibit the drug efflux pump via P-glycoprotein 1 (P-gp). Alkyl gallates were not extruded from the cells, however this effect was dependent on their alkyl chain lengths [58]. Results reported by Calcabrini et al. [59] showed that dodecyl gallate reduced cell survival and promoted cell cycle alterations in MCF7 (human breast cancer), MCF7 ADR (human breast cancer multidrug-resistant) and MDA-MB-231 cells (mutant p53 breast cancer). An up-regulation of p21 (a potent cyclin-dependent kinase inhibitor that function as a regulator of cell cycle) was observed in all three breast cancer cell lines, followed by apoptotic cell death. Both p21 up-regulation and apoptosis induction were regulated, at least in part, by ERK1/2 activation. Interestingly, a stronger cytotoxic effect was observed on multidrug-resistant MCF7 ADR cells as well as on MDA-MB-231 carrying mutations in the p53 gene, which, in theory, should be more resistant to cytotoxic treatments and apoptosis. The induction of death caused by dodecyl gallate in MCF7 cells may be directly related to the ability of this gallate to inhibit the Pgp.

In vitro studies performed with GA and its derivatives showed that these compounds exhibit significant antitumoral property, being interesting to conduct *in vivo* studies to confirm these findings. Based on the reports related to GA and its derivatives Locatelli et al (2012) proposed by which pathways the gallates might act as anticancer agents.

3. Anti-inflammatory Activity of Gallic Acid

Inflammation is a defense reaction of the organism placed to eliminate or limit the spread of harmful agents. There are many components of an inflammatory response that may contribute to the associated symptoms and harmful effects to tissues [60]. It involves a complex web of intracellular cytokine signals, which activate monocytes and/or macrophages releasing a variety of inflammatory mediators such as tumor necrosis factor-α (TNF-α), interleukin-1β (IL-1β), interleukin-6 (IL-6), ROS, prostaglandin E2 (PGE2) and nitric oxide (NO) [61, 62].

Moreover, the MAPKs and the transcription factor NF-kB are important mediators of extracellular and cellular responses. Some important MAPKs of mammalian cells include the extracellular signal regulated kinase (ERK), p38 and c-jun N-terminal kinase (JNK). These mediators play an important role in the regulation of cellular responses to pro-inflammatory molecules, particularly TNF-α, IL-1 β and IL-6 [63, 64].

TNF-α provides cellular signals that result in the activation of NF-κB, which is a key factor for regulating gene expression of enzymes and cytokines associated with inflammation, for example iNOS, COX-2, TNF-α, IL-1β and IL-6 [61, 65, 66]. NF-κB controls the expression of numerous genes involved in immune and inflammatory responses and cellular proliferation, actuality involved in various diseases [67]. Because of their importance in inflammatory gene expression, NF-κB is a target for the treatment of various inflammatory diseases, and most anti-inflammatory drugs have been shown to inhibit expression of inflammatory cytokines by inhibiting activation via NF-κB [61, 65].

It has been demonstrated that inhibition of the production and the biological activity of iNOS, COX-2, TNF-α and IL-1β by selective inhibitors result in significant improvement in the development of various inflammatory diseases [68]. Similarly to many natural products, GA is outstanding, given its anti-inflammatory actions by suppressing pro-inflammatory cytokines and chemokines as COX-2 [68, 69].

Increased expression of COX-2 appears to be involved in cancer development by promoting cell division and inhibition of apoptosis [70]. Among many factors related to tumor progression, COX-2 is being considered as an important factor involved in cancer inflammation [71]. Induction of apoptosis appears to be directly related to inhibition of COX preventing the release of cytochrome *c* and caspase activation and PARP cleavage. Chandramohan et al [70] demonstrated that GA (25-50 μM) reduced levels of COX-2 and the results were comparable to celecoxib (80 -160 mM).

GA is already known to have antiproliferative and antitumor activity in lung cancer, prostate cancer and leukemia. However, the molecular basis for the activity in inhibiting the growth and proliferation of cancer cells is still unclear. Recently, it was reported that a potent anti-HAT (histone acetyltransferase) activity of GA [67, 72, 73].

The acetylation of p65 regulates the biological action of NF-κB, including the activation of the transcription, DNA binding activity and IκBα assembly. It has been reported that GA inhibited the activation of NF-κB-dependent p65 acetylation and production of inflammatory markers. The low acetylation rate of p65 resulted in a complete loss of function of NF-κB, indicating that its acetylation is necessary for the signaling pathway mediated by NF-κB. Thus, studies suggested that interference in the selective acetylation of p65 with small molecules as GA might be considered as a new class of anti-inflammatory drugs [67].

In the early stages of atherosclerosis, increased adhesion molecule expression and leukocyte adhesion to vascular endothelium are common events. This scenario happens with the activation of endothelial cells by pro-inflammatory molecules, including TNF-α, which is critical for the beginning of the process. The atherosclerosis event is also characterized by an increased expression of endothelial adhesion molecules such as ICAM-1 and VCAM-1, observed in animal models of inflammation and in human atherosclerotic plaques [61]. Phenolic compounds like GA have been reported to inhibit the expression of endothelial leukocyte adhesion on the surface of vascular endothelium covering inflamed atherosclerotic lesions [65, 74]. Additionally, it has been reported that GA caused a decrease in the expression of ICAM-1 and VCAM-1 in endothelial cells [75].

The 4-O-methylgallic acid (4-OMGA) is one of the main metabolites of GA and, *in vivo*, its level in plasma and urine is rapidly increased after oral administration of GA [76]. Therefore, 4-OMGA has been extensively investigated by its anti-inflammatory effect and its molecular mechanism of action in endothelial cells (HUVEC) stimulated by TNF-α. The metabolite reduced the expression of ICAM-1 and VCAM-1 as well as monocyte adhesion to HUVEC treated with TNF-α by inhibiting the activity of NF-κB [65]. The structure of the 4-OMGA is similar to the GA ester of epigallocatechin-3-gallate (EGCG), which suppresses inflammatory gene expression by inhibiting the activation of the signaling pathway related to NF-κB [61]. The authors also demonstrated that 4-OMGA downregulated the expression of inflammatory enzymes (iNOS and COX-2) and cytokines (TNF-α and IL-1β) by suppressing the activation of NF-κB, as a result of elimination of intracellular ROS and inhibition of IKK activation. This inhibitory effect has important implications for the development of potential drugs for the treatment of inflammatory diseases, such as atherosclerosis and endotoxemia [61, 65].

Recent studies suggested anti-inflammatory effects of GA on rheumatoid arthritis (RA). RA is an inflammatory disease characterized by abnormal synovial hyperplasia and inflammatory cell infiltration in the synovial membrane, causing inflammation and destruction of joints. When synovial inflammation is initiated, synovial cells of rheumatoid arthritis (RA FLS) resist the apoptotic process to proliferate and activate inflammatory cells to produce several pro-inflammatory mediators, including cytokines and chemokines, which results in exacerbation of synovitis and the destruction of the integrity of gasket [77]. Kuppan et al. [78] showed that treatment with GA decreases the expression of TNF-α and IL-6 from human monocytes, and Kim et al. [79] found that GA inhibited the mast cell-derived from inflammation by suppressing pro-inflammatory cytokines [78, 79]. Yoon et al [69] studied the effect of GA on RA and found that the expression levels of a number of important pro-inflammatory mediators of RA FLS were significantly inhibited by the treatment.

Gallic acid was also proposed for the treatment of Alzheimer's disease (AD). AD is the most common form of dementia, and is characterized by progressive decline of cognitive function. The activation of NF-κB as a central event of inflammation is a common feature of many neurodegenerative diseases. In the brains of patients with AD, activation of NF-κB is observed predominantly in neurons and glial cells in areas with senile plaques containing amyloid beta (Ab). In addition, the Ab also appeared to activate NF-κB, which eventually leads to increased cytokine production in neurons and glial cells [80, 81]. Kim et al [80] found that treatment with GA suppresses the activation of NF-κB induced by Ab and cytokine production in microglia through the hypoacetylation, which finally led to the reduction of neurotoxicity induced by Ab. Also there was a restorative effect of GA in Ab-induced cognitive dysfunction. Further, treatment with GA blocks neuronal cell death through the downregulation of the cytokine expression levels and *in vivo* acetylation of NF-κB.

Table 1. Antioxidant, antitumoral and anti-inflammatory activities of gallic acid (AG)

Actions		
Antioxidant	Antitumoral	Anti-inflammatory
↑H_2O_2; ↑OH; ↑ O_2^-; ↓ GSH; ↓ $\Delta\Psi_m$; ↑Release of cytochrome c; ↑ catalase expression; ↑ DNA damage;	↑ Caspase 3; ↑ Caspase 9; ↑ Ca^{2+}; ↓ expression and activity of ADAM17; ↓ p-Erk; ↓ p-Akt; ↑ p21; ↑ p27; ↓ Cyclin D1/2/3/E; ↑ Bax; ↑ FasL; ↓ BcL-2; ↓ tyrosine BCR/ALB kinase phosphorylation; ↑Release of cytochrome c; ↓ NF-kB activity; ↑ DNA damage; ↑ PARP cleavage;	↓ NF-kB activity ↓ COX-1; ↓ COX-2; ↓ p65 acetylation; ↓ HAT; ↓ ICAM-1; ↓ VCAM-1; ↓ TNF-α; ↓ IL-1β; ↓ IL-6; ↓ iNOS;

FasL, death receptor ligand; ADAM, disintegrin and metalloproteinase domains; ErK, extracellular signal-regulated kinases; AkT/PKB, protein kinase B; COX, cyclooxygenase; NF-kB nuclear factor kappa-light-chain-enhancer of activated B cells; ICAM, intercellular adhesion molecule; VCAM, vascular adhesion molecule; PARP, poly ADP-ribose polymerase; BCL-2, anti-apoptotic protein; Bax, pro-apoptotic member of the Bcl-2 protein family; GSH reduced glutathione; p21, cell cycle regulator protein; p65, transcription factor p65 (RELA); DNA, deoxyribonucleic acid; TNF-α, tumor necrosis factor-α; IL-1β, interleukin-1β; IL-6, interleukin-6; HAT, histone acetyltransferase; iNOS, Inducible nitric oxide synthase; ΔΨm, mitochondrial membrane potential.

Despite the effect of GA in inflammatory diseases such AR and AD, the allergic reaction was additionally studied regarding GA treatment. Mast cells, which are constituted of essentially all the organs and tissues, are important mediators of allergic reaction. They, when activated, might produce histamine as well as a wide variety of other inflammatory mediators. Although, these cytokines have a beneficial effect on host defense, they may cause pathological states when over-expressed. Therefore, reduction of TNF-α and IL-6 released by mast cells is a key indicator of decreased in allergic inflammatory symptoms [79]. Additionally, studies have shown that GA reduced the release of histamine from basophil leukemia cells of mice and suppresses production of pro-inflammatory cytokines in murine peritoneal macrophages. Furthermore, GA, as well as epigallocatechin-3-gallate, and the aflavin-3, 3'-digallate inhibited the activation of MAPK. These data substantiated the inhibitory effects of GA on allergic inflammatory responses.

In an attempt to summarize the findings described, a Table with antitumoral, anti-inflammatory and antioxidant properties of GA was elaborated (Table 1) showing the different mechanisms of induction of apoptosis, oxidative stress defenses and anti-inflammatory processes.

REFERENCES

[1] Grundhöfer, P.; Niemetz, R.; Schilling, G.; Gross, G. G., Biosynthesis and subcellular distribution of hydrolyzable tannins. *Phytochemistry* 2001, 57, 915-27.

[2] Bucher, D. L., *Antioxidants for fish oils*. Fishery Market News, 1945; Vol. 7.

[3] Thompson, P. E.; Moore, A. M.; reinertson, J. W., *Antimalarial activity* of *b-resorcylic acid and analogs and reversal by p-hydroxybenzoic acid. Antibiotics and Chemotherapy* 1953, 3, 399-408.

[4] Mahadevan, A.; Reddy, M. K., The effect of phenolic compounds on growth, polygalacturonase production and activity of Fusarium oxysporum f. vasinfectum. *Netherlands Journal of Plant Pathology* 1968, 74, 87-90.

[5] Florov, A. F.; Mishenkova, E. L., Inhibitory effects of some preparations from higher plants on reproduction of influenza virus in vitro and in vivo. *Mikrobiol Zh* 1970, 32, 628-33.

[6] Kawada, M.; Ohno, Y.; Ri, Y.; Ikoma, T.; Yuugetu, H.; Asai, T.; Watanabe, M.; Yasuda, N.; Akao, S.; Takemura, G.; Minatoguchi, S.; Gotoh, K.; Fujiwara, H.; Fukuda, K., Anti-tumor effect of gallic acid on LL-2 lung cancer cells transplanted in mice. *Anticancer Drugs* 2001, 12, 847-52.

[7] Ohno, T.; Inoue, M.; Ogihara, Y., Cytotoxic activity of gallic acid against liver metastasis of mastocytoma cells P-815. *Anticancer Res* 2001, 21, 3875-80.

[8] Yang, F.; Oz, H. S.; Barve, S.; de Villiers, W. J.; McClain, C. J.; Varilek, G. W., The green tea polyphenol (-)-epigallocatechin-3-gallate blocks nuclear factor-kappa B activation by inhibiting I kappa B kinase activity in the intestinal epithelial cell line IEC-6. *Mol Pharmacol* 2001, 60, 528-33.

[9] van der Heijden, C. A.; Janssen, P. J.; Strik, J. J., Toxicology of gallates: a review and evaluation. *Food Chem Toxicol* 1986, 24, 1067-70.

[10] Locatelli, C.; Carvalho, D. R.; Mascarello, A.; de Cordova, C. A.; Yunes, R. A.; Nunes, R. J.; Pilati, C.; Creczynski-Pasa, T. B., Antimetastatic activity and low systemic toxicity of tetradecyl gallate in a preclinical melanoma mouse model. *Invest New Drugs* 2011.

[11] Davies, K. J., Oxidative stress, antioxidant defenses, and damage removal, repair, and replacement systems. *IUBMB Life* 2000, 50, 279-89.

[12] Valko, M.; Leibfritz, D.; Moncol, J.; Cronin, M. T.; Mazur, M.; Telser, J., Free radicals and antioxidants in normal physiological functions and human disease. *Int J Biochem Cell Biol* 2007, 39, 44-84.

[13] Sakagami, H.; Satoh, K.; Kochi, M., Comparative study of the antitumor action between sodium 5,6-benzylidene-L-ascorbate and sodium ascorbate (minireview). *Anticancer Res* 1997, 17, 4451-2.

[14] Inoue, M.; Sakaguchi, N.; Isuzugawa, K.; Tani, H.; Ogihara, Y., Role of reactive oxygen species in gallic acid-induced apoptosis. *Biological & Pharmaceutical Bulletin* 2000, 23, 1153-1157.

[15] You, B. R.; Park, W. H., Gallic acid-induced lung cancer cell death is related to glutathione depletion as well as reactive oxygen species increase. *Toxicol In Vitro* 2010, 24, 1356-62.

[16] Ow, Y. Y.; Stupans, I., Gallic acid and gallic acid derivatives: Effects on drug metabolizing enzymes. *Current Drug Metabolism* 2003, 4, 241-248.

[17] Yeh, C. T.; Ching, L. C.; Yen, G. C., Inducing gene expression of cardiac antioxidant enzymes by dietary phenolic acids in rats. *J Nutr Biochem* 2009, 20, 163-71.

[18] Yeh, C. T.; Yen, G. C., Induction of hepatic antioxidant enzymes by phenolic acids in rats is accompanied by increased levels of multidrug resistance-associated protein 3 mRNA expression. *J Nutr* 2006, 136, 11-5.

[19] Ferk, F.; Chakraborty, A.; Jäger, W.; Kundi, M.; Bichler, J.; Mišík, M.; Wagner, K. H.; Grasl-Kraupp, B.; Sagmeister, S.; Haidinger, G.; Hoelzl, C.; Nersesyan, A.; Dušinská, M.; Simić, T.; Knasmüller, S., Potent protection of gallic acid against DNA oxidation: results of human and animal experiments. *Mutat Res* 2011, 715, 61-71.

[20] Chen, C.; Tang, H. R.; Sutcliffe, L. H.; Belton, P. S., Green tea polyphenols react with 1,1-diphenyl-2-picrylhydrazyl free radicals in the bilayer of liposomes: direct evidence from electron spin resonance studies. *J Agric Food Chem* 2000, 48, 5710-4.

[21] Hashimoto, T.; Kumazawa, S.; Nanjo, F.; Hara, Y.; Nakayama, T., Interaction of tea catechins with lipid bilayers investigated with liposome systems. *Biosci Biotechnol Biochem* 1999, 63, 2252-5.

[22] Lu, Z.; Nie, G.; Belton, P. S.; Tang, H.; Zhao, B., Structure-activity relationship analysis of antioxidant ability and neuroprotective effect of gallic acid derivatives. *Neurochem Int* 2006, 48, 263-74.

[23] Saeki, K.; Yuo, A.; Isemura, M.; Abe, I.; Seki, T.; Noguchi, H., Apoptosis-inducing activity of lipid derivatives of gallic acid. *Biol Pharm Bull* 2000, 23, 1391-4.

[24] Serrano, A.; Palacios, C.; Roy, G.; Cespon, C.; Villar, M. L.; Nocito, M.; Gonzalez-Porque, P., Derivatives of gallic acid induce apoptosis in tumoral cell lines and inhibit lymphocyte proliferation. *Archives of Biochemistry and Biophysics* 1998, 350, 49-54.

[25] Fiuza, S. M.; Gomes, C.; Teixeira, L. J.; Girão da Cruz, M. T.; Cordeiro, M. N.; Milhazes, N.; Borges, F.; Marques, M. P., Phenolic acid derivatives with potential anticancer properties--a structure-activity relationship study. Part 1: methyl, propyl and octyl esters of caffeic and gallic acids. *Bioorg Med Chem* 2004, 12, 3581-9.

[26] Sakaguchi, N.; Inoue, M.; Ogihara, Y., Reactive oxygen species and intracellular Ca2+, common signals for apoptosis induced by gallic acid. *Biochemical Pharmacology* 1998, 55, 1973-1981.

[27] Isuzugawa, K.; Inoue, M.; Ogihara, Y., Ca2+-dependent caspase activation by gallic acid derivatives. *Biological & Pharmaceutical Bulletin* 2001, 24, 844-847.

[28] Jacobi, H.; Hinrichsen, M. L.; Wess, D.; Witte, I., Induction of lipid peroxidation in human fibroblasts by the antioxidant propyl gallate in combination with copper(II). *Toxicology Letters* 1999, 110, 183-190.

[29] Kobayashi, H.; Oikawa, S.; Hirakawa, K.; Kawanishi, S., Metal-mediated oxidative damage to cellular and isolated DNA by gallic acid, a metabolite of antioxidant propyl gallate. *Mutation Research-Genetic Toxicology and Environmental Mutagenesis* 2004, 558, 111-120.

[30] Locatelli, C.; Leal, P. C.; Yunes, R. A.; Nunes, R. J.; Creczynski-Pasa, T. B., Gallic acid ester derivatives induce apoptosis and cell adhesion inhibition in melanoma cells: The relationship between free radical generation, glutathione depletion and cell death. *Chem Biol Interact* 2009, 181, 175-84.

[31] Meyskens, F. L.; Buckmeier, J. A.; McNulty, S. E.; Tohidian, N. B., Activation of nuclear factor-kappa B in human metastatic melanoma cells and the effect of oxidative stress. *Clinical Cancer Research* 1999, 5, 1197-1202.

[32] Locatelli, C.; Rosso, R.; Santos-Silva, M. C.; de Souza, C. A.; Licinio, M. A.; Leal, P.; Bazzo, M. L.; Yunes, R. A.; Creczynski-Pasa, T. B., Ester derivatives of gallic acid with potential toxicity toward L1210 leukemia cells. *Bioorganic & Medicinal Chemistry* 2008, 16, 3791-3799.

[33] de Cordova, C. A. S.; Locatelli, C.; Assuncao, L. S.; Mattei, B.; Mascarello, A.; Winter, E.; Nunes, R. J.; Yunes, R. A.; Creczynski-Pasa, T. B., Octyl and dodecyl gallates induce oxidative stress and apoptosis in a melanoma cell line. *Toxicology in Vitro* 2011, 25, 2025-2034.

[34] Feng, Q.; Kumagai, T.; Nakamura, Y.; Uchida, K.; Osawa, T., Correlation of antimutagenic activity and suppression of CYP1A with the lipophilicity of alkyl gallates and other phenolic compounds. *Mutat Res* 2003, 537, 101-8.

[35] Inoue, M.; Suzuki, R.; Sakaguchi, N.; Li, Z.; Takeda, T.; Ogihara, Y.; Jiang, B. Y.; Chen, Y., Selective induction of cell death in cancer cells by gallic acid. *Biol Pharm Bull* 1995, 18, 1526-30.

[36] Ohnishi, T.; Salerno, J. C., Conformation-driven and semiquinone-gated proton-pump mechanism in the NADH-ubiquinone oxidoreductase (complex I). *FEBS Lett* 2005, 579, 4555-61.

[37] Shi, L.; Pardini, R. S., Effect of NDGA on beef heart mitochondria and EMT6 mouse mammary carcinoma cells. *Res Commun Mol Pathol Pharmacol* 1995, 90, 235-54.

[38] Gut, I.; Nedelcheva, V.; Soucek, P.; Stopka, P.; Tichavská, B., Cytochromes P450 in benzene metabolism and involvement of their metabolites and reactive oxygen species in toxicity. *Environ Health Perspect* 1996, 104 Suppl 6, 1211-8.

[39] Ho, H. H.; Chang, C. S.; Ho, W. C.; Liao, S. Y.; Wu, C. H.; Wang, C. J., Anti-metastasis effects of gallic acid on gastric cancer cells involves inhibition of NF-kappaB activity and downregulation of PI3K/AKT/small GTPase signals. *Food Chem Toxicol* 2010, 48, 2508-16.

[40] Lu, Y.; Jiang, F.; Jiang, H.; Wu, K.; Zheng, X.; Cai, Y.; Katakowski, M.; Chopp, M.; To, S. S., Gallic acid suppresses cell viability, proliferation, invasion and angiogenesis in human glioma cells. *Eur J Pharmacol* 2010, 641, 102-7.

[41] Weng, C. J.; Yen, G. C., Chemopreventive effects of dietary phytochemicals against cancer invasion and metastasis: phenolic acids, monophenol, polyphenol, and their derivatives. *Cancer Treat Rev* 2012, 38, 76-87.

[42] Libra, M.; Scalisi, A.; Vella, N.; Clementi, S.; Sorio, R.; Stivala, F.; Spandidos, D. A.; Mazzarino, C., Uterine cervical carcinoma: role of matrix metalloproteinases (review). *Int J Oncol* 2009, 34, 897-903.

[43] Cheng, J. C.; Chou, C. H.; Kuo, M. L.; Hsieh, C. Y., Radiation-enhanced hepatocellular carcinoma cell invasion with MMP-9 expression through PI3K/Akt/NF-kappaB signal transduction pathway. *Oncogene* 2006, 25, 7009-18.

[44] Javelaud, D.; Poupon, M. F.; Wietzerbin, J.; Besançon, F., Inhibition of constitutive NF-kappa B activity suppresses tumorigenicity of Ewing sarcoma EW7 cells. *Int J Cancer* 2002, 98, 193-8.

[45] Park, J. M.; Kim, A.; Oh, J. H.; Chung, A. S., Methylseleninic acid inhibits PMA-stimulated pro-MMP-2 activation mediated by MT1-MMP expression and further tumor invasion through suppression of NF-kappaB activation. *Carcinogenesis* 2007, 28, 837-47.

[46] Hennessy, B. T.; Smith, D. L.; Ram, P. T.; Lu, Y.; Mills, G. B., Exploiting the PI3K/AKT pathway for cancer drug discovery. *Nat Rev Drug Discov* 2005, 4, 988-1004.

[47] Ho, H. H.; Chang, C. S.; Ho, W. C.; Liao, S. Y.; Lin, W. L.; Wang, C. J., Gallic acid inhibits gastric cancer cells metastasis and invasive growth via increased expression of RhoB, downregulation of AKT/small GTPase signals and inhibition of NF-κB activity. *Toxicol Appl Pharmacol* 2013, 266, 76-85.

[48] Lu, X.; Lu, D.; Scully, M.; Kakkar, V., ADAM Proteins-Therapeutic Potential in Cancer. *Current Cancer Drug Targets* 2008, 8, 720-732.

[49] Kheradmand, F.; Werb, Z., Shedding light on sheddases: role in growth and development. *Bioessays* 2002, 24, 8-12.

[50] Wildeboer, D.; Naus, S.; Sang, Q.-X. A.; Bartsch, J. W.; Pagenstecher, A., Metalloproteinase disintegrins ADAM8 and ADAM19 are highly regulated in human primary brain tumors and their expression levels and activities are associated with invasiveness. *Journal of Neuropathology and Experimental Neurology* 2006, 65, 516-527.

[51] Bernhaus, A.; Fritzer-Szekeres, M.; Grusch, M.; Saiko, P.; Krupitza, G.; Venkateswarlu, S.; Trimurtulu, G.; Jaeger, W.; Szekeres, T., Digalloylresveratrol, a new phenolic acid derivative induces apoptosis and cell cycle arrest in human HT-29 colon cancer cells. *Cancer Lett* 2009, 274, 299-304.

[52] Nocentini, G., Ribonucleotide reductase inhibitors: new strategies for cancer chemotherapy. *Crit Rev Oncol Hematol* 1996, 22, 89-126.

[53] Horvath, Z.; Saiko, P.; Illmer, C.; Madlener, S.; Hoechtl, T.; Bauer, W.; Erker, T.; Jaeger, W.; Fritzer-Szekeres, M.; Szekeres, T., Synergistic action of resveratrol, an ingredient of wine, with Ara-C and tiazofurin in HL-60 human promyelocytic leukemia cells. *Exp Hematol* 2005, 33, 329-35.

[54] Madlener, S.; Illmer, C.; Horvath, Z.; Saiko, P.; Losert, A.; Herbacek, I.; Grusch, M.; Elford, H. L.; Krupitza, G.; Bernhaus, A.; Fritzer-Szekeres, M.; Szekeres, T., Gallic acid inhibits ribonucleotide reductase and cyclooxygenases in human HL-60 promyelocytic leukemia cells. *Cancer Letters* 2007, 245, 156-162.

[55] Liang, Y. C.; Tsai, S. H.; Chen, L.; Lin-Shiau, S. Y.; Lin, J. K., Resveratrol-induced G2 arrest through the inhibition of CDK7 and p34CDC2 kinases in colon carcinoma HT29 cells. *Biochem Pharmacol* 2003, 65, 1053-60.

[56] Tang, X. M.; Sun, Y. J.; Half, E.; Kuo, M. T.; Sinicrope, F., Cyclooxygenase-2 overexpression inhibits death receptor 5 expression and confers resistance to tumor

necrosis factor-related apoptosis-inducing ligand-induced apoptosis in human colon cancer cells. *Cancer Research* 2002, 62, 4903-4908.

[57] Rizzo, M. T., Cyclooxygenase-2 in oncogenesis. *Clinica Chimica Acta* 2011, 412, 671-687.

[58] Kitagawa, S.; Nabekura, T.; Kamiyama, S.; Takahashi, T.; Nakamura, Y.; Kashiwada, Y.; Ikeshiro, Y., Effects of alkyl gallates on P-glycoprotein function. *Biochemical Pharmacology* 2005, 70, 1262-1266.

[59] Calcabrini, A.; Garcia-Martinez, J. M.; Gonzalez, L.; Julian Tendero, M.; Agullo Ortuno, M. T.; Crateri, P.; Lopez-Rivas, A.; Arancia, G.; Gonzalez-Porque, P.; Martin-Perez, J., Inhibition of proliferation and induction of apoptosis in human breast cancer cells by lauryl gallate. *Carcinogenesis* 2006, 27, 1699-1712.

[60] Arfan, M.; Amin, H.; Khan, N.; Khan, I.; Saeed, M.; Khan, M. A.; Fazal-ur-Rehman, Analgesic and anti-inflammatory activities of 11-O-galloylbergenin. *J Ethnopharmacol* 2010, 131, 502-4.

[61] Na, H. J.; Lee, G.; Oh, H. Y.; Jeon, K. S.; Kwon, H. J.; Ha, K. S.; Lee, H.; Kwon, Y. G.; Kim, Y. M., 4-O-Methylgallic acid suppresses inflammation-associated gene expression by inhibition of redox-based NF-kappaB activation. *Int Immunopharmacol* 2006, 6, 1597-608.

[62] Chang, Y. H.; Lee, S. T.; Lin, W. W., Effects of cannabinoids on LPS-stimulated inflammatory mediator release from macrophages: involvement of eicosanoids. *J Cell Biochem* 2001, 81, 715-23.

[63] Azzolina, A.; Bongiovanni, A.; Lampiasi, N., Substance P induces TNF-alpha and IL-6 production through NF kappa B in peritoneal mast cells. *Biochim Biophys Acta* 2003, 1643, 75-83.

[64] Beyaert, R.; Cuenda, A.; Vanden Berghe, W.; Plaisance, S.; Lee, J. C.; Haegeman, G.; Cohen, P.; Fiers, W., The p38/RK mitogen-activated protein kinase pathway regulates interleukin-6 synthesis response to tumor necrosis factor. *EMBO J* 1996, 15, 1914-23.

[65] Lee, G.; Na, H. J.; Namkoong, S.; Jeong Kwon, H.; Han, S.; Ha, K. S.; Kwon, Y. G.; Lee, H.; Kim, Y. M., 4-O-methylgallic acid down-regulates endothelial adhesion molecule expression by inhibiting NF-kappaB-DNA-binding activity. *Eur J Pharmacol* 2006, 551, 143-51.

[66] Ghosh, S.; May, M. J.; Kopp, E. B., NF-kappa B and Rel proteins: evolutionarily conserved mediators of immune responses. *Annu Rev Immunol* 1998, 16, 225-60.

[67] Choi, K. C.; Lee, Y. H.; Jung, M. G.; Kwon, S. H.; Kim, M. J.; Jun, W. J.; Lee, J.; Lee, J. M.; Yoon, H. G., Gallic acid suppresses lipopolysaccharide-induced nuclear factor-kappaB signaling by preventing RelA acetylation in A549 lung cancer cells. *Mol Cancer Res* 2009, 7, 2011-21.

[68] Jung, H. J.; Kim, S. J.; Jeon, W. K.; Kim, B. C.; Ahn, K.; Kim, K.; Kim, Y. M.; Park, E. H.; Lim, C. J., Anti-inflammatory activity of n-propyl gallate through down-regulation of NF-κB and JNK pathways. *Inflammation* 2011, 34, 352-61.

[69] Yoon, C. H.; Chung, S. J.; Lee, S. W.; Park, Y. B.; Lee, S. K.; Park, M. C., Gallic acid, a natural polyphenolic acid, induces apoptosis and inhibits proinflammatory gene expressions in rheumatoid arthritis fibroblast-like synoviocytes. *Joint Bone Spine* 2012.

[70] Chandramohan Reddy, T.; Bharat Reddy, D.; Aparna, A.; Arunasree, K. M.; Gupta, G.; Achari, C.; Reddy, G. V.; Lakshmipathi, V.; Subramanyam, A.; Reddanna, P., Anti-leukemic effects of gallic acid on human leukemia K562 cells: downregulation of

COX-2, inhibition of BCR/ABL kinase and NF-κB inactivation. *Toxicol In Vitro* 2012, 26, 396-405.

[71] Wang, M. T.; Honn, K. V.; Nie, D., Cyclooxygenases, prostanoids, and tumor progression. *Cancer Metastasis Rev* 2007, 26, 525-34.

[72] Jagan, S.; Ramakrishnan, G.; Anandakumar, P.; Kamaraj, S.; Devaki, T., Antiproliferative potential of gallic acid against diethylnitrosamine-induced rat hepatocellular carcinoma. *Mol Cell Biochem* 2008, 319, 51-9.

[73] Madlener, S.; Illmer, C.; Horvath, Z.; Saiko, P.; Losert, A.; Herbacek, I.; Grusch, M.; Elford, H. L.; Krupitza, G.; Bernhaus, A.; Fritzer-Szekeres, M.; Szekeres, T., Gallic acid inhibits ribonucleotide reductase and cyclooxygenases in human HL-60 promyelocytic leukemia cells. *Cancer Lett* 2007, 245, 156-62.

[74] Frankel, E. N.; Kanner, J.; German, J. B.; Parks, E.; Kinsella, J. E., Inhibition of oxidation of human low-density lipoprotein by phenolic substances in red wine. *Lancet* 1993, 341, 454-7.

[75] Hidalgo, M.; Martin-Santamaria, S.; Recio, I.; Sanchez-Moreno, C.; de Pascual-Teresa, B.; Rimbach, G.; de Pascual-Teresa, S., Potential anti-inflammatory, anti-adhesive, anti/estrogenic, and angiotensin-converting enzyme inhibitory activities of anthocyanins and their gut metabolites. *Genes Nutr* 2012, 7, 295-306.

[76] Shahrzad, S.; Aoyagi, K.; Winter, A.; Koyama, A.; Bitsch, I., Pharmacokinetics of gallic acid and its relative bioavailability from tea in healthy humans. *J Nutr* 2001, 131, 1207-10.

[77] Yan, J.; Chen, Y.; He, C.; Yang, Z. Z.; Lü, C.; Chen, X. S., Andrographolide induces cell cycle arrest and apoptosis in human rheumatoid arthritis fibroblast-like synoviocytes. *Cell Biol Toxicol* 2012, 28, 47-56.

[78] Kuppan, G.; Balasubramanyam, J.; Monickaraj, F.; Srinivasan, G.; Mohan, V.; Balasubramanyam, M., Transcriptional regulation of cytokines and oxidative stress by gallic acid in human THP-1 monocytes. *Cytokine* 2010, 49, 229-34.

[79] Kim, S. H.; Jun, C. D.; Suk, K.; Choi, B. J.; Lim, H.; Park, S.; Lee, S. H.; Shin, H. Y.; Kim, D. K.; Shin, T. Y., Gallic acid inhibits histamine release and pro-inflammatory cytokine production in mast cells. *Toxicol Sci* 2006, 91, 123-31.

[80] Kim, M. J.; Seong, A. R.; Yoo, J. Y.; Jin, C. H.; Lee, Y. H.; Kim, Y. J.; Lee, J.; Jun, W. J.; Yoon, H. G., Gallic acid, a histone acetyltransferase inhibitor, suppresses β-amyloid neurotoxicity by inhibiting microglial-mediated neuroinflammation. *Mol Nutr Food Res* 2011, 55, 1798-808.

[81] Mattson, M. P., Emerging neuroprotective strategies for Alzheimer's disease: dietary restriction, telomerase activation, and stem cell therapy. *Exp Gerontol* 2000, 35, 489-502.

In: Handbook on Gallic Acid
Editors: M. A. Thompson and P. B. Collins
ISBN: 978-1-62618-921-8

Chapter 11

BIOTECHNOLOGICAL PRODUCTION OF GALLIC ACID

P. Aguilar-Zárate, M. L. Chávez-González, R. Rodríguez-Herrera and C. N. Aguilar*
Department of Food Science and Technology, School of Chemistry, Universidad Autónoma de Coahuila, Venustiano, Caranza, Colonia República Oriente, Saltillo, Coahuila, México

ABSTRACT

Gallic acid (3, 4, 5-trihydroxybenzoic acid), is pytochemical compound occurring as a free molecule or as part of tannic acid molecule in plant tissue. Gallic acid has many significant biological activities such as antioxidant, antiinflmatory, antigungal,, and carcinogenic properties, it has attracted considerable interest Tannic acid biodegradation is the key condition for the gallic acid production. Biodegradation of tannic acid can be controlled by enzymatic or microbiological ways, giving the possibility for developing biotechnological processes for the production of gallic acid. Several studies on biotechnological production of gallic acid have generated significant achievements in recent years, interest in the search for new routes to produce more "*clean*" compounds has turned the interest of researchers for fermentation processes, biotransformation rich substrates tannins by microorganisms has led to high yields of gallic acid, however, these new areas of opportunity still present challenges which need studies involve more factors in the microbial production of gallic acid is necessary to make efforts to deepen scale up production studies. In addition, methods for higher recovery and purification of this important compound are necessary.

Keywords: Tannic acid, biodegradation, gallic acid, solid and submerged fermentation, bioprocess

---

* Corresponding author: C. N. Aguilar. Departament of Food Science and Technology. School of Chemistry. Universidad Autónoma de Coahuila. Venustiano Caranza. Colonia República Oriente. CP. 25280. Saltillo, Coahuila, México. E-mail: cristobal.aguilar@uadec.edu.mx.

Introduction

Gallic acid (3, 4, 5-trihydroxybenzoic acid), is an organic substance occurring in many plants as a free molecule or as part of tannic acid molecule (Figure 1) (Bajpai and Patil, 2008). Gallic acid was obtained by first time from a precipitate of gall-nuts solution, that was called essential salts of gall or gall-nut salts (Haslam, 1992). One of the most important activities is due to the phenolic hydroxyl group in their structure that acts as a hydrogen donor, and they can effectively scavenge free radicals (Nakamura *et al*, 2012).

Since was found to have many significant biological activities such as antioxidant, antiinflmatory, antigungal,, and carcinogenic properties, it has attracted considerable interest (Daneshfar *et al.* 2008).

It is mainly used in the pharmaceutical industry for manufacturing trimethoprim an antimicrobial agent (Kar and Banerjee, 2000). It can be used in the manufacturing of ordinary writing inks and dyes, as photographic developer, in tannery industry for homogenization of tannins, for the production of pyrogallol and gallic acid esters among other compounds (Banerjee *et al.*, 2005).

Gallic acid is also used in food industries as a substrate for the chemical or enzymatic synthesis of propyl gallate, a potent antioxidant and food preservative (Fernandez-Lorente *et al.*, 2011). Gallic acid has important biological activities as antioxidant, antibacterial, antiviral, analgesic, etc. Its antioxidant activity allows it to act as an anti apoptotic agent and to protect the cells of oxidation damage (Treviño-Cueto *et al.*, 2007). Gallic acid show citotoxic activity against cancer cells, without harming normal cells (Mondal *et al.,* 2001).

Due to its several interesting properties and commercial applications gallic acid is a compound of great interest to both pharmaceutical and chemical industries. The world wide annual demand of gallic acid is 8000 tonnes approximately (Lokeswari *et al.*, 2010) and the natural occurrence is restricted. Hence, new ways to obtain gallic acid from different sources has been developed.

Sources of Gallic Acid

Tannins are polyphenolic compounds widely distributed in nature. They have molecular weight higher than 500, reaching values above 20000 kDa (Belmares *et al.*, 2004). Tannins are divided into four major groups: gallotannins, ellagitannins, condensed tannins, and complex tannins. Gallotannins differ from the other groups by the presence of gallic acid esterified to a glucose core. To maintain the binding capability, gallotannins must have more than two gallic acid constituents in their chemical structure. Gallotannins can be easily hydrolyzed by different ways to obtain gallic acid. Gallic acid is one of the structural constituents of some hydrolysable tannins, such as tannic acid (Aguilar *et al.*, 2007).

Gallic acid is usually obtained from different parts (barks, needles, heartwood, grasses, seeds and flowers) of vascular plants, by chemical synthesis (alkali or acid hydrolysis) or by the transformation of substrates using microorganisms. Gallic acid and its esters are industrially important chemicals used in the food and pharmaceutical industry (synthesis of propyl gallate and trimethoprim) (Daneshfar *et al.* 2008). Below some of the sources of gallic acid are described.

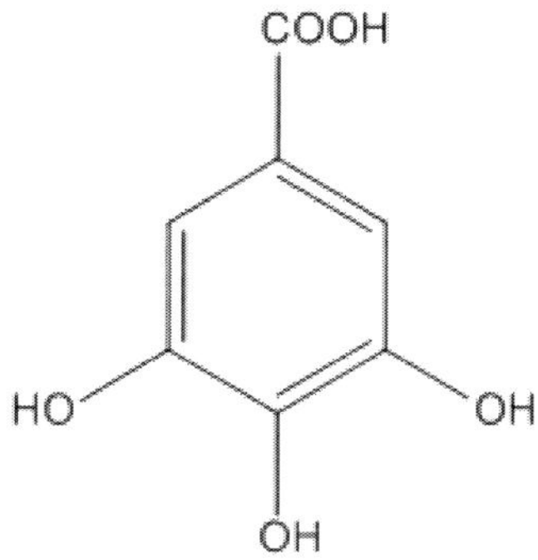

Figure 1. Gallic acid.

Plants

Plant kingdom has a wide chemical diversity as consequence of the natural adaptation and evolution. Plants are significant sources of different natural products that have important functions on them. That compounds have broader ecological roles and may participate in plant growth and development. Recent information suggests that some bioactive compounds may not serve as simply chemical barriers. Phytochemicals have several roles in defense-related signaling process.

Tannins are the fourth most abundant plants constituents after cellulose, hemicellulose and lignin (Das Mohapatra *et al.*, 2006). They play an important role in the immunity, protecting vulnerable parts of plants from microbial attack. Gallic acid is the main production of gallotannins hydrolysis, and it is present in some leaves, cortexes, fruits and they are polymerized or in free form (Martínez-Vázquez, *et al.*, 2011). Production of gallic acid from tannin-containing plant materials (gallotannins) has been reported by a number of workers.

One of the most studied plants is *Camellia sinensis*. This is used in the elaboration of green and black tea. It is well known its higher content of phenolic compounds. Tea polyphenols constitute the major portion of the soluble ingredients and are also the essential components of tea which have physiological functions. Tea catechines are the primary polyphenols in the tea (Hara *et al.*, 1995).

The essential components include epigallocatechine gallate, epigallocatechine, epicatechin gallate, epicatechin, gallocatechin and catechin (Lu *et al.*, 2009). Content of polyphenolic compounds are greater in green tea than black tea because of the latter is completely fermented. The fermentation of green tea results on the hydrolysis of polyphenols. This phenomenon increases the content of gallic acid in black tea as reported by Lu and Chen (2008) and Lu *et al.* (2009). They demonstrated the increase of gallic acid content and hydrolysis of epigallocatechin gallate and epicatechin gallate in green tea leaves infusion after an enzymatic treatment. Chandini *et al.*, (2011) enhance the recovery of polyphenols and gallic acid in black tea using the same technique.

American plants have also been used in the elaboration of beverages and in traditional medicine. Leaves of hojasen (*Fluorensia cernua*) and gobernadora (*Larrea tridentata*) are used to treat digestive disorders, rheumatism, venereal diseases, sores, bronchitis, chicken pox and common cold (Ventura *et al.*, 2009).

Both plants have a high content of hydrolysable tannins and condensed tannins (Belmares *et al.*, 2004). Production of phenolic antioxidants present in tannin-rich shrubs mainly derived from hydrolysable tannins has been demonstrated (Ventura *et al.*, 2008). Treviño-Cueto *et al.* (2007) reported a high content of hydrolysable tannins in gobernadora leaves. They have obtained an accumulation of 0.33 g l^{-1} of gallic acid using a microbial tannins transformation. Ventura *et al.* (2009) evaluated the use of hojasen and gobernadora to produce gallic acid. Biotransformation of tannins to gallic acid reaches a 93% in gobernadora and 89% in hojasen. Considering the acceptable levels of tannins in these plants, they could be an alternative for the solution to the worldwide demand of gallic acid.

Some other plants as cascalote (*Caesalpinia cacalaco*) is a Mexican tree and it is an excellent source of phenolics and tannic acid used in Mexican tannery industry the gallic acids is content in a 410.5 g/L (Veloz-García *et al.*, 2010).

Agro-industrial residues and forest products are generally consider the best source of tannin-rich substrate (Das Mohapatra *et al.*, 2006). Hence many plants and industrial by products have been used as substrates for gallic acid synthesis (Table 1).

Das Mohapatra *et al.* (2006) tested different forest plant extracts rich in tannins for the production of gallic acid. Lokeswari and Raju (2007) proposed a method for bioconversion of tannins obtained from myrobalan (*Terminalia chebula*) into gallic acid. Gallotannins concentration was estimated to be 38% w/v therefore high content of gallic acid (7.68 mg/mL) was obtained. A yield of 8.1 mg/mL of gallic acid was also obtained from *Terminalia belerica* under optimal conditions (Patil *et al.*, 2011).

Application of agro-industrial residues is certainly economical and it also reduces environmental pollution. Tea stalks are a novel and less exploited agro-residue. It is an interesting substrate for gallic acid production due to its abundant content of tannins (Wang *et al.*, 2012). An industrial waste such as the Indian rosewood (*Dalebergia* sissoo) sawdust which is a residue of the furniture making has been utilized for gallic acid production (Beniwal *et al.*, 2012). A tannins content of approximately 2.1 mg/g allowed releasing 5.4 mg/g ds of gallic acid. In nature gallic acid is found polymerized or joined to a sugar core and rarely can be found as a free form. Therefore, the employ of mechanisms to release this compound are necessary.

Chemical Synthesis

At the present gallic acid is produced industrially by acid hydrolysis of natural occurring gallotannins.

Due to the high cost, low yield of desired product and production of large toxic effluent by acid hydrolysis an enzyme based eco-friendly technology for gallic acid production is urgently required (Beniwal *et al.*, 2010).

Some efforts to produce gallic acid was carried out, Kambourakis Frost (2000) reported the conversion of 3-Dehydroshikimic acid into gallic acid likely proceeds via initial enolization (catalyzed by inorganic phosphate) of an α-hydroxycarbonyl and oxidation (Cu^{2+}) of the resulting enediol.

Table 1. Plants and agro-industrial residues used as potential sources of gallic acid

Plant / Agro-industrial residue	Reference
Caesalpinia digyna	Banerjee *et al.*, 2005
Coffee husk	Sabu *et al.*, 2006
Tamarind seed powder	Sabu *et al.*, 2006
Acacia auriculiformis	Das Mohapatra *et al.*, 2006
Casuarina equisetifolia	Das Mohapatra *et al.*, 2006
Psidium guazava	Das Mohapatra *et al.*, 2006
Anacardium occidentale	Das Mohapatra *et al.*, 2006
Delonix regia	Das Mohapatra *et al.*, 2006
Eucalyptus tereticornis	Das Mohapatra *et al.*, 2006
Casia fistula	Das Mohapatra *et al.*, 2006
Ficus benghalensis	Das Mohapatra *et al.*, 2006
Terminalia chebula	Lokeswari and Raju, 2007
Larrea tridentata Cov.	Treviño-Cueto *et al.*, 2007
Creosote bush	Treviño-Cueto, *et al.*, 2007
Syzygium cumini	Kumar *et al.*, 2007
Phyllanthus emblica	Yadav *et al.*, 2008
Zyzyphus Mauritania	Yadav *et al.*, 2008
Syzygium sp.	Yadav *et al.*, 2008
Acacia nilotica	Yadav *et al.*, 2008
Olive mill wastewater	Rodríguez *et al.*, 2009
Wine distillery wastewater	Rodríguez *et al.*, 2009
Flourencia cernua	Ventura *et al.*, 2009
Camellia sinensis	Lu *et al.*, 2009
Terminalia belerica Roxb.	Patil *et al.*, 2011
Eugenia cuspidate	Selwal and Selwal, 2012
Tea stalks	Wang *et al.*, 2012
Indian rosewood sawdust	Beniwal *et al.*, 2012
Coal wastewater	Ramakrishnan and Surampalli, 2013

Fermentation

Fermentation is a process where a complex molecule is transformed into another simple. It refers to various processes involving limited biochemical changes brought about by microorganisms and their enzymes. Consequently, microbial metabolic potentials are widely exploited for industrial processes.

Fermentation as an effective way to obtain gallic acid through clean and controlled processes, has been used various kinds of both liquid and solid state fermentation, where plants are used as support and substrate for metabolic reactions, there are some studies where have been reported the use of bacteria, fungi and yeasts (Banerjee and Pati, 2007).

Normally tannins inhibit the growth of a number of microorganisms, resist microbial attack and are recalcitrant to biodegradation. Despite the antimicrobial properties of tannins, many fungi, bacteria and yeasts are quite resistant to tannins, and can grow and develop on them (Bath *et al.*, 1998). This ability to grow up in this kind of substrate is due the production of tannase enzyme or tannin acyl hydrolase (E.C. 3.1.20), tannase catalyses the breakdown of ester bonds in hydrolysable tannins such as tannic acid (Aguilar *et al.*, 2007; Ordoñez *et al.* 2011).

The mechanism by which occurs gallic acid production is by the catalytic action of the enzyme to breakdown of ester bonds (Figure 2) from hydrolysable tannins such as tannic acid, methyl gallate, ethyl gallate, *n*-propylgallate and isoamyl gallate (Bajpai and Patil, 1997; Aguilar *et al.* 2007). In some cases although tannase production can occur in the absence of tannic acid. (Aguilar *et al.* 2007). The action of this enzyme on tannic acid, TAH gives glucose and gallic acid as final reaction products (Aguilar Gutiéerez-Sánchez, 2001).

TAH can be obtained from plant, animal an microbial sources. From plant sources, the enzyme is present in tannin-rich vegetables mainly in their fruits, leaves, branches and barks of trees lime. As for animal sources, TAH and be extracted from bovine intestine and from the ruminal mucous. Also, it has been reported that some insects produce the enzyme during the larval state. The most important sources to obtain the enzyme is by microbial way, because the produced enzymes are more stable than similar one obtained from other sources(Aguilar Gutiérrez-Sánchez, 2001), also tannase produced from microbial sources find immense applications at various industrial processes due to its stability and availability (Natarajan and Rajendran, 2009). Microorganisms have been reported to produce the enzyme tannase during the fermentation of tannins (Banerjee and Pati, 2007).

MICROBIAL PRODUCTION OF GALLIC ACID

Microbes rich in tannase carry out tannin hydrolysis of tannin rich substrates efficiently producing tannase and gallic acid. Generally they use the central core of tannins as carbon source. The concentration of tannase and gallic acid production depends upon the properties such as enzymes of microorganisms, substrate and fermentation conditions (Srivastava and Kar, 2011).

The main hindrance in the development of a successful bioconversion process is the sensitivity of the microorganisms to tannic acid and the oxidation of the unused tannic acid (Bajpai and Patil, 2008). However, tannase and gallic acid production has been reported widely in various microorganisms (Figure 3).

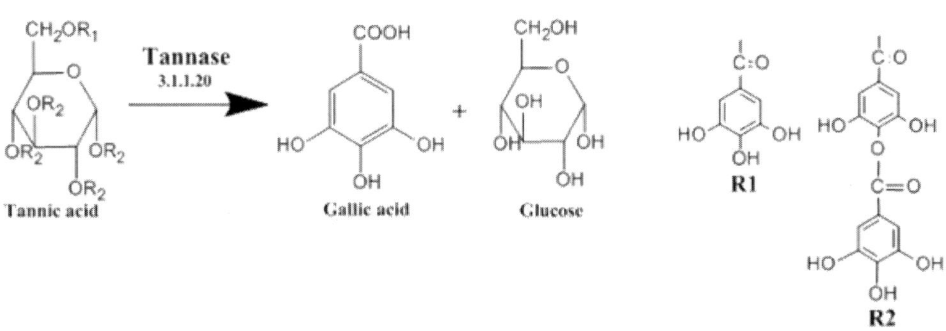

Figure 2. Enzymatic hydrolysis of tannic acid by tannase and release of gallic acid and glucose.

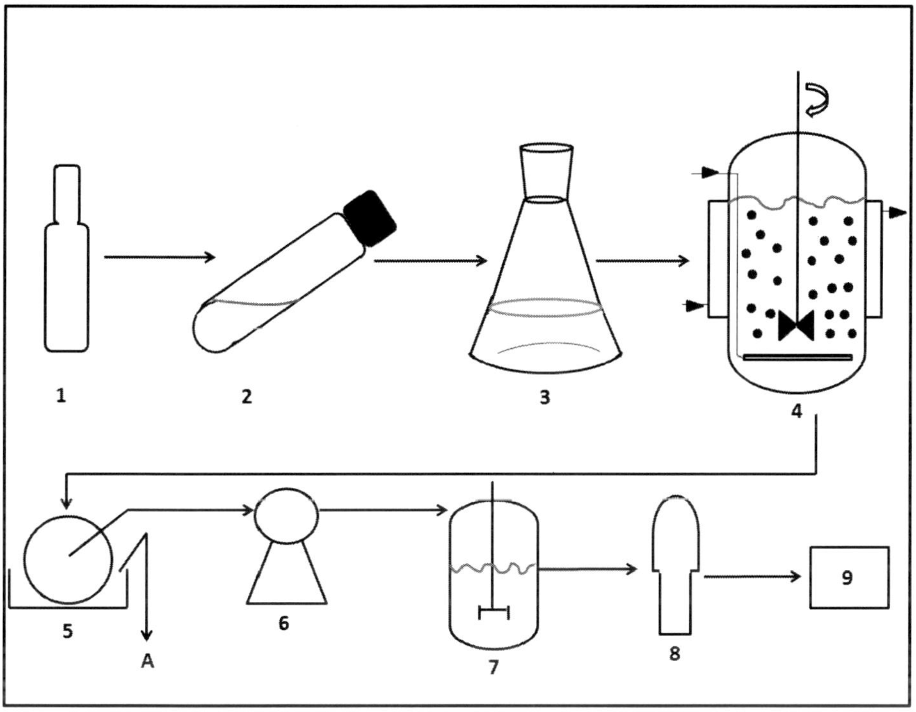

Figure 3. Process production of gallic acid by fermentation. 1. Conserved strain, 2. Slant, 3. Seed culture, 4. Fermenter, 5. Filter rotor, 6. Extraction centrifuge, 7. Crystallizer, 8. Dryer, 9. Final product (gallic acid). A. Biomass.

Bacterial

Literature on production of gallic acid and tannase using bacterial strains is limited. Lewis *et al.* (1969) were the first in report a bacterial strain (*Achomobacter* sp.) capable of degrading tannins as sole carbon source. Deschamps *et al.* 1983 isolated a group of bacteria able to grow in a media supplemented with tannic acid. However few bacteria are known to produce tannase and gallic acid and include certain species of *Bacilli, Corynebacterium* sp., *Lactobacillus* sp., and *Serratia* sp. (Rodríguez *et al.*, 2008; Milva *et al.*, 2010). Most of the bacterial strains capable of degrading tannins isolated till date are anaerobic or facultative anaerobic. However, the process based on anaerobic bacteria is slow (Chowdhury *et al.*, 2007; Beniwal *et al.*, 2010).

In recent years, investigations have centered in the search and isolation of new bacterial strains with the capability to produce high titers of tannase and to synthesize gallic acid (table 2).

Over the past decade bacterial strains mainly the genres *Bacillus* and *Lactobacillus* have been reported as tannin degrader. Osawa *et al.*, (2000) were the first in report the isolation of a *Lactobacillus* strain capable of degrading tannins. *Lactobacillus plantarum* is a lactic acid bacteria with high tannase activity (Vaquero *et al.*, 2004). Lactic acid bacteria play a vital role in hydrolysis of tannins present in food and intestines due to their ability to hydrolyze and therefore, decrease the absorption of these compounds to intestine cells. Considering the above, lactic acid bacteria have been tested as tannase and gallic acid producer.

Table 2. Tannin-degrading bacteria

Microorganism	Reference
Achomobacter sp	Lewis *et al.*, (1969)
Bacillus plumilus	Deschamps *et al.*, (1983)
Bacillus polymixa	Deschamps *et al.*, (1983)
Corinebacterium sp.	Deschamps *et al.*, (1983)
Klebsiella pneumoniae	Deschamps *et al.*, (1983)
Pseudomonas solanaceaum	Deschamps *et al.*, (1983)
Citrobacter freundii	Kumar *et al.*, (1999)
Lactobacillus plantarum	Osawa *et al.*, (2000)
Lactobacillus paraplantarum	Osawa *et al.*, (2000)
Lactobacillus pentosus	Osawa *et al.*, (2000)
Bacillus licheniformis	Mondal *et al.*, (2000)
Bacillus cereus	Mondal *et al.*, (2001)
Lactobacillus plantarum	Ayed y Hamdi (2002)
Selenomonas ruminantium	Belmares *et al.*, (2004)
Lactobacillus paraplantarum	Nishitani *et al.*, (2004)
Lactobacillus acidophilus	Nishitani *et al.*, (2004)
Lactobacillus pentosus	Nishitani *et al.*, (2004)
Lactobacillus animalis	Nishitani *et al.*, (2004)
Lactobacillus murinus	Nishitani *et al.*, (2004)
Lactobacillus faecalis	Nishitani *et al.*, (2004)
Lactobacillus acidilactici	Nishitani *et al.*, (2004)
Lactobacillus pentosaceaus	Nishitani *et al.*, (2004)
Enterococcus faecalis	Goel *et al.*, (2005)
Lactobacillus sp. ASR-S1	Sabu *et al.*, (2006)
Pediococcus pentosaceus	Guzmán-López, *et al.*, (2009)
Lactobacillus buchneri	Guzmán-López, *et al.*, (2009)
Lactobacillus hilgardii	Guzmán-López, *et al.*, (2009)
Weisella confusa	Guzmán-López, *et al.*, (2009)
Pseudomonas aeruginosa	Selwal *et al.*, (2010)
Serratia ficaria	Belur *et al.*, (2010)
Serratia marcescens	Belur *et al.*, (2010)
Microbacterium terregens	Belur *et al.*, (2010)
Providencia rettgeri	Belur *et al.*, (2010)
Bacillus sphaericus	Raghuwanshi *et al.*, (2011)
Bacillus massieliensis	Belur *et al.*, (2012)

Bacterial production of gallic acid and tannase has been carried out mainly in submerged fermentation (Belur *et al.*, 2010). Higher production of tannase and gallic acid have been obtained by submerged fermentation due to bacteria prefer this process.

However Sabu *et al.* (2006) carried out solid state fermentation using *Lactobacillus* sp. and agro-industrial wastes as solid substrate. To enhance the gallic acid production by bacteria, molecular modifications in bacterial strains have been carried out. Muir *et al.*, 2011 cloned the gene responsible for gallic acid synthesis from plants on *E. coli*.

They over expressed the gene for enzyme shikimate dehydrogenase and demonstrated the production in vitro of gallic acid.

Production of enzymes and microbial compounds through fermentation are influenced by physical and chemical environment of cultural conditions such as quality and quantity of carbon and nitrogen source, presence of inducers, pH, temperature, aeration, etc. For effective bioprocess, it is necessary to optimize culture conditions and composition of media (Das Mohapatra *et al.*, 2009).

Beniwal *et al.*, 2010 optimized the process for tannase and gallic acid production by *Enterobacter cloacae*. They used the strategy to optimize each parameter (incubation time and temperature; pH; inoculum; substrate concentration; sugar, phosphate and nitrogen sources) independent of the others reaching a maximal production of 3.47 mg/mL for gallic acid and 0.38 U/mL of tannase.

Conventional optimization procedures involve altering one independent variable at the time keeping the others constant. These procedures are time consuming, require more experimental and cannot provide information about the mutual interactions (Rao *et al.*, 2008). Therefore statistical tools and experimental designs help to gain more information about optimization.

Taguchi design of experiments is an orthogonal array that consists in the study of any given system by a set of independent variables (factors) over a specific region of interest (levels) (Taguchi, 1986). This approach facilitates the identification of individual factors and the relationship between variables (Eriksson *et al.*, 1990). Das Mohapatra *et al.*, 2009 applied this methodology to optimize the production process for tannase using a strain of *Bacillus licheniformis* demonstrating an increase of 2.8-fold compared with the unoptimized condition.

Response Surface Methodology (RSM) is an efficient tool for optimization of physicochemical parameters. It is an experimental design upper than traditional designs and has the advantage of determine the influence of variables over the process and reach high yields under the best possible economic conditions. Raghuwanshi *et al.* (2011) reported the use of a face-centered central composite design, a type of RSM, for the optimization of tannase and gallic acid production by *Bacillus sphaericus*.

They reach the higher bacterial tannase activity (16.5 U/L) which was 13.6-fold higher than unoptimized conditions and a higher conversion (90.8%) of tannic acid to gallic acid. Thus literature review suggests that optimal conditions can vary considerably depending on the microorganism, type of fermentation, physicochemical conditions, and experimental process.

Fungus

Gallic acid production by fungus has been studied in various culture media, the most studied and fungal species that have shown greater ability of biotransformation are the following species: *Aspergillus*, *Penicillium*, *Fusarium* and *Trichoderma* (Aguilar *et al.*, 2007). The production of gallic acid through fungus has been carried out by tannase action (Riul *et al.*, 2013) and this enzyme that can be produced in the presence of tannic acid by various filamentous fungi, (Ramírez-Coronel, *et al.* 2003). The mechanism by which occurs gallic acid production by fungus is the same that reported by bacteria and yeast.

Studies on tannase production by fungus have been carried out in submerged and solid-state cultures. Depending on the strain and the culture conditions the enzyme is induced and expressed with different levels of activity, showing different production patterns (Aguilar *et al.*, 2007).

Production of tannase has been used in the production of gallic acid have been biosynthesized in fungal submerged culture (Gaaathon et al. 1989; Kar and Banerjee 2000; García-Nájera *et al.* 2002) and solid state culture (Kar and Banerjee 2000). Studies have reported interesting advantages between the tannase produced by solid state culture in relation with that produced by submerged culture (Aguilar *et al.*, 2001; Viniegra-González *et al.*, 2003). In those, attractive advantages were indicated: the high-production titles, the extracellular nature of the enzyme, and the stability towards wide pH and temperature ranges (Aguilar *et al.*, 2007). Also has been reported studies in modified solid state culture (Kar *et al.* 2002) using microencapsulated tannase in chitosan and as free enzyme (Yu *et al*, 2004) and by mycelium-bon tannase in organic solvent lime 1-propanol (Yu and Li 2006). This is last is an important contribution to discover the new reaction of tannase in nonconventional media because the industrial biocatalysis has grown considerably in recent years for its potential applications in fine-chemical, agrochemical, perfumery, flavor pharmaceutical, and drug industries.

A large number of fungi have been reported with the ability to produce the enzyme tannase, like: *A. japonicus*, *A. gallomyces*, *A. awamori*, *A. versicolor*, *A. flavus*, *A. caespitosum*, *A. oryzae*, *A. aureus*, *A. fischeri*, *A. rugulosus*, *A, terreus*, *A. oryzae*, *A. niger*, *A. fumigatus*, *A. foetidus*, *A. aculeatus*, *A. awamori nakazawa*, *A. ruber*, *R. oryzae*, *P. variable*, *P. atramentosm*, *Lenzites elegans P. notatum*, *P. islandicum*, *P. chrysogenum*, *P. digitatum*, *P. acrellanum*, *P. carylophilum*, *P. citrinum*, *P. charlessi*, *P. glaucum*, *P. crustosum*, *P. restrictum*, *P. glabrum*, *Trichoderma viride*, *T. hamatum*, *T. harzianum*, *Fusarium solani*, *F. oxysporium*, *Mucor sp. Paecilomyces variotii*, *Rhizopus oryzae*, *Cryphonectria parasitica*, *Heliocostylum sp.*, *Cunninghamela sp.*, *Syncephalastrum reacemosum*, *Neurospora crassa* (Kumar *et al.*, 2007; Goncalves *et al.*, 2011; Jana *et al.* 2012; Chávez-González *et al.*, 2012).

May be highlighted as good producers of versatiles tannases that act efficiently on different types of tannins. Tannase production by fungi has reported using both submerged and solid-state fermentations in the presence and absence of an inducer (Goncalves *et al.* 2011).

Another form to obtain gallic acid is when the reaction substrate is the methyl ester of tannic acid, TAH gives gallic acid and methanol as final products (Aguilar Gutiérrez-Sánchez, 2001).

One way more to obtain gallic acid is prepared by the degalloylation of an epigallocatechin this phenomen has been reported by Battestin *et al.* using extract from green tea and tannase from *Paecilomyces variotii* (2008). Enzymatic synthesis of gallic acid esters from gallic acid and alcohols in organic solvents was studied using microencapsulated tannase from *A. niger* (Yu *et al.*, 2004).

There are reports about the use of natural sources rich in tannins to release gallic acid by fermentation, some examples of studies is the use of *Larrea tridentata* plant this by culture on solid medium using the fungus *Aspergillus niger* Aa-20 (Treviño-Cueto *et al.*, 2007). In fact Gobernadora is a potential source of gallic acid by solid state culture, reported an accumulation of 0.33 g/L^{-1} of gallic acid during the first 48 h of culture.

With a tannic acid concentration of 45 g/L and agitation under optimum conditions for gallic acid production was reached a maximum of 40.3 gl^{-1} at 24 h of fermentation by *Aspergillus awamori* (Seth, M. Chand, S. 2000).

The fungal culture produced high yields of extracellular tannase under solid-state fermentation (SSF) using different agro forest residues such as Amla leaves (*Phyllanthus emblica*), Ber leaves (*Zyzphus mauritina*), Jamun leaves (*Syzygium cumini*), Jamoa leaves (*Syzygium sp.*) and Keekar leaves (*Acacia niloticaI*). Among differentsubstrates used, Jamun leaves yielded maximal extra-cellular production of tannase (Manjit *et al.* 2008).

Also have been used inert supports in the gallic acid production to avoid recover some other compounds, product of heterogeneity of matrix natural sources.

Exists some studies where are investigate the effects of polyurethane matrices in a solid state culture on *Aspergillus niger*-induced tannase and gallic acid production and they found discontinuous matrix of foam polyurethane favoured the gallic acid accumulation; 7.64 g/L was obtained during the first 12 h of the culture, whereas the continuous and semi-discontinuous matrices produced 0.7 and 0.86 g/L of gallic acid, respectively. The gallic acid accumulation was proportional to the tannase biosynthesis in the production systems (Treviño *et al.* 2007). Among yeast that are involved in the production of tannase are species like as: *Arxula adeninivorans*, *Candida* sp., *Cadida utilis, Saccharomyces cerevisiae*, *Mycotorula japonica*, *Pichia* spp., *Debaryomyces hansenii* (Chávez-González *et al.*, 2012)

CONCLUDING REMARKS

Research on microbial production of gallic acid have generated significant achievements in recent years, interest in the search for new routes to produce more "*clean*" compounds has turned the interest of researchers for fermentation processes, biotransformation rich substrates tannins by microorganisms has led to high yields of gallic acid, however, these new areas of opportunity still present challenges which need studies involve more factors in the microbial production of gallic acid is necessary to make efforts to deepen scale up production studies.

In addition, methods for higher recovery and purification of this important compound are necessary.

ACKNOWLEDGMENTS

Authors P. Aguilar-Zarate and M. Chávez-González thank CONACYT for the financial support.

REFERENCES

Aguilar, C. N., Rodríguez, R., Gutiérrez-Sánchez, G., Augur, C., Favela-Torres, E., Prado-Barragán, L. A., Ramírez-Coronel A., and Contreras-Esquivel, J. C. (2007). Microbial tannases: advances and perspectives. *Applied Microbiology and Biotechnology*. 76, 47-59.

Aguilar, C. N. and Gutiérrez-Sánchez. (2001). Review: Sources, Properties, Applications and Potential uses of Tannin Acyl Hydrolase. *Food Sci. Tech. Int.* 7 (5) 373-382.

Ayed, L. y Hamdi, M. (2002). Culture conditions of tannase production by *Lactobacillus plantarum*. *Biotechnology Letters*. 24:1763-1765

Bajpai, B., Pail, S. (1997). Introduction of tannin acyldrolase (EC 3.1.1.20) activity in some members of fungi imperfecti. *Enzyme Microb. Technol.* 20:612-14

Bajpai, B. and Patil, S. (2008). A new approach to microbial production of gallic acid. *Brazilian Journal of Microbiology*. 39, 708-711.

Banerjee, D. and Pati, B. R. (2007). Optimization of tannase production by *Aureobasidium pullulans* DBS66. *Journal of Microbiology and Biotechnology*. 17(6), 1049-1053.

Banerjee, R., Mukherjee, G., Patra, K. C. (2005). Microbial transformation of tannin-rich substrate to gallic acid through co-culture method. *Bioresource Technology*. 96:949-953.

Battestin, V., Macedo, G. A., De Freitas, V. A. P. (2008). Hydrolysis of epigallocatechin gallate using a tannase from *Paecilomyces variotii*. *Food Chemistry*. 108. 228-233.

Bhat, T. K., Singh, B., Sharma, O. P. (1998). Microbial degradation of tannins-A current perspective. *Biodegradation*. 9:343-357.

Belmares, R., Contreras-Esquivel, J. C., Rodriguez-Herrera, R., Ramirez-Coronel, A. y Aguilar, C. N. (2004). Microbial production of tannase: an enzyme with potential use in food industry. *Lebensmittel-Wissenschaft Technologie* 37:857-864.

Belur, P. D., Gopal, M., Nirmala, K. R. y Basavaraj, N. (2010). Production of novel cell-associated tannase from newly isolated Serratia ficaria DTC. *Journal of Microbiology and Biotechnology*. 20(4), 732–736.

Beniwal, V., Chhokar, V., Singh, N., and Sharma, J. (2010). Optimization of process parameters for the production of tannase and gallic acid by *Enterobacter cloacae* MTCC 9125. *Journal of American Science*. 6 (8), 389-397.

Beniwal, V., Goel, G., Kumar, A., and Chhokar, V. (2012). Production of tannase through solid state fermentation using Indian Rosewood (*Dalbergia Sissoo*) sawdust –a timber industry waste. *Annals of Microbiology*. In press. DOI 10.1007/s13213-012-0508-6

Chandini, S. K., Rao, L. J., Gowthaman, M. K., Haware, D. J., and Subramanian, R. (2011). Enzymatic treatment to improve the quality of black tea extracts. *Food Chemistry*. 127, 1039-1045.

Chávez-González, M., Rodríguez-Durán, L. V., Balagurusamy, N., Prado-Barragán, A., Rodríguez, R., Contreras, J. C., Aguilar, C. N. (2012). Biotechnological Advances and Challenge of Tannase: An Overview. *Food Bioprocess Technol*. 5:445-459.

Chowdhury, S. P., Khana, S., Verma, S. C., and Tripathi, A. K. (2007). Molecular diversity of tannic acid degrading bacteria isolated from tannery soil. *Journal of Applied Biochemistry*. 97, 1210-1219.

Daneshfar, A., Ghaziaskar, H. S. and Homayoun, N. (2008). Soulubility of gallic acid in methanol, ethanol, water and ethyl acetate. *Journal of Chemical and Engeenering. Data*. 53 (3): 776-778.

Das Mohapatra, P., Mondal, K. C. and Pati, B. R. (2006). Production of tannase through submerge fermentation of tannin-containing plant extracts by *Bacillus licheniformis* KBR6. *Polish Journal of Microbiology*. 55(4), 297-301.

Das Mohapatra, P. K., Maity, C., Rao, R. S., Pati, B. R., and Mondal, K. C. (2009). Tannase production by *Bacillus licheniformis* KBR6: optimization of submerged culture conditions by Taguchi DOE methodology. *Food Research International*. 42, 430-435.

Deschamps, A. M., Otuk, G. y Lebeault, J. M. (1983). Production of tannase and degradation of chestnut tannin by bacteria. *Journal of Fermentation Technology*. 61:55-59.
Deschamps, A. M., Otuk, G. y Lebeault, J. M. (1983). Production of tannase and degradation of chestnut tannin by bacteria. *Journal of Fermentation Technology* 61:55-59.
Eriksson, K. E. L., Blanchette, R. A. and Ander, P. (1990). Microbial and enzymetic degradation of chestnut tannin by bacteria. *Microbial and Enzymatic Degradation of Wood and Wood Components*. Springer-Verlag, Berlin.
Fernández-Lorente, G., Bolivar, J. M., Rocha-Martin, J., Curiel, J. A., Muñoz, R., de las Rivas, B., Carrascosa, A. V., and Guisan, J. M. (2011). Synthesis of propyl gallate by transesterification of tannic acid in aqueous media catalyzed by immovilised derivatives of tannase from *Lactobacillus plantarum*. *Food Chemistry*. 128, 214-217.
Gaathon, A., Gross, Z., Rozhanski, M. (1989). Propyl gallate: enzymatic synthesis ina reverse micelle system. *Enzyme Microb. Technol.* 11:604-609.
García-Nájera, J. A., Contreras-Esquivel, J. C., Rodríguez-Herrera, R., Prado-Barragán, L. A., Aguilar, C. N. (2002). Fungal production of 3,4,5-trihydroxybenzoic acid in submerged culture (in Spanish). *Proceeding of the third international symposium on bioprocess engineering*, IBT-UNAM, Cuernavaca, Morelos, Mexico.
Goel, G., Puniya, A. K., Aguilar, C. N., and Singh, K. (2005). Interaction of gut microflora with tannins in feeds. *Naturewissenchaften*. 92, 947-503.
Goncalves, H. B., Riul, A. J., Terenzi, H. F., Jorge, J. A. Guimaraes, L. H. S. (2011). Extracellular tannase from *Emericella nidulans* showing hypertoletance to temperature and organic solvents. *Journal of Moelcular Catalysis B: Enzymatic*. 71:29:35.
Guzmán-López, O., Loera, O., Parada, J. L., Castillo-Morales, A., Martínez-Ramirez, C., Augur, C., Gaime-Perraud, I. y Saucedo-Castañeda, G. (2009). Microcultures of lactic acid bacteria: Characterization and selection of strains, optimization of nutrients and gallic acid concentration. *Journal of Industrial Microbiology and Biotechnology* 36:11-20.
Hara, Y., Luo, S. J., Wickremasinghe, R. L., and Yamannishi, T. (1995). Chemical composition of tea. *Food Research International*. 11(3), 457-471.
Haslam, E. (1992). Gallic acid and its metabolites. In: Hemingway, R. W. and Laks, P. E. (Eds.) *Plant polyphenols*. Plenum Press, New York, 169-194.
Jana, A., Maity, C., Halder, S. K., Mondal, K. C., Pati, B. R., Mohapatra, P. K. D. (2012). Tannase Production by *Penicillium purpurogenum* PAF6 in Solid State Fermentation of Tannin-Rich Plant Residues Following OVAT and RSM. *Applied Biochemistry and Biotechnology*. DOI 10.1007/s 12010-012-9547-5.
Kambourakis, S., Frost, J. W. (2000). Synthesis of Gallic Acid: Cu^{2+} Mediated Oxidation of 3-Dehydroshikimic Acid. *J. Org. Chem.* 65: 6904-6909.
Kar, B., Banerjee, R., 2000. Biosynthesis of tannin acyl hydrolase from tannin-rich forest residue under different fermentation conditions. *J. Ind. Microbiol. Biotechnol.* 25(1):29-38.
Kar, B., Banerjee, R., Bhattacharyya, B. C. (2002). Optimization of physicochemical parameters for gallic acid production by evolutionary operation-factorial design techinique. *Process Biochem.* 37:1395-1401.
Kumar, R. A., Gunasekaran, P. and Lakshmanan, M. (1999). Biodegradation of tannic acid by *Citrobacter freundii* isolated from a tannery effluent. *Journal of Basic Microbiology*. 39, 161-168.

Kumar, R., Sharma, J. y Singh, R. (2007). Production of tannase from *Aspergillus ruber* under solid-state fermentation using jamun (*Syzygium cumini*) leaves. *Microbiological Research* 162(4): 384–390.

Lewis, J. A. y Starkey, R. L. (1969). Decomposition of Plant Tannins by Some Soil Microorganisms. *Soil Science.* 107(4):235 – 241.

Lokeswari, N. and Raju, K. J. (2007). Optimization of gallic acid production from terminalia chebula by *Aspergillus niger*. *E-Journal of Chemistry*. 4(2), 287-293.

Lokeswari, N., Sriramireddy, D., Pola, S., and Bobbarala, V. (2010). Production of gallic acid using mutant strain of *Aspergillus oryzae*. *Journal of Pharmacy Research*. 3(6), 1402-1406.

Lu, M. J. and Chen, C. (2008). Enzymatic modification by tannase increases the antioxidant activity of green tea. *Food Research International*. 41, 130-137.

Lu, M. J., Chu, S. C., Yan, L., and Chen, C. (2009). Effect of tannase treatment on protein-tannin aggregation and sensory attributes of green tea infusion. *LWT-Food Science and Technology*. 42, 338-342.

Manjit., Yadav, A., Aggarwal, N. K., Kumar, K., and Kumar, A. (2008). Tannase production by Aspergillus fumigatus MA under solid-state fermentation. *World Journal of Microbiology and Biotechnology*. 24: 3023-3030.

Martínez-Vazquez, D. G., Aguilera-Carbó, A., Reyes-Vega, M. L., Rodríguez-Herrera, R., and Aguilar, C. N. (2011). Secondary plant metabolites and phytochemical defense. In: Rodriguez-Herrera, R., Aguilar, C. N., Simpson-Williampson, J. K., and Gutierrez-Sánchez, G. (Eds.) Phytopatology in the omics era. Research Signpost, Kerala, India.

Milva, P., Lucia, R. L., Roberto, A., Alessandro, E., Guido, P., Monia, R., Arianna, L., Antonio, F., Simone, G., and Silvano, E. F. (2010). Tannic acid degradation by bacterial strains *Serratia* sp. and *Pantoea* sp. isolated from olive mill wastes mixtures. *International Biodeterioration and Biodegradation*.64, 73-80.

Mondal, K. C., Samanta, S., Giri, S., and Pati, B. R. (2001). Distribution of tannic acid degrading microorganisms in the soil and comparative study of tannase from two fungal strains. *Acta Microbiologica Polonica*. 50 (1), 75-82.

Mondal, K. C., Banerjee, D., Banerjee, R. y Pati, B. R. (2001). Production and characterization of tannase from *Bacillus cereus* KBR9. *Journal of General and Applied Microbiology* 47:263-267.

Mondal, K. C., Banerjee, R. y Pati, B. R. (2000). Tannase production by *Bacillus licheniformis*. *Biotechnology Letters* 22: 767-769.

Muir, R. M., Ibañez, A. M., Uratsu, S. L., Ingham, E. S., Leslie, C. A., McGranaham, G. H., Batra, N., Goyal, S., Joseph, J., Jammis, E. D., and Dandekar, A. M. (2011) Mechanism of gallic acid biosynthesis in bacteria (*E. coli*) and walnut (*Junglans regia*). *Plant Molecular Biology*. 75, 555-565.

Nakamura, K., Yamada, Y., Ikai, H., Kanno, T., Sasaki, K., and Niwano, Y. (2012). Bactericidal action of photoirradiated gallic acid via reactive oxygen species formation. *Journal Agriculture and Food Chemistry*. 60: 10048-10054.

Natarajan, K., Rajendran, A. (2009). Effect of fermentation parameters on extra cellular tannase production by *Lactobacillus plantarum* MTCC 1407. *Electronic Journal of Chemistry*. 6(4), 979-984.

Nishitani, Y., Sasaki, E., Fujisawa, T., and Osawa, R. (2004). Genotypic analyses of *Lactobacilli* with a range of tannase activities isolated from human feces and fermented foods. *Systematic and Applied Microbiology*. 27, 109-117.

Ordoñez, R. M., Colombo, L., Alberto, M., Isla, M. I. (2011). Production of tannase from wood-degradinng ungus using as substrate plant resides: purification and characterization. *World Journal Microbiology and Biotechnoogy*. 27: 2325-2333.

Osawa, R., Kuroiso, K., Goto, S. y Shimizu, A. (2000). Isolation of tannin degrading lactobacilli from human and fermented foods. *Applied Environmental Microbiology*. 66:3093-3097

Patil, D. B., Das, S. B., Das Mohapatra, P. K., and Nag, A. (2011). Physico-chemical studies and optimization of gallic acid production from the seed coat af *Terminalia belerica* Roxb. *Annals of Microbiology*. 61, 649-654.

Raghuwanshi, S., Dutt, K., Gupta, P., Swati, M., and Saxena, R. K. (2011). *Bacillus sphaericus*: the highest bacterial tannase producer with potential for gallic acid synthesis. *Journal of Bioscience and Bioengineering*. 111 (6), 635-640.

Ramakrishnan, A. and Surampalli, R. Y. (2013). Performance of anaerobic hybrid reactors for the treatment of complex phenolic wastewater with biogas recirculation. *Bioresource Technology*. 129, 26-32.

Rao, R. S., Kumar, G. C., Prakasham, S. R., and Hobbs, P. J. (2008). The Taguchi methodology as a statistical tool for biotechnological applications: A critical appraisal. *Biotechnology Journal*. 3, 510-523.

Riul, A. J., Goncalves, H. B., Jorge, J. A., Guimaraes, L. E. S. (2013). Characterization of a glucose-and solvent-tolerant extracellular tannase from *Aspergiilus phoenicis*. *Journal of Molecular Catalysis B: Enzymatic*. 85-86. 126-133.

Rodríguez, H., Curiel, J. A., Landete, J. M., de las Rivas, B., López de Felipe, F., Gómez-Cordovés, C., Mancheño, M. A., and Muñoz, R. (2009). Food phenolics and lactic acid bacteria. *International Journal of Food Microbiology*. 132, 79-90.

Ramírez-Coronel, M. A., Viniegra-González, G., Darvil, A., Augur, G. (2003). A novel tannase from *Aspergillus niger* with β-glucosidase activity. *Microbiology*. 149. 2941-2946.

Rodríguez, H., de las Rivas, B., Gómez-Cordovés, C., and Muñoz, R. (2008). Degradation of tannic acid by cell free extracts of *Lactobacillus plantarum*. *Food Chemistry*. 107, 664-670.

Sabu, A., Augr, C., Swatu, C., and Pandey, A. (2006). Tannase production by *Lactobacillus* sp. ASR-S1 under solid-state fermentation. *Procces Biochemistry*. 41, 575-580.

Selwal, M. K. and Selwal, K. K. (2012). High-level tannase production by *Penicillium antramentosum* KM using agro residues under submerged fermentation. *Annals of Microbioology*. 62, 139-148.

Selwal, M. K., Yadav, A., Selwal, K. K., Aggarwal, N. K., Gupta, R. y Gautam, S. K. (2010). Optimization of cultural conditions for tannase production by *Pseudomonas aeruginosa* IIIB 8914 under submerged fermentation. *World Journal of Microbiology and Biotechnology* 26:599-605

Seth, M. Chand, S. (2000). Biosynthesis of tannase andhydrolysis of tannins to gallic acid by *Aspergillus awamori*-optimisation of process parameters. *Process Biochemistry*. 36. 39-44.

Srivastava, A. and Kar, R. (2011). Production of gallic acid by immobilized *Aspergillus niger* using polyurethane foam as solid support. *International Journal of Food Engineering*. Vol. 7, Iss. 6, Article 1.

Taguchi, G. (1986). Introduction to quality engineering. *UNIPUB/Kraus International*, White Plains.

Treviño-Cueto, B., Luis, M., Contreras-Esquivel, J. C., Rodríguez, R., Aguilera, A., and Aguilar, C. N. (2007). Gallic acid and tannase accumulation during fungal solid state cultura of a tannin-rich desert plant (*Larrea tridentate* Cov.). *Bioresource Technology*. 98, 721-724.

Vaquero, I., Marcobal, A. and Muñoz, R. (2004). Tannase activity by lactic acid bacteria isolated from grape must and wine. *International Journal of Food Microbiology*. 96, 199-204.

Veloz-García, R., Martín-Martínez, R., Veloz-Rodríguez, R., Roríguez-Guerra, R., Torres-Pacheco, I., González-Chavira, M. M., Anaya-López, J. L., Guevara-Olvera, L., Feregrino-Pérez, A. A., Loarca-Piña, G., Guevara-González, R. G. (2010). Antimicrobial activitis of cascalote (*Caesalpinia cacalaco*) phenolics-containing extract against fungus *Colletotrichum lindemuthianum*.

Ventura, J., Belmares, R., Aguilera-Carbó, A., Gutierrez-Sánchez, G., and Rodríguez-Herrera, R. (2008). Fungal biodegradation of tannins from cresote bush (*Larrea tridentata*) and tar bush (*Fluorensia cernua*) for gallic and ellagic acid production. *Food Technology and Biotechnology*. 46, 211-215.

Ventura, J., Gutierrez-Sánchez, G., Rodríguez-Herrera, R., and Aguilar, C. N. (2009). Fungal culturas of tar bush and cresote bushfor production of two phenolic antioxidants (pyrocatechol and gallic acid). *Folia Microbiologica*. 54(3), 199-203.

Viniegra-González, G., Favela-Torres., Aguilar,C. N., Romero-Gomez, S. J., Díaz-Godínez, G., Augur, C. (2003). Advantages of fungal enzyme production in solid state over liquid fermetation systems. *Biochem. Eng. J.* 13:157-167.

Wang, F., Ni, H., Cai, H. N., and Xiao, A. N. (2012). Tea stalks –a novel agro-residue for the production of tannase under solid state fermentation by *Aspergillus niger* JMU-TS528. *Annals of Microbiology*. In press. DOI 10.1007/s13213-012-0541-5.

Yadav, M. A., Aggarwal, N. K., Kumar, K., and Kumar, A. (2008). Tannase production by *Aspergillus fumigates* MA under solid-state fermentation. *World Journal of Microbiology and Biotechnology*. 24, 3023-3030.

Yu, X., Li, Y., Wu, D. (2004). Enzymatic synthesis of gallic acid esters using microencapsulated tannase: effect of organic solvent and enzymes specificity. *Journal of Molecular Catalysis B*. 30. 69-73.

Yu, X. W., Li, Y. Q. (2006). Kinetics and thermodynamics of syntehesis of propyl gallate by mycelium-bound tannase from *Aspergillus niger* in organic solvent. *J. Mol. Catal. B. Enzym*. 30 (2):69-73.

In: Handbook on Gallic Acid
Editors: M. A. Thompson and P. B. Collins
ISBN: 978-1-62618-921-8

Chapter 12

GALLIC ACID AS A USEFUL INDICATOR OF THE ANTIOXIDANT CAPACITY OF LIQUID FOODS?

F. J. Barba, M. J. Esteve and A. Frigola*
Department of Nutrition and Food Chemistry, Universitat de València, Burjassot, Spain

ABSTRACT

In terms of public health, drinking fruit and vegetable juices may well be as effective as consuming whole fruits and vegetables with regard to reducing the risk of chronic disorders. Many of their health benefits can be attributed to phenolic compounds. Phenolic compounds are aromatic metabolites of plants secondary metabolism that have a common structure with an aromatic ring with at least one hydroxyl group, which provides the ability to neutralize reactive species, helping the body to protect itself from oxidative stress. Phenols contribute to fruits' color and taste and have been described as possessing anticarcinogenic and antimutagenic activity. Furthermore, they have been shown to be good contributors to the total antioxidant capacity of foods, resulting in a beneficial effect to human health.

Due to the growing popularity of phenolic antioxidant over the past two decades, an increasing interest in determining the antioxidant activities exhibited by phenolic compounds and their derivatives should also be noted. In this line, special attention has been given to gallic acid. Gallic acid is a naturally abundant plant phenolic compound which is present in liquid foods derived from fruits and vegetables and has been found to exhibit antioxidative properties. Therefore, at this stage of development, there is a need to establish the relationship between phenolic compounds, including gallic acid, with the antioxidant capacity and total phenolic compounds of these food products.

Keywords: Gallic acid, phenolic profile, total phenolic compounds, antioxidant capacity

* E-mail: ana.frigola@uv.es. Phone: +34 963544955, Fax: +34 963544954.

INTRODUCTION

The growing interest in fruit and vegetable-based products is particularly due to the health benefits connected with their antioxidant properties, which are attributable to the ascorbic acid, carotenoids and phenolic compounds that they contain (Garcia-Salas et al., 2010; Cartea et al., 2011; Palafox et al., 2012).

Phenolic compounds or polyphenols form a large group of chemical substances considered as secondary metabolites of plants, with various chemical structures and activities. Chemically, phenolic compounds are substances that possess an aromatic ring and a benzene ring with one or more hydroxide groups including functional derivatives (esters, methyl esters, glycosides, etc.) (Sahidi and Naczk, 2004; Craft et al., 2012). In plants, they appear in conjugated form with one or more glucose residues joined to the hydroxyl groups, although in some cases direct junctions may be formed between a glucose molecule and an aromatic carbon. Consequently, the most common form in which they are found in nature is as glycosides, which are soluble in water and organic solvents. They may also be found joined to carboxylic acids, organic acids, amines, lipids and other phenolic compounds (Bravo, 1998).

Polyphenols are the most abundant antioxidants in our diet, since the average daily intake is about 1 g, which is almost 10-fold the intake of vitamin C, 100-fold the intake of vitamin E, and 500-fold the intake of carotenoids (Scalbert and Williamson, 2000). Foods and beverages rich in polyphenols may have a large potential regarding prevention of diseases. This is the case of fruits and vegetables associated with the prevention of stroke and cancers (Dai et al., 2006; Ruxton et al., 2006).

Most of the articles in the published literature use the Folin-Ciocalteu´s assay for quantifying total phenolic content of samples. The concentration of phenolic compounds is then reported as gallic acid, chlorogenic acid and catequin equivalents (Vermerris and Nicholson, 2007).

In general, gallic acid is used as standard based on the availability of a stable and pure substance, and because it is less expensive than other options. In addition, the response to gallic acid has been shown to be equivalent to most other phenolics in different fruits on a mass basis. In this line, Waterhouse et al. (2001) tested the stability of gallic acid standard solutions and they lose less than 5% of their value over two weeks when refrigerated and kept tightly closed.

Moreover, Escarpa and González (2001) studied the calibration curves for six polyphenolic compound standards, representative of all polyphenolic structures in different fruit samples (Fresh red wine pomace (*Vitis Vinifera*), apple varieties (*Malus domestica*, Golden and Reineta), pears (*Pyrus comunis*), and vegetables (fresh green beans (*Phaseolus vulgaris*) from two varieties (1 and 2) and dried lentils (*Lens culinaris*). They found a satisfactory correlation within the linear range of all investigated compounds. They proved the viability of the discussed method to determine the mentioned compounds in real extracts. As far as sensitivity (represented by calibration slope) was concerned, different values were obtained for each studied polyphenolic structure. In fact, sensitivity decreased in the following order: gallic acid > (+)-catechin > chlorogenic acid > rutin. Caffeic acid exhibited an identical slope as that found for gallic acid, while (−)-epicatechin showed the sensitivity identical to that found for (+)-catechin. Based on these results the total polyphenolic content was determined as a function of the standard employed in the calibration curves.

Nevertheless, in some articles, (+)-catechin has been used in order to determine total polyphenolic. For example, some apple samples allowed securing the same values when employing chromatographic methods, probably, because these phenolic structures were dominant in the studied samples (Escarpa and González, 2001).

The mode of action of antioxidants is complex and may be highly dependent upon a wide range of variables within a system (Craft et al., 2012). Previous studies in the published literature have been conducted with specific liquid foods evaluating the level of correlation between total phenolics content and observed antioxidant activity; the results of which have shown great variation (Barba et al., 2010; Barba et al., 2013; Di Majo et al., 2008; Sulaiman et al., 2011). Overall, these authors found that in sources in which a strong correlation is observed, it is typically concluded that phenolics are largely responsible for the antioxidant activities seen within the samples. However, in sources in which strong correlations are not observed, they found that there were significant amounts of antioxidants other than the measured phenolics present in the system, and that the specific phenolic species present in the system could not be quantified properly through the total phenol assay (Zulueta et al., 2009; Barba et al., 2013).

Thus, although the phenolics content assay may in many cases provide indication of the potential antioxidant capacity of a food product, it should not be considered as an accurate assessment and there is a need to evaluate individual phenolic compounds (Craft et al., 2012).

In this line, special attention has been given to gallic acid. Gallic acid (3, 4, 5-trihydroxybenzoic acid) is a naturally abundant plant phenolic acid of the benzoic family, which is synthesized from l-phenylalanine (and l-tyrosine) in plants. This process is commonly referred to as phenylpropanoid metabolism (De la Rosa et al., 2010). This compound is commonly found in liquid foods derived from fruits and vegetables, and since it was found to exhibit antioxidative properties, it has attracted considerable interest (Strlic et al., 2002; Kelebek et al., 2009). The relationship between phenolic bioactive compounds, their antioxidant capacity (Palafox et al., 2011) and the health benefits are well established. However, information about phenolic acids, especially gallic acid, and their interactions on the antioxiodant capacity is scarce.

Numerous in vitro studies have been conducted to evaluate total antioxidant capacity (TAC) of food products. So far, however, there is no official standardized method, and therefore, it is recommended that each evaluation should be made with various oxidation conditions and different methods of measurement (Karadag et al., 2009; Zulueta et al., 2009; Barba et al., 2013). The methods for measuring antioxidant capacity are basically classified into two groups, depending on the reaction mechanism: methods based on hydrogen atom transfer (HAT) and methods based on electron transfer (ET). Oxygen radical antioxidant capacity (ORAC) and Trolox equivalent antioxidant capacity (TEAC) assays are the most popularly used methods (Berker et al., 2012; Huang et al., 2005). Nowadays, the most widely used methods for measuring antioxidant activity are those that involve the generation of radical species and the presence of antioxidants determining the scavenging of these radicals (Arnao et al., 2001; Szydłowska-Czerniak et al., 2011).

The aim of the present study was to evaluate the relationship between the different phenolics, including gallic acid, with total antioxidant capacity and total phenolic compounds of the different liquid foods.

METHODS

Total Phenols

Total phenols were determined according to the method reported by Georgé et al. (2005), with some modifications. Briefly, 10 mL of sample were homogenized with 50 mL of a mixture of acetone/water (7/3, v/v) for 30 min. Mixture supernatants were then recovered by filtration (Whatman nº 2, England) and constituted the raw extracts (REs). REs (2 mL) were settled on an Oasis cartridge (Waters). Interfering water-soluble components (reducing sugars, ascorbic acid) were recovered with 2 x 2 mL of distilled water. The recovered volume of the washing extract (WE) was carefully measured. In order to eliminate vitamin C, heating was carried out on the washing extract (3 mL) for 2 h at 85 °C and led to the heated washing extract (HWE). All extracts (RE, WE, and HWE) were submitted to the Folin-Ciocalteu method, adapted, and optimized (Carbonell-Capella et al., 2013): Sodium carbonate solution (3 mL) 2 % (w/v) and 100 μL of Folin–Ciocalteau reagent were added to an aliquot of 100 μL of sample. The mixture was incubated for 1 h at room temperature. Absorbance was measured at 765 nm.

Liquid Chromatographic Analysis of Phenolic Profile

High performance liquid chromatography (HPLC) analysis was performed in accordance to Kelebek et al. (2009), with some modifications. Samples were filtered through a Sep-Pak® cartridge (a reverse-phase C-18 cartridge; Millipore, MA, USA) which retains phenolic compounds. The cartridges were previously activated with 10 mL of methanol (MeOH) and 10 mL of water. Every 10 mL of sample was eluted with 2 mL of MeOH, and all methanolic fractions were collected, filtered through a 0.45 μm membrane filter Millex-HV13 (Millipore) and then analysed by liquid chromatography. The LC system consisted of two isocratic pumps (Prostar 210, Varian Inc, California, USA) with degasser (Degassit, MetaChem, USA), column thermostat (Prostar 510, Varian) and UV-vis detector (Varian Inc, California, USA). The whole LC system was operated by a Varian STAR Chromatography Workstation Ver. 6.0 (Varian Inc, California, USA). Luna PFP(2) precolumn (guard column); Luna 100 PFP(2) column (5 μm, 150 x 4.6 mm) (Phenomenex, Spain). The mobile phase consisted of two solvents: Solvent A, water/formic acid (95:5; v/v) and Solvent B, acetonitrile/solvent A (60:40; v/v). Gallic acid was eluted under the following conditions: 1 mL/min flow rate and the temperature was set at 40 °C, isocratic conditions from 0 to 10 min with 0% B, gradient conditions from 0% to 15% B in 20 min, from 15% to 22% B in 45 min, from 22% to 100% B in 15 min, from 100% to 0% B in 5 min, followed by washing and reconditioning the column. Chromatograms were recorded at 280 nm. Identification of gallic acid was carried out by using authentic standard and by comparing the retention times, while quantification was performed by external calibration with standard.

Trolox Equivalent Antioxidant Capacity (TEAC) Assay

The Trolox Equivalent Antioxidant Capacity (TEAC) test was conducted in accord to the method proposed by Ou et al. (2001) adapted, and optimized by Barba et al. (2013). The method was based on the capacity of antioxidants to inhibit the radical cation 2,2-azino-bis(3-ethylbenzothiazoline6-sulphonate) (ABTS), which has a characteristic long-wavelength absorption spectrum, showing a maximal peak at 734 nm. The ABTS radical cation is formed by the interaction of ABTS (7 mM) with $K_2S_2O_8$ (2.45 mM).

Oxygen Radical Absorbance Capacity (ORAC) Assay

The oxygen radical absorbance capacity (ORAC) assay used, with fluorescein as the "fluorescent probe", was that described by Barba et al. (2013). The automated ORAC assay was carried out on a Wallac 1420 VICTOR^2 multilabel counter (Perkin-Elmer, USA) with fluorescence filters, for an excitation wavelength of 485 nm and an emission wavelength of 535 nm. The measurements were made in plates with 96 white flat bottom wells (Sero-Wel, BibbySterilin Ltd., Stone, UK). The reaction was performed at 37 °C, as the reaction was started by thermal decomposition of AAPH in 75 mM phosphate buffer (pH 7.0). Triplicate analyses were performed for each sample.

SAMPLES

In order to determine the contribution of phenolic compounds, mainly gallic acid, to antioxidant capacity of liquid foods, different fruit-based products were prepared and analysed in the laboratory. Moreover, some commercial samples were obtained from a local supermarket (Valencia, Spain).

With regard to samples prepared in the laboratory, all the juices were extracted after appropriate washing and hygienization of the fruits. Orange, lime and lemon juices were extracted with a conventional juicer. In addition, litchi, mango, papaya, pitaya blackcurrant, black grape, blueberry, plum, pomegranate and prickly pear juices were squeezed in a blender (Moulinex, Spain). Then the pulp was removed.

The orange juice–milk beverage (OJM) was prepared according to the method of Zulueta et al. (2009) by mixing 50% (v/v) of orange juice with the pulp removed, 20% (v/v) of UHT skimmed milk, and 30% (v/v) of distilled water. Sugar (7.5% w/v), citric acid (0.1% w/v), and high-methoxyl citrus pectin (0.3% w/v) were added as sweetener, preservative, and homogenizer of the samples, respectively. Solid ingredients were placed in water in the weight proportions indicated. The beverage was prepared just before use.

In addition, some commercial samples of orange juice, orange juice from concentrate, orange nectar, pineapple nectar and açaí juice were purchased from a local supermarket (Valencia, Spain). Triplicate analyses were performed for each sample.

Statistical Analysis

All the determinations were performed by triplicate. An analysis of variance (ANOVA) was applied to the results obtained in order to verify whether there were significant differences in the parameters studied in relation to sample analyzed, and to ascertain possible interactions between factors (differences at $p<0.05$ were considered significant). Where there were differences, an LSD test was applied to indicate the samples in which differences were observed. A multiple regression analysis was performed to study the influence of phenolic profile to antioxidant capacity (the results are shown in the significant cases, $p<0.05$). A study was conducted with the aim of determining whether there were correlations between a pair of variables (Pearson´s test). To gain insight into the data structure a multivariate analysis called principal component analysis (PCA) to describe the relations between variables and observations. All statistical analyses were performed using SPSS® (Statistical Package for the Social Sciences) v.20.0 for Windows (SPSS Inc., Chicago, USA).

Results

Until the date, several authors have demonstrated the high correlation of total phenolic compounds with the total antioxidant capacity of fruit-based products (Khanizadeh et al., 2008; Barba et al., 2013). However, there is a lack of information about the relative contribution of each phenolic compound.

In the present research study, we analyze the total phenolic compounds, antioxidant capacity of the samples (TEAC and ORAC assay) as well as the phenolic profile of 19 fruit-based beverages from different origin, in order to establish the relationship between them.

The total phenolic content of the samples analyzed in the present research work are shown in Table 1. Among all the samples studied, black grape juice had the highest content of total phenolic compounds (1734.75±37.97 mg GAE/L). Moreover, total phenolic content was also important in some dark fruit juices such as pomegranate (965.10±4.95 mg GAE/L), blackcurrant (1452.65±45.75 mg GAE/L), açaí (1325.10±4.67 mg GAE/L) and blueberry (1543.12±25.13 mg GAE/L), which have been considered traditionally an excellent source of phenolic compounds (Gil et al., 2000; Howard et al., 2003; Szajdek & Borowska, 2008; González-Molina et al., 2009; Mousavinejad et al., 2009; Vázquez-Araujo et al., 2010; Barba et al., 2011; Barba et al., 2012). However, pitaya juice (an exotic fruit juice) had the lowest phenolic content (47.35±1.34 mg GAE/L). Overall, the total phenolic content of the samples studied in the present work were lower than those obtained by other authors in juices from açaí (1900.0 mg GAE/L) (Seeram et al., 2008); black currant (2698.6-2813.1 mg GAE/L) (Mitić et al., 2011); black grape (2182.1-2258.8 mg GAE/L); blueberry (2350.0 mg GAE/L) (Seeram et al., 2008); orange-based beverages (595.6-684.2 mg GAE/L) (Klimczak et al., 2007; Seeram et al., 2008); papaya (410.0-650.0 mg GAE/L) (Özkan et al., 2011), pitaya (92.1-133.4 mg GAE/L) (Esquivel et al., 2007); and pomegranate (2487.5-3800.0) (Gil et al., 2000; Seeram et al., 2008).

Table 1. Total phenolic compounds (TPC), gallic acid and protocatechuic acids content of different liquid foods

		Benzoic acids	
Sample	TPC (mg GAE/L)	Gallic acid (mg/L)	Protocatechuic acid (mg/L)
Açaí juice commercial	1325.10±4.67	15.66±0.81	-
Black currant juice	1452.65±45.75	98.60±2.88	5.26±0.20
Black grape juice	1734.75±37.97	34.48±1.39	-
Blueberry juice	1543.12±25.13	156.83±5.08	-
Lemon juice	369.86±17.39	0.56±0.08	-
Lime juice	165.55±8.70	-	-
Litchi juice	299.20±4.67	5.22±0.57	0.98±0.04
Mango juice	84.90±2.26	0.31±0.01	0.04±0.01
Orange juice	337.17±28.81	3.87±0.12	1.11±0.06
Orange juice commercial	214.65±5.87	1.37±0.01	0.66±0.06
Orange juice from concentrate	432.03±7.13	5.56±0.08	1.35±0.06
Orange juice-milk	146.74±5.03	2.57±0.23	0.50±0.05
Orange néctar commercial	166.39±5.90	0.99±0.01	0.53±0.04
Papaya juice	105.23±0.47	0.16±0.02	-
Pineapple nectar commercial	196.53±3.04	1.20±0.01	1.20±0.01
Pitaya juice	47.35±1.34	2.04±0.01	0.31±0.02
Plum juice	340.10±22.06	0.63±0.11	122.55±2.81
Pomegranate juice	965.10±4.95	6.92±0.20	-
Prickly pear juice	279.95±3.04	6.57±0.02	0.06±0.01

Table 2. Cinnamic acids content (mg/L) of different liquid foods

Sample	Caffeic acid (mg/L)	Chlorogenic acid (mg/L)	*p*-coumaric acid (mg/L)	Ferulic acid (mg/L)	Cinnamic acid (mg/L)
Açaí juice commercial	3.59±0.04	-	-	6.36±0.20	5.41±0.40
Black currant juice	1.73±0.23	-	37.16±1.36	4.63±0.09	10.75±0.26
Black grape juice	0.20±0.01	0.75±0.04	1.64±0.05	2.64±0.45	0.71±0.15
Blueberry juice	2.34±0.31	6.80±0.49	2.32±0.12	3.86±0.24	0.13±0.04
Lemon juice	1.76±0.28	3.67±0.16	4.18±0.04	5.29±0.19	-
Lime juice	2.33±0.05	2.26±0.08	3.59±0.04	1.26±0.06	-
Litchi juice	-	-	13.25±1.38	0.68±0.28	0.11±0.01
Mango juice	0.11±0.04	5.87±0.07	1.37±0.18	3.02±1.37	-
Orange juice	5.68±0.18	9.91±0.21	3.85±0.10	20.88±0.38	-
Orange juice comercial	2.76±0.01	9.65±0.19	13.33±0.02	69.08±0.52	-
Orange juice from concentrate	18.70±0.28	-	0.85±0.01	-	-
Orange juice-milk	3.76±0.18	4.87±0.13	1.92±0.08	10.43±0.17	-
Orange néctar commercial	2.27±0.01	10.58±0.15	5.09±0.01	-	-
Papaya juice	-	1.62±0.04	-	0.53±0.04	0.60±0.04
Pineapple néctar commercial	0.01±0.01	-	-	2.49±0.01	6.46±0.08
Pitaya juice	-	12.69±0.31	1.09±0.20	0.06±0.01	-
Plum juice	2.38±0.21	8.29±0.20	2.17±0.07	2.93±0.12	0.40±0.02
Pomegranate juice	-	-	23.50±1.29	3.68±0.20	6.53±0.64
Prickly pear juice	2.03±0.03	9.37±0.58	2.34±0.11	3.29±0.13	0.16±0.01

In the last two decades, most of the phenolic compounds have attracted the interest of scientific community due to its beneficial properties. In this work we studied the content in benzoic acids (gallic and protocatechuic acids), cinnamic acids (caffeic, chlorogenic, *p*-coumaric, ferulic and trans-cinnamic acids), and flavanoids (apigenin, catechin, rutin, quercetin, naringenin, and hesperidin). The individual contents of each phenolic compound in the samples studied are shown in Tables 1-3. As can be seen, the concentration of each phenolic compound in the studied samples differs from the food matrix since phenolic amount depends on variety, growing conditions and processing among other factors. Gallic acid was identified and quantified in all the samples studied except in the lime juice. In the line of the results obtained for TPC, dark fruit juices had also the higher content of gallic acid (from 6.92±0.20 to 156.83±5.08 mg/L), being especially important the content found in blueberry (156.83±5.08 mg/L) and black currant (98.60±2.88 mg/L) juices.

Table 3. Flavanoids content (mg/L) of different liquid foods

	Flavones	Flavanols	Flavonols		Flavanones	
Sample	Apigenin (mg/L)	Catechin (mg/L)	Rutin (mg/L)	Quercetin (mg/L)	Naringenin (mg/L)	Hesperidin (mg/L)
Açaí juice	80.27±1.82	31.60±1.46	0.28±0.05	22.18±0.87	-	-
Black currant juice	40.63±1.32	32.92±1.09	2.44±0.27	37.76±1.14	-	16.73±1.65
Black grape juice	83.50±1.31	6.04±0.18	1.27±0.09	19.56±1.48	114.08±2.20	32.65±1.20
Blueberry juice	51.18±1.46	41.89±0.46	15.67±0.77	7.63±0.09	-	-
Lemon juice	5.17±0.06	-	-	1.38±0.04	3.00±0.21	156.63±1.73
Lime juice	3.20±0.02	-	-	1.17±0.06	-	154.86±2.33
Litchi juice	30.83±0.69	2.97±0.46	0.72±0.17	25.06±1.25	-	8.75±0.70
Mango juice	-	33.52±0.62	0.18±0.05	23.17±0.86	-	-
Orange juice	29.19±0.81	-	-	5.73±0.16	11.93±0.05	79.50±1.36
Orange juice commercial	74.28±0.16	-	-	4.73±0.23	11.26±0.03	88.49±0.23
Orange juice from concentrate	-	-	-	9.12±0.22	48.90±0.57	210.05±13.51
Orange juice-milk	14.87±0.13	-	-	2.84±0.08	0.97±0.04	48.68±0.16
Orange néctar commercial	44.28±0.23	-	-	3.72±0.09	10.97±0.01	43.02±0.04
Papaya juice	25.93±0.27	-	-	4.61±0.11	-	-
Pineapple néctar commercial	-	1.27±0.11	0.63±0.01	-	24.83±0.03	-
Pitaya juice	4.51±0.36	2.24±0.47	-	0.71±0.10	-	0.15±0.01
Plum juice	107.99±3.59	3.70±0.18	1.11±0.06	57.47±1.04	-	-
Pomegranate juice	155.18±4.19	61.97±2.32	0.87±0.07	31.58±0.08	-	43.18±1.07
Prickly pear juice	62.06±2.79	32.55±0.69	5.07±0.37	38.91±1.38	-	37.84±0.64

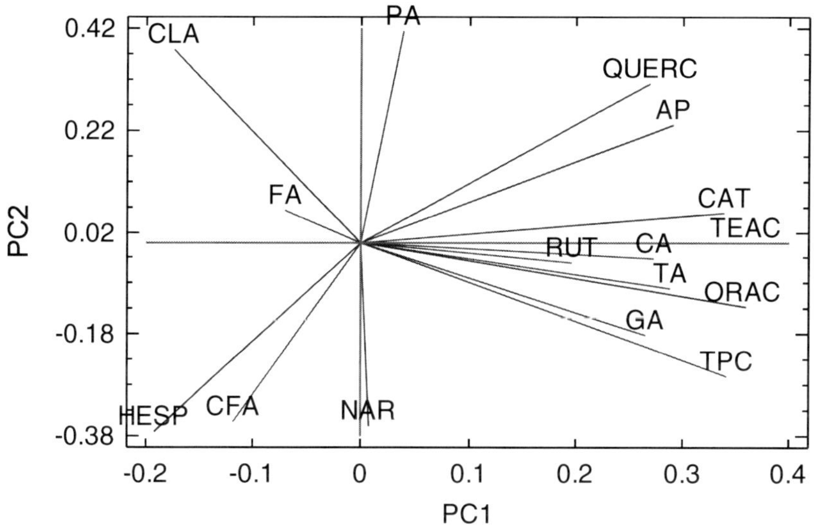

Figure 1. Principal components plot of the samples studied. AP (apigenin). CAFA (caffeic acid). CAT (catechin). CLA (chlorogenic acid). CA (*p*-coumaric acid). FA (ferulic acid). GA (gallic acid). HESP (hesperidin). NAR (naringenin). PA (protocatechuic acid). QUERC (quercitin). RUT (rutin). TA (*trans*-cinnamic acid). TPC (total phenolic compounds). TEAC (trolox equivalent antioxidant capacity). ORAC (oxygen radical antioxidant capacity).

Table 4. Eigenvectors, principal components for the antioxidant capacity, total phenolic compounds and phenolic profile of the liquid foods studied

	1	2	3	4	5	6
Total phenolics	0.341566	-0.265018	0.115881	0.188569	-0.000456946	-0.180796
TEAC	0.361567	-0.000405494	-0.164599	0.0177698	-0.0135145	-0.102607
ORAC	0.359035	-0.125487	-0.0621134	0.0851363	0.0541302	-0.0189218
Gallic acid	0.264994	-0.18203	0.474292	0.0663944	0.0817427	0.131402
Protocatechuic acid	0.0383098	0.414937	-0.166923	0.459177	-0.0846023	0.228637
Catechin	0.338327	0.0576257	0.123643	-0.124372	0.0351574	0.152405
Caffeic acid	-0.11751	-0.353639	-0.117953	0.368898	0.186539	0.467044
Chlorogenic acid	-0.172956	0.379505	0.313575	0.128752	0.309064	-0.0574426
Coumaric acid	0.27214	-0.0329438	-0.207404	-0.303518	0.358455	0.173884
Ferulic acid	-0.0697929	0.0639781	-0.030769	-0.0285828	0.749409	-0.296437
Transcinnamic	0.287741	-0.0911438	-0.213615	-0.352419	-0.0392176	0.111921
Rutin	0.196115	-0.0393826	0.586467	0.179044	0.0437131	0.132023
Naringin	0.00722382	-0.362263	-0.136135	0.401197	-0.128032	-0.536393
Quercetin	0.2685	0.313471	-0.22219	0.273733	-0.066586	0.232004
Hesperidin	-0.191954	-0.372539	-0.19601	0.161552	0.297671	0.307812
Apigenin	0.29093	0.230185	-0.188192	0.249621	0.220244	-0.240563

TEAC: Trolox equivalent antioxidant capacity. ORAC: Oxygen radical antioxidant capacity.

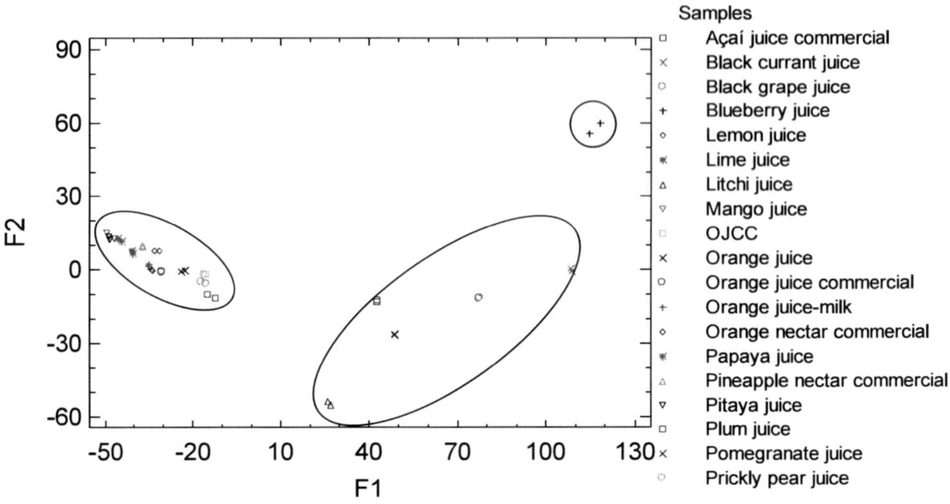

Figure 2. Principal component score plot of the variables with function 1 (F1) and function 2 (F2) based on the antioxidant capacity (TEAC and ORAC assays), total phenolic compounds and phenolic profile of the samples studied.

The different composition can be a useful tool for classifying the groups and to discriminate the origin of the liquid foods. In this line, a principal component analysis (PCA) was carried out using the whole set of data in order to discover the similarities or differences among the liquid foods studied and to specify the main parameters, which could differentiate the samples (Table 4). Apart from TPC, TEAC and ORAC, the main components responsible for the differences in liquid foods were gallic acid, catechin, coumaric acid, transcinnamic acid, quercetin and apigenin (Table 4, Figure 1). The projection of the beverages on the first two principal components (84.52% of the total variability) showed that the citric based beverages and exotic fruit juices were clearly separated from the berry juices. Citric based beverages and exotic fruit juices were located in the negative part of the function 2 and the berry juices were placed in the positive part of the function 1 (Figure 2).

Total phenolic compounds (TPC) determined with Folin-Ciocalteu´s assay has been used widely as a routine measurement for quantifying the phenolic content of foods (Craft et al., 2012). Although gallic acid has been widely used as a phenolic standard for the determination of TPC, there is a need to study the correlation of TPC with each phenolic compound, in order to facilitate the selection of the best phenolic standard for the different food matrix studied. Therefore, a Pearson´s test was performed to study the relationship between each phenolic compound and total phenolic compounds in the samples analysed in the present research (Table 5). A positive significant correlation ($p<0.05$) was found between TPC with apigenin, catechin, *p*-coumaric acid, gallic acid, naringenin, rutin and trans-cinnamic acid, obtaining the stronger relationship between TPC and gallic acid ($r= 0.7262$, $p= 0.0000$).

Moreover, correlation between total phenolic compounds measured using the modified Folin-Ciocalteu´s assay with the sum up of each phenolic compound was studied. As can be observed in the equations below, a good relationship between TPC and the sum up was found for exotic and citric-based fruit beverages (Eq. 1) while the correlation was not clear when dark fruit juices were studied (Eq. 2).

Table 5. Correlations ($p<0.05$) of phenolic compounds in the samples studied with total phenolic compounds (TPC), TEAC and ORAC assays

Phenolic compounds	TEAC (mM TE)	ORAC (mM TE)	TPC (mg GAE/L)
Apigenin	0.7673 (p=0.0000)	0.7672 (p=0.0000)	0.4810 (p=0.0022)
Caffeic acid	-	-	-
Catechin	0.7248 (p=0.0000)	0.4451 (p=0.0051)	0.5415 (p=0.0004)
Chlorogenic acid	-	-	-
p-Coumaric acid	0.5012 (p=0.0013)	0.5014 (p=0.0013)	0.3441 (p=0.0344)
Ferulic acid	-	-	-
Gallic acid	-	0.5441 (p=0.0004)	0.7262 (p=0.0000)
Hesperidin	-	-	-
Naringenin	-	-	0.4035 (p=0.0120)
Protocatechuic	-	-	-
Quercetin	0.4661 (p=0.0015)	0.4962 (p=0.0015)	
Rutin	-	0.3285 (p=0.0440)	0.4933 (p=0.0017)
trans-cinnamic	0.5003 (p=0.0024)	0.5000 (p=0.0014)	0.4811 (p=0.0022)

TEAC: trolox equivalente antioxidant capacity. ORAC: oxygen radical antioxidant capacity.

$$\text{Sum up (mg/L)} = 57.429+1.024*\text{TPC (mg GAE/L)} \quad (R^2= 42.965, \text{ standard error} = 138.237) \tag{1}$$

$$\text{Sum up (mg/L)} = 595.456+0.020*\text{TPC (mg GAE/L)} \quad (R^2= 0.172, \text{ standard error} = 239.129) \tag{2}$$

Table 6. Total antioxidant capacity (mM trolox equivalent (TE)) of different liquid foods

Sample	TEAC (mM TE)	ORAC (mM TE)
Açaí juice commercial	12.97±0.04	22.93±0.05
Black currant juice	14.28±0.40	38.80±0.33
Black grape juice	17.26±0.01	28.31±0.36
Blueberry juice	16.34±0.40	32.44±0.17
Lemon juice	3.36±0.29	6.63±0.10
Lime juice	2.56±0.19	3.78±0.46
Litchi juice	16.94±0.71	38.31±0.36
Mango juice	4.75±0.75	2.78±0.86
Orange juice	4.86±0.11	11.51±0.08
Orange juice commercial	4.74±0.16	10.75±0.08
Orange juice from concentrate commercial	6.59±0.53	12.67±0.16
Orange juice-milk	3.54±0.12	9.49±0.09
Orange néctar commercial	3.57±0.35	10.20±0.19
Papaya juice	7.31±0.78	6.11±0.48
Pineapple néctar commercial	8.77±0.10	5.31±0.13
Pitaya juice	2.59±0.37	3.73±0.40
Plum juice	11.48±0.21	15.63±0.59
Pomegranate juice	41.23±0.52	26.14±0.14
Prickly pear juice	10.26±0.98	14.40±0.23

TEAC: trolox equivalent antioxidant capacity. ORAC: oxygen radical antioxidant capacity.

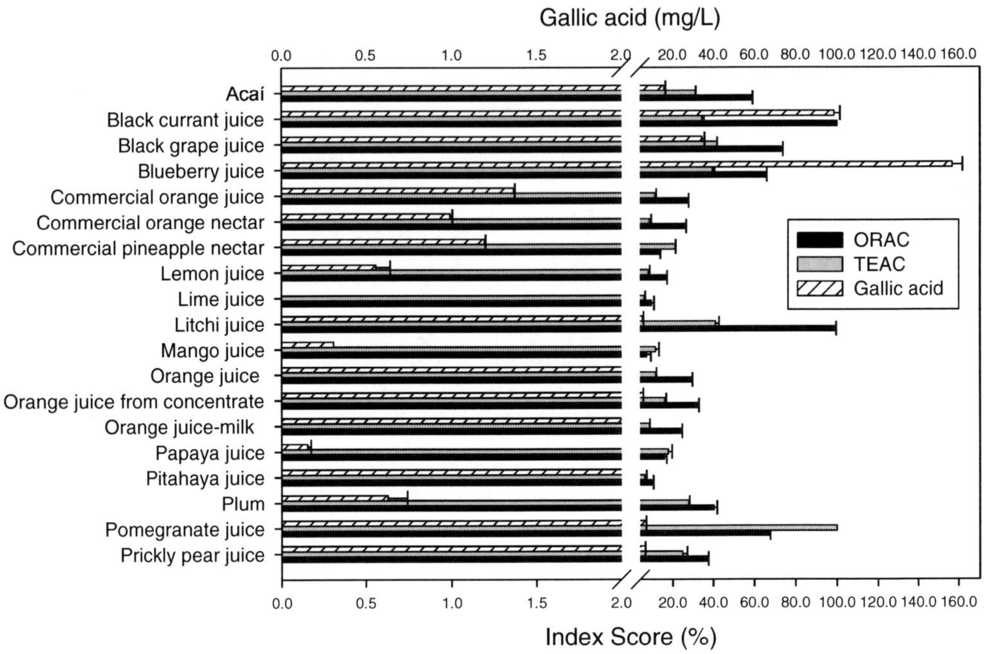

Figure 3. Antioxidant potency (index score) and gallic acid content (mg/L) for each studied liquid food.

where, Sum up is the sum of each phenolic measured in the samples, TPC is the total phenolic content measured using Folin-Ciocalteu´s assay and GAE stands for gallic acid equivalents.

Folin-Cioalteu´s assay is based on the reduction of phosphor-wolframate-phosphomolybdate complex by phenolics to a blue reaction product (Bonoli et al., 2004; Kyoung et al., 2005). Thus, some phenolic compounds (i.e. anthocyanins) that react when Folin-Ciocalteu´s is used were not analyzed by the chromatographic analysis in the present study.

This fact may explain the differences observed depending on food matrix. Antioxidant capacity measured using TEAC assay ranged from 2.56±0.19 mM trolox equivalent (TE) (lime juice) to 41.23±0.52 mM TE (pomegranate juice) (Table 6). With regard to ORAC values, the higher value was found for black currant juice (38.80±0.33 mM TE) while the lower value was obtained when pitaya juice was analyzed (3.73±0.40 mM TE), which had also the lower total phenolic content. As can be observed in Table 6, the magnitude of the antioxidant capacity in the different samples differs from the applied method. This fact makes difficult the comparison of the results obtained with the different antioxidant capacity assays. For this reason, an overall antioxidant potency composite index was determined by assigning all assays an equal weight, ascribing an index value of 100 to the best score for each test, and then calculating an index score = [(sample score/best score)×100]; the average of the values of all tests for each beverage was then used for the antioxidant index (Seeram et al. 2008). Overall, it should be noted that when ORAC method was applied, the index scores were higher than those obtained by TEAC assay except for commercial pineapple nectar, as well as mango, papaya and pomegranate juices (Figure 3). This can be related with the nature and chemical structures of the different compounds included in food products that confer this antioxidant capacity. The higher antioxidant index corresponded to blackcurrant juice and pomegranate juices for ORAC and TEAC assays, respectively. As can be observed black currant juice is a good source of phenolic compounds (1452.65±45.75 mg GAE/L), especially gallic acid (98.60±2.88 mg/L), while pomegranate had also a high content in vitamin C which one of the main contributors to antioxidant capacity when TEAC assays is used.

A Pearson´s test was conducted in order to establish the relationship of each phenolic compound with the antioxidant capacity, measured with TEAC and ORAC assays (Table 5). With regard to gallic acid, the correlation test revealed a positive significant ($p<0.05$) correlation between gallic acid with ORAC ($r = 0.9858$) while a non-significant relationship was found with TEAC assay. To verify the efficiency of each method in the measurement of the antioxidant capacity of gallic acid, we prepared concentration curves for gallic acid standard (0.30–1.30 mg/mL). Gallic acid showed antioxidant capacity when both TEAC and ORAC assays were used. However, when ORAC method was employed for measuring the antioxidant capacity, a higher increase in ORAC values was found when gallic acid concentration was higher in comparison to TEAC values. Linear regression analysis was significant in all the cases ($p<0.05$) as can be seen in Figure 4. However, same concentration produces a different increase in total antioxidant capacity, obtaining slopes of 45.56 and 30.25 $mg \cdot mL^{-1} \cdot mM\ TE^{-1}$ for ORAC and TEAC, respectively. This fact can be explained by the fact that the ORAC method is a reaction based on the transfer of H atoms, and groups present in gallic acid (3OH) may be better represented by this assay. In addition, it must be pointed out that the ORAC method is the only one so far that combines the total inhibition time and the percentage of the free radical damage by the antioxidant into a single quantity, ensuring that, by the end of the process, all the antioxidants present in the sample have reacted with the radicals generated (Zulueta et al., 2009). Therefore, it can be concluded that in samples with high concentration of gallic acid, measurement of the antioxidant capacity using TEAC assay is not appropriately evaluated.

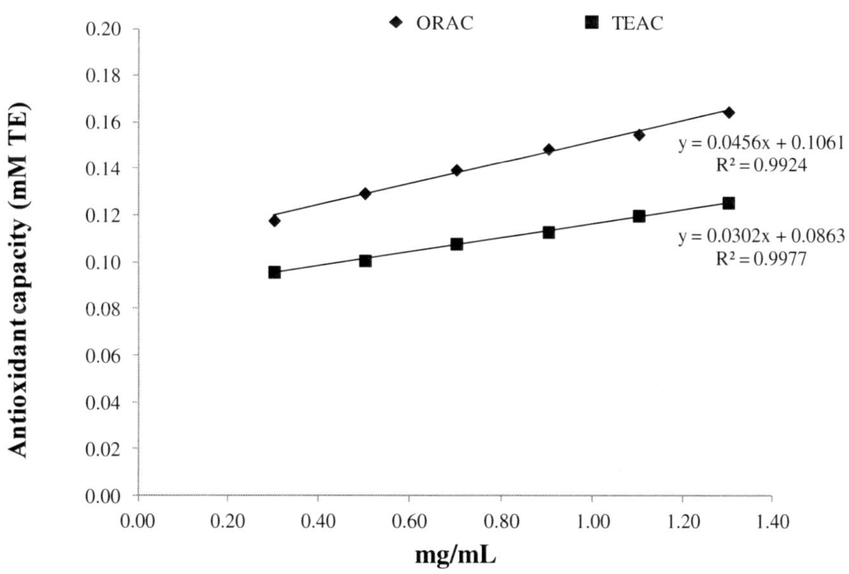

Figure 4. Antioxidant capacity of gallic acid reference substance evaluated by the TEAC and ORAC methods. TE (trolox equivalents).

ACKNOWLEDGMENTS

This research project was supported by the Spanish Ministry of Science and Technology and European Regional Development Funds (AGL2010-22206-C02-01).

REFERENCES

Arnao, M. B., Cano, A. & Acosta, M. (2001). The hydrophilic and lipophilic contribution to total antioxidant activity. *Food Chemistry, 73,* 239-244.

Barba, F. J., Esteve, M. J. & Frígola, A. (2010). Ascorbic acid is the only bioactive that is better preserved by high hydrostatic pressure than by thermal treatment of a vegetable beverage. *Journal of Agricultural and Food Chemistry, 58,* 10070−10075.

Barba, F. J., Esteve, M. J. & Frígola, A. (2011). Physicochemical and nutritional characteristics of blueberry juice after high pressure processing. *Food Research International, In Press,* DOI:10.1016/j.foodres.2011.02.038.

Barba, F. J., Jäger, H., Meneses, N., Esteve, M. J., Frígola, A. & Knorr, D. (2012). Evaluation of quality changes of blueberry juice during refrigerated storage after high-pressure and pulsed electric fields processing. *Innovative Food Science and Emerging Technologies, 14,* 18-24.

Barba, F. J., Esteve, M. J., Tedeschi, P., Brandolini, V. & Frigola, A. (2013). A comparative study of the analysis of antioxidant activities of liquid foods employing spectrophotometric, fluorometric, and chemiluminescent methods. *Food Analytical Methods, 6,* 317-327.

Berker, K. I., Demirata, B. & Apak, R. (2012). Determination of total antioxidant capacity of lipophilic and hydrophilic antioxidants in the same solution by using ferric–ferricyanide assay. *Food Analytical Methods, 3,* 154-168.

Bonoli, M., Verardo, V., Marconi, E. & Fiorenza, C. M. (2004). Antioxidant phenols in barley (*Hordeum vulgare* L.) flour: comparative spectrophotometric study among extraction methods of free and bound phenolic compounds. *Journal of the Science Food Chemistry, 84,* 5195- 5200.

Bravo, L. (1998). Polyphenol: chemistry, dietary sources, metabolism and nutritional significance. *Nutrition Reviews, 56 (11),* 317-333.

Carbonell-Capella, J. M., Barba, F. J., Esteve, M. J. & Frígola, A. (2013). High pressure processing of fruit juice mixture sweetened with *Stevia rebaudiana* Bertoni: Optimal retention of physical and nutritional quality. *Innovative Food Science & Emerging Technologies, In Press, Accepted Manuscript.* DOI: 10.1016/j.ifset.2013.01.011.

Cartea, M. E., Francisco, M., Soengas, P. & Velasco, P. (2011). Phenolic compounds in *Brassica* vegetables. *Molecules, 16,* 251-280.

Craft, B. D., Kerrihard, A. L., Amarowicz, R. & Pegg, R. B. (2012). Phenol-based antioxidants and the in vitro methods used for their assessment. *Comprehensive Reviews in Food Science and Food Safety, 11,* 148-173.

Dai, Q., Borenstein, A. R., Wu, Y., Jackson, J. C. & Larson, E. B. (2006). Fruit and vegetable juices and Alzheimer's disease: the Kame project. *American Journal of Medicine, 119,* 751-759.

De la Rosa, L. A., Alvarez-Parrilla, E. & González-Aguilar, G. A. (2010). *Fruit and vegetable phytochemicals. Chemistry, nutritional value and stability.* Iowa: Blackwell Publishing.

Di Majo, D., La Guardia, M., Giammanco, S., La Neve, L. & Giammanco, M. (2008). The antioxidant capacity of red wine in relationship with its polyphenolic constituents. *Food Chemistry, 111,* 45–49.

Escarpa, A. & González, M. A. (2001). Approach to the content of total extractable phenolic compounds from different food samples by comparison of chromatographic and spectrophotometric methods. *Analytica Chimica Acta, 427,* 119–127.

Esquivel, P., Stintzing, F. C. & Carle, R. (2007). Phenolic compound profiles and their corresponding antioxidant capacity of purple pitaya (*Hylocereus sp.*) Genotypes. *Zeitschrift für Naturforsch, 62,* 636-644.

Garcia-Salas, P., Morales-Soto, A., Segura-Carretero, A. & Fernández-Gutiérrez, A. (2010). Phenolic-compound-extraction systems for fruit and vegetable samples. *Molecules, 15,* 8813-8826.

Georgé, S., Brat, P., Alter, P. & Amiot, M. J. (2005). Rapid determination of polyphenols and vitamin C in plant-derived products. *Journal of Agricultural and Food Chemistry, 53,* 1370–1373.

Gil, M. I., Tomás-Barberán, F. A., Hess-Pierce, B., Holcroft, D. M. & Kader, A. A. (2000). Antioxidant activity of pomegranate juice and its relationship with phenolic composition and processing. *Journal of Agricultural and Food Chemistry, 48,* 4581–9.

González-Molina, E., Moreno, D. A. & García-Viguera, C. (2009). A new drink rich in bioactives combining lemon and pomegranate juices. *Food Chemistry, 115,* 1364–1372.

Howard, L. R., Clark, J. R. & Brownmiller, C. (2003). Antioxidant capacity and phenolic content in blueberries as affected by genotype and growing season. *Journal of the Science of Food and Agriculture, 83,* 1238–1247.

Huang, D., Ou, B. & Prior, R. L. (2005). The chemistry behind antioxidant capacity assays. *Journal of Agricultural and Food Chemistry, 53,* 1841–1856.

Karadag, A., Ozcelik, B. & Saner, S. (2009). Review of methods to determine antioxidant capacities. *Food Analytical Methods, 2,* 41–60.

Kelebek, H., Selli, S., Canbas, A. & Cabaroglu, T. (2009). HPLC determination of organic acids, sugars, phenolic compositions and antioxidant capacity of orange juice and orange wine made from a Turkish cv. Kozan. *Microchemical Journal, 91,* 187-192.

Klimczak, I., Małecka, M., Szlachta, M. & Gliszczyńska-Świgło, A. (2007). Effect of storage on the content of polyphenols, vitamin C and the antioxidant activity of orange juices. *Journal of Food Composition and Analysis, 20,* 313–322.

Khanizadeh, S., Tsao, R., Rekika, D., Yang, R., Charles, M. T. & Vasantha-Rupasinghe, H. P. (2008). Polyphenol composition and total antioxidant capacity of selected apple genotypes for processing. *Journal of Food Composition and Analysis, 21(5),* 396-401.

Kyoung, Ch. O., Kim, D., Smith, N., Schroeder, D., Taek, H. J. & Yong, L. Ch. (2005). Daily consumption of phenolics and total antioxidant capacity from fruit and vegetables in the American diet. *Journal of the Science of Food and Agriculture, 85,* 1715–1724.

Mitić, M. N., Obradović, M. V., Kostić, D. A., Nasković, D. Č. & Micić, R.J. (2011). Phenolics content and antioxidant capacity of commercial red fruit juices. *Hemijska Industrija, 65(5),* 611–619.

Mousavinejad, G., Emam-Djomeh, Z., Rezaei, K., Hossein, M. & Khodaparast, H. (2009). Identification and quantification of phenolic compounds and their effects on antioxidant activity in pomegranate juices of eight Iranian cultivars. *Food Chemistry, 115,* 1274–1278.

Ou, B., Hampsch-Woodill, M. & Prior, R. L. (2001). Development and validation of an improved oxygen radical oxygen assay using fluorescein as the fluorescent probe. *Journal of Agricultural and Food Chemistry, 49,* 4619–4626.

Özkan, A., Gübbük, H., Güneş, E. & Erdoğan, A. (2011). Antioxidant capacity of juice from different papaya (*Carica papaya* L.) cultivars grown under greenhouse conditions in Turkey. *Turkish Journal of Biology, 35,* 619-625.

Palafox-Carlos, H., Ayala-Zavala, J. F. & González-Aguilar, G. A. (2011). The role of dietary fiber in the bioaccessibility and bioavailability of fruit and vegetable antioxidants. *Journal of Food Science, 76,* 6–15.

Palafox-Carlos, H., Gil-Chávez, J., Sotelo-Mundo, R. R., Namiesnik, J., Gorinstein, S. & González-Aguilar, G. A. (2012). Antioxidant interactions between major phenolic compounds found in 'Ataulfo' mango pulp: chlorogenic, gallic, protocatechuic and vanillic Acids. *Molecules, 17,* 12657-12664.

Ruxton, C. H. S., Gardner, E. J. & Walker, D. (2006). Can pure fruit and vegetable juices protect against cancer and cardiovascular disease too? A review of the evidence. *International Journal of Food Science and Nutrition, 57,* 249-272.

Scalbert, A. & Williamson, G. (2000). Dietary intake and bioavailability of polyphenols. *Journal of Nutrition, 130,* 2073S-2085S.

Seeram, N. P., Aviram, M., Zhang, Y., Henning, S. M., Feng, L., Dreher, M. & Heber, D. (2008). Comparison of antioxidant potency of commonly consumed polyphenol-rich beverages in the United States. *Journal of Agricultural and Food Chemistry, 56,* 1415–1422.

Shahidi, F. & Naczyk, M. *Phenolics in food and nutraceuticals.* Boca Ratón, Fl: CRC Press LLC; 2004.

Strlic, M., Radovic, T., Kolar, J. & Pihlar, B. (2002). Anti- and pro-oxidative properties of gallic acid in Fenton-like systems. *Journal of Agricultural and Food Chemistry, 50,* 6313–6317.

Sulaiman, S. F, Yusoff, N. A. M., Eldeen, I. M., Seow, E. M., Sajak, A. A. B. & Supriatno, O. K. L. (2011). Correlation between total phenolic and mineral contents with antioxidant activity of eight Malaysian bananas (*Musa sp.*). *Journal of Food Composition and Analysis, 24,* 1-10.

Szajdek, A. & Borowska, E. J. (2008). Bioactive compounds and health-promoting properties of berry fruits: a review. *Plant Foods for Human Nutrition, 63,* 147–56.

Szydłowska-Czerniak, A., Tułodziecka, A. & Szłyk, E. (2011). Determination of antioxidant capacity of unprocessed and processed food products by spectrophotometric methods. *Food Analytical Methods, 5,* 807–813.

Vázquez-Araújo, L. Chambers, E., Adhikari, K. & Carbonell-Barrachina, A. A. (2010). Sensory and physicochemical characterization of juices made with pomegranate and blueberries, blackberries, or raspberries. *Journal of Food Science, 75(7),* 398-404.

Vermerris, W. & Nicholson, R. Phenolic compound biochemistry. New York: Springer; 2007.

Villaño, D., Fernández-Pachón, M. S., Troncoso, A. M. & García-Parrilla, M. C. (2005). Comparison of antioxidant activity of wine phenolic compounds and metabolites in vitro. *Analytica Chimica Acta, 538,* 391–398.

Waterhouse, A.L. Determination of Total Phenolics, In: Wrolstad, R.E. editors. Current Protocols in Food Analytical Chemistry. New York: John Wiley & Sons Inc.; 2001; I1.1.1-I1.1.8.

Zulueta, A., Esteve, M. J. & Frígola, A. (2009). ORAC and TEAC assays comparison to measure the antioxidant capacity of food products. *Food Chemistry, 114(1),* 310–316.

In: Handbook on Gallic Acid
Editors: M. A. Thompson and P. B. Collins

ISBN: 978-1-62618-921-8

Chapter 13

ANTIOXIDANT AND CANCER PROTECTIVE EFFECTS OF GALLIC ACID *IN VITRO* AND *IN VIVO*

Franziska Ferk*[1], *Armen Nersesyan*[1], *Miroslav Mišík*[1], *Karl-Heinz Wagner*[2] *and Siegfried Knasmüller*[1]
[1]Institute of Cancer Research, Department of Medicine I, Medical University of Vienna, Austria
[2]Department of Nutritional Sciences, University of Vienna, Austria

ABSTRACT

Recent findings of our group indicate that a small amount of gallic acid (GA 12 mg/P/d) protects humans against DNA damage and induces the activities of antioxidant enzymes. Furthermore, we saw in animal studies that similar doses prevent damage of the genetic material in inner organs of rats and the induction of preneoplastic hepatic foci by γ-radiation. These promising findings stimulated us to evaluate the current knowledge on ROS and cancer protective properties of the phenolic. Approximately a dozen *in vitro* studies have been published which concern the antioxidant properties of GA; with high concentrations sometimes pro-oxidant effects were observed. Also a few *in vivo* results which were obtained with laboratory rodents are available which indicate GA induces ROS protective enzymes possibly *via* induction of the transcription factor Nrf2. In regard to the cancer protective properties of the compound, it is notable that *in vitro* experiments showed that GA inhibits cell cycle progression at G0/G1 and also at G2/M, (via modulation of the transcription of cyclines) and induces apoptosis in various human cancer cell lines (possible modes of action are inhibition of the synthesis of DNA bases, induction of NfκB, depletion of glutathione and release of ROS). Furthermore, was found that the phenolic decreases the motility and adhesion of cancer cells, inhibits metastasis and prevents angiogenesis *via* inhibition of VEGF. Also data from animal experiments (xenograft models with mice) show that the compound prevents the tumour growth. Most of these findings were obtained with doses which are substantially higher as the daily intake of GA *via* the diet which is in Middle Europe in the rage of 18 mg/P/d; however some findings were seen with lower levels that can be easily reached after increase of the

* Corresponding Author: Siegfried Knasmüller. E-Mail: siegfried.knasmueller@meduniwien.ac.at; Tel: +43 1 40160 57561; Fax: +43 1 40160 957500.

consumption by dietary measures. Taken together the current data indicate that GA is an extremely potent antioxidant in humans and may protect against malignant diseases.

1. Introduction

The promising results which we obtained with gallic acid (GA) in human and animal studies (Ferk et al., 2011, Chakraborty et al., 2009) stimulated us to evaluate the current knowledge on the antioxidant and cancer protective properties of this phenolic compound which is contained in numerous plant derived foods and also released in the digestive tract from gallotanins and other derivatives that are contained in vegetables and beverages.

Our findings from a human intervention trial (chapter 2.2) suggest that low amounts which are consumed *via* the diet prevent oxidative DNA damage. Since instability of the genetic material plays a key role in the aetiology of cancer (Ames 1993, Shaughnessy and DeMarini, 2009) our findings suggest that GA may protect humans against malignant disorders. In this context, it is notable that it was shown in a number of studies, that GA causes cycle arrest of cancer cells, induces programmed cell death (apoptopsis) and prevents metastasis and angiogenesis; these findings are summarised in chapters 3, 4 and 5. The last chapters of this article concern results of studies in which cancer protective properties of the phenolic were investigated in laboratory rodents and statements on the relevance of the *in vitro* and *in vivo* findings in regard to beneficial effects in humans.

2. Antioxidant Properties of GA

Phenolics and as well as other plant derived compounds protect against ROS (reactive oxygen species) *via* two direct molecular mechanisms, namely direct scavenging and induction of antioxidant enzymes (Knasmüller et al., 2008).

2.1. *In Vitro* Effects

Table 1 shows examples for studies which were conducted with mammalian and human cell lines. It can be seen that the phenolic causes protective effects in experimental models which reflect the inactivation of radicals (e.g. in the ferric reducing antioxidant assay), reduces the intracellular levels of radicals (2,2-diphenyl-1-picrylhydrazyl test), prevents the oxidation of lipids in cell membranes, and decreases the oxidation of glutathione.

Not all *in vitro* systems reflect the induction of antioxidant enzymes; however findings which were obtained with a human vascular endothelial cell line (HUVEC) show, that GA induces ROS protective enzymes such as superoxide dismutase (SOD) and catalase (CAT) (Yang et al., 2011). Furthermore, it was found in experiments with rat liver microsomes, that exposure to high concentrations causes an increase of the activity of GST which catalyses the detoxification of xenobiotics but is also involved in the activation of ROS (Shinno et al., 2005, Knasmüller et al., 2008).

Table 1. Antioxidant properties of GA *in vitro*

Tests system	Cell lines	Dose (μM)	Exposure time (h)	Effects (LOEL)/ Remarks	Ref.
FRAP	Rat gastric mucosal cells	5-20	1	GA > quercetin > ascorbic acid FRAP ↑ (10 μM)	Pal et al., 2010
MDA	Human peripheral blood lymphocytes	50-200	18	MDA ↓ (100 μM)	Sohi et al., 2003
GSH/GSSG	B16 melanoma cells	5-50	24	GSH ↑ (5 μM), GSSG ↓ (5 μM)	Kim 2007
GST	Rat liver microsomes	1000	0.83	GST ↑ (5-fold)	Shinno et al., 2005
DPPH TBARS GPx GR GSH GST	Human erythrocytes	10-100	4	The strongest DPPH effect was seen with 34.95 μM GA, it decreased in the order GA > caffeic acid > ferulic > acid protocatechuic acid. GA (10 μM)caused the strongest inhibitory effect of formationTBARS GPx ↔ GR, GSH, GST ↔	Hsieh et al., 2005
DPPH MDA	Human erythrocytes	50	0.5	The DPPH scavenging activity was lower than vitamin C in the order vitamin C > GA > Trolox > green tea extract. GA ↓ (294 μM) caused inhibition of MDA formation.	Hseu et al., 2008
DCFH-DA TBARS SOD catalase	Human umbilical vein endothelial cells (HUVEC)	29.3	2	Cells were exposed to AAPH (15 mM) in presence and absence of GA. DCFH-DA and TBARS decreased to basal levels. SOD ↑ (2-fold) catalase ↑	Yang et al., 2011
Ratio GSH/GSSG	Human monocytes (THP-1)	10	24	Cells were incubated with high glucose (25 mM) with and without GA. GSH/GSSG ↑ (not significant with GA alone) GSH/GSSG ↑ (GA with high glucose)	Kuppan et al., 2010
DCFH-DA TBARS GSH (intracellular) GSSG (extracellular)	Rat hepatocytes	100	3	Cells were isolated via perfusion and exposed to CHP (120 μM) with and without GA. GA had ↑ activity than aqueous sumach extract in regard to intracellular ROS and TBARS. GA prevented CHP-induced intracellular GSH depletion and ↑ the absorption of extracellular GSSG.	Pourahmad et al., 2010
DCFH-DA TBARS GSH (intracellular) GSSG (extracellular)	Rat hepatocytes	100	3	Cells were isolated via perfusion and exposed to CHP (120 μM) with and without GA. GA was a potent scavenger of ROS induced by CHP. Mango extract was more effective than GA in ↓ of TBARS GA reduced CHP induced intracellular GSH depletion and ↑ GSSG levels.	Pourahmad et al., 2010
DPPH MDA	Human chronic myelogenous leukemia cell line (K562)	6-600 1200-4800	24	DPPH scavenging activity was seen with the lowest dose of 1 μg/mL but not with the highest dose. The most pronounced inhibition was seen with the highest dose (72% ↓)	Abdelwahed et al., 2007

Table 1. (Continued)

Tests system	Cell lines	Dose (μM)	Exposure time (h)	Effects (LOEL)/ Remarks	Ref.
TEAC ORAC PST (P and M)	Human platelet cells	6.7	0.33	GA had highest scavenging capacity and the highest ORAC value among the hydroxy benzoic acid and cinnamic acid derivatives. PST (of both forms) ↑	Yeh and Yen 2003

Abbreviations: AAPH, 2,2´-azo-bis(2-amidinopropane) hydrochloride; CHP, cumene hydroperoxide; CYP3A, Cytochrome P450; DPPH, 2,2-diphenyl-1-picrylhydrazyl; DCF-DA, 2´,7´-dichlorofluorescin-diacetat; DCFH-DA, 2´,7´-dihydrofluorescein-diacetate; FRAP, ferric reducing antioxidant; GR, glutathione reductase; GSH, reduced glutathione; GSSG, oxidized glutathione; GST, glutathione *S*-transferase; MDA, malondialdehyde; ROS, reactive oxygen species; PST, platelet sulfotransferase; TEAC, trolox equivalent antioxidant capacity; SOD, superoxide dismutase; TBARS, thiobarbituric acid reactive substances.

Findings which we obtained in experiments with peripheral human lymphocytes are shown in Figure 1. We measured in this experiment the impact of exposure of the cells to GA on ROS (H_2O_2) induced DNA instability in single cell gel electophoresis assays (SCGE) which are based on the determination of DNA migration in an electric field (Tice et al., 2000) under standard conditions. It can be seen, that the phenolic caused dose-dependent protective effects in the range between 0.03 and 0.15 μg/mL (0.17-0.85 μM).

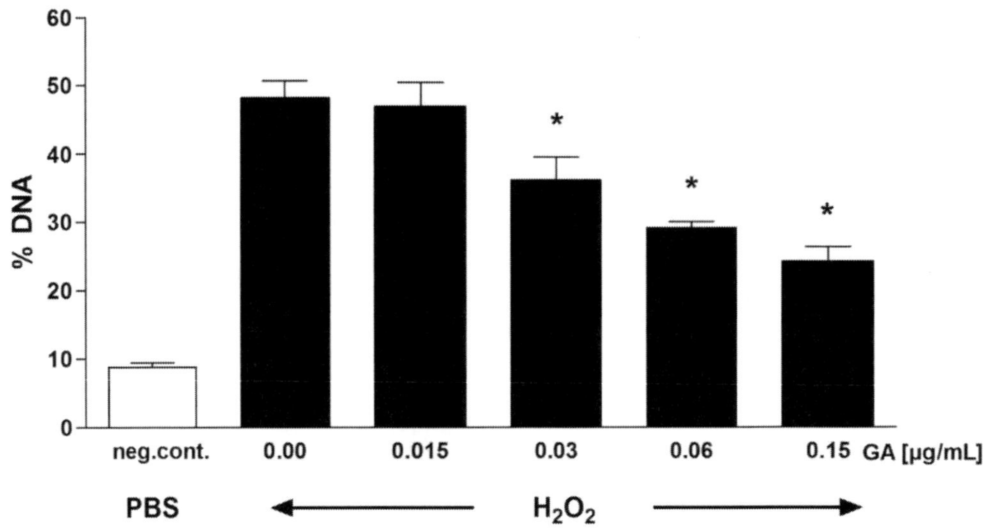

Figure 1. ROS-protective effects of GA in human lymphocytes. The cells were treated for 60 min with different concentrations of GA (0.015–0.15 μg/mL). At the end of the exposure period H_2O_2 (75 μM) was added for 5 min. Subsequently DNA-migration was monitored under standard conditions. Per experimental point, three cultures were made in parallel and from each culture 50 cells were evaluated with a computer aided image analysis system (Comet Assay IV, Perceptive Instruments, UK). Stars indicate statistical significance (*$p < 0.05$, ANOVA). For a description of the method see Ferk et al., 2011).

The antioxidant properties of GA which were seen in *in vitro* experiments were similar to those observed with other common phenolics. For example, the protection against ROS (H_2O_2) induced DNA migration in human lymphocytes was in the same range as that detected with quercetin (Ferk et al., unpublished); Soobrattee and coworkers (2005) compared the antioxidant properties of GA with those of butylated hydroxyanisole (BHA) in the TEAC and FRAP assays and found more or less identical activities.

2.2. Antioxidant Effects of GA in Laboratory Rodents and in Human

Results of a few investigations show that the acid is also protective in animals and humans. For example, it was found that it prevents liver damage in experimental models in which ROS play an indirect role: such as inflammations and liver injury which were induced by carbon tetrachloride and diethylnitrosamine (Tung et al. 2009; Jadon et al., 2007; Anand et al., 1997; Jagan et al., 2008). In the later case, it is known that radicals play a key role in the induction of cellular damage and cancer (Teufelhofer et al., 2005).

Evidence for induction of the antioxidant defence system was also obtained in a number of other *in vivo* studies. In rat brains, a significant increase of thiobarbituric acid reactive substances (TBARS), SOD, CAT and GSH was seen after intraperitoneal administration of GA (20 mg/kg b.w./d) for 7 days (Nabavi et al., 2012). Similar observations were made in inner organs such as liver, intestine and colon in 1,2-dimethylhydrazine treated rats after intragastric intubation with GA (Giftson et al., 2010). In an experimental model of aging with mice, significant enhancement of the activities of CATand GPx were seen in whole blood and in hepatic tissue while MDA levels in liver, brain and kidney were reduced after oral treatment with GA (80 mg/kg b.w./d) over 30 days (Li et al., 2005).

In our own experiments, we supplemented drinking water with GA and gave it over a period of 3 days to rats (3/group), subsequently we analysed the extend of DNA damage in single cell gel electrophoresis assays. In an additional experiment, the animals were exposed in a ^{60} Co γ-source. It is known, that this form of radiation causes DNA damage partly due to intracellular formation of ROS which are released as a consequence of the cleavage of water molecules. The results are summarised in Figure 2. Significant protective effects were seen in all organs in regard to endogenous formation of oxidised DNA bases (i.e. of purines and pyrimidines), furthermore, we observed also reduction of radiation induced DNA damage under standard conditions (which reflect formation of single and double strand breaks).

According to our knowledge, only one study has been published which concerns the antioxidant effects of GA in humans (Ferk et al., 2011). The participants consumed in this intervention trial GA in drinking water over a period of 3 days. Subsequently, peripheral lymphocytes were isolated and analysed for comet formation under standard conditions with lesion-specific enzymes (formamidopyrimidine-DNA glycosylase and endonuclease III) enzymes and after ROS (H_2O_2) treatment. These enzymes enable to quantify oxidatively damaged purines and pyrimidines. Furthermore, markers were analysed which are indicative for the redox status (malondialdehyde, MDA and the total antioxidant capacity, TAC). Also the activities of different antioxidant enzymes (SOD; glutathione peroxidise, GPx; overall glutathione *S*-tranferase, GST; glutathione *S*-tranferase α, GST-α and glutathione *S*-tranferase π, GST-π) were measured before and after the intervention phase. The results are summarised in Figure. 3. It can be seen that GA prevented the endogenous formation of DNA bases and protected the cells against ROS (H_2O_2) induced DNA damage, while no significant alterations of makers were observed which are commonly used to assess the antioxidant status in humans. It is likely that the protection of the genetic material is due to an increase of the activities of antioxidant enzymes: The levels of three important enzymes which are involved in the detoxification of radicals namely SOD, GPx and GST-π were significantly elevated after GA consumption.

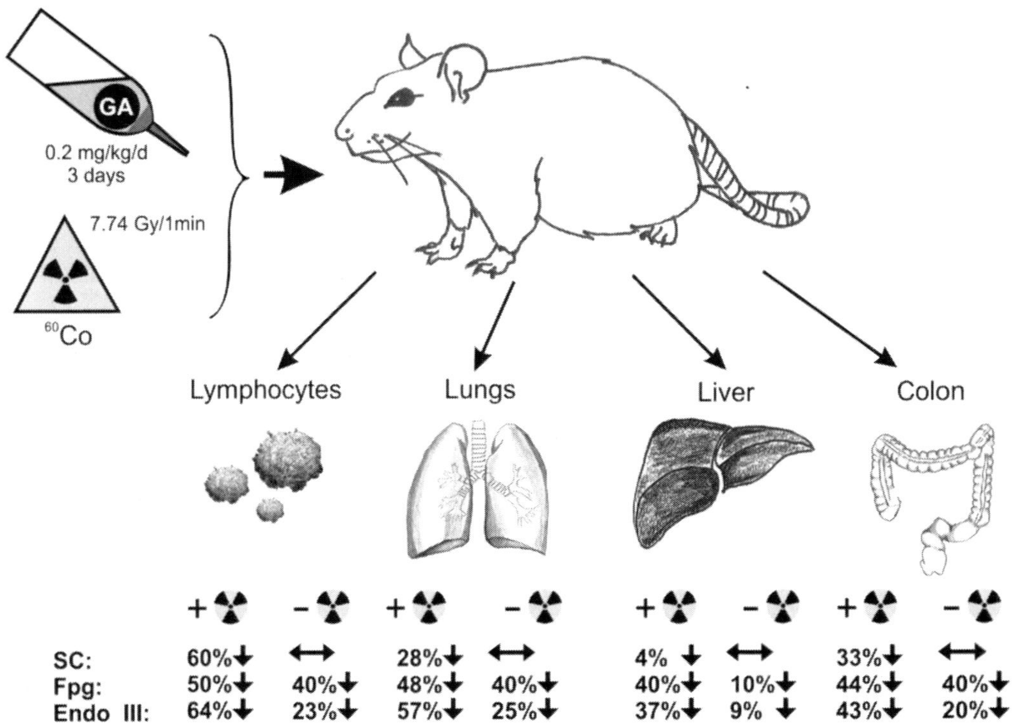

Figure 2. Protective effects of GA in rats. The animals (3/group) received the phenolic in drinking water (0.2 mg/kg bw/day) for a period of 3 days; half of the gruops were exposed in a ^{60} Co gamma source to 7.74 Gy. Control animals received normal tap water. The nuclei were isolated by homogenisation as described by Sasaki et al. (2002), DNA migration was measured as described in detail in the paper of Ferk et al. (2011). From each organ, 3 slides were made and 50 cells were evaluated per slide. Numbers indicate in % the reduction of comet formation (i.e. of %DNA in tail) compared to the corresponding controls. ↑ induction, ↓ reduction, ↔ no effects. Abbreviations: Endo III, endonuclease III, Fpg, formamido-pyrimidine-DNA glycosylase, SC, standard conditions.

In this context it is notable that induction of antioxidant enzymes was also seen in earlier investigations in rodents and humans with a number of other dietary constituents for example with different phenolics, allyslulfides (Gudi and Singh 1991, Smith and Yang 2000), coffee specific diterpenoids and also with beverages such as coffee and teas (Cavin et al., 2002, Hoelzl et al., 2009; Bichler et al., 2007). It was postulated that these effects are due to activation of the transcription factor Nrf2 which controls numerous genes that are involved in the detoxification of xenobiotics and in antioxidant defence (Kensler et al., 2010). It is notable that Yeh and Yen published in 2006 a study in which they showed that oral administration of GA to rats (100 mg/kg b.w.) over 14 days causes activation of Nrf2 in liver.

In the same study, an increase of the activities of the antioxidant enzymes such as SOD, GPx and CATin the livers and small intestines was observed. In a subsequent experiment by the same authors, induction of Nrf2 by GA was seen in the heart muscle of rats, also this effect was paralleled by increases of SOD, GPx, CATand of the oxygen radical absorbance capacity (ORAC) in heart tissue (Yeh et al., 2009).

Table 2. Effects of GA on cell division

Cell lines	Inhibitory dose (μM)	Exposure time (h)	Methods	Remarks/ molecular mechanisms	Ref.
Human bladder transitional carcinoma cells (TSGH-8301)	30	24	FC WB IB	Cell cycle arrest at G2/M via Chk2-mediated phosphorylation of Cdc25C at Ser-216. Cdk1 ↓, Cyclin B1 ↓, Cdc25C ↓ Induction of Cip1/p21 protein expression	Ou et al., 2010
Human prostate carcinoma cells (DU 145)	50	6	FC IB IB	Cell cycle arrest at G2/M via Chk2-mediated phosphorylation of Cdc25C at Ser-216. Phosphohistone H2A.X (a marker of DNA damage) ↑ Induction of Cip1/p21 protein expression	Agarwal et al., 2006
Human prostate carcinoma cells (DU 145)	5.8	24	FACS IB	Cell cycle arrest at G2/M Chk1 ↑, Chk2 ↑, Cdc25C ↓	Chen et al., 2009
Human promyelocytic leukemia cells (HL-60)	50	24	FC WB	Cell cycle arrest at G0/G1 via ↓ of Cyclin D and E p21 ↑, ↑ p27.	Yeh et al., 2011
Human cervix adenocarcinoma cells (HeLa)	70	24	WB IB	Phosphorylation of ERK ↓, JNK ↑, p38 ↔	You & Park 2011
Human prostate cancer cells (PC-3)	100	24	FC RT-PCR	ATM ↓, ART ↓ (checkpoint kinases, activated by double-stranded DNA breaks) p53 ↓	Liu et al., 2011
Human myeloid leukemia cells (K562)	50	24	FC WB	Cell cycle arrest at G2/M p21 ↑, p27 ↑, Cyclin D ↓, Cyclin E ↓	Chandramohan et al., 2012
Human breast cancer cells (MCF-7)	11.7	24	FC WB	Cell cycle arrest at G2/M p27 (Kip1) ↑, p21 (Cip1) ↑, Cyclin A ↓, Cyclin B1 ↓, Skp2 ↓	Hsu et al., 2011
Human promyelocytic leukemia cells (HL-60)	20	24	FACS	Cell cycle arrest at G2/M	Madlener et al., 2007

Abbreviations: Cdk1, cyclin-dependent kinases; Chk2, checkpoint kinase 2; FACS, fluorescence activated cell sorting; FC, flow cytometry; IB, immunoblotting ; RT-PCR, real time polymerase chain reaction; WB, western blotting.

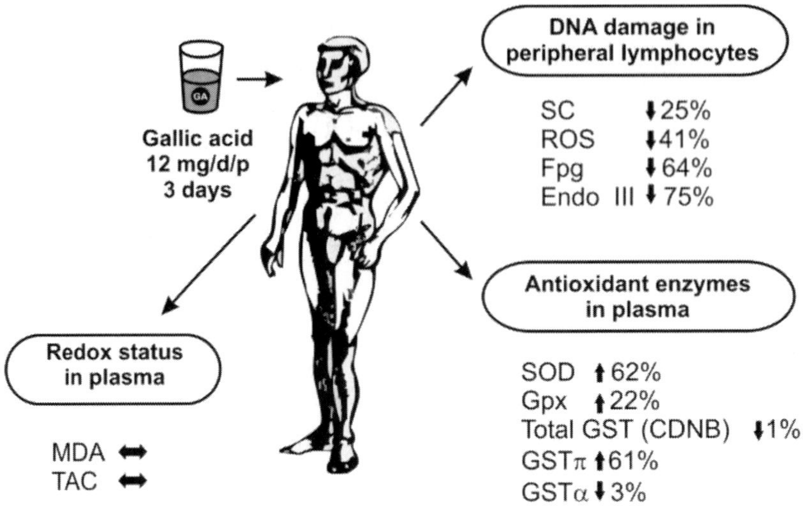

Figure 3. Impact of consumption of GA on the antioxidant status in humans. The participants (8/group) consumed GA in the drinking water or a placebo (non supplemented water) over a period of 3 days. Subsequently, blood was collected and DNA damage measured in SCGE experiments, additionally also parameters were determined which reflect the redox status and antioxidant enzymes were monitored with established methods (for details see Chakaborty et al., 2009; Ferk et al., 2011). Numbers indicate in % the differences between effects seen at the beginning and at the end of the intervention. In addition, a placebo group was included and no alterations of the different parameters were detected (data not shown). ↑ induction, ↓ reduction, ↔ no effects. Abbreviations: Endo III, endonuclease III, Fpg, formamido-pyrimidine-DNA glycosylase, GPx, glutathione peroxidase; overall GST, glutathione S-tranferase; GST-α, glutathione S-tranferase α; GST-π, glutathione S-tranferase π; MDA, malondialdehyde; ROS, reactive oxygen species (released by H_2O_2 treatment); SC, standard conditions; SOD, superoxide dismutase; TAC, total antioxidant capacity.

3. Inhibition of the Division of Cancer Cells

The impact of GA on cell division was investigated in a number of studies, some of them are listed in Table 2. It can be seen that two checkpoints of the cycle are affected namely G2/M and G0/G1. The molecular mechanisms are quite complex and include inhibition of the transcription and/or synthesis of cyclines as well as inhibition of other proteins which are involved in cell cycle progression. It is notable, that significant effects were seen in the majority of experiments at dose levels which are in the range between 30 -100 μM, the most sensitive cell line is obviously MCF-7, which was responsive at doses ≥ 12 μM.

4. Impact of GA on Programmed Cell Death

Numerous studies provide evidence that GA causes apoptosis human and animal derived cell lines. It can be seen in Table 3 that the phenolic induces apoptosis through both, caspase-dependent and caspase-independent pathways.

Table 3. Effect of GA on apoptosis and cellular changes associated with programmed cell death

Cell lines	LOEL (μM)	Exposure time (h)	Methods	Remarks/ molecular mechanisms	Ref.
Human chronic myeloid leukemia cells (K562)	50	24	SDS/Proteinase K/ RNAse A extraction FACS Calibur ™ flow cytometry FACS WB EIA	DNA fragmentation ↑ $\Delta\Psi_m$ ↓ Apoptosis ↑ PARP ↑, FASL ↑, caspase-3 ↑, caspase-8 ↑, Bax ↑, Bcl-2 ↓, BCR/ABL kinase ↓, COX-2 ↓, Cytochrome c ↑ NF-κB ↓	Chandramohan et al., 2012
Human promyelocytic Leukemia cells (HL-60)	50	6	FC WB RT-PCR	$\Delta\Psi_m$ ↓, Ca^{2+} ↑ caspase-3 ↑, caspase-8 ↑, caspase-9 ↑, AIF ↑, AIF ↑, cytochrome c ↑, Endo G ↑, FAS ↑, FASL ↑, BID ↑, Bax ↑, Bcl-2 ↓, caspase-4 ↑	Yeh et al., 2011
Lymph-node carcinoma cells from the prostate (LNCaP)	8.8	12 /24	ELISA, TU ELISA WB	Apoptosis ↑ (12 h) DNA fragmentation ↑ (24 h) PARP ↑, caspase-3 ↑ (12 h), caspase-9 ↑ (24 h)	Reddivari et al., 2010
Prostate cancer cells (PC-3)	8.8	24	ELISA, TU ELISA	Apoptosis ↑ DNA fragmentation ↑ Caspase independent cell death mechanisms were seen in PC-3 cells due to androgen independency of the cells.	Reddivari et al., 2010
Human prostate cancer cells (DU145)	50	24	FC	Annexin V-FITC positive cells ↑ (75%) Caspase independent cell death mechanisms were seen in DU145 due to androgen independency of the cells.	Kaur et al., 2009
Human prostate cancer cells (DU145)	100	8	FC SDS-PAGE and IB	$\Delta\Psi_m$ ↓, cytochrome c release caspase-3 ↑, caspase-9 ↑, Bax ↑, Bcl-2 ↓, PARP ↑	Chen et al., 2009
Human prostate cancer cells (DU145)	50	24	FC SDS-PAGE and IB	Apoptosis ↑ Caspase-3 ↑, caspase-9 ↑, PARP ↑	Agarwal et al., 2006
Human peripheral lymphocytes	50	18	FACStar flow cytometry DNA ladder SDS-PAGE and IB FC	Apoptosis ↑ DNA fragmentation ↑ Bcl-2 ↔ 100 μM of GA protect lymphocytes from PS externalisation. (PS externalisation is a specific marker for apoptosis)	Sohi et al., 2003
Human melanoma cells (A375.S2)	250	48	FC WB RT-PCR	Apoptosis ↑, $\Delta\Psi_m$ ↓ Caspase-3 ↑, caspase-8 (12 h) ↑, caspase-9 ↑ Bcl-2 ↑ (12 h), Bax ↔, Bid ↑, caspase-4 ↔, Fas ↑, FasL ↑ (12 h), cytochrome c ↑, Endo G ↑, AIF ↑	Lo et al., 2010
Human lung adenocarcinoma cells (A549)	200	1	DNA-fragmentation assay FC Colorometric assay	DNA fragmentation ↓ $\Delta\Psi_m$ ↓ Caspase-3 ↑	Maurya et al., 2011

Table 3. (Continued)

Cell lines	LOEL (μM)	Exposure time (h)	Methods	Remarks/ molecular mechanisms	Ref.
Human pulmonary adenocarcinoma cells (Calu-6, A549)	50 100	24	FACStar flow cytometry FC	Annexin V-FITC positive cells ↑ (Calu-6, at 50 μM, A549 at 100 μM) $\Delta\Psi_m$ ↓	You et al., 2011
Human pancreatic cancer cell lines (CFPAC-1, MiaPaCa-2)	14.7	48	FACStar flow cytometry FC analysis Fluorometric assay FC	Annexin V-FITC positive cells ↑ $\Delta\Psi_m$ ↓ Caspase-3 ↑ , caspase-8 ↔ , caspase-9 ↑ Bax	Liu et al., 2012
Human cervix adenocarcinoma (HeLa)	164.3	4	Fluorometric assay	caspase-9 ↓ (67.8%)	Erol-Dayi et al., 2012
Human cervix adenocarcinoma (HeLa)	70	24	FC FACStar flow cytometry	$\Delta\Psi_m$ ↓ annexin V-FITC positive cells ↑ (45 %)	You et al., 2011
Human monocytic cells (U937)	5.8	24	FC RT-PCR	DNA fragmentation ↑ NF-κB ↑	Kim et al., 2011
Human osteosacroma cells (U-20S)	50	24	FACSalibur flow cytomerty Colorimetric analysis	Annexin V-FITC positive cells ↑ Caspase-3 ↑ (1.5-fold), caspase-8 ↔ , caspase-9 ↑ (2-fold)	Liang et al., 2012
Human gastric carcinoma cells (AGS)	3.5	24	IB	NF-κB ↔	Ho et al., 2013
Human promyelocytic Leukemia cells (HL-60)	60	24	Microscopy (Hoechst/propidium Iodid staining) EIA	% Apoptosis ↑ COX 1 ↓ , COX 2 ↓	Madlener et al., 2007

Abbreviations: AIF, apoptosis inducing factor; EIA, enzyme immuno assay; ELISA, enzyme linked immunosorbent assay; Endo G, endonuclease G; FAS, fatty acid synthase; FC, flow cytometry; annexin V-FITC, annexin V-fluorescin isothiocyanate; IB, immunoblotting; LOEL, lowest effect level ; $\Delta\Psi_m$, mitochondrial membrane potential; PS, phospatidylserine; TU, TUNEL assay; WB, western blotting.

Investigations concerning the underlying molecular mechanisms show that GA increases the transcription of caspases which play a key role in the onset of programmed cell death. Furthermore, also up-regulation of pro-apoptotic proteins such Bax and down-regulation of anti-apoptotic proteins (e.g. Bcl-2) play a role. In many studies, alterations of the mitochondrial membrane potential and release of cytochrom c from mitochondria was observed, indicating that these mechanisms are of particular importance (Lo et al., 2010).

GA causes a number of phenomena which are assumed to lead to programmed cell death. (i) Induction of NF-κB transcription *via* increase of IκB. The transcription factor plays a key role in inflammations and is involved in the onset of programmed cell death (Park et al., 2007). (ii) Induction of PARP and (iii) nucleotide pool imbalances which were observed in human promyelocytic cells (Yeh et al., 2011). (iv) Also down-regulation of the Rho family of small GTPases, which are key regulators of processes like cell motility and cell cycle progression may be involved. (v) It is known from a number of investigations that GA acts as a pro-oxidant (as many other phenolic compounds) and releases reactive oxygen radicals.

Such effects were seen in different experimental models with GA at dose levels ≥ 50 μM (Arouma et al., 1993; Yen et al., 2002; Russell et al., 2012) and it is possible that ROS formation accounts for the increase of the apoptosis rates in high dose experiments. (vi) The depletion of GSH is often a consequence of ROS production and known to play a key role in the onset of programmed cell death (Yang and Park 2010). However, it was postulated that modulation of the GSH levels caused by GA not due to disturbances of the redox status but a consequence of the reduced expression of γ-glutamylcysteine synthetase (γ-GCS), an enzyme with catalyses the rate limiting step in the synthesis of the tripeptide (Locatelli et al., 2009).

5. Inhibition of Cell Migration, Metastasis and Angiogenesis

Table 4 summarizes the results of *in vitro* studies which concern the impact of the phenolic on invasive growth of cancer cells and on metastasis.

Observations with human glioblastoma cells showed that treatment with GA causes a dose-dependent down-regulation of metalloproteinase 17 (ADAM 17). The ADAM family belongs to a group of Zn-dependent metalloproteinases (MMP), ADAM 17 is involved in the proteolysis of collagens of the extracellular matrix and it is assumed that it plays an important role in the invasive growth of cancer cells (Ho et al., 2010). In the same study, it was also found that P13K/Akt as well as Ras/MAPK are down-regulated and similar observations were made in experiments with osteosacroma and prostate cancer cells. The first pathway plays a role in the metastasis of different tumours, for example in melanoma cells. Activated Akt can bind to the IκB kinase (IκK) which has an impact on the inhibition of NfκB which was seen in some *in vitro* experiments.

GA treatment of different cell lines led also to decreased expression of Ras-homologous (Rho) GTPases which play a role in cellular functions associated with metastasis (Weng and Yen 2012). RhoA down-regulation (which was observed in gastric, prostate and melanoma cells) promote like other GTPases family members (such as Ras) invasive growth and metastasis, while RhoB, which was inhibited in stomach-derived cells, has possibly a tumour suppressive role.

In many trials, it was found that GA reduces the expression of several of matrix metalloproteinases (MMPs). These zinc-dependent enzymes play a role in the invasion, metastasis and angiogenesis of tumour cells.

In some of these mechanistic studies, additional experiments were conducted in which the impact of GA in cell motility was investigated. For example, human gastric carcinoma cells (AGS) were studied in a wound healing assay by Ho and co-workers (2010), and reduced migration was observed with low GA concentrations (2-5 μM). Also in studies with Matrigel coated cell culture chambers, clear inhibition of migration was seen with melanoma (A375.S2) cells with 10-fold higher exposure concentrations (Lo et al. 2011).

The adhesive ability of metastatic tumour cells is another critical factor in the extravasation and in the formation of new tumour foci. The impact of GA on this process was investigated in experiments with a murine melanoma cell line (B16F10) and clear inhibitory effects were seen with GA at dose levels ≥ 25 μM (Locatelli et al., 2009).

Table 4. Effects of GA on cell mobility, invasion and metastasis

Cell lines	LOEL (μM)	Exposure time (h)	Methods	Remarks/ molecular mechanisms	Ref.
Human gastric carcinoma cells (AGS)	2	48	GZ WB FM	MMP-9 ↓, MMP2 ↓ NF-κB ↓, Ras ↓, cdc42 ↓, Rac1 ↓, RhoA ↓, RhoB ↓, P13K ↓, p38 MAPK ↓ F-actin ↓	Ho et al., 2010
Human glioblastoma (U87, U251n)	234.7	24	WB	ADAM 17 ↓, p-AKT ↓, p-Erk ↓ P13K/Akt ↓,Ras/MAPK ↓	Lu et al., 2010
Human vascular endothelial cell line (HUVEC)	29.3	24	WB	MMP-9 ↓, MMP2 ↓	Hseu et al., 2011
Human melanoma cells (A375.S2)	100	24	GZ RT-PCR WB	MMP-9 ↓, MMP2 ↓ MMP-9 ↓, MMP2 ↓, RhoA ↓ Ras ↓, ERK1/2 ↓, p38 ↓, MMP-2 ↓, MMP-9 ↓	Lo et al., 2011
Human prostate cancer cells (PC-3)	25	24	GZ WB RT-PCR	MMP-9 ↓, MMP2 ↓ MMP-9 ↓, MMP2 ↓, ERK1/2 ↓, pAKT ↓, NF-κB ↓, P13K/AKT ↑ RhoA ↓, FAK ↓	Liu et al., 2011
Human chronical myeloid leukemia cells (K562)	10	24	WB GZ	MMP-9 ↓, MMP2 ↓ (Bcr-Abl-negativ)	Chen et al., 2012
Human acut myeloid leukemia cells (U937)	10	24	WB GZ	MMP-9 ↔, MMP-2 ↔ (Bcr-Abl-positiv)	Chen et al., 2012
Human osteosacroma cell lines (U-2 OS)	20	24	GZ WB RT-PCR	MMP-9 ↓, MMP-2 ↓ MMP-9 ↓, MMP-2 ↓, P13K ↔, AKT ↔, p38 ↓, ERK1/2 ↓, NF-κB ↓, MAPK ↓ MMP-9 ↓, MMP-2 ↓	Liao et al., 2012

Abbreviations: FM: Fluorescence microscopy GZ, gelatine zymography; LOEL ,lowest effective dose ; MMPs, matrix metalloproteinases; RT-PCR, real time polymerase chain reaction; WB, western blotting.

A group from Taiwan (Hseu et al., 2011) studied the anti-cancer effects of *Toona sinensis*, a plant which is used in vegetarian cuisine but also as a herbal remedy to treat cancer and inflammations. The authors assumed that GA is one of the active principles and demonstrated that the extract of the plant and also the phenolic itself inhibit the induction of cell proliferation by the vasoendothelial growth factor (VEGF) in human umbilical vein endothelial cells (HUVECs) and also in a human vascular endothelial cell line (EA.hy 926). This effect was paralleled by suppression of MMP-9 and MMP-2 which are activated by VEGF. These collagenases play an important role in tumour invasion as they break down basement membrane compounds. The effects of GA were seen with relatively low concentrations (29.3 μM). Subsequent results obtained in chicken embryo chorioallantoic membrane (CAM) assays showed that GA is also effective in preventing angiogenesis *in vivo* (Hseu et al., 2011).

Table 5. Effects of GA in cancer models

Species/Strain	Cancer models/organs	Dose (mg per kg b.w.) treatment route duration (d)	Results	Ref.
C57Black mice male	Mouse lung cancer cell lines (LL-2 cells)	400 (drinking water) 28	9.2% ↓ of the size of tumor in GA-group 12,5% ↓ in average weights of the tumor in GA-group Apoptotic cells (in TUNEL-positive cells) 2.6-fold ↑	Kawada et al., 2001
TRAM mice male	Animals developed spontaneous progressive stages of prostatic disease with 2 stages.	1200 and 4000 drinking water 140	Histopathological analyses (H & E staining) show a marked difference in the area covered by less aggressive tumors in the GA supplemented group (63%, 67%) compared with the positive controls. Apoptotic cells (in TUNEL-positive cells) were 3-fold ↑ Immunoblot analyses of protein levels of: Cdk2 (96%, 100%) ↓ , Cdk4 (72%, 90%) ↓ , Cdk6 (69%, 74%) ↓ , Cdc2 (100% for both conc.) ↓, Cyclin A (80%, 70%) ↓, Cyclin B1 (50%, 62%) ↓, Cyclin E (41%, 52%) ↓. Changes in the expression levels of: Cip1/p21 ↔ , Kip1/p27 ↔	Raina et al., 2008
BALB/c mice male	Murine leukemia cells (WEHI-3)	40 and 80 (gavage) 21 d	↓ in the weights of spleens and livers of GA treated group (spleen and liver enlargement are characteristic for WEHI-3 leukemia). GA promotes macrophage phagocytosis from peripheral cells, but decreased macrophage phagocytosis in isolated peritoneal cells at 80 mg/kg b.w.	Ho et al., 2009
BALB/c $^{nu/nu}$ female	Human lung carcinoma cells (NCI-H460)	20 i.p. 34	40.4% ↓ of tumor size and 40% ↓ of tumor weights after GA administration caspase-3 ↑ , caspase-8 ↑ , caspase-9 ↑	Ji et al., 2009
BALB/c $^{nu/nu}$ male	Human prostate cancer cells (DU145, 22Rv1)	1200 and 4000 (drinking water) 42	↓ tumor size: DU145: 25%, 47.5%, 22Rv1: 28.5%, 62% Dose dependent anti-proliferative effects (in PCNA-positive cells) were seen: DU145: 12%, 22.5% ↓ , 22Rv1: 10%, 24% ↓ Anti-apoptotic effects (in TUNEL-positive cells) were induced in a dose dependent manner. DU145:1.9-fold and 2.25-fold ↑ , 22Rv1: 1.6-fold and 1.8-fold ↑ Anti-angiogenic effects were monitored by CD31-immunohistochemical staining (in CD31 positive cells). DU145: 26%, 39% ↓ , 22Rv1: 24%, 29% ↓	Kaur et al., 2009
Him-OFA male	Liver	0.2 (drinking water) 8	43% ↓ of radiation-induced preneoplastic lesions (GST-P^{+}) in the liver	Ferk et al., 2011
BALB/c $^{nu/nu}$ female	Human osteosarcoma cells (MNNG/HOS)	1200 and 4000 (drinking water) 5	1.2-fold and 1.6-fold ↓ of tumor volume Anti-proliferative effects (in PCNA-positive cells) were seen in a dose dependent manner. 75% and 95% ↓ 25% and 50% ↓ of anti-angiogenic effects were monitored by CD31-immunohistochemical staining (in CD31 positive cells).	Liang et al., 2012
BALB/ c $^{nu/nu}$ male	Human gastric carcinoma cells (AGS)	1000 and 2000 (drinking water) 30	1.5-fold and 2.4-fold ↓ of tumor volume, 2-fold and 3-fold ↓ of tumor weights	Ho et al., 2013

6. Results of Animal Studies

The findings obtained in *in vivo* experiments with laboratory rodents are summarized in Table 5. Protective effects such as reduction of tumour growth and size as well as anti-angiogenic properties and increased apoptosis rat were seen in several xenograft mouse models. Furthermore, it is notable that a significant reduction of the formation of preneoplastic hepatic foci (which are considered to be precursors of liver tumours), was observed in a study with rats. All these investigations, except the radiation experiment were conducted with high doses which exceed the human uptake *via* food consumption (see below).

7. Does GA Protect Humans against Cancer?

The evaluation of the literature shows that GA prevents the formation of cancer cells (*via* protection of the genetic material) furthermore, protective effects were seen on other stages of cancer development, i.e. inhibition of division of cancer cells, induction of programmed cell death and prevention of metastasis and angiogenesis (see previous chapters).The different modes of action are summarized in Figure 4.

The results obtained in the animal experiments support the assumption that protection against cancer may also take place in humans. To elucidate under which conditions chemoprotective effects can be expected in humans, the doses which are required to cause effects in animal models and the concentrations which were used in *in vitro* experiments have to be compared with the amounts which are consumed *via* the diet and with the plasma levels in humans.

The amounts which were given in the xenograft studies to rodents in chow or drinking water are in almost all cases equivalent to uptake of several grams/person/day. These levels are three to four orders of magnitude higher compared to the daily uptake which is consumed by populations in Central Europe (18 mg/P/d, Radkte et al., 1998).

The most promising data in regard to beneficial effects come from intervention trials in which a relatively small dose (12 mg/P/d) was found to be sufficient to cause protection against oxidative DNA damage, which was paralleled by an increase of the activity of antioxidant enzymes (Ferk et al., 2011). However, it is still a matter of debate if antioxidant effects are directly associated with cancer protective properties (Collins 2013). In this context it is notable that the same dose which we used in the human study was not only effective in regard to prevention of DNA damage in different inner organs of rat but also in regard to inhibition of the formation of radiation induced precursors of liver tumours.

A number of studies has been performed which concern the bioavailability of GA. The levels which were seen in human plasma after intake of 50 mg /P/d of the pure the compound or of an equivalent amount in black tea were around 1.8-2.1 μM/L (Manach et al 2005). In our experiments (Ferk et al., 2011) we detected after intake of 12 mg/P/d a maximal concentration of 0.7 μM in the serum. The doses which were effective in most of the *in vitro* experiments are about 10-100 fold higher than the levels detected in the human studies (see Tables 2 and 3). However, in some investigations, for example with human monocytic cells (Kim et al., 2011) and human prostate carcinoma cells (Chen et al., 2009) promising results in regard to induction of apoptosis and inhibition of cell division were seen already with concentrations between 5 and 10 μM. The effective levels which were obtainedin *in vitro* studies concerning the antioxidant properties of GA were in a similar range (see Table 1) and as mentioned above, also inhibition of migration of gastric cells was detectedwith low doses.

In this context it is notable that that certain plant derived foods and beverages contain high amounts of the phenolic, for example mangos up to 440 mg/100 g d.w., red wine up to 5 mg/100 mL and teas up to 39 mg/100 mL (Barreto et al, 2008; Landraut et al., 2001; Lin et al., 1998). Therefore, the daily uptake levels can be increased substantially by dietary measures.

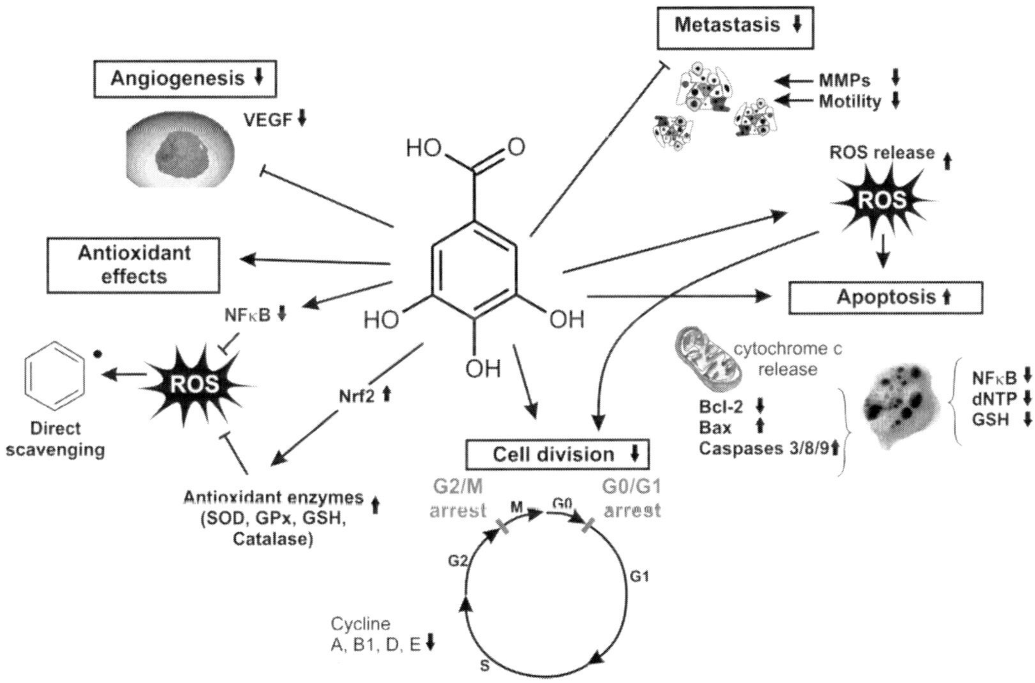

Figure 4. Properties of GA which are associated with prevention of formation of cancer cells and growth of tumours.

Comparisons of the DNA protective effects of GA with those of other dietary compounds (e.g. other phenolics) indicate, that the acid is more effective than other common antioxidants contained in the human diet (Moller and Loft 2002). The high antioxidant activity of GA which was seen in the human intervention trial and also animal experiments may be due to its extraordinary efficient absorption which is by far higher than that of other phenolics (Manach et al., 2005), furthermore, GA acts not only as a direct radical scavenger but also indirectly *via* induction of Nrf2.

Taken together, the evaluation of currently available data shows that GA is one of the most promising antioxidants and cancer protective compounds in the human diet. The findings obtained in animal models suggest that this compound is active at different molecular levels and indicate that it may be not only useful in the prevention of DNA damage and malignant diseases but become also relevant in regard to its putative therapeutic properties. However, substantial research will be required for a better understanding of the cancer protective potential of this promising compound.

ACKNOWLEDGMENTS

The study was supported by a grant from the "Herzfeldersche Familienstiftung" (to SK).

REFERENCES

Abdelwahed A, Bouhlel I, Skandrani I, Valenti K, Kadri M, Guiraud P, et al. Study of antimutagenic and antioxidant activities of gallic acid and 1,2,3,4,6-pentagalloylglucose from Pistacia lentiscus. Confirmation by microarray expression profiling. *Chem Biol Interact* 2007;165:1-13.

Agarwal C, Tyagi A, Agarwal R. Gallic acid causes inactivating phosphorylation of cdc25A/cdc25C-cdc2 via ATM-Chk2 activation, leading to cell cycle arrest, and induces apoptosis in human prostate carcinoma DU145 cells. *Mol Cancer Ther* 2006;5:3294-3302.

Ames BN. DNA damage from micronutrient deficiencies is likely to be a major cause of cancer. *Mutat Res* 2001;475:7-20.

Anand KK, Singh B, Saxena AK, Chandan BK, Gupta VN, Bhardwaj V. 3,4,5-Trihydroxy benzoic acid (gallic acid), the hepatoprotective principle in the fruits of Terminalia belerica-bioassay guided activity. *Pharmacol Res* 1997;36:315-321.

Aruoma OI, Evans PJ, Kaur H, Sutcliffe L, Halliwell B. An evaluation of the antioxidant and potential pro-oxidant properties of food additives and of trolox C, vitamin E and probucol. *Free Radic Res Commun* 1990;10:143-157.

Barreto JC, Trevisan MT, Hull WE, Erben G, de Brito ES, Pfundstein B, et al. Characterization and quantitation of polyphenolic compounds in bark, kernel, leaves, and peel of mango (Mangifera indica L.). *J Agric Food Chem* 2008;56:5599-5610.

Bichler J, Cavin C, Simic T, Chakraborty A, Ferk F, Hoelzl C, et al. Coffee consumption protects human lymphocytes against oxidative and 3-amino-1-methyl-5H-pyrido[4,3-b]indole acetate (Trp-P-2) induced DNA-damage: results of an experimental study with human volunteers. *Food Chem Toxicol* 2007;45:1428-1436.

Cavin C, Holzhaeuser D, Scharf G, Constable A, Huber WW, Schilter B. Cafestol and kahweol, two coffee specific diterpenes with anticarcinogenic activity. *Food Chem Toxicol* 2002;40:1155-1163.

Chakraborty A, Ferk F, Simic T, Brantner A, Dusinska M, Kundi M, et al. DNA-protective effects of sumach (Rhus coriaria L.), a common spice: Results of human and animal studies. *Mutat Res* 2009;661:10-17.

Chandramohan Reddy T, Bharat Reddy D, Aparna A, Arunasree KM, Gupta G, Achari C, et al. Anti-leukemic effects of gallic acid on human leukemia K562 cells: downregulation of COX-2, inhibition of BCR/ABL kinase and NF-kappaB inactivation. *Toxicol In Vitro* 2012;26:396-405.

Chen HM, Wu YC, Chia YC, Chang FR, Hsu HK, Hsieh YC, et al. Gallic acid, a major component of Toona sinensis leaf extracts, contains a ROS-mediated anti-cancer activity in human prostate cancer cells. *Cancer Lett* 2009;286:161-171.

Chen YJ, Chang LS. Gallic acid downregulates matrix metalloproteinase-2 (MMP-2) and MMP-9 in human leukemia cells with expressed Bcr/Abl. *Mol Nutr Food Res* 2012;56:1398-1412.

Collins AR. Kiwifruit as a modulator of DNA damage and DNA repair. *Adv Food Nutr Res 2013*;68:283-299.

Erol-Dayi O, Arda N, Erdem G. Protective effects of olive oil phenolics and gallic acid on hydrogen peroxide-induced apoptosis. *Eur J Nutr* 2012;51:955-960.

Ferk F, Chakraborty A, Jager W, Kundi M, Bichler J, Misik M, et al. Potent protection of gallic acid against DNA oxidation: results of human and animal experiments. *Mutat Res* 2011;715:61-71.

Giftson JS, Jayanthi S, Nalini N. Chemopreventive efficacy of gallic acid, an antioxidant and anticarcinogenic polyphenol, against 1,2-dimethyl hydrazine induced rat colon carcinogenesis. *Invest New Drugs* 2010;28:251-259.

Gudi VA, Singh SV. Effect of diallyl sulfide, a naturally occurring anti-carcinogen, on glutathione-dependent detoxification enzymes of female CD-1 mouse tissues. *Biochem Pharmacol* 1991;42:1261-1265.

Ho CC, Lin SY, Yang JS, Liu KC, Tang YJ, Yang MD, et al. Gallic acid inhibits murine leukemia WEHI-3 cells in vivo and promotes macrophage phagocytosis. *In Vivo* 2009;23:409-413.

Ho HH, Chang CS, Ho WC, Liao SY, Wu CH, Wang CJ. Anti-metastasis effects of gallic acid on gastric cancer cells involves inhibition of NF-kappaB activity and downregulation of PI3K/AKT/small GTPase signals. *Food Chem Toxicol.* 2010; 48:2508-2516.

Ho HH, Chang CS, Ho WC, Liao SY, Lin WL, Wang CJ. Gallic acid inhibits gastric cancer cells metastasis and invasive growth via increased expression of RhoB, downregulation of AKT/small GTPase signals and inhibition of NF-kappaB activity. *Toxicol Appl Pharmacol* 2013;266:76-85.

Hoelzl C, Knasmuller S, Wagner KH, Elbling L, Huber W, Kager N, et al. Instant coffee with high chlorogenic acid levels protects humans against oxidative damage of macromolecules. *Mol Nutr Food Res* 2010;54:1722-1733.

Hseu YC, Chang WH, Chen CS, Liao JW, Huang CJ, Lu FJ, et al. Antioxidant activities of Toona Sinensis leaves extracts using different antioxidant models. *Food Chem Toxicol* 2008;46:105-114.

Hseu YC, Chen SC, Lin WH, Hung DZ, Lin MK, Kuo YH, et al. Toona sinensis (leaf extracts) inhibit vascular endothelial growth factor (VEGF)-induced angiogenesis in vascular endothelial cells. *J Ethnopharmacol* 2011;134:111-121.

Hsieh CL, Yen GC, Chen HY. Antioxidant activities of phenolic acids on ultraviolet radiation-induced erythrocyte and low density lipoprotein oxidation. *J Agric Food Chem* 2005;53:6151-6155.

Hsu JD, Kao SH, Ou TT, Chen YJ, Li YJ, Wang CJ. Gallic acid induces G2/M phase arrest of breast cancer cell MCF-7 through stabilization of p27(Kip1) attributed to disruption of p27(Kip1)/Skp2 complex. *J Agric Food Chem* 2011;59:1996-2003.

Jadon A, Bhadauria M, Shukla S. Protective effect of Terminalia belerica Roxb. and gallic acid against carbon tetrachloride induced damage in albino rats. *J Ethnopharmacol* 2007;109:214-218.

Jagan S, Ramakrishnan G, Anandakumar P, Kamaraj S, Devaki T. Antiproliferative potential of gallic acid against diethylnitrosamine-induced rat hepatocellular carcinoma. *Mol Cell Biochem* 2008;319:51-59.

Ji BC, Hsu WH, Yang JS, Hsia TC, Lu CC, Chiang JH, et al. Gallic acid induces apoptosis via caspase-3 and mitochondrion-dependent pathways in vitro and suppresses lung xenograft tumor growth in vivo. *J Agric Food Chem* 2009;57:7596-7604.

Kaur M, Velmurugan B, Rajamanickam S, Agarwal R, Agarwal C. Gallic acid, an active constituent of grape seed extract, exhibits anti-proliferative, pro-apoptotic and anti-tumorigenic effects against prostate carcinoma xenograft growth in nude mice. *Pharm Res* 2009;26:2133-2140.

Kawada M, Ohno Y, Ri Y, Ikoma T, Yuugetu H, Asai T, et al. Anti-tumor effect of gallic acid on LL-2 lung cancer cells transplanted in mice. *Anticancer Drugs* 2001;12:847-852.

Kensler TW, Wakabayashi N. Nrf2: friend or foe for chemoprevention? *Carcinogenesis* 2010;31:90-99.

Kim YJ. Antimelanogenic and antioxidant properties of gallic acid. *Biol Pharm Bull* 2007;30:1052-1055.

Kim NS, Jeong SI, Hwang BS, Lee YE, Kang SH, Lee HC, et al. Gallic acid inhibits cell viability and induces apoptosis in human monocytic cell line U937. *J Med Food* 2011; 14:240-246.

Knasmüller S, Nersesyan A, Mišík M, Gerner C, Mikulits W, Ehrlich V, et al. Use of conventional and -OMICS based methods for health claims of dietary antioxidants: a critical overview. *Br J Nutr* 2008;99, E-Suppl. 1:ES3-ES52.

Kuppan G, Balasubramanyam J, Monickaraj F, Srinivasan G, Mohan V, Balasubramanyam M. Transcriptional regulation of cytokines and oxidative stress by gallic acid in human THP-1 monocytes. *Cytokine* 2010;49:229-234.

Landrault N, Poucheret P, Ravel P, Gasc F, Cros G, Teissedre PL. Antioxidant capacities and phenolics levels of French wines from different varieties and vintages. *J Agric Food Chem* 2001;49:3341-3348.

Li L, Ng TB, Gao W, Li W, Fu M, Niu SM, et al. Antioxidant activity of gallic acid from rose flowers in senescence accelerated mice. *Life Sci* 2005;77:230-240.

Liang CZ, Zhang X, Li H, Tao YQ, Tao LJ, Yang ZR, et al. Gallic acid induces the apoptosis of human osteosarcoma cells in vitro and in vivo via the regulation of mitogen-activated protein kinase pathways. *Cancer Biother Radiopharm* 2012;27:701-710.

Liao CL, Lai KC, Huang AC, Yang JS, Lin JJ, Wu SH, et al. Gallic acid inhibits migration and invasion in human osteosarcoma U-2 OS cells through suppressing the matrix metalloproteinase-2/-9, protein kinase B (PKB) and PKC signaling pathways. *Food Chem Toxicol* 2012;50:1734-1740.

Lin J-K, Lin C-L, Liang Y-C, Lin-Shiau S-Y, Juan IM. Survey of Catechins, Gallic Acid, and Methylxanthines in Green, Oolong, Pu-erh, and Black Teas. *Journal of Agricultural and Food Chemistry* 1998;46:3635-3642.

Liu KC, Ho HC, Huang AC, Ji BC, Lin HY, Chueh FS, et al. Gallic acid provokes DNA damage and suppresses DNA repair gene expression in human prostate cancer PC-3 cells. *Environ Toxicol* 2011; 1-9.

Liu Z, Li D, Yu L, Niu F. Gallic acid as a cancer-selective agent induces apoptosis in pancreatic cancer cells. *Chemotherapy* 2012;58:185-194.

Lo C, Lai TY, Yang JH, Yang JS, Ma YS, Weng SW, et al. Gallic acid induces apoptosis in A375.S2 human melanoma cells through caspase-dependent and -independent pathways. *Int J Oncol;* 2010; 37:377-385.

Lo C, Lai TY, Yang JS, Yang JH, Ma YS, Weng SW, et al. Gallic acid inhibits the migration and invasion of A375.S2 human melanoma cells through the inhibition of matrix metalloproteinase-2 and Ras. *Melanoma Res* 2011; 21:267-273.

Lu Y, Jiang F, Jiang H, Wu K, Zheng X, Cai Y, et al. Gallic acid suppresses cell viability, proliferation, invasion and angiogenesis in human glioma cells. *Eur J Pharmacol* 2010;641:102-107.

Locatelli C, Leal PC, Yunes RA, Nunes RJ, Creczynski-Pasa TB. Gallic acid ester derivatives induce apoptosis and cell adhesion inhibition in melanoma cells: The relationship between free radical generation, glutathione depletion and cell death. *Chem Biol Interact* 2009;181:175-184.

Madlener S, Illmer C, Horvath Z, Saiko P, Losert A, Herbacek I, et al. Gallic acid inhibits ribonucleotide reductase and cyclooxygenases in human HL-60 promyelocytic leukemia cells. *Cancer Lett* 2007;245:156-162.

Manach C, Williamson G, Morand C, Scalbert A, Remesy C. Bioavailability and bioefficacy of polyphenols in humans. I. Review of 97 bioavailability studies. *Am J Clin Nutr* 2005;81:230S-242S.

Maurya DK, Nandakumar N, Devasagayam TP. Anticancer property of gallic acid in A549, a human lung adenocarcinoma cell line, and possible mechanisms. *J Clin Biochem Nutr* 2011;48:85-90.

Moller P, Loft S. Oxidative DNA damage in human white blood cells in dietary antioxidant intervention studies. *Am J Clin Nutr* 2002;76:303-310.

Nabavi SF, Habtemariam S, Jafari M, Sureda A, Nabavi SM. Protective role of gallic acid on sodium fluoride induced oxidative stress in rat brain. Bull Environ Contam Toxicol 2012;89:73-77.

Ou TT, Wang CJ, Lee YS, Wu CH, Lee HJ. Gallic acid induces G2/M phase cell cycle arrest via regulating 14-3-3beta release from Cdc25C and Chk2 activation in human bladder transitional carcinoma cells. *Mol Nutr Food Res* 2010;54:1781-1790.

Pal C, Bindu S, Dey S, Alam A, Goyal M, Iqbal MS, et al. Gallic acid prevents nonsteroidal anti-inflammatory drug-induced gastropathy in rat by blocking oxidative stress and apoptosis. *Free Radic Biol Med* 2010;49:258-267.

Park JM, Kim A, Oh JH, Chung AS. Methylseleninic acid inhibits PMA-stimulated pro-MMP-2 activation mediated by MT1-MMP expression and further tumor invasion through suppression of NF-kappaB activation. *Carcinogenesis* 2007;28:837-847.

Pourahmad J, Eskandari MR, Shakibaei R, Kamalinejad M. A search for hepatoprotective activity of fruit extract of Mangifera indica L. against oxidative stress cytotoxicity. *Plant Foods Hum Nutr* 2010;65:83-89.

Pourahmad J, Eskandari MR, Shakibaei R, Kamalinejad M. A search for hepatoprotective activity of aqueous extract of Rhus coriaria L. against oxidative stress cytotoxicity. *Food Chem Toxicol* 2010;48:854-858.

Radtke J, Linseisen J, Wolfram G. [Phenolic acid intake of adults in a Bavarian subgroup of the national food consumption survey]. *Z Ernahrungswiss* 1998;37:190-197.

Raina K, Rajamanickam S, Deep G, Singh M, Agarwal R, Agarwal C. Chemopreventive effects of oral gallic acid feeding on tumor growth and progression in TRAMP mice. *Mol Cancer Ther* 2008;7:1258-1267.

Reddivari L, Vanamala J, Safe SH, Miller JC, Jr. The bioactive compounds alpha-chaconine and gallic acid in potato extracts decrease survival and induce apoptosis in LNCaP and PC3 prostate cancer cells. *Nutr Cancer* 2010;62:601-610.

Russell LH, Jr., Mazzio E, Badisa RB, Zhu ZP, Agharahimi M, Oriaku ET, et al. Autoxidation of gallic acid induces ROS-dependent death in human prostate cancer LNCaP cells. *Anticancer Res;* 2012; 32:1595-1602.

Sasaki YF, Kawaguchi S, Kamaya A, Ohshita M, Kabasawa K, Iwama K, et al. The comet assay with 8 mouse organs: results with 39 currently used food additives. *Mutat Res* 2002;519:103-119.

Shaughnessy DT, DeMarini DM. Types and consequences of DNA damage. 2009, In: *Chemoprevention of Cancer and DNA damage by Dietary factors.* Wiley-Blackwell, Germany,pp.21-33.

Shinno E, Shimoji M, Imaizumi N, Kinoshita S, Sunakawa H, Aniya Y. Activation of rat liver microsomal glutathione S-transferase by gallic acid. *Life Sci* 2005;78:99-106

Smith TJ, Yang CS. Effect of organosulfur compounds from garlic and cruciferous vegetables on drug metabolism enzymes. *Drug Metabol Drug Interact* 2000;17:23-49.

Sohi KK, Mittal N, Hundal MK, Khanduja KL. Gallic acid, an antioxidant, exhibits antiapoptotic potential in normal human lymphocytes: A Bcl-2 independent mechanism. *J Nutr Sci Vitaminol* (Tokyo) 2003;49:221-227.

Soobrattee MA, Neergheen VS, Luximon-Ramma A, Aruoma OI, Bahorun T. Phenolics as potential antioxidant therapeutic agents: mechanism and actions. *Mutat Res* 2005;579:200-213.

Teufelhofer O, Parzefall W, Kainzbauer E, Ferk F, Freiler C, Knasmuller S, et al. Superoxide generation from Kupffer cells contributes to hepatocarcinogenesis: studies on NADPH oxidase knockout mice. *Carcinogenesis* 2005;26:319-329.

Tice RR, Agurell E, Anderson D, Burlinson B, Hartmann A, Kobayashi H, et al. Single cell gel/comet assay: guidelines for in vitro and in vivo genetic toxicology testing. *Environ Mol Mutagen* 2000;35:206-221.

Tung YT, Wu JH, Huang CC, Peng HC, Chen YL, Yang SC, et al. Protective effect of Acacia confusa bark extract and its active compound gallic acid against carbon tetrachloride-induced chronic liver injury in rats. *Food Chem Toxicol* 2009;47:1385-1392.

Weng CJ, Yen GC. Chemopreventive effects of dietary phytochemicals against cancer invasion and metastasis: phenolic acids, monophenol, polyphenol, and their derivatives. *Cancer Treat Rev*; 2012; 38:76-87.

Yang HL, Chen SC, Lin KY, Wang MT, Chen YC, Huang HC, et al. Antioxidant activities of aqueous leaf extracts of Toona sinensis on free radical-induced endothelial cell damage. *J Ethnopharmacol* 2011;137:669-680.

Yeh CT, Yen GC. Effects of phenolic acids on human phenolsulfotransferases in relation to their antioxidant activity. *J Agric Food Chem* 2003;51:1474-1479.

Yeh CT, Yen GC. Induction of hepatic antioxidant enzymes by phenolic acids in rats is accompanied by increased levels of multidrug resistance-associated protein 3 mRNA expression. *J Nutr* 2006;136:11-15.

Yeh CT, Ching LC, Yen GC. Inducing gene expression of cardiac antioxidant enzymes by dietary phenolic acids in rats. *J Nutr Biochem* 2009;20:163-171.

Yeh RD, Chen JC, Lai TY, Yang JS, Yu CS, Chiang JH, et al. Gallic acid induces G(0)/G(1) phase arrest and apoptosis in human leukemia HL-60 cells through inhibiting cyclin D and E, and activating mitochondria-dependent pathway. *Anticancer Res* 2011; 31:2821-2832.

Yen GC, Duh PD, Tsai HL. Antioxidant and pro-oxidant properties of ascorbic acid and gallic acid. *Food Chem* 2002; 79:307-313.

Yang YM, Park WH. p38 inhibitor enhances growth inhibition and death in gallic acid-treated endothelial cells. *Int J Mol Med*;26:295-300.

You BR, Park WH. The effects of mitogen-activated protein kinase inhibitors or small interfering RNAs on gallic acid-induced HeLa cell death in relation to reactive oxygen species and glutathione. *J Agric Food Chem* 2011;59:763-771.

You BR, Kim SZ, Kim SH, Park WH. Gallic acid-induced lung cancer cell death is accompanied by ROS increase and glutathione depletion. *Mol Cell Biochem* 2011;357:295-303.

In: Handbook on Gallic Acid
Editors: M. A. Thompson and P. B. Collins ISBN: 978-1-62618-921-8

Chapter 14

Application of Spectroscopic Techniques for the Study of Gallic Acid Autoxidation

***Goran M. Nikolić*[1]*, Aleksandar M. Veselinović*[1]
and Ružica S. Nikolić[2]**
[1]Department of Chemistry, Faculty of Medicine, University of Niš, Serbia
[2]Department of Chemistry, Faculty of Natural Sciences and Mathematics, University of Niš, Serbia

Abstract

Gallic acid is a polyphenolic compound widely distributed in plants ether in the free form or as a part of more complex polyphenolic compounds. It has a range of biological activities but is best known for its antioxidative activity. All the biological effects of gallic acids in living cells may be connected to the ease of its oxidation or/and autoxidation. Various methods may be used for the study of such processes but spectroscopic methods seem to be especially well suited for such purpose. Out of many spectroscopic methods used for the study of oxidation/autoxidation of various phenolic compounds UV/Vis spectrophotometry and electron spin resonance (ESR) have some particular advantages. UV/Vis spectrophotometry enables direct study of the oxidation/autoxidation of polyphenolic compounds but it lacks the specificity of other spectroscopic techniques. This disadvantage may be overcomed by the application of various chemometric techniques for spectra processing. UV/Vis spectrophotometry is also one of the most economic spectroscopic techniques found in virtually all analytical laboratories and is readily used for detection in liquid chromatography, usually as a HPLC-DAD (high performance liquid chromatography with diode array detection). ESR spectroscopy allows direct detection and characterization of transient free radical species formed in the course of polyphenolic compounds transformations under various conditions. In this chapter we present the results of the spectroscopic studies of the gallic acid autoxidation under various conditions with special emphasis on the results obtained by the UV/Vis spectrophotometry with chemometric spectral processing and ESR spectroscopy and their application for a better understanding of mechanisms of gallic acid transformation and its action upon living cells.

Keywords: Gallic acid, autoxidation, UV/Vis spectrophotometry, electron spin resonance

INTRODUCTION

Polyphenols are a class of plant-derived natural products attracting much attention by both scientific community and general public because of their presence and abundance in foodstuffs and beverages whose consumption has been claimed to have beneficial effects for human health (Quideau et al., 2011). They present a wide variety of compounds and although they are abundant micronutritients in our diet their bioavailability differs greatly. Gallic acid (3,4,5-trihydroxybenzoic acid) is one of the most well-absorbed polyphenols (Manach et al., 2005) and it also has a wide range of biological activities. Recent reports indicated the protective effects of gallic acid in various pathological states induced by different agents like carbon tetrachloride (Jadon, Bhadauria, and Shukla, 2007), paracetamol (Mahaboob et al., 2010), lindane (Padma et al., 2011), 6-hydroxydopamine (Mansouri et al., 2013), and sodium fluoride (Nabavi et al., 2013). Gallic acid and its derivatives also influence drug metabolizing enzymes (Ow and Stupans, 2003), but most importantly they proved to be promising for the prevention of cardiovascular diseases (Morton et al., 2000) and cancer (Locatelli, Filippin-Monteiro, and Creczynski-Pasa, 2013; Verma, Singh, and Mishra, 2013).

Numerous studies have shown that polyphenols exert their biological activity by scavenging free radicals and reactive oxygen species thus supplementing the action of cellular antioxidants like glutathione, glutathione peroxidase, or superoxide dismutase (Quideau et al., 2011). This is also the case with gallic acid (Inoue et al., 2000) and this scavenging activity is connected with the presence of pyrogallol moiety in its structure. Beneficial biological effects of gallic acid are the consequence of its antioxidative properties and regarding the radical scavenging capacity in some systems it compares favorably to some common antioxidants like ascorbic acid, cysteine, and Trolox (Lo Scalzo, Todaro, and Rapisarda, 2012). However, the ease of oxidation or autoxidation of pyrogallol moiety is responsible for either antioxidative or prooxidative properties of gallic acid in various systems (Strlič et al., 2002) - as in the case of other polyphenols containing pyrogallol or catechol moieties (Quideau et al., 2011). Therefore, it is very important to know the exact mechanisms of gallic acid autoxidation under various conditions, since some recent studies showed that the autoxidation of gallic acid was directly responsible for some of its biological activities (Gil-Longo and González-Vázquez, 2010; Russel L. H. Jr. et al., 2012).

AUTOXIDATION OF GALLIC ACID

Oxidation and autoxidation of gallic acid were studied extensively for the variety of reasons. Measurement of oxygen uptake by gallic acid during the autoxidation in highly alkaline solution (Tulyathan, Boulton, and Singleton, 1989) or results obtained in studies aimed at water purification by various methods like photochemical treatment (Benitez et al., 2005; Lucas et al., 2008), electrochemical removal (Boye et al., 2006), and radiolytic degradation (Melo et al., 2009) have no direct biological significance but may give some insight into the behavior of gallic acid in aqueous solutions. Some useful data about the antioxidant activity of gallic acid and gallates were obtained by the electrochemical methods (Gunckel et al., 1998; Abdel-Hamid and Newair, 2011) and pulse radiolysis studies (Dwibedy et al., 1999), while for the quantitative determination of gallic acid in various studies mostly high performance liquid chromatography (HPLC) was employed.

UV-visible (UV/Vis) spectrophotometry, as relatively simple and readily accessible spectroscopic method, was extensively used for studying various transformations of gallic acid including evaluation of the antioxidant and prooxidant actions of gallic acid and its derivatives (Arouma et al., 1993), effect of pH on its stability in aqueous solutions (Friedman and Jürgens, 2000), enzymatic and chemical oxidation (Muñoz-Muñoz et al., 2009) and complexation with metal ions (Andjelković et al., 2006; Lu, Li, and Lu, 2009). However, given the ubiquitous presence of free radicals in all the processes relevant for the biological activities of gallic acid, electron spin resonance (ESR) spectroscopy with its unique capability for direct detection and characterization of free radicals seems to be especially well suited for the study of gallic acid autoxidation. For that reason we used ESR spectroscopy in combination with UV/Vis spectrophotometry for our study of gallic acid autoxidation in aqueous solutions.

UV/Vis Spectophotometry

Over the long period of time UV/Vis spectrophotometry was the main tool for studying the structure and reactions of natural phenolic compounds and even now, although it lacks the specificity of some other spectroscopic techniques, spectrophotometry remains very useful for such kind of research, especially when interactions with metal ions (Malešev and Kuntić, 2007) or the influence of pH in solution are concerned (Friedman and Jürgens, 2000). In their study of the effect of pH on the stability of various plant phenolic compounds in aqueous solution, including gallic acid, Friedman and Jürgens (2000) concluded that some naturally occurring polyphenolic compounds which are part of our diet are damaged when exposed to high pH which was closely related to their chemical structure. Although it was obvious that structural changes, as evidenced by spectral changes, were the consequence of the autoxidation of these compounds, no attempt was made to characterize these reactions based on the spectrophotometric data.

We tried to extract more information about structural changes of gallic acid by applying various chemometric techniques to the UV/Vis spectra obtained during the autoxidation in alkaline aqueous solutions. Most simple approach was to use graphical analysis by the matrix method (Coleman, Varga, and Mastin, 1970) which is capable of identifying equilibrium between absorbing species in solution even in the cases where no isobestic points appear in the spectra (Nikolić, Premović, and Nikolić, 1998). UV/Vis spectra of gallic acid in aqueous solution at pH 10 recorded after various time intervals are presented in Figure 1.

During the period of 1 hour there was a continuous absorbance increase around 250 nm and a continuous absorbance decrease around 300 nm and at the same time new absorption bands/shoulders appeared at higher wavelengths. The presence of isobestic points in the spectra recorded in time interval from 1 min to 45 min suggested that there was equilibrium of two absorbing species in solution and this was confirmed by applying the test for the two absorbing species with restrictions (Nikolić et al., 2010).

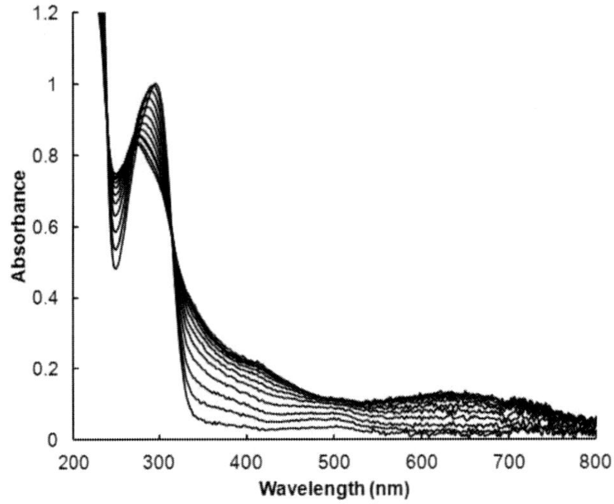

Figure 1. UV/Vis spectra of 0.3 mM gallic acid in aqueous solution at pH 10 at various time intervals after NaOH addition.

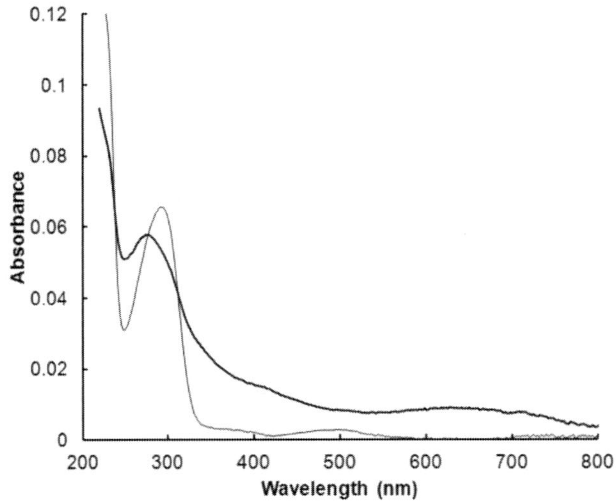

Figure 2. Principal spectral components of the UV/Vis spectra recorded during the gallic acid autoxidation in aqueous solution at pH 10.

Further chemometric analysis of these UV/Vis spectra was performed by the use of multivariate curve resolution-alternating least squares (MCR-ALS) method (Tauler, 1995; Jaumot et al., 2005) which we have already successfully applied for the study of pyrogallol autoxidation in weakly alkaline aqueous solutions (Veselinović, Nikolić, and Nikolić, 2012). Application of MCR-ALS method allowed the separation of the two principal spectral components and they are shown in Figure 2.

One of the spectral components obtained by the MCR-ALS method clearly represents partially deprotonated gallic acid while the second one we tentatively attributed to some kind of gallic acid dimer, since the dimerization of gallic acid during its autoxidation or oxidation in aqueous solutions at high pH values was already reported in the literature (Tulyathan, Boulton, and Singleton, 1989; Oniki and Takahama, 2004). In order to confirm these assignments we used the HPLC-DAD technique. Two well resolved peaks appeared in the chromatogram of the solution of gallic acid undergoing autoxidation at pH 10 for 45 min (Nikolić et al., 2011b) and based on their retention times and corresponding spectra we arrived at the same conclusions. However, one minor peak appearing in the chromatogram indicated that maybe some small quantity of ellagic acid was also formed under these conditions which is in accordance to the results of gallic acid autoxidation study at very high pH values (Tulyathan, Boulton, and Singleton, 1989).

At lower pH values autoxidation of gallic acid in pure aqueous solutions proceeds at much slower rate, but our previous results have shown that the autoxidation of gallic acid at pH 8.5 was greatly influenced by the presence of Mg(II) ions (Nikolić et al., 2011a; Nikolić et al., 2011b). Not only that the presence of Mg(II) ions increased the autoxidation rate, but it also changed the mechanism of autoxidation as it is evidenced by both UV/Vis spectral analysis and even more by HPLC-DAD (Nikolić et al., 2011b). Therefore, we decided to study the autooxidation of gallic acid in weakly alkaline aqueous solution at pH 7.4 (which is the most relevant for the processes in biological systems) in the presence of various alkaline-earth metal ions (some of which also having biological importance). Analysis of UV/Vis spectra recorded during the autoxidation of gallic acid at pH 7.4 in the presence of alkaline-earth cations by the MCR-ALS method revealed two principal spectral components in all cases, the very similar one being attributable to the partially deprotonated (and probably metal ion complexed) gallic acid and the second one corresponding to the product of autoxidation. However, while second principal spectral components were very similar in the cases of gallic acid autoxidation in the presence of Ca(II), Sr(II), and Ba(II) ions, its appearance in the case of gallic acid autoxidation in the presence of Mg(II) ions was very different. Second principal spectral components obtained by the MCR-ALS analysis of UV/Vis spectra recorded during the autoxidation of gallic acid at pH 7.4 in the presence of Mg(II) and Ca(II) ions are shown in Figure 3.

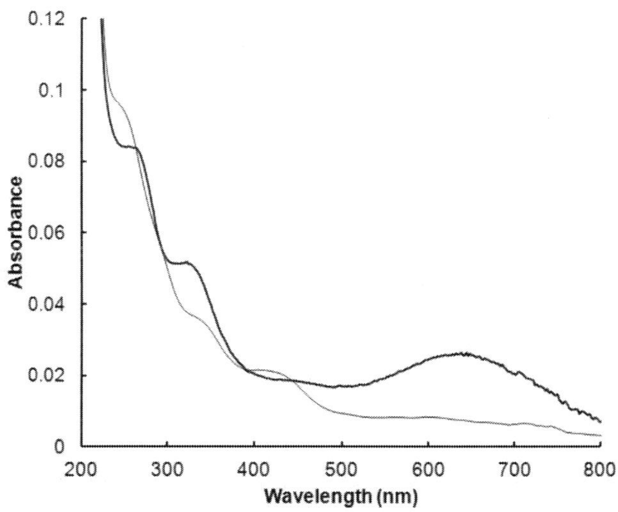

Figure 3. Second principal spectral components obtained by the MCR-ALS analysis of UV/Vis spectra recorded during the autoxidation of gallic acid at pH 7.4 in the presence of Mg(II) (thin line) and Ca(II) ions (thick line).

The difference of the effects of Mg(II) and Ca(II) ions on the autoxidation of gallic acid is not in accordance to the results obtained in the study of the effects of alkaline-earth metal ions on the catecholate oxidation in which case the influence of Mg(II) and Ca(II) ions was similar and differed from the influence of Sr(II) and Ba(II) ions (Lebedev et al., 2007). This fact is worth further study since the presence of both Mg(II) and Ca(II) ions is ubiquitous in most of the biological systems where gallic acid exerts its activity.

Electron Spin Resonance (ESR) Spectroscopy

Electron spin resonance (ESR) spectroscopy (alternatively called electron paramagnetic resonance or EPR) is a technique which can be used for studies of the oxidative status of biological systems and the antioxidative properties of different compounds in various ways (Spasojević et al., 2011; Yu et al., 2011). Since the early reports that ESR spectroscopy can be conveniently used for the detection and characterization of semiquinone free radicals formed during the autoxidation of many phenolic compounds (Adams, Blois, and Sands, 1958) this technique has found many applications for studying the systems in which such processes are important. Primary free radicals obtained by the autoxidation of gallic acid in aqueous solutions have also been studied in detail by ESR spectroscopy (Yoshioka et al., 2003; Eslami et al., 2010). At pH > 10 a characteristic three line ESR spectrum with 1 : 2 : 1 line intensity ratio was obtained indicating the presence of two equivalent protons in the structure of radical with the hyperfine splitting constant value of 1.07 G. However, ESR spectrum of buffered aqueous solution of gallic acid recorded at pH 9, although rather noisy, indicated the presence of free radical with three nonequivalent protons in its structure (Eslami et al., 2010). No ESR spectra were reported for radicals obtained by the autoxidation of gallic acid in aqueous solutions below pH 9.

ESR spectroscopy was also used for a direct detection of characteristic gallic acid radical in complex systems like lipid peroxidation system (Milić, Djilas, and Čanadanović-Brunet, 1998), extracts of some medicinal plants (Mouhajir et al., 2001) and mixture of green tea polyphenols (Severino et al., 2009); for studying interactions of gallic acid with other antioxidants (Sakagami and Satoh, 1996; Nogaki et al., 1998) or paramagnetic metal ions (Severino et al., 2011); for characterization of secondary radical(s) obtained by the oxidation of gallic acid with $K_4[Fe(CN)_6]$ at high pH values (Oniki and Takahama, 2004), and for studying properties and interactions of other antioxidants possessing pyrogallol moiety (Pasanphan, Buettner, and Chirachanchai, 2010; Perron et al., 2011). Some ESR studies of proanthocyanidins and gallate esters employed the spin stabilization technique (Bors, Michel, and Stettmaier, 2000) which was firstly reported for the study of complexation of diamagnetic metal ions with *ortho*-benzosemiquinone (Eaton, 1964). Spin stabilization technique is especially well suited for the study of *ortho*-semiquinone type radicals because it enables their detection and characterization in neutral and even slightly acidic aqueous solutions since good resolution low noise ESR spectra may be obtained in favorable cases. It was extensively used for the ESR study of redox reactions of many biologically active compounds with catechol moiety in their structure like catecholamines (Kalyanaraman, 1990) and flavonoids (van Acker et al., 1996), but was seldom applied to the study of compounds with pyrogallol moiety in structure (Nikolić et al., 1998; Bors, Michel, and Stettmaier, 2000; Perron et al., 2011).

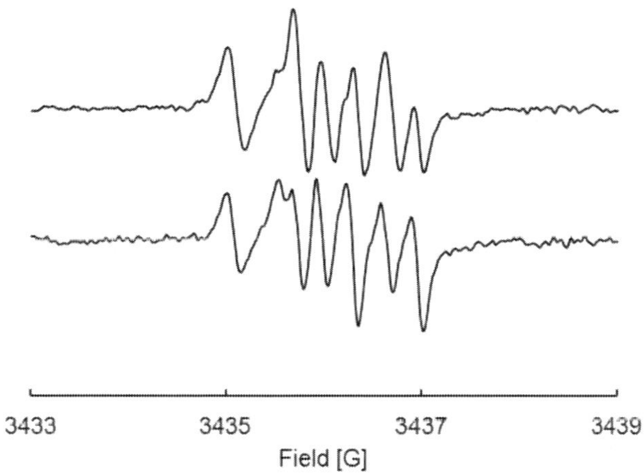

Figure 4. ESR spectra of radicals obtained durig the autoxidation of gallic acid in aqueous solutions at pH 8.5 in the presence of Mg(II) ions at the gallic acid concentrations of 5 mM (upper trace) and 20 mM (lower trace).

We employed spin stabilization technique for the room temperature ESR study of gallic acid autoxidation in weakly alkaline aqueous solutions in the presence of Mg(II) ions (Nikolić et al., 2011a). ESR spectra obtained for two various concentrations of gallic acid in solution at pH 8.5 recorded in less than 3 min after mixing gallic acid solutions with Tris buffer containing Mg(II) ions are given in Figure 4.

Asymmetric appearance of the ESR spectra shown in Figure 4. clearly indicates that more than one radical species was formed at the initial stage of gallic acid autoxidation in the presence of Mg(II) ions at pH 8.5. After prolonged autoxidation time (up to 2 hours) general appearance of ESR spectra did not change but overall signal intensity decreased and spectra became more noisy. Attempts to obtain good quality ESR spectra at lower pH values or in the absence of Mg(II) ions did not give satisfactory results. The structure of free radicals formed during the autoxidation of gallic acid in aqueous solutions at pH 8.5 in the presence of Mg(II) ions was resolved by using the computer simulation of ESR spectra. A reasonably good computer simulations of the ESR spectra were obtained by assuming the existence of two free radical species in solution and the results of computer simulation for the ESR spectrum recorded for the 5 mM gallic acid solution are shown in Figure 5.

Individual components of computer simulated spectrum correspond to one free radical with two equivalent protons (triplet with signal intensity ratio 1 : 2 :1) and one free radical with two nonequivalent protons (slightly overlapping quartet with signal intensity ratio 1 : 1 : 1 : 1). It is clear that the first radical presents Mg(II) ion spin stabilized primary radical of gallic acid. The much smaller hyperfine coupling constant value of 0.60 G in comparison with the value of 1.07 G reported in the literature for the anion radical of gallic acid obtained during its autoxidation at pH 11 (Eslami et al., 2010) is in accordance with similar changes of the hyperfine constant values for *ortho*-benzosemiquinone anion radical upon the spin stabilization with Mg(II) ions (Eaton, 1964). The secondary radical obtained in this study has a structure with two nonequivalent protons whose hyperfine coupling constant values are 0.94

G and 0.67G. This structure differs greatly from the structure of secondary radical formed by the oxidation of gallic acid with $K_4[Fe(CN)_6]$ at high pH values and identified by its ESR spectrum to be a C-O-C dimer of gallic acid (Oniki and Takahama, 2004). Proposed structures of two radicals identified in our experiment are shown in Figure 6.

Computer simulation of experimental ESR spectrum obtained for the gallic acid autoxidation in aqueous solution at pH 8.5 in the presence of Mg(II) ions at gallic acid concentration of 20 mM has also given satsfactory results with the same two radicals and the main change was in the relative abundance of two radicals at different concentrations of gallic acid. With the increase of gallic acid concentration the proportion of primary radical also increased (from about 25% at 5 mM to slightly over 50% at 20 mM).

A significant result of this ESR study is the fact that the presence of Mg(II) ions greatly increased the stability of free radicals formed during the autoxidation of gallic acid. This is important because prolonged lifetime of radicals may give them more opportunity to exibit cytotoxic prooxidant activity. The spin stabilization of radicals by metal ions also has potential implications for the etiology of diseases associated with increased accumulations of metals (Sakihama et al., 2002).

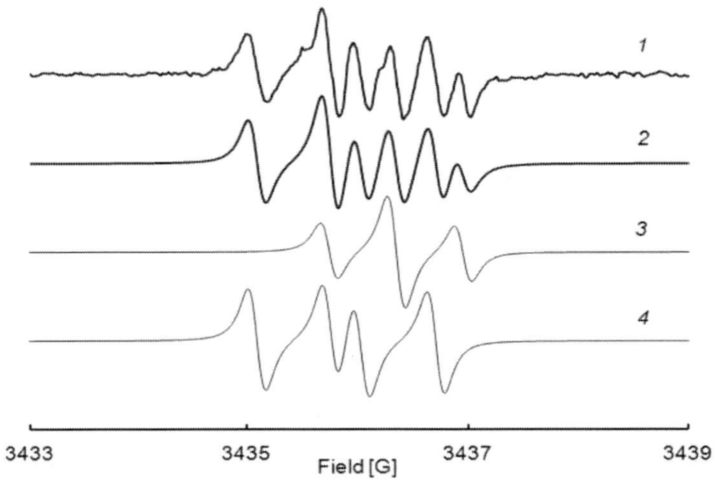

Figure 5. Experimental ESR spectrum of gallic acid at pH 8.5 in aqueous solution containing Mg(II) ions (1), computer simulated spectrum (2), and individual components of computer simulated spectrum (3 and 4).

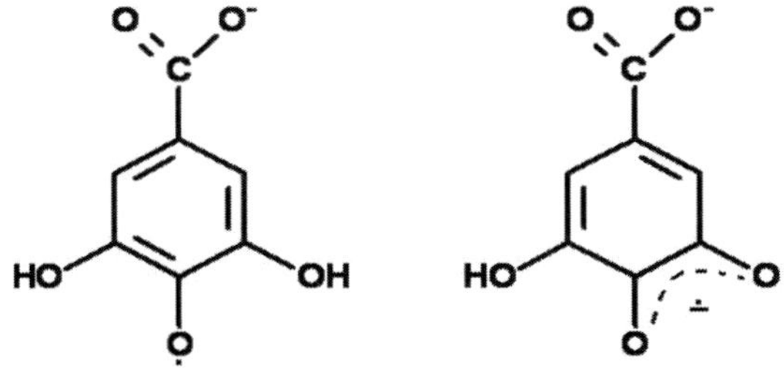

Figure 6. Proposed structures of primary (left side) and secondary (right side) free radicals formed during the autoxidation of gallic acid in aqueous solutions at pH 8.5 in the presence of Mg(II) ions.

Conclusion

UV/Vis spectrophotometry and ESR spectroscopy were used to study the autoxidation of gallic acid in aqueous solutions under various conditions. It was demonstrated that the presence of diamagnetic metal ions has great influence on the mechanism of gallic acid autoxidation. The presence of metal ions not only increased the autoxidation rate of gallic acid but also increased the lifetime of free radicals formed in this process by the effect of spin stabilization and this may have a profound influence on the biological activity of gallic acid.

Acknowledgments

This work was supported by the Ministry of Education, Science and Technological Development of the Republic of Serbia under the project No. 172044.

References

Abdel-Hamid, R. & Newair, E. F. (2011). Electrochemical behavior of antioxidants: I. Mechanistic study on electrochemical oxidation of gallic acid in aqueous solutions at glassy-carbon electrode. *Journal of Electroanalytical Chemistry*, *657*, 107-112.

Adams, M., Blois, S. M. Jr. & Sands, R. H. (1958). Paramagnetic resonance spectra of some semiquinone radicals. *Journal of Chemical Physics*, *28*, 774-776.

Andjelković, M., van Camp, J., de Meulenaer, B., Depaemelaere, G., Socaciu, C., Verloo, M. & Verhe, R. (2006). Iron-chelation properties of phenolic acids bearing catechol and galloyl groups. *Food Chemistry*, *98*, 23-31.

Arouma, O. I., Murcia, A., Butler, J. & Hallivell, B. (1993). Evaluation of the antioxidant and prooxidant actions of gallic acid and its derivatives. *Journal of Agricultural and Food Chemistry*, *41*, 1880-1885.

Benitez, F. J., Real, F. J., Acero, J. L., Leal, A. I. & Garcia, C. (2005). Gallic acid degradation in aqueous solutions by UV/H_2O_2 treatment, Fenton's reagent and photo-Fenton system. *Journal of Hazardous Materials B*, *126*, 31-39.

Bors, W., Michel, C. & Stettmaier, K. (2000). Electron paramagnetic resonance studies of radical species of proanthocyanidins and gallate esters. *Archives of Biochemistry and Biophysics*, *374*, 347-355.

Boye, B., Brillas, E., Buso, A., Farnia, G., Flox, C., Giomo, M. & Sandonà, G. (2006). Electrochemical removal of gallic acid from aqueous solutions. *Electrochimica Acta*, *52*, 256-262.

Coleman, J. S., Varga, L. P. & Mastin, S. H. (1970). Graphical methods for determining the number of species in solution from spectrophotometric data. *Inorganic Chemistry*, *9*, 1015-1020.

Dwibedy, P., Dey, G. R., Naik, D. B., Kishore, K. & Moorthy, P. N. (1999). Pulse radiolysis studies on redox reactions of gallic acid: one electron oxidation of gallic acid by gallic acid-OH adduct. *Physical Chemistry Chemical Physics*, *1*, 1915-1918.

Eaton, D. R. (1964). Complexing of metal ions with semiquinones. An electron spin resonance study. *Inorganic Chemistry*, *3*, 1268-1271.

Eslami, A. C., Pasanphan, W., Wagner, B. A. & Buettner, G. R. (2010). Free radicals produced by the oxidation of gallic acid: An electron paramagnetic resonance study. *Chemistry Central Journal*, *4*, 15.

Friedman, M. & Jürgens, H. S. (2000). Effect of pH on the stability of plant phenolic compounds. *Journal of Agricultural and Food Chemistry*, *48*, 2101-2110.

Gil-Longo, J. & González-Vázquez, C. (2010). Vascular pro-oxidant effects secondary to the autoxidation of gallic acid in rat aorta. *Journal of Nutritional Biochemistry*, *21*, 304-309.

Gunckel, S., Santander, P., Cordano, G., Ferreira, J., Munoz, S., Nunez-Vergara, L. J. & Squella, J. A. (1998). Antioxidant activity of gallates: an electrochemical study in aqueous media. *Chemico-Biological Interactions*, *114*, 45-59.

Inoue, M., Sakaguchi, N., Isuzugawa, K., Tani, H. & Ogihara, Y. (2000). Role of reactive oxygen species in gallic acid-induced apoptosis. *Biological & Pharmaceutical Bulletin*, *23*, 1153-1157.

Jadon, A., Bhadauria, M. & Shukla, S. (2007). Protective effect of *Terminalia belerica* Roxb. and gallic acid against carbon tetrachloride induced damage in albino rats. *Journal of Ethnopharmacology*, *109*, 214-218.

Jaumot, J., Gargallo, R., de Juan, A. & Tauler, R. (2005). A graphical user-friendly interface for MCR-ALS: a new tool for multivariate curve resolution in MATLAB. *Chemometrics and Intelligent Laboratory Systems*, *76*, 101-110.

Kalyanaraman, B. (1990). Characterization of *o*-semiquinone radicals in biological systems. *Methods in Enzymology*, *186*, 333-343.

Lebedev, A. V., Ivanova, M. V., Timoshin, A. A. & Ruuge, E. K. (2007). Effect of group II metal cations on the catecholate oxidation. *ChemPhysChem*, *8*, 1863-1869.

Locatelli, C., Filippin-Monteiro, F. B. & Creczynski-Pasa, T. B. (2013). Alkyl esters of gallic acid as anticancer agents: A review. *European Journal of Medicinal Chemistry*, *60*, 233-239.

Lo Scalzo, R., Todaro, A. & Rapisarda, P. (2012). Methods used to evaluate the peroxyl ($ROO^{\cdot}$) radical scavenging capacities of four common antioxidants. *European Food Research and Technology*, *235*, 1141-1148.

Lu, L-L., Li, Y-H. & Lu, X-Y. (2009). Kinetic study of the complexation of gallic acid with Fe(II). *Spectrochimica Acta Part A*, *74*, 829-834.

Lucas, M. S., Dias, A. A., Bezzera, R. M. & Peres, J. A. (2008). Gallic acid photochemical oxidation as a model compound of winery wastewaters. *Journal of Environmental Science and Health Part A*, *43*, 1288-1295.

Malešev, D. & Kuntić, V. (2007). Investigation of metal-flavonoid chelates and the determination of flavonoids *via* metal-flavonoid complexing reactions. *Journal of the Serbian Chemical Society*, *72*, 921-939.

Mahaboob, K. R., Evan, P. S., Segu, R. R., Pranatharthiharan, P., Patel, S., Mandal, N., Mishra, P. P. & Samuel, J. (2010). Hepatoprotective nad antioxidant effect of gallic acid in paracetamol-induced liver damage in mice. *Journal of Pharmacy and Pharmacology*, *62*, 638-643.

Manach, C., Williamson, G., Morand, C., Scalbert, A. & Rémésy, C. (2005). Bioavailability and bioefficacy of polyphenols in humans. I. Review of 97 bioavailability studies. *American Journal of Clinical Nutrition*, *81*, 230S-242S.

Mansouri, M. T., Farbood, Y., Sameri, M. J., Sarkaki, A., Naghizadeh, B. & Rafeirad, M. (2013). Neuroprotective effects of oral gallic acid against oxidative stress induced by 6-hydroxydopamine in rats. *Food Chemistry*, *138*, 1028-1033.

Melo, R. P., Leal, J. P., Takács, E. & Wojnárovits, L. (2009). Radiolytic degradation of gallic acid and its derivatives in aqueous solution. *Journal of Hazardous Materials*, *172*, 1185-1192.

Milić, B. Lj., Djilas, S. M. & Čanadanović-Brunet, J. M. (1998). Antioxidative activity of phenolic compounds on the metal-ion breakdown of lipid peroxidation system. *Food Chemistry*, *61*, 443-447.

Morton, L. W., Cacceta, R. A-A., Puddey, I. B. & Croft, K. D. (2000). Chemistry and biological effects of dietary phenolic compounds: Relevance to cardiovascular disease. *Clinical and Experimental Pharmacology and Physiology*, *27*, 152-159.

Mouhajir, F., Pedersen, J. A., Rejdali, M. & Towers, G. H. N. (2001). Phenolics in Maroccan medicinal plant species as studied by electron spin resonance spectroscopy. *Pharmaceutical Biology*, *39*, 391-398.

Muñoz-Muñoz, J. L., García-Molina, F., García-Ruiz, P. A., Arribas, E., Tudela, J., García-Cánovas, F. & Rodríguez-López, J. N. (2009). Enzymatic and chemical oxidation of trihydroxylated phenols. *Food Chemistry*, *113*, 435-444.

Nabavi, S. F., Nabavi, S. M., Habtemariam, S., Moghaddam, A. H., Sureda, A., Jafari, M. & Latifi, A. M. (2013). Hepatoprotective effect of gallic acid isolated from *Peltiphyllum peltatum* against sodium fluoride-induced oxidative stres. *Industrial Crops and Products*, *44*, 50-55.

Nikolić, G. M., Nikolić, R. S., Milić, B. Lj. & Čanadanović-Brunet, J. M. (1998). Application of electron spin resonance method for studying the antioxidative properties of natural phenolic compounds. *Acta Facultatis Medicae Naissensis*, *15*, 183-188. (in Serbian)

Nikolić, G. M., Premović, P. I. & Nikolić, R. S. (1998). Spectrophotometric study of catechol oxidation by aerial O_2 in alkaline aqueous solutions containing Mg(II) ions. *Spectroscopy Letters*, *31*, 327-333.

Nikolić, G. M., Veselinović, A. M.., Nikolić, R. S., Mitić, Ž. J. & Živković J. V. (2010). *Spectrophotometric study of gallic acid autoxidation in alkaline aqueous solutions*. Proceedings of the 10th International Conference on Fundamental and Applied Aspects of Physical Chemistry, Belgrade, Serbia, 188-190.

Nikolić, G. M., Veselinović, A. M.., Nikolić, R. S. & Mitić, S. S. (2011a). Spectroscopic study of Mg(II) ion influence on the autoxidation of gallic acid in weakly alkaline aqueous solutions. *Russian Journal of Physical Chemistry A*, *85*, 2270-2273.

Nikolić, G., Veselinović, A., Mitić, Ž. & Živanović, S. (2011b). HPLC-DAD study of gallic acid autoxidation in alaline aqueous solutions and the influence of Mg(II) ion. *Scientific Journal of the Faculty of Medicine in Niš*, *28*, 219-224.

Nogaki, A., Satoh, K., Iwasaka, K., Takano, H., Takahama, M., Ida, Y. & Sakagami, H. (1998). Radical intensity and cytotoxic activity of curcumin and gallic acid. *Anticancer Research*, *18*, 3487-3491.

Oniki, T. & Takahama, U. (2004). Free radicals produced by the oxidation of gallic acid and catechin derivatives. *Journal of Wood Science*, *50*, 545-547.

Ow, Y-Y. & Stupans, I. (2003). Gallic acid and gallic acid derivatives: Effects on drug metabolizing enzymes. *Current Drug Metabolism*, *4*, 241-248.

Padma, V. V., Sowmya, P., Felix, T. A., Baskaran, R. & Poornima, P. (2011). Protective effect of gallic acid against lindane induced toxicity in experimental rats. *Food and Chemical Toxicology*, *49*, 991-998.

Pasanphan, W., Buettner, G. R. & Chirachanchai, S. (2010). Chitosan gallate as a novel potential polysaccharide antioxidant: an EPR study. *Carbohydrate Research*, *345*, 132-140.

Perron, N. R., García, C. R., Pinzón, J. R., Chaur, M. N. & Brumaghim, J. L. (2011). Antioxidant and prooxidant effects of polyphenol compounds on copper-mediated DNA damage. *Journal of Inorganic Biochemistry*, *105*, 745-753.

Quideau, S., Deffieux, D., Douat-Casassus, C. & Pouységu, L. (2011). Plant polyphenols: Chemical properties, biological activities, and synthesis. *Angewandte Chemie International Edition*, *50*, 586-621.

Russel, L. H. Jr., Mazzio, E., Badisa, R. B., Zhu, Z-P., Agharahimi, M., Oriaku, E. T. & Goodman, C. B. (2012). Autoxidation of gallic acid induces ROS-dependent death in human prostate cancer LNCaP cells. *Anticancer Research*, *32*, 1595-1602.

Sakihama, Y., Cohen, M. F., Grace, S. C. & Yamasaki, H. (2002). Plant phenolic antioxidant and prooxidant activities: phenolics-induced oxidative damage mediated by metals in plants. *Toxicology*, *177*, 67-80.

Sakagami, H. & Satoh, K. (1996). The interaction between two antioxidants, sodium ascorbate and gallic acid: radical intensity and apoptosis induction. *Anticancer Research*, *16*, 1231-1234.

Severino, J. F., Goodman, B. A., Kay, C. W. M., Stolze, K., Tunega, D., Reichenauer, T. G. & Pirker, K. F. (2009). Free radicals generated during oxidation of green tea polyphenols: Electron paramagnetic resonance spectroscopy combined with density functional theory calculations. *Free Radical Biology & Medicine*, *46*, 1076-1088.

Severino, J. F., Goodman, B. A., Reichenauer, T. G. & Pirker, K. F. (2011). Is there a redox reaction between Cu(II) and gallic acid? *Free Radical Research*, *45*, 115-124.

Spasojević, I., Mojović, M., Ignjatović, A. & Bačić, G. (2011). The role of EPR spectroscopy in the studies of the oxidative status of biological systems and the antioxidative properties of various compounds. *Journal of the Serbian Chemical Society*, *76*, 647-677.

Strlič, M., Radovič, T., Kolar, J. & Pihlar, B. (2002). Anti- and prooxidative properties of gallic acid in Fenton-type systems. *Journal of agricultural and Food Chemistry*, *50*, 6313-6317.

Tauler, R. (1995). Multivariate curve resolution applied to second order data. *Chemometrics and Intelligent Laboratory Systems*, *30*, 133-146.

Tulyathan, V., Boulton, R. B. & Singleton, V. L. (1989). Oxygen uptake by gallic acid as a model for similar reactions in wines. *Journal of Agricultural and Food Chemistry*, *37*, 844-849.

van Acker, S. A. B. E., de Groot, M. J., van den Berg, D-J., Tromp, M. N. J. L., den Kelder, G. D-O., van der, Vijgh, W. J. F. & Bast, A. (1996). A quantum chemical explanation of the antioxidant activity of flavonoids. *Chemical Research in Toxicology*, *9*, 1305-1312.

Verma, S., Singh, A. & Mishra, A. (2013). Gallic acid: Molecular rival of cancer. *Environmental Toxicology and Pharmacology*, *35*, 473-485.

Veselinović, A. M., Nikolić, R. S. & Nikolić, G. M. (2012). Application of multivariate curve resolution-alternating least squares (MCR-ALS) for resolving pyrogallol autoxidation in weakly alkaline aqueous solutions. *Central European Journal of Chemistry*, *10*, 1942-1948.

Yoshioka, H., Ohashi, Y., Fukuda, H., Senba, Y. & Yoshioka, H. (2003). Spectral simulation of the ESR spectra of polyphenol radicals formed by reaction with hydoxyl radical. *The Journal of Physical Chemistry A*, *107*, 1127-1132.

Yu, C., Yongijan, W., Jian, W., Chan, S. & Ao, Y. (2011). Antioxidant activity studies using electron paramagnetic resonance methods. *Progress in Chemistry*, *23*, 1959-1972. (in Chinese)

In: Handbook on Gallic Acid
Editors: M. A. Thompson and P. B. Collins
ISBN: 978-1-62618-921-8

Chapter 15

GALLIC ACID BIOAVAILABILITY IN HUMANS

Andriana C. Kaliora, Panagiotis T. Kanellos and Nick Kalogeropoulos*
Laboratory of Chemistry-Biochemistry-Physical Chemistry of Foods, Department of Science of Nutrition–Dietetics, Harokopio University, Athens, Greece

ABSTRACT

Polyphenols are extensively metabolized either in tissues, once they are absorbed through the gut barrier or, for the non-absorbed fraction and the fraction re-excreted in the bile, by the colonic microflora. All polyphenols are conjugated in our body to form O-glucuronides, sulphate esters and O-methyl ethers. The formation of anionic derivatives by conjugation with glucuronides and sulphate groups facilitates their urinary and biliary excretion and explains their rapid elimination.

Animal studies first revealed the possible metabolic fate of gallic acid. In rats, rabbits and chickens the major urinary metabolite is 4-O-methylgallic acid, followed by pyrogallol (conjugated and unconjugated). Small amounts of conjugated 2-O-methylpyrogallol were also detected in rats. In sheep, resorcinol glucuronide is the major product of gallic acid metabolism and unconjugated pyrogallol and resorcinol comprise minor urinary metabolites.

Moreover, recent *in vitro* studies have reported phenolic acids to be main metabolites of anthocyanins after fecal fermentation. However, fragmentation of anthocyanins to phenolic acids in humans has not been studied in detail. One recently identified anthocyanin metabolite is 3-O-methylgallic acid, which is presumably the metabolite of petunidin-3-glucoside, but possibly also a demethylation product of malvidin-3-glucoside.

The few existing studies addressing the bioavailability of gallic acid in humans revealed that, compared to other polyphenols, this phenolic compound is extremely well absorbed. The most common metabolite of gallic acid identified in human biological fluids is 4-O-methylgallic acid, which was found to increase significantly in human

* Corresponding author: Nick Kalogeropoulos. Tel.: +30 2109549251; fax: +30 2109577050; E-mail address: nickal@hua.gr.

plasma within the first 4 h after consumption of a single dose of red wine or dealcoholized red wine. Plasma concentrations of free and glucuronidated forms of gallic acid and 4-O-methylgallic acid also increased after ingestion of 50 mg pure gallic acid, while gallic acid was rapidly absorbed after human administration with tea or acidum gallicum. Concerning the excretion of gallic acid, many studies have identified gallic acid and its metabolites in urinary samples. For example, after acute consumption of 3 cups of black tea, gallic acid was not detected in urine of subjects. However, three gallic acid methyl ethers, 4-O-methylgallic acid, 3-O-methylgallic acid and 3,4-O-dimethylgallic acid, were identified. A recent study on the bioavailability of phenolic compounds present in Corinthian raisins (*Vitis vinifera*) –containing 115.7 ng gallic acid per g- indicated that gallic acid is bioavailable in humans, as its concentrations in plasma increased one hour after raisins consumption. Overall, gallic acid concentrations in plasma were found to range between 2.2±0.9 to 9.9±4.6 (SEM) μg/L.

1. Introduction

Phenolic compounds, abundant in plant products, have the characteristic structure of at least one aromatic ring with one or more hydroxyl groups attached. Over 8000 phenolic structures have been reported and they are widely dispersed throughout the plant kingdom– many occurring in foods.

Phenolic compounds range from simple, low molecular weight, single aromatic ring compounds to the large and complex tannins and derived polyphenols. They can be classified by the number and arrangement of their carbon atoms and are commonly found conjugated to sugars and organic acids. Phenolics occurring naturally in healthy plant tissue can be classified into two groups, the flavonoids (i.e. the flavonol quercetin) and the non-flavonoids (i.e. caffeic acid or resveratrol or cyanidin) (Table 1).

Table 1. Characteristic flavonoid and non-flavonoid phenolic compounds

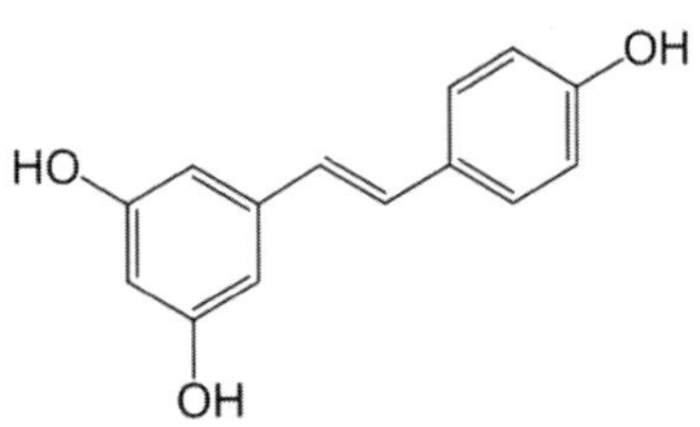 resveratrol (stilbene)	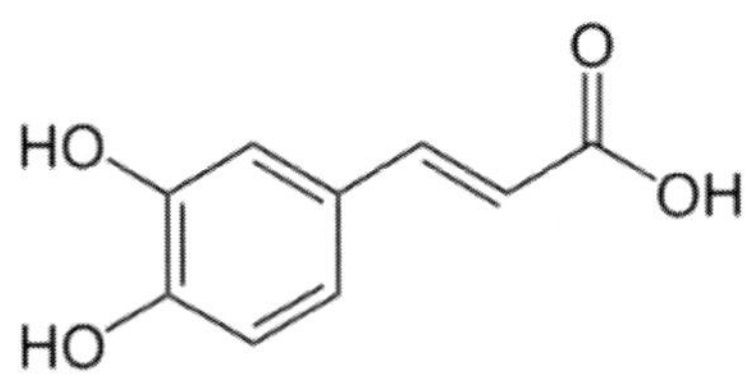 caffeic acid (phenolic acid)
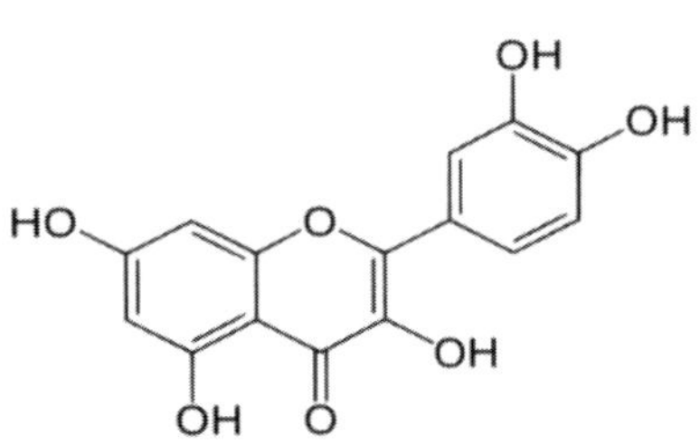 quercetin (flavonol)	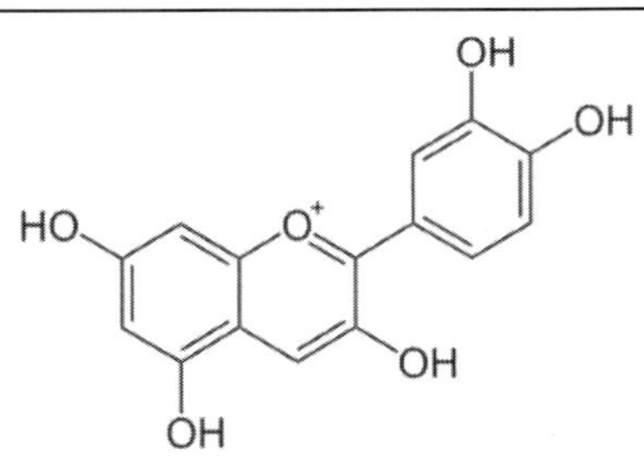 cyanidin (anthocyanin)

Traditionally processed foods and beverages, such as black tea, matured red wine, coffee and cocoa, may contain phenolic transformation products that are best described as 'derived polyphenols'. Tannins are the active ingredients of traditional plant extracts used to convert hides to leather and occur widely in foods and beverages but at concentrations too low to tan hides.

Flavonoids are polyphenolic compounds comprising 15 carbons, with two aromatic rings connected by a three carbon bridge, hence C6–C3–C6. They are the most abundant of the phenolics and are found throughout the plant kingdom.

The main non-flavonoids of dietary significance are the C6–C1 phenolic acids, most notably gallic acid, which is the biosynthetic precursor of hydrolysable tannins, the C6–C3 hydroxycinammates and their conjugated derivatives, and the polyphenolic C6–C2–C6 stilbenes.

The most important food sources of phenolic compounds are fruit and vegetables, green and black tea, red wine, coffee, chocolate and extra virgin olive oil.

Herbs and spices, nuts, and algae are also potentially relevant sources of polyphenols, depending on culinary habits.

2. Gallic Acid

2.1. Biosynthesis

Gallic acid is synthesized via the shikimate pathway, a seven step metabolic route used by bacteria, fungi, algae, parasites and plants for the biosynthesis of aromatic amino acids (phenylalanine, tyrosine, and tryptophan).

This pathway is not found in animals and in humans, hence the products of this pathway represent essential amino acids that must be obtained from the animal's diet.

The first enzyme involved is the shikimate kinase, an enzyme that catalyzes the ATP-dependent phosphorylation of shikimate to form shikimate 3-phosphate.

Biosynthesis of 3-dehydroquinate from phosphoenolpyruvate and erythrose-4-phosphate

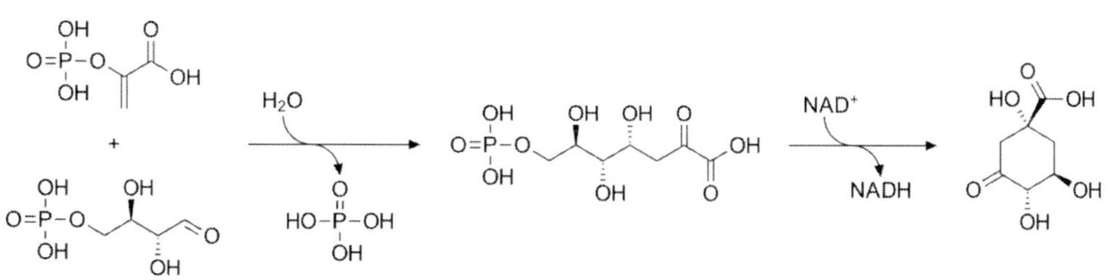

Biosynthesis of 3-dehydroshikimate acid from 3-dehydroquinate

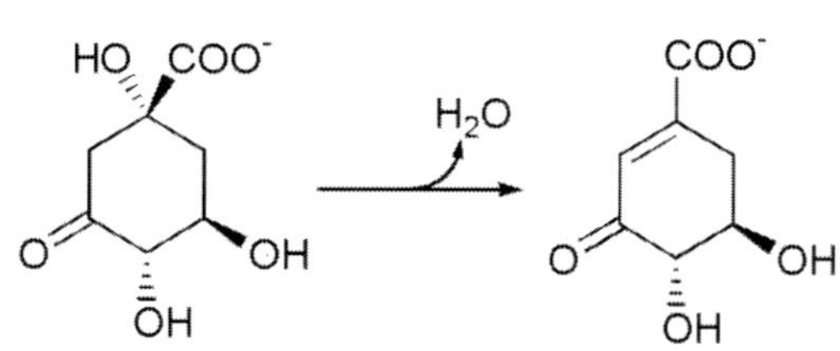

3-Dehydroshikimate under the action of the enzyme shikimate dehydrogenase is transformed in 3,5-didehydroshikimate and finally rearranged to gallic acid

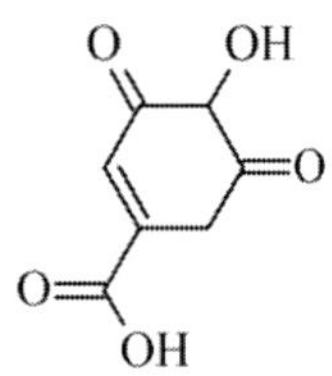

3,5-Didehydroshikimate

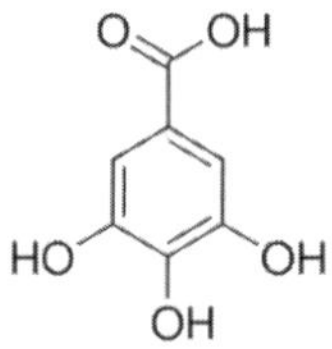

Gallic acid

2.2. Natural Occurrence in Plants/Foods

Gallic acid is the commonest phenolic acid, and occurs widely as complex sugar esters in gallotannins such as 2-O-digalloyltetra-O-galloyl-glucose but these are found only to a limited extent in dietary components. Non-sugar galloyl esters in grapes, wine, mangoes, green tea and black tea are the major source of gallic acid in the human diet.

3. Bioavailability of Gallic Acid

3.1. The Concept of Bioavailability

Bioavailability is defined as the amount of an ingested nutrient that is absorbed and available for physiological functions. Bioavailability is dependent on digestion, release from the food matrix, absorption by intestinal cells, and transportation to body cells. *Bioaccessibility* is the amount of an ingested nutrient that is potentially available for absorption. It is dependent only on digestion and release from the food matrix.

3.2. *In Vitro* Studies

In vitro screening methods for the determination of nutrient bioaccessibility and bioavailability from foods have been developed mostly during the last decades.

These methods can provide useful information, especially when the vast number of factors that potentially affect nutrient absorption is taken into consideration.

In vitro bioavailability methods could provide knowledge on possible interactions between nutrients and/or food components, the effects of luminal factors (including pH and enzymes), food preparation and processing practices, nature of the food matrix, on either micronutrient absorbability (a component of bioavailability) or on the potential for a nutrient to be absorbed (bioaccessibility).

Even if *in vitro* methods have many advantages compared to human or animal studies (less expensive, faster, better control of experimental variables) (Sandberg, 2005), they cannot be substituted for *in vivo* studies.

There are principally four *in vitro* methods for measuring bioaccessibility and/or bioavailability: solubility, dialyzability, or a gastrointestinal model for bioaccessibility, and the Caco-2 models for bioavailability (Table 2).

The main *in vitro* bioavailability method that has been used repeatedly to measure the bioaccessibility of polyphenols is *in vitro* solubility. This method has been used to test the bioaccessibility of various polyphenols in a plethora of foods.

Fogliano et al. (2011) examined the *in vitro* digestion of the water-insoluble cocoa fraction (WICF) with gastrointestinal enzymes as well as its bacterial fermentation in a human colonic model system. Experiments were carried out to investigate bioaccessibility and biotransformation of WICF polyphenols, respectively. In this study *in vitro* digestion solubilized 38.6% of WICF with pronase and Viscozyme L treatments releasing 51% of the total phenols from the insoluble material. Regarding biotransformation flavanols were converted into phenolic acids, including gallic acid, by the microbiota following a concentration gradient resulting in high concentrations of 3-hydroxyphenylpropionic acid (3-HPP) in the last gut compartment.

In the study of López de Lacey et al. (2012) who examined the bioaccessibility of polyphenols in green tea-agar films, most of the release of gallic acid occurred in the gastric phase of digestion, whereas little or no increase was observed in the duodenal phase.

Table 2. Characteristics of *in vitro* methods for measuring bioaccessibility and/or bioavailability

In vitro method	End point	Advantages	Limitations
Solubility	Bioaccessibility	Simple to do Relatively inexpensive Easy to conduct, equipment in all labs	Sometimes not a reliable indicator of bioavailability Cannot assess rate of uptake or absorption or transport kinetics Cannot measure nutrient or food component competition at the site of absorption
Dialyzability	Bioaccessibility	Simple to do Relatively inexpensive Easy to conduct, equipment in all labs	Cannot assess rate of uptake or absorption or transport kinetics Cannot measure nutrient or food component competition at the site of absorption
Gastrointestinal models	Bioaccessibility Bioavailability when coupled to intestinal cells	Incorporates many digestion parameters (peristalsis, churning, body temperature, etc.,) Allows the collection of digest at any step of the digestive system	Expensive Few validation studies
Caco-2 cell model	Bioavailability	Allows the study of nutrient or food component competition at the site of absorption	Requires trained personnel with knowledge of cell culture methods

Adjusted by Etcheverry et al., 2012.

The recovery from the film samples was significantly lower, irrespectively of the presence of gelatin. However, the interactions established between gallic acid and gelatin in this study seemed to be weak, allowing, therefore, an easy release of the phenolic compound. Consequently, in bioaccessibility experiments, the release of gallic acid is highly dependent on the matrix.

Recent *in vitro* studies have reported phenolic acids to be main metabolites of anthocyanins after fecal fermentation (Aura et al., 2005). However, fragmentation of anthocyanins to phenolic acids in humans has not been studied in detail.

One recently identified anthocyanin metabolite is 3-O-methylgallic acid, which is presumably the metabolite of petunidin-3-glucoside, but possibly also a demethylation product of malvidin-3-glucoside (Forester and Waterhouse, 2008).

3.3. Animal Studies

The use of animals in research is essential to the development of new and more effective methods for diagnosing and treating diseases that affect both humans and animals. Animal studies aiming at examining the bioavailability of phenolic compounds have been also conducted (Table 3).

Konishi et al. (2004) examined the absorption of orally administered gallic acid in rats to obtain serum pharmacokinetic profiles and to investigate their intestinal absorption characteristics *in vivo*. Specifically, rats were administered 100 μmol/kg body weight of gallic acid, and both gallic acid and its metabolites were subsequently quantified with a highly selective and sensitive colorimetric detection method using high-performance liquid chromatography-electrochemical detection. Gallic acid was shown to be slowly absorbed, with a tmax for intact gallic acid of 60 min and a Cmax of 0.71 μmol/L and the area under the curve for intact gallic acid was calculated from the serum concentration profile in the portal vein to be 42.6 μmol min L^{-1}. These findings are in good agreement with the results obtained in vitro using a Caco-2 cell system (Deprez et al., 2001). In a recent study, Ferruzzi et al. (2009) examined the bioavailability and brain deposition of a grape seed polyphenolic extract (GSPE) in Sprague Dawley rats.

Table 3. Studies of gallic acid bioavailability in Sprague Dawley rats fed on different nutritional sources of the phenolic acid

Supplementation	Identified compounds	Biological samples	Method	Reference
Gallic acid (100 μmol/kg body weight)	Gallic acid 4-Methyl gallic acid	Serum	HPLC-electrochemical detection	Konishi et al., 2004
Grape seed polyphenolic extract (GSPE) (50, 100, and 150 mg)	Gallic acid	Plasma	LC-UV-MS	Ferruzzi et al., 2009
Concentrated cranberry powder (3.3, 6.6, and 33 mg/kg of diet)	Gallic acid	Urine	LC-MS/MS	Prior et al., 2010

Plasma pharmacokinetic response of major GSPE phenolic components was measured following intragastric gavage of 50, 100, and 150 mg GSPE per kg body weight. Liquid chromatography-mass spectrometry (LC-MS) analysis identified gallic acid in plasma of rats gavaged acutely with GSPE. Additionally, 4-O-methylgallic acid and other metabolites of GSPE, such as 3-methylcatechin, were identified as circulating phenolic constituents. Cmax for individual GSPE constituents and their metabolites increased in a dose-dependent fashion (with increasing GSPE oral dose). Repeated daily exposure to GSPE was found to significantly increase bioavailability (defined as plasma area under the curve, AUC, 0-8 h) of gallic acid by 198% relative to animals receiving only a single acute GSPE dose. This study suggests that brain deposition of gallic acid is affected by repeated dosing of GSPE. The objective of another study by Prior et al. (2010) was to identify and quantify the urinary excretion of 19 phenolic acids and their conjugates in rats fed on three doses of a concentrated cranberry powder (3.3, 6.6, and 33 mg/kg of diet). The basic diet used contained very low amounts of any polyphenolic compounds. Among the phenolic acids studied, gallic acid was excreted in the urine in concentrations of 0.1-2 μg/mg creatinine, a relative low one compared to other phenolic acids.

3.4. Human Studies of Gallic Acid Bioavailability

There are only a few published studies concerning the bioavailability of gallic acid in humans (Table 4). Shahrzad and Bitsch (1998) examined the metabolism of gallic acid in human body by developing two methods for the identification and determination of gallic acid and its phenolic metabolites in human plasma and urine after oral administration of 50 mg gallic acid. Determinations were carried out by reversed-phase high-performance liquid chromatography, using UV detection and involving isocratic elution. One of these methods enables the simultaneous separation and determination of gallic acid, 4-O-methylgallic acid, pyrogallol, 2-O-methylpyrogallol and resorcinol in biological fluids with a long analysis time (57 min). The second method was developed to determine gallic acid and 4-O-methylgallic acid in a shorter analysis time (25 min).

However, after oral administration of 50 mg gallic acid, both methods failed to show metabolites other than 4-O-methylgallic acid and gallic acid in plasma and urine.

In 2001, the same research team conducted a study to determine gallic acid pharmacokinetics and relative bioavailability in healthy humans, by employing acidum gallicum tablets that contained 10% gallic acid and 90% glucose or a black tea brew that contained 93% of gallic acid in free form. After the administration of a single oral dose of acidum gallicum tablets or tea (each containing 0.3 mmol gallic acid) to 10 healthy volunteers, gallic acid was rapidly absorbed and eliminated with mean half-lives of 1.19±0.07 and 1.06±0.06 h and mean maximum concentrations of 1.83±0.16 and 2.09±0.22 μmol/L (plasma), respectively.

After oral administration of the tablets or black tea, 36.4±4.5 and 39.6±5.1% of the gallic acid dose were present in urine in the form of gallic acid and of its metabolite 4-O-methylgallic acid, respectively. The relative bioavailability of gallic acid from tea compared with that from the tablets was 1.06±0.26, showing that its bioavailability was independent of the matrix, either tea or tablets. Caccetta et al. (2000) examined whether certain phenolic acids can be detected in the circulation after red wine consumption.

Table 4. Studies of gallic acid bioavailability in humans administrated with different nutritional sources of the phenolic acid

Supplementation	Identified compounds	Biological samples	Method	Reference
Gallic acid (50 mg)	Gallic acid 4-Methyl gallic acid	Plasma Urine	Reversed-phase HPLC-UV detection	Shahrzad and Bitsch, 1998
Acidum gallicum tablets (10% gallic acid and 90% glucose) or black tea	Gallic acid Gallic acid 4-Methyl gallic acid	Plasma Urine	HPLC mobile phase	Shahrzad et al., 2001
Red wine	4-Methyl gallic acid	Plasma	GC-MS	Caccetta et al., 2000
Red wine	Gallic acid 4-Methyl gallic acid 3-O Methyl gallic acid	Plasma	HPLC-MS	Cartron et al., 2003
Polyphenol-rich juice	O-Methylgallic acid-O-sulfates	Plasma, urine	HPLC-MS	Mullen et al., 2010
Concord grape juice	Gallic acid	Ileal effluent	HPLC-MS	Stalmach et al., 2012
Black tea	4-Methyl gallic acid 3-O Methyl gallic acid 3,4-O-Dimethylgallic acid	Urine	GC-MS	Hodgson et al., 2000
Tea	4-Methyl gallic acid	Urine	GC-MS	Hodgson et al., 2004
Six polyphenol-rich beverages	Gallic acid	Urine	HPLC-electrospray ionization-MS	Ito et al., 2005
Black tea and wine	Gallic acid 4-Methyl gallic acid	Urine	HPLC-electrospray ionization-MS-MS	Mennen et al., 2006

Twelve healthy male nonsmokers consumed red wine, phenol-stripped red wine, dealcoholized red wine, or water, each at a separate visit, in random order and 1 week apart, and plasma phenolic acids were measured by gas chromatography-mass spectrometry. The metabolite of gallic acid, 4-O-methylgallic acid, increased significantly ($P<0.025$) after consumption of red wine and dealcoholized red wine compared with water or phenol-stripped red wine. Cartron et al. (2003) sought to evaluate the effect of one single intake (300 mL) of red wine on plasma phenolics over a 24-h time period following the intake. In the first part, blood samples were collected just before and after wine consumption. In the second part, subjects received the 3 types of wine successively, only at the mealtime, over a 3-week periods separated by a 3-week wash out.

Blood samples were drawn in fasting condition before and after each 3-week wine consumption period. The peak of plasma antioxidant capacity appeared at 3-4 h following the single intake of red wine, not coinciding with the peak of plasma gallic acid (earlier to 3-4 h, 1.5 μmol/L), yet coinciding with that of catechin.

In plasma, the predominating form of gallic acid was its 4-O-methylated derivative, but also the 3-O-methyl derivative appeared as minor derivative. Mullen et al. (2010) developed a polyphenol-rich juice drink as a potential approach to increase intake of dietary polyphenols and examined the bioavailability of flavonoids and other phenolic compounds. Ten healthy humans consumed 350 mL of the polyphenol-rich juice drink, after which plasma and urine samples were collected over a 0-24 h period. Applying high-performance liquid chromatography - mass spectrometry analysis resulted in the identification of a total of 13 metabolites in plasma and a total of 20 metabolites in urine, including O-methylgallic acid-O-sulfates as metabolites of gallic acid. Following this study, Stalmach and co-workers (2011) examined the bioavailability of the phenolic compounds in 350 mL of Concord grape juice following acute intake by eight healthy volunteers. Plasma and urine were collected over 0-24 h and analyzed for parent compounds and metabolites. In total, 41 compounds, principally metabolites, were identified; however, gallic acid was not detected in both plasma and urine.

In another study, Stalmach et al. (2012) aimed to compare the *in vitro* gastrointestinal stability of phenolic compounds in Concord grape juice with recoveries in ileal fluid after the ingestion of the juice by ileostomists. Of the 18±1 μmol of gallic acid ingested through the consumption of 350 mL of Concord grape juice, 8.5±1.7 μmol of gallic acid were recovered in ileal effluent, representing 47% of intake. The juice was also ingested by healthy subjects with an intact functioning colon. Peak plasma concentrations of several phenolic compounds were determined, however the authors failed to quantify gallic acid concentration nor in plasma neither in urine.

Apparently, the gallic acid administered with Concord grape juice was absorbed, shown by the recoveries in ileal fluid; however its metabolites were not targeted and detected in plasma and urines. Hodgson et al. (2000) conducted several studies to identify urinary gallic acid metabolites with potential to serve as markers of black tea intake.

In an initial study, the concentrations of nine compounds, assessed by gas chromatography-mass spectrometry, were found to increase in urine after consumption of 3 cups of black tea over 3 h. A subsequent study employed a controlled crossover design in which 10 subjects consumed 5 cups of black tea per day or water for 4 weeks in random order. Twenty-four hour urine samples were collected at the end of each period. Of the 9 candidate compounds identified in the initial study, only 3 were present at higher concentrations in urine of all the 10 subjects during tea-drinking in comparison to water-drinking periods.

These compounds were all identified as methyl ether derivatives of gallic acid, namely 4-O-methylgallic acid, 3-O-methylgallic acid, and 3,4-O-dimethylgallic acid. It is suggested that these compounds have the potential to be used as markers of black tea intake. Another study from Hodgson et al. (2004) explored the relationships of tea intake with 24 h urinary excretion of 4-O-methylgallic acid in human subjects. The relationship of long-term usual (111 participants) and contemporaneously recorded current (344 participants) tea intake with 24 h urinary excretion of 4-O-methylgallic acid was assessed in two populations. 4-Methyl gallic acid was related to usual ($r=0.50$, $P<0.001$) and current ($r=0.57$, $P<0.001$) tea intake, indicating that 4-methyl gallic acid is a good biomarker for black tea-derived polyphenol exposure. Ito et al. (2005) examined a fast method suitable for the analysis of polyphenols in urine, selected as potential biomarkers of intake.

This method was applied to estimate polyphenol recovery after ingestion of six different polyphenol-rich beverages.

Among many other polyphenols, gallic acid was quantified in human urine by high-performance liquid chromatography coupled with electrospray ionisation mass - mass spectrometry with a run time of 6 min per sample after ingestion of green tea or grape-skin extract or cocoa beverage or coffee or grapefruit juice or orange juice. Levels of urinary excretion suggested that gallic acid could be used as specific biomarker to evaluate the consumption of its nutritional sources. Mennen et al. (2006) evaluated the associations between the intake of polyphenol-rich foods and the urinary excretion of several phenolic compounds and examined whether these compounds could be used as biomarkers of intake. Fifty-three participants collected 24 h urine and a spot urine sample and filled a dietary record covering a 2 d period. Thirteen polyphenols and metabolites, including gallic acid and 4-O-methylgallic acid were measured using high-performance liquid chromatography-electrospray ionization- mass spectrometry- mass spectrometry.

In spot samples the combination of fruits and/or fruit juices was positively correlated to gallic acid and 4-O-methylgallic acid ($r=0.24-0.44$, $P<0.05$). Further, black tea and wine consumption were positively correlated with gallic and 4-O-methylgallic acids ($r=0.37-0.54$, $P<0.001$). When Russell et al. (2009) examined the bioavailability of phenolic acids after consumption of 750 g of Scottish strawberries, gallic acid or its metabolites were not detected in the plasma of the four participants.

Most recently, our research team looked for the postprandial effect of raisin administration in plasma antioxidant capacity of ten healthy subjects. Raisins examined were Corinthian Currants, produced exclusively in Greece. They are most widely consumed dried fruits in Greece obtained by black grapes naturally dried without the addition of any substitute. The content of Corinthian raisins in phytochemicals was determined by gas chromatography - mass spectrometry.

A total of 25 phytochemicals were identified and quantified in raisins. Ten healthy volunteers who consumed 144 g of raisins -containing 16.7 μg of gallic acid- were subjected to blood collection 4 hours after consumption. In obtained samples total polyphenols and serum oxidation resistance significantly increased after the first hour of raisin consumption. Increase in serum oxidation resistance 1 h after ingestion was positively correlated to increase in total phenolic compounds in plasma, meaning that increase was due to the absorbed and bioavailable polyphenolic compounds. When quantification of individual phenolic compounds in plasma was carried out, plasma concentration of gallic acid peaked during the first hour after ingestion, while its concentrations ranged from 2.2 ± 0.9 to 9.9 ± 4.6 (SEM) μg/L.

Conclusion

Although gallic acid and its derivatives are ubiquitous in food items of plant origin, it's *in vivo* bioavailability in humans has not been studied extensively. Studies carried out so far indicate that gallic acid is well absorbed and in most cases its metabolites together with the parent molecule are detected in body fluids. However, in some surveys gallic acid or its metabolites were not detected either in blood or in urine. Available *in vivo* experimental data for humans indicate that the kinetics of gallic acid uptake results in plasma peak concentrations 1-2 hours after ingestion.

REFERENCES

Aura, A-M., Martin-Lopez, P., O'Leary, K. A., Williamson, G., Oksman-Caldentey, K-M., Poutanen, K., Santos-Buelga, C. In vitro metabolism of anthocyanins by human gut microflora. *Eur. J. Nutr.* 2005, 44 (3):133-142.

Caccetta, R. A., Croft, K. D., Beilin, L. J., Puddey, I. B. Ingestion of red wine significantly increases plasma phenolic acid concentrations but does not acutely affect ex vivo lipoprotein oxidizability. *Am. J. Clin. Nutr.* 2000, 71 (1):67-74.

Cartron, E., Fouret, G., Carbonneau, M. A., Lauret, C., Michel, F., Monnier, L., Descomps, B., Léger, C. L. Red-wine beneficial long-term effect on lipids but not on antioxidant characteristics in plasma in a study comparing three types of wine--description of two O-methylated derivatives of gallic acid in humans. *Free Radic. Res.* 2003, 37(9):1021-1035.

Deprez, S., Mila, I., Huneau, J. F., Tome, D., Scalbert, A. Transport of proanthocyanidin dimer, trimer, and polymer across monolayers of human intestinal epithelial Caco-2 cells. *Antioxid. Redox. Signal.* 2001, 3:957-967.

Etcheverry, P., Grusak, M. A., Fleige, L. E. Application of in vitro bioaccessibility and bioavailability methods for calcium, carotenoids, folate, iron, magnesium, polyphenols, zinc, and vitamins B6, B12, D, and E. *Front Physiol.* 2012, 3:317.

Ferruzzi, M. G., Lobo, J. K., Janle, E. M., Whittaker, N., Cooper, B., Simon, J. E., Wu, Q-L., Welch, C., Ho, L., Weaver, C., Pasinetti, G. M. Bioavailability of gallic acid and catechins from grape seed polyphenol extract is improved by repeated dosing in rats: Implications for treatment in Alzheimer's Disease. *J. Alzheimers Dis.* 2009, 18(1):113-124.

Fogliano, V., Corollaro, M. L., Vitaglione, P., Napolitano, A., Ferracane, R., Travaglia, F., Arlorio, M., Costabile, A., Klinder, A., Gibson, G. In vitro bioaccessibility and gut biotransformation of polyphenols present in the water-insoluble cocoa fraction. *Mol. Nutr. Food Res.* 2011, 55 Suppl. 1: S44-S55.

Forester, S. C., Waterhouse, A. L. Identification of Cabernet Sauvignon anthocyanin gut microflora metabolites. *J. Agric. Food Chem.* 2008, 56 (19):9299-9304.

Hodgson, J. M., Chan, S. Y., Puddey, I. B., Devine, A., Wattanapenpaiboon, N., Wahlqvist, M. L., Lukito, W., Burke, V., Ward, N. C., Prince, R. L., Croft, K. D. Phenolic acid metabolites as biomarkers for tea- and coffee-derived polyphenol exposure in human subjects. *Br. J. Nutr.* 2004, 91(2): 301-306.

Hodgson, J. M., Morton, L. W., Puddey, I. B., Beilin, L. J., Croft, K. D. Gallic Acid Metabolites Are Markers of Black Tea Intake in Humans. *J. Agric. Food Chem.* 2000, 48: 2276-2280.

Ito, H., Gonthier, M. P., Manach, C., Morand, C., Mennen, L., Rémésy, C., Scalbert, A. Polyphenol levels in human urine after intake of six different polyphenol-rich beverages. *Br. J. Nutr.* 2005, 94(4):500-509.

Konishi, Y., Hitomi, Y., Yoshioka, E. Intestinal Absorption of p-Coumaric and Gallic Acids in Rats after Oral Administration. *J. Agric. Food Chem.* 2004, 52:2527-2532.

López de Lacey, A. M., Giménez, B., Pérez-Santín, E., Faulks, R., Mandalari, G., López-Caballero, M. E., Montero, P. Bioaccessibility of green tea polyphenols incorporated into an edible agar film during simulated human digestion. *Food Research International* 2012, 48:462-469

Mennen, L. I., Sapinho, D., Ito, H., Bertrais, S., Galan, P., Hercberg, S., Scalbert, A. Urinary flavonoids and phenolic acids as biomarkers of intake for polyphenol-rich foods. *Br. J. Nutr.* 2006, 96(1):191-198.

Mullen, W., Borges, G., Lean, M. E., Roberts, S. A., Crozier, A. Identification of metabolites in human plasma and urine after consumption of a polyphenol-rich juice drink. *J. Agric. Food Chem.* 2010, 58(4):2586-95.

Prior, R. L., Rogers, T. R., Khanal, R. C., Wilkes, S. E., Wu, X., Howard, L. R. Urinary Excretion of Phenolic Acids in Rats Fed Cranberry. *J. Agric. Food Chem.* 2010, 58:3940-3949.

Russell, W. R., Scobbie, L., Labat, A., Duthie, G. G. Selective bio-availability of phenolic acids from Scottish strawberries. *Mol. Nutr. Food Res.* 2009, 53 Suppl. 1:S85-S91.

Sandberg, A. S. Methods and options in vitro dialyzability; benefits and limitations. *Int. J. Vitam. Nutr. Res.* 2005,75:395-404.

Shahrzad, S., Aoyagi, K., Winter, A., Koyama, A., Bitsch, I. Pharmacokinetics of Gallic Acid and Its Relative Bioavailability from Tea in Healthy Humans. *J. Nutr.* 2001, 131(4):1207-1210.

Shahrzad, S., Bitsch, I. Determination of gallic acid and its metabolites in human plasma and urine by high-performance liquid chromatography. *J. Chromatogr. B Biomed. Sci. Appl.* 1998, 705:87-95.

Stalmach, A., Edwards, C. A., Wightman, J. D., Crozier, A. Gastrointestinal stability and bioavailability of (poly)phenolic compounds following ingestion of Concord grape juice by humans. *Mol. Nutr. Food Res.* 2012, 56(3):497-509.

Stalmach, A., Edwards, C. A., Wightman, J. D., Crozier, A. Identification of (poly)phenolic compounds in concord grape juice and their metabolites in human plasma and urine after juice consumption. *J. Agric. Food Chem.* 2011, 59(17):9512-9522.

In: Handbook on Gallic Acid
Editors: M. A. Thompson and P. B. Collins
ISBN: 978-1-62618-921-8

Chapter 16

Gallic Acid and Its Derivatives: Occurrence and Identification in High Altitude Edible and Medicinal Plants

J. Mauricio Peñarrieta,[1,*] Patricia Mollinedo,[1] Enzo Aliaga-Rossel,[1,2] José L. Vila[1] and José A. Bravo[1]

[1]Instituto de Investigaciones en Productos Naturales, Carrera de Ciencias Químicas, Universidad Mayor de San Andrés, La Paz, Bolivia
[2]Instituto de Ecología, Universidad Mayor de San Andrés, La Paz, Bolivia

Abstract

Phenolic compounds are molecules that have one or more hydroxyl groups bound to an aromatic ring. Together with vitamins, phenolic compounds are considered important dietary antioxidants. Thousands of phenolic compounds are found in plants, and they are classified into different types of functional groups. Phenolic compounds play a number of metabolic roles in plants, in growth and reproduction, and in protection against pathogens and external stress, UV radiation and predators for instance.

Gallic acid is trihydroxybenzoic acid, a phenolic acid, also known as 3,4,5-trihydroxybenzoic acid. The presence of this phenolic compound and its derivatives is extensive in nature and it can be found in many plants and foods like fruits, vegetables and cereals.

Bolivia is currently considered among few countries with a high bio-diversity in the world. This diversity is evident from many species of food plants found in so contrasting geographical regions like the Andean mountains or the Amazon flood plain. In particular, the Andean region has a variety of endemic and adapted foods like roots, fruits, cereals and pseudo-cereals all growing at high altitude (2000 to 4200 m.a.s.l.). The presence of ancient native populations such as the Aymara, Quechua and Guarani ethnic groups contributed through the centuries to present with their own and particular *savoir-faire* to the adaptation and use of these plants.

[*] E-mail. mauricio.penarrieta@gmail.com.

As a sample, this part of the world has given to the world's kitchen two of the most celebrated foods in mankind's history -potato and corn- in their many varieties. Endemic Bolivian non-conventional crops have increased dramatically their importance during recent years due to the introduction of Andean foods onto the international market, never as widely open as nowadays. A typical example is quinoa -*Chenopodium quinoa*-, a protein-rich Andean grain currently available and a very common product in European markets.

Bolivia offers a unique environment for researches on high solar UV radiation levels and related phenolics' contents in plants. The altiplano region particularly, is a *sui generis* site (because of its high altitude), where populations and food plants grow in an abundant variety and quantity with a non-negligible percentage in the per-year national statistics.

This chapter presents the results of gallic acid and its derivatives found in food and high altitude plants. These results include an EIMS and NMR spectroscopic approach for gallic acid.

INTRODUCTION

Bolivia is currently considered among few high bio-diversity countries in the world. This diversity is evident from many species of food plants found in so contrasting geographical regions like the Andean mountains or the Amazon flood plain. In particular, the Andean region has a variety of endemic and adapted foods like roots, fruits, cereals and pseudo-cereals all growing at high altitude (2000 to 4200 m.a.s.l.). The presence of ancient native populations such as the Aymara, Quechua and Guarani ethnic groups contributed through the centuries to present with their own and particular *savoir-faire* to the adaptation and use of these plants.

As a sample, this part of the world has given to the world's kitchen two of the most celebrated foods in mankind's history -potato and corn- in their many varieties.. Endemic Bolivian non-conventional crops have increased dramatically their importance during recent years due to the introduction of Andean foods onto the international market, never as widely open as nowadays. A typical example is quinoa -*Chenopodium quinoa*-, a protein-rich Andean grain currently available and a very common product in European markets.

Bolivia offers a unique environment for researches on high solar UV radiation levels and related phenolics' contents in plants. The altiplano region particularly, is a *sui generis* site (because of its high altitude), where populations and food plants grow in an abundant variety and quantity with a non-negligible percentage in the per-year national statistics.

This chapter presents the results of gallic acid and its derivatives found in food and high altitude plants. These results include an EIMS and NMR spectroscopic approach for gallic acid.

PLANT ADAPTATIONS TO HIGH ALTITUDE

High altitude is characterized by severe and frequently life-threatening abiotic conditions. Plants at high altitude suffer several adapting processes to survive extreme environmental conditions, among others, night-time low temperature and daylight UV high radiation related to high altitude, strong winds, arid and damp atmospheric fluctuations, as well as dramatic changes in barometric pressure provoking changes in diverse biological processes like for instance the metabolic rate [1].

However, all species are capable of growing successfully thanks to highly specialized physiological processes. These mechanisms affect the chemical response of the plant, from the synthesis of special lipids that modify cell membranes to anti-freeze carbohydrates, ultraviolet filters, potent antioxidants and free radical quenchers characteristic of these plants and less common in low altitude vegetal species [1,2].

Temperature and other climate conditions change rapidly at high altitudes and plants developed specific adaptation mechanisms by means of seasonal phenology and morphology or by variable growth form preferences. Such mechanisms include the increasing in quantity of the solutes of exposed tissues to avoid freezing, preventing thus ice-derived super-cooling crystallization. The water availability can eventually vary in this geographical location and plants exhibit high desiccation tolerance with an increased stomata control as well as the exudates of waxy substances on the leaves that seals moisture. About noon, when the highest temperature is usually recorded, stems or rhizomes may extend deeper into the soil in order to absorb as much water as possible. Roots and tubercles are adapted for storing food [1]. Ultraviolet radiation increases when increasing altitude, another stress factor in high plateau's plants. However, to the best of our knowledge it is still uncertain if the growth and development of plants is notably affected by UV radiation [1, 2].

The studied plants were collected in a 2,650 to 4,200 m altitude range above sea level, representing species differently adapted depending on the specific level of elevation, radiation and temperature. These are valuable species for the local indigenous communities in the area due to their food and traditional medicinal qualities.

GALLIC ACID

Gallic acid belongs to the chemical phenolic group known as hydroxybenzoic acids. Hydroxybenzoic acids are compounds with a carboxylic group (acid group) and hydroxyl groups (one or more) in an aromatic ring, as exemplified in Figure 1. They are present in plants and foods such as fruits, vegetables and cereals. Many studies have reported the biological activity of hydroxybenzoic acids; for example, protocatechuic acid for the protective effect against liver lesions in animal models [3].

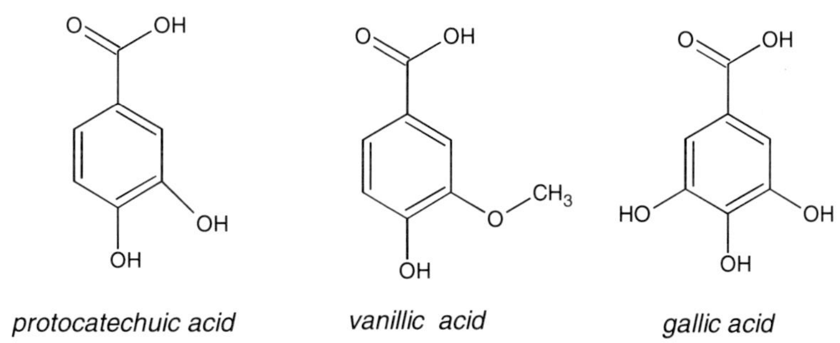

Figure 1. Hydroxybenzoic acids.

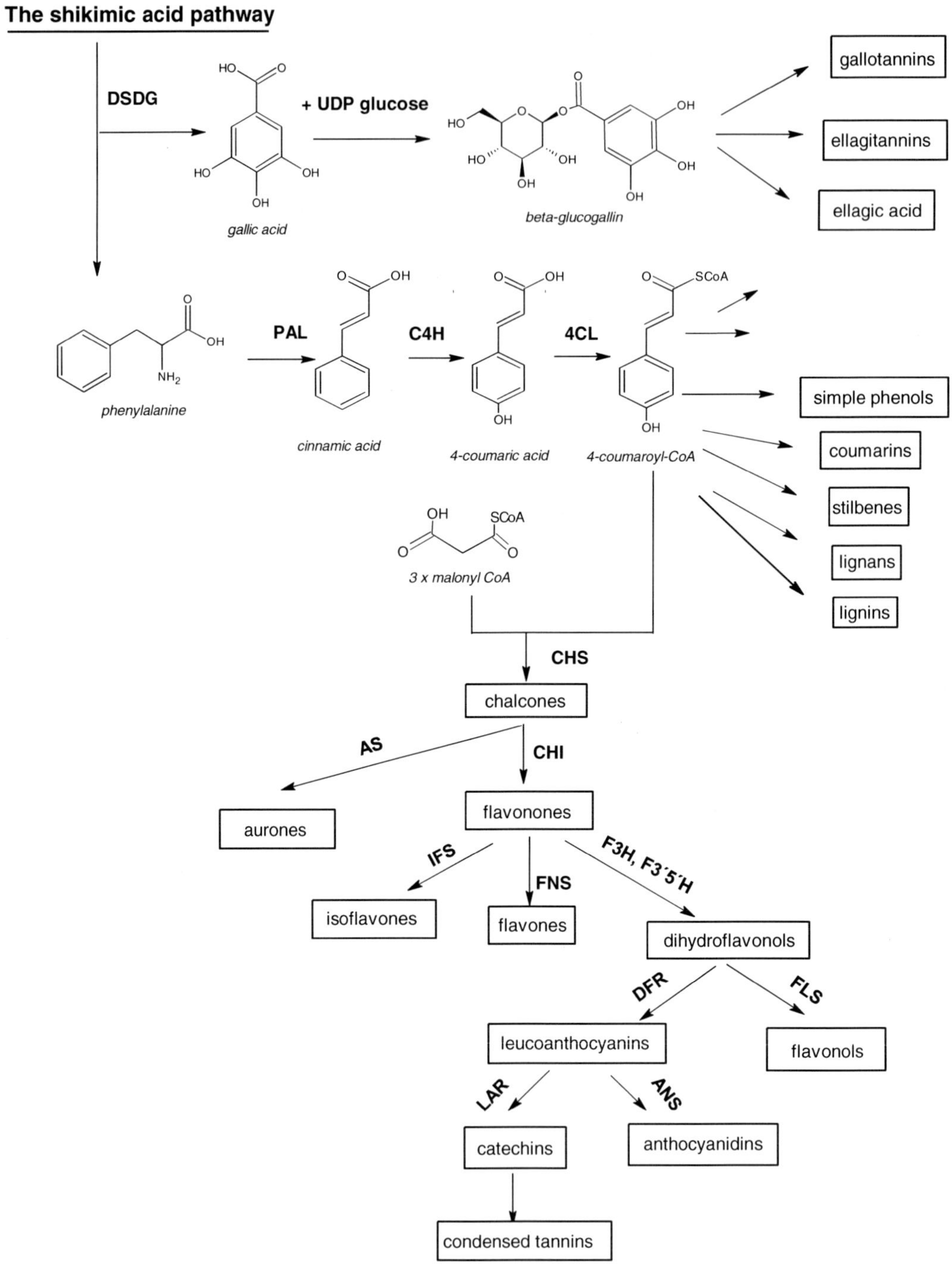

Figure 2. Biosynthesis of phenolic compounds (adapted from [4,5].) AN = anthocyanin synthetase; AS = aurone synthetase; CHI = chalcone isomerase; CHS chalcone synthetase C4H = cinnamic acid 4-hydroxylase;. DFR = dihydroflavonol-4-reductase; DSDG= dehydroshikimate dehydrogenase; IFS = isoflavone synthetase; F3´5´H = flavonoid-3´,5´-hydroxylase; F3H = flavonoid-3-hydroxylase; FLS = flavonol synthetase; FN = flavone synthetase; LAR = leucoanthocyanidin reductase;.4CL= CoA ligase.

Biosynthesis of Phenolic Compounds

It is important to know phenolic compounds' biosynthesis to understand their presence, particularly gallic acid and its derivatives, in plants. Phenolic compounds are formed in plants by the so-called shikimic acid pathway. The enzymatic process in the plant begins with the production of phenylalanine and tyrosine.

Phenylalanine is the substrate of phenylalanine ammonia-lyase, a key enzyme that leads to the formation of most phenolic compounds. Other phenolic compounds such as gallic acid and its derivatives are formed from the 3-dehydroshikimate via dehydroshikimate dehydrogenase. The central metabolism of the phenyl propanoid pathway starts when phenylalanine ammonia-lyase converts phenylalanine in cinnamic acid and this is hydroxylated by cinnamic acid 4-hydroxylase. Finally, 4-coumaric acid is esterified by coenzyme A by 4-coumaric acid:CoA ligase. Flavonoid biosynthesis continues via the phenylpropanoid pathway, starting with the formation of chalcones through a reaction catalysed by chalcone synthetase using 4-coumaric acid:CoA and malonyl-CoA as substrates. The different flavonoid groups are formed in specific enzymatic reactions, as shown in Figure 2 [4, 5].

Gallic Acid and Derivatives in Bolivian Food

Regarding the studies about the presence and contents of gallic acid and its derivatives in food plants growing at high altitude above sea level, it is worth to mention that they were found in several Andean crops.

Gallic acid was found in the Andean crops known as oca (*Oxalis tuberosa*) and ulluco (*Ullucus tuberosum*) species growing at 2000-4500 m above sea level. These are part of traditional food in Bolivia, Peru, Colombia and Ecuador. Oca is also found in Venezuela and in northern Chile and Argentina. At the end of the 19th century oca was introduced in New Zealand where it's a popular food plant under the name of New Zealand yam, a fact revealing its nutritional worldwide importance. [6]

In other studies, gallic acid was identified and quantified in potatoes collected in an altitudinal range of 3790 and 4000 meters above sea level (m.a.s.l.). This resulted in one of the major phenolic compounds present in potato peels [7]. Gallic acid derivatives were found in wild strawberries (*Fragaria vesca*) collected between 2650 and 3300 m.a.s.l. in the Bolivian valleys [8]. The presence of Gallic acid in foods growing at high altitude can be linked to the influence of the solar UV radiation. There are several reports on the relationship between UV radiation and the content of phenolic compounds, as early mentioned [9,10].

Gallic Acid and Derivatives in Medicinal Plants

In the last decades more than 60 phenolic compounds were found in medicinal Bolivian plants [11]. The gallic acid derivatives were mostly found in the plants of the genus *Piper* and are reported for uses in the treatment of malaria and other parasitic diseases. [12].

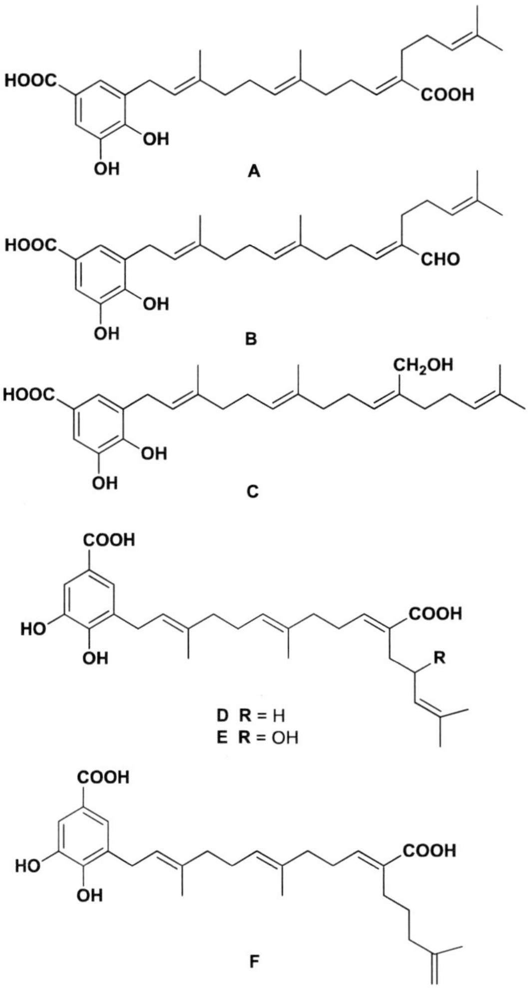

Figure 3. Gallic acid derivatives found in medicinal plants growing in Bolivia.

The *Piper* genus can be found from 1500 to 2500 m.a.s.l. elevations. Among the compounds identified we can mention the following, arieianoic acid (A), arieianal (B) and arieianol (C) (Figure 3). These molecules were isolated and identified in *Piper heterophyllum* [11]. 3-[(2E,6E,10E)-11-carboxy-3,7,15-trimethyl-2,6,10,14-hexadecatetraenyl)-4,5-dihydroxybenzoic acid (D), 3-[(2E,6E,10E)-11-carboxy-13-hydroxy-3,7,15-trimethyl-2,6,10,14-hexadecatetraenyl]-4,5-dihydroxybenzoic acid (E) and 3-[(2*E*,6*E*,10*E*)-11-carboxy-14-hydroxy-3,7,15-trimethyl-2,6,10,15-hexadecatetraenyl]-4,5-dihydroxybenzoic acid (F) were isolated from *Piper aduncum* [12] (Figure 3).

Gallic Acid, Spectrometric Approach

Six solvent systems of varying suitability are reported for the thin-layer chromatographic separation of gallic acid and other simple phenolics. The solvent system chloroform-ethyl acetate-acetic acid (50:50:1) can be economically used. Detection is carried out using iodine vapor, ferric chloride reagent, ferric chloride-ferricyanide reagent and vanillin-sulfuric acid reagent.[14] Gallic acid is soluble in acetone, ethyl acetate, 6 parts of alcohol, 2 parts of boiling water, slightly soluble in cold water but insoluble in benzene or chloroform.[15]

Spectrometric Approaches Include IR, UV Spectrometry as Well as NMR and EIMS

The NMR Characterization of Gallic Acid

In spite of the low solubility in chloroform, this solvent in its deuterated form ($CDCl_3$) was used to treat 30 mg of gallic acid (AG-1) at room temperature (20 ºC) adding CD_3OD by dropping until obtaining the more homogeneous mixture possible under the scope of a rational methanol-D_4 added volume. Filtration was applied to have a homogeneous solution filtrate which was transferred to a 5 mm NMR tube. More effective but more expensive deuterated solvents for a better dissolving of gallic acid are available as ethylacetate-D_8 or acetone-D_6. The NMR description of AG-1 will be based on the molecular ion peak appearing at *m/z* 170 from EIMS data, corresponding to a $C_7H_6O_5$ molecular formula (molecular ion peak at *m/z* 169.96; see Figure 12). The NMR spectra of AG-1 were recorded in an Avance 300 MHz Bruker spectrometer at the Department of Chemical Sciences of Universidad Mayor de San Andres at La Paz.

The ^{1}HNMR Spectrum

The experimental ^{1}HNMR spectrum of AG-1 (Figure 4) shows three signals from higher to lower field. The first signal is a wide and high singlet at δ_H 3.67 corresponding to CD_3OD polluting water (solvent). The next signal at lower field (leftwards) is a fine and intense singlet at δ_H 7.05 followed by a short fine singlet at δ_H 7.28 corresponding to $CDCl_3$ polluting protio-chloroform (solvent).

Due to the absence of TMS internal standard, the ^{1}H spectrum has been calibrated using this signal ($CHCl_3$, δ_H 7.28). Hence the only signal corresponding to proton resonances in AG-1 is the fine and intense singlet at δ_H 7.05. This chemical shift value can be referred to tabled data [16] (Tables B15, B20) where such δ_H resonance value corresponds to ethylenic, HN-amide, or aromatic protons, mainly. Unfortunately because of the use of deuterated solvents where a rapid H↔D interchange occurs, the resonances of the acidic protons HO- are not visibly present in the ^{1}HNMR spectrum.

The ^{13}CNMR and DEPT 135-Experiment Spectra

Figures 5 and 6 respectively show the ^{13}C and DEPT 135-experiment spectra of AG-1. The remarkable signals in these spectra are as follows. From higher to lower field we have the two first big signals calling us for attention appearing as a multiplet centered at 49.0 (Table

C260 [16]) and as an intense triplet centered at δ_C 77.1 (Table C255 [16]). These are well-known signals corresponding to solvents CD_3OD and $CDCl_3$, respectively.

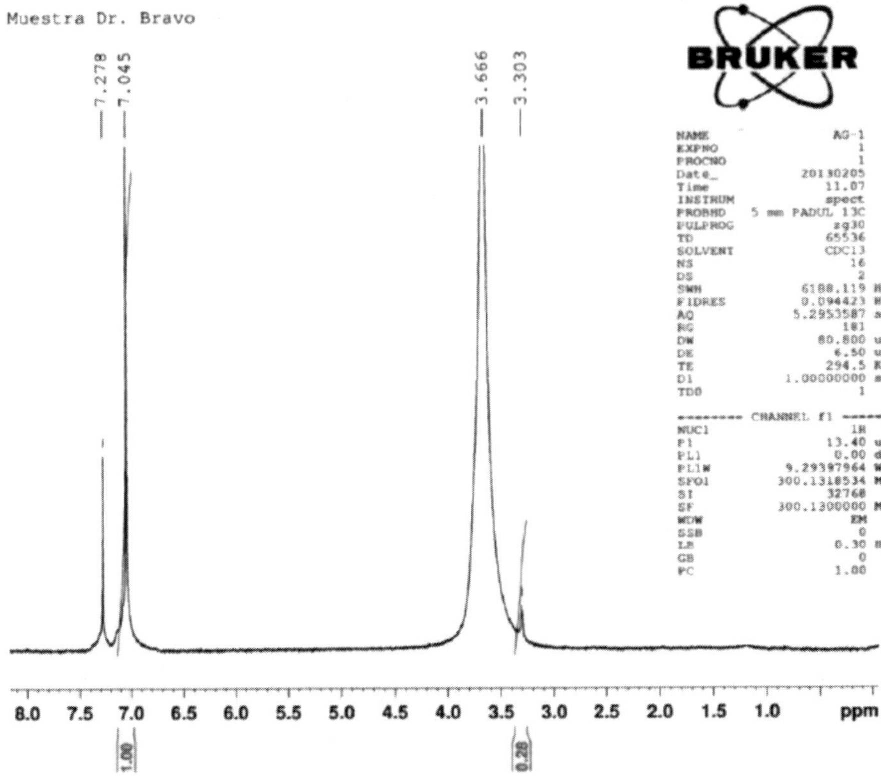

Figure 4. ^{1}HNMR spectrum of AG-1 (300 MHz, $CDCl_3/CD_3OD$).

Being deuterium a magnetically active nucleus with an angular momentum quantum number I=1 and a directional quantum number m= I, I-1, I-2, the split pattern due to each deuterium directly bonded to a carbon nucleus is a triplet (D→C indirect spin-spin coupling).

All according to the total number of 3 possible orientations or m values (1, 0 and -1).[17] This is why the $CDCl_3$ carbon nucleus signal appears as a triplet and the CD_3OD carbon nucleus as a septet. Toward lower field from the chloroform triplet on, sequentially and in an increasing chemical shift values order we find 5 carbon signals at δ_C 109.4, 120.8, 137.7, 144.5 and 169.4. This last value corresponds to a carbonyl carbon (Table B10 [16]).

The only rationalized possibility for such δ_C value, according to the literature (Table B10 [16]) and the molecular formula of AG-1, is an α,β-unsaturated carboxylic acid function, the amide, acyl chloride or acetate functions being excluded. Aldehydes and ketones resonate at much lower field. The δ_C range for these nuclei is according to tabled data (Table B10 [16]) about δ_C 200 and δ_C 205 to 220 respectively. The other four carbons all ranging *grosso modo* in the δ_C 109 – 144 interval, could be according to tabled data (Table B10 [16]), belonging to a variety of organic functions namely pyridine, aromatic, vinylic, and ethylenic carbons. From the molecular formula and calculating the index of hydrogen deficiency [18] or I = 7(C) - 6/2(H) + 1 = 5, we know that we have five insaturations and/or cycles in all for AG-1. Hence the N° double bonds/N° cycles ratio could have the following forms: 1/4, 2/3, 3/2, 4/1. The carbon spectrum (Figure 5) shows five signals corresponding to double bonds carbons. One of the double bonds carbons is the C=O (δ_C 169.4) with four double bond carbons left. The 1/4 ratio is not feasible because we have more than 1 double bond. The 2/3 is not feasible either since this means the presence of carbonyl and only one more double bond (2 double bond carbons), and we have four double bond carbons (two double bonds) besides C=O. The 3/2 ratio means one carbonyl and two more double bonds (4 double bond carbons) and two cycles. This choice means that for instance we have a 1,3-diene-portion in a six-membered ring giving a total of 5 double bond carbons (including the C=O) and two not double bond carbons remaining left. This implies the presence of ^{13}CNMR signals resonating at field higher than the solvent signal at δ_C 77.1, a fact contrary to the spectral data. A similar argument can be adduced for a five-membered ring. We can even propose a 4-membered cyclobutadiene ring (two C=C plus the carbonyl or three double bonds) where the two unemployed carbons in the first cycle cannot form another cycle to accomplish with the I=5 value obtained from the molecular formula, being the form 3/2 discarded as well. Finally the only possibility for the N° double bonds/N° cycles ratio is 4/1. This means a carbonyl (an extracyclic double bond) and 3 double bonds plus one cycle or I=5. This possibility contemplates 3 conjugated double bonds (C=C-C=C-C=C), one cycle and one carbonyl. The only possibility is a benzene ring substituted somewhere by a carbonyl or 4 double bonds and one cycle (3C=C + C=O + ⬡ = 5).

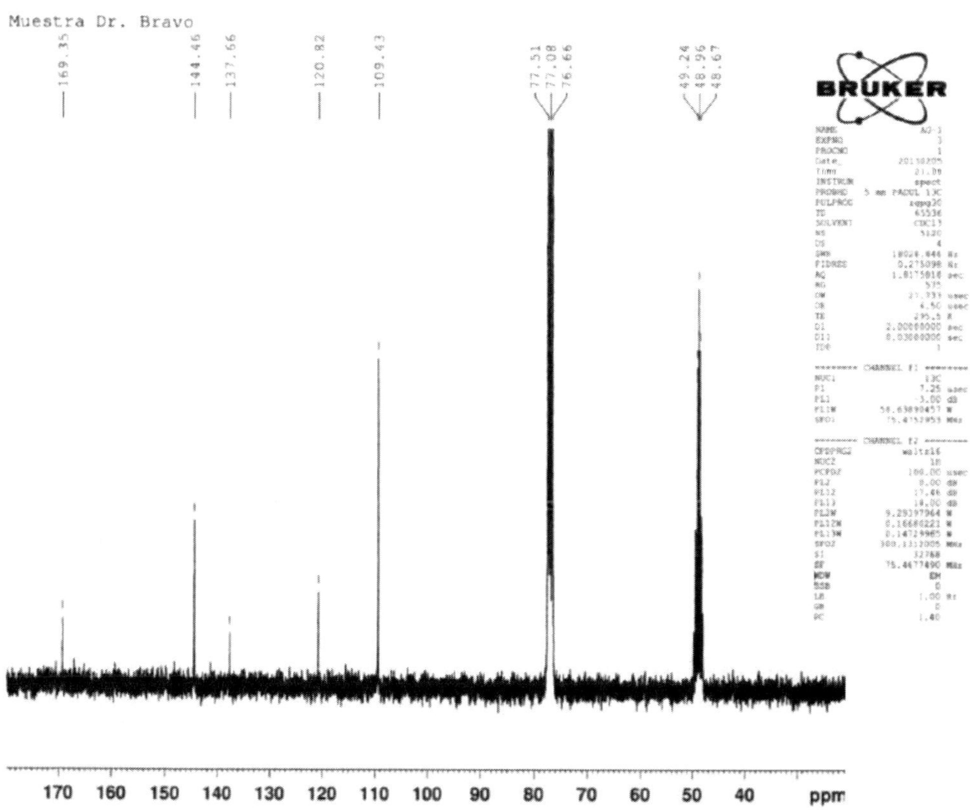

Figure 5. ^{13}CNMR spectrum of AG-1 (75.5 MHz, $CDCl_3/CD_3OD$).

The HSQC and HMBC-Experiments Spectra

Figure 6 shows that AG-1 aromatic carbons can be classified as one methine carbon (δ_C 109.4, see upper spectrum) and three more non-hydrogenated or quaternary carbons (δ_C 120.8, 137.7, 144.5 see downer spectrum). Figure 8 is a direct ^1J-couplings heteronuclear chart run under the HSQC program. This chart exhibits an only crosspeak connecting carbon signal at δ_C 109.4 to proton signal at δ_H 7.05. Summing up; crossing information of the so far analyzed spectral data with the molecular formula ($C_7H_6O_5$) gave us the 7 carbon units assigned as one carbonyl carbon plus 6 benzene carbons. With respect to the 5 oxygen units, we have two oxygen atoms imbibed in the carboxylic acid residue. This forces us to employ the remaining 3 oxygen atoms as 3OH aromatic ring substituents. The six hydrogen units from the molecular formula are assigned one to the carboxylic acid hydroxyl group and 3 for the benzene ring OH substituents giving four from 6 hydrogen atoms. The 2 aromatic protons left have the very same chemical shift value. This means that the singlet at δ_H 7.05 (Figure 1) integrates for two protons that are chemically and magnetically equivalents. This means too that at δ_C 109.4 we have resonating two aromatic carbons with the same δ_C value. In other words the molecule must be symmetrical. These elements can be visualized in Figure 7.

Let us now focus on the establishment of long range heteronuclear couplings in AG-1 or the $^{2\text{ or }3}J_{H\rightarrow C}$-couplings. We don't have so many proton signals available. In fact the only one is the 2H-singlet at δ_H 7.05. A good start point in this kind of analysis is the carbonyl carbon that for AG-1 resonates at δ_C 169.4. Figure 9 is the HMBC experiment 2D-chart of AG-1. Looking at the carbon projection at δ_C 169.4 and rightwards we can readily find an only doublet-shape crosspeak for this carbonyl carbon connecting it to our only proton 2H-singlet at δ_H 7.05. This means that both of our aromatic protons are close enough to C=O to be coupled to.

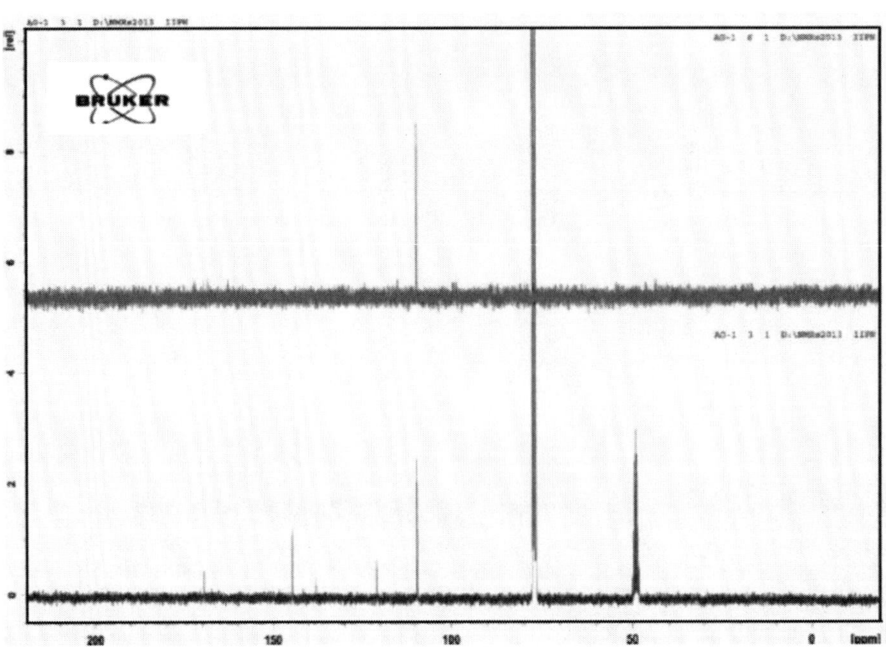

Figure 6 ^{13}CNMR spectrum (blue) contrasted to DEPT 135 experiment spectrum (red) of AG-1 (75.5 MHz, $CDCl_3/CD_3OD$).

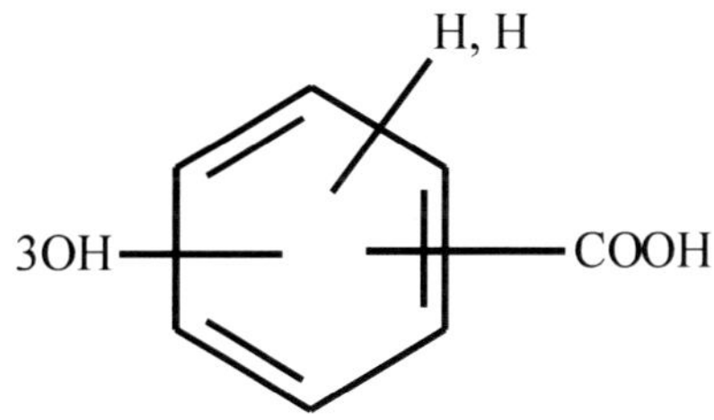

Figure 7. AG-1 structural elements.

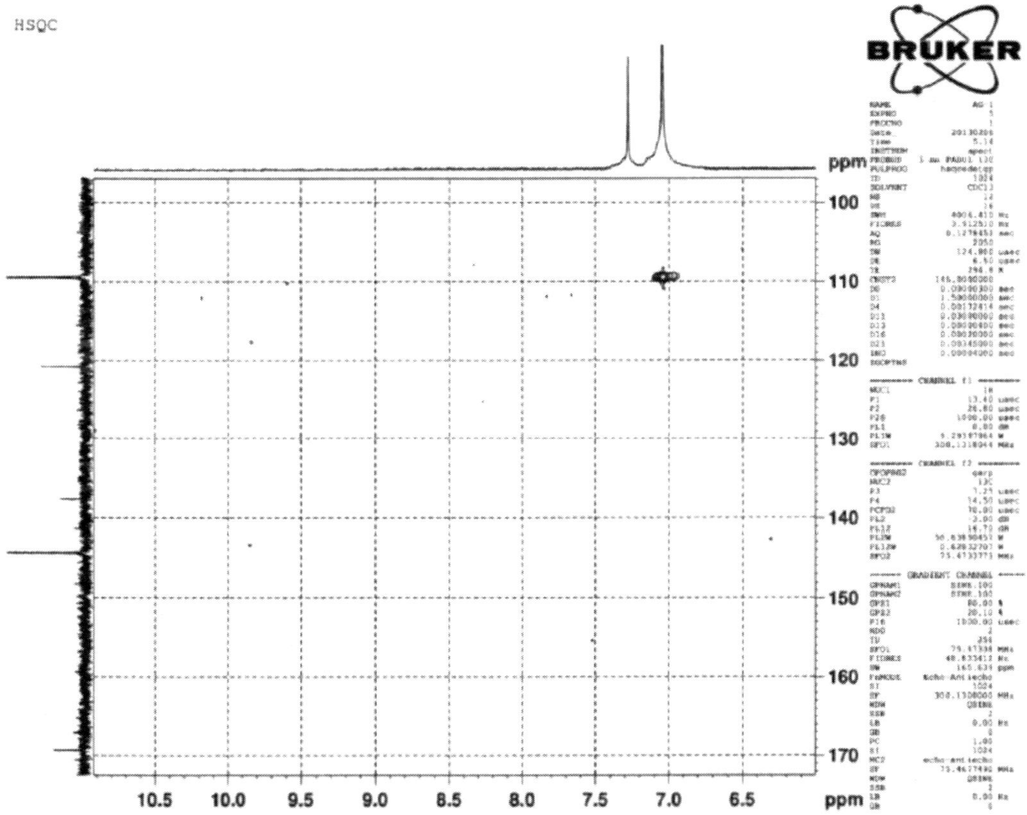

Figure 8. HSQC experiment spectrum of AG-1 for ^{1}J-coupling establishments (300 MHz, 75.5 MHz, $CDCl_3/CD_3OD$).

From the structural elements enounced above (Figure 7) we can place the carbonyl as one of the substituents in the benzene ring defining so the C-1 position in the 6-membered ring. This obliges to the proposal of a $^{3}J_{H\rightarrow C}$-coupling for this crosspeak (See Figure 9 A).

Such remote heteronuclear correlation can provide us with the global or final structure for AG-1, being both equivalent protons forcedly placed in positions 2 and 6 into the aromatic cycle. This settling is respectful of the symmetrical disposition of both protons and indeed with their chemical and magnetic equivalence qualities.

Placing both aromatic protons over positions 3 and 5 is also respectful of these qualities and their symmetrical positioning in the aromatic ring (see Figure 9 B), however the 4-bond distance from H to C=O prevents any H→C coupling. Hence, the only possible settling for the aromatic protons is position 2 and 6 (Figure 9 A). This situation forces us to locate the three water residues (3OH) over positions 3 to 5 (Figure 9 C).

The remaining question now is the peak-assignment for the quaternary carbons in the oxygenated positions as well as in the quaternary supporting the carboxylic acid residue. The assignment in the carbon spectrum (Figure 5) started with peak at δ_C 169.4 for the carbonyl of the carboxylic acid extracyclic residue, followed by the assignment of the aromatic methine carbons at δ_C 109.4 to C-2 and C-6. From Figure 5 we can establish the following facts to be considered before assigning peaks. A.- The quaternary signals are δ_C 120.8, 137.7 and 144.5. B.- The more deshielded a nucleus the higher δ_C is assigned. C.- The more intense carbon peaks could bear two chemically equivalent nuclei. D.- The quaternaries resonate at lower field than methine carbons (congruent with the already assigned C-2,6 at δ_C 109.4 in comparison with all three quaternary signals). E.- Quaternaries are shorter than methines (also congruent with the C-2,6 signal). F.- Oxygenated quaternary carbons resonate at lower field than non oxygenated quaternary carbons. Since according to AG-1 structure (Figure 9 A) we need three types of quaternaries being the first the non oxygenated C-1, the second type the oxygenated chemically equivalent C-3,5 carbons and the third the oxygenated C-4, we can under the just mentioned premises *a priori* assign signals at δ_C 120.8 to C-1, δ_C 137.7 to C-4 and δ_C 144.5 to C-3,5. Under the scope of the preceding enounced facts, we look for useful correlations in the HMBC spectrum (Figure 10). The presence of the doublet-shape crosspeak correlating δ_C 137.7 (supposedly C-4) to the proton signal at δ_H 7.05, would represent the $^3J_{H\rightarrow C}$-coupling H-2,6→C-4, being thus C-4 signal at δ_C 137.7 (see Figure 9 D).

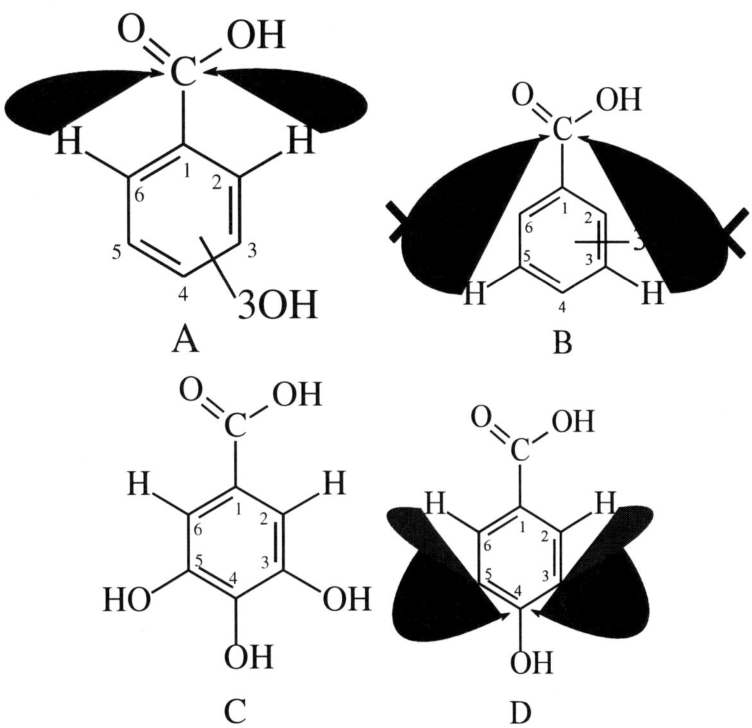

Figure 9. Structural definitions for AG-1 derived from ^3J-coupling establishments in the HMBC spectrum (300 MHz, 75.5 MHz, $CDCl_3/CD_3OD$).

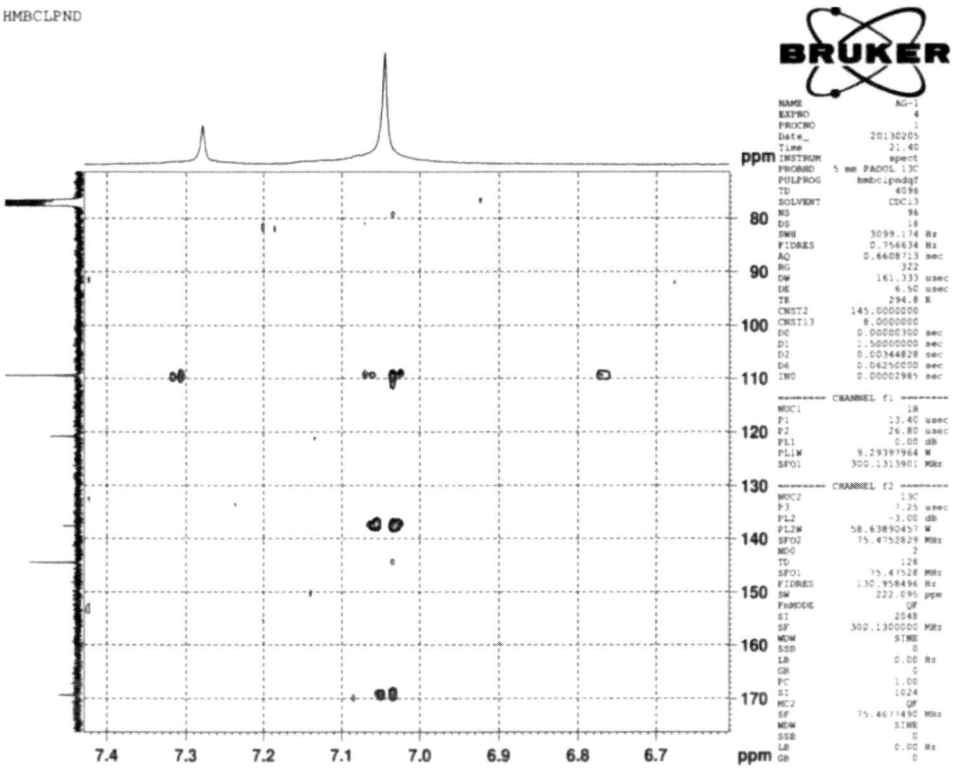

Figure 10. HMBC experiment spectrum of AG-1 (300 MHz, 75.5 MHz, $CDCl_3/CD_3OD$).

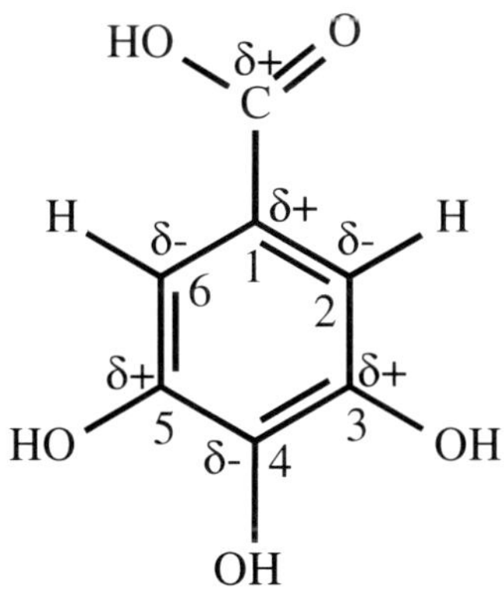

Figure 11. Carbonyl-provoked electronic densities; shielding and deshielding effects are congruent with the chemical shift on the basis of chemical shift values mainly; AG-1.

The remaining carbon signals to assign is the oxygenated C-3,5 and the non-oxygenated C-1. From the AG-1 already deduced structure, we notice the chemical and magnetic equivalence of C-3 and C-5 since a free rotation operation makes possible the positional interchange of such nuclei in this symmetrical molecule. The last, even so weak but assignable crosspeak left, appears at δ_C 144.5 (supposedly C-3,5) corresponding to $^2J_{H\rightarrow C}$-coupling H-2,6→C-3,5 (Figure 9 D).

A much more concentrated sample of AG-1 with a higher number of scans should show more intense remote correlations. Nevertheless, these two HMBC-correlations conduct to ambiguous assignments. By the way, the carbonyl group provokes some partial electronic densities inside the aromatic ring (Figure 11). These densities are congruent with shielding-deshielding effects giving rise to the chemical shift assignments done above on the basis of chemical shift values mainly and not taking into account the ambiguous assignment derived from the HMBC approach. The full assignment done in this analysis is congruent with published data on gallic acid [19] and elemicin, [20] a gallic acid derivative. The following table resumes the NMR data for AG-1, gallic acid

Table 1. NMR data of Gallic Acid , 300 and 75.5 MHz ($CDCl_3/CD_3OD$), δ in ppm

Atom	1H	13C
1	-	120.8
2	7.05 s	109.4
3	-	144.5
Atom	1H	13C
4	-	137.7
5	-	144.5
6	7.05 s	109.4
C=O	-	169.4

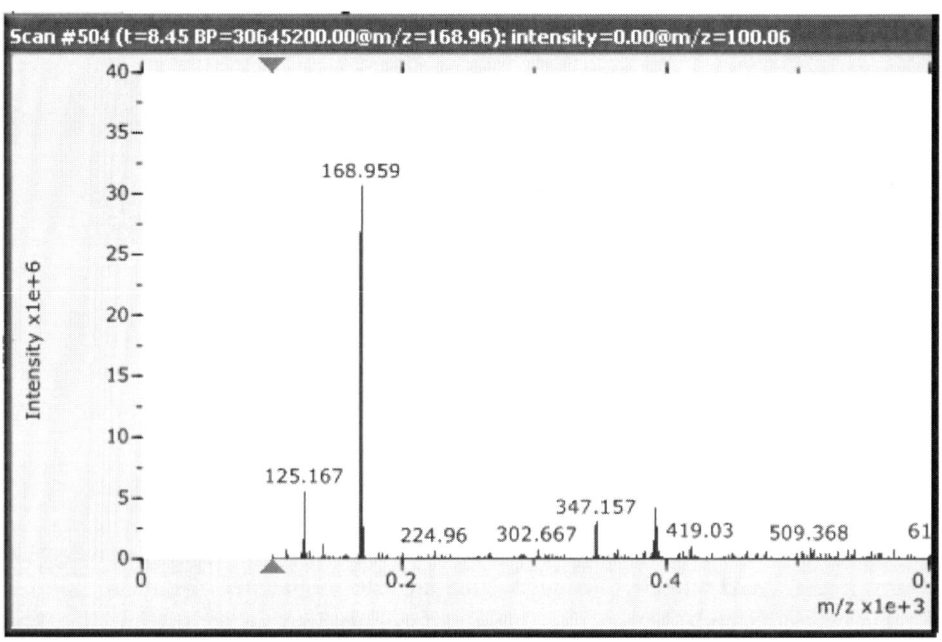

Figure 12. HPLC-MS/EI spectrum of AG-1.

Mass Spectroscopy of Gallic Acid HPLC-MS/EI

The EI-mass spectrum of gallic acid shows a molecular ion peak at *m/z* 169.96. (Fig.12). Favoured modes of fragmentation involve the loss of a hydrogen atom at *m/z* 168.96, the loss of a carbon monoxide unit and a formyl radical at *m/z* 151.96 and 150.96 respectively, and finally the oxidative loss of carbon dioxide unit at *m/z* 125.167.

GC and HPLC coupled to Mass spectra were obtained at the School of Food Science - University of Idaho, USA.

REFERENCES

[1] Larcher, W. (2003). *Physiological Plant Ecology: Ecophysiology and Stress Physiology of Functional Groups*. Berlin:Springer.

[2] Tranquillini, W. (1964). The Physiology of Plants at High Altitudes. *Annual Review of Plant Physiology*, 15, 345-362.

[3] Lui, C.L. Wang, J.M. Chu, C.Y. Cheng, M.T. and Tseng T.H. (2002). In vivo protective effect of protocatechuic acid on tertbutylhydroperoxide-induced rat hepatotoxicity. *Food Chemistry and Toxicology*, 40, 635-641.

[4] Paiva, N. (2000). An introduction to the biosynthesis of chemicals used in plant-microbe communication. *Journal of Plant Growth Regulation*, 19,131-143.

[5] Anttonen, M.J. (2007). Evaluation of means to increase the content of bioactive phenolic compounds in soft fruits. Doctoral thesis, University of Kuopio, Kuopio, Finland.

[6] Peñarrieta, J.M.(2009). *Antioxidants in Bolivian food plants*. Doctoral Thesis. Lund University, Lund,Sweden.

[7] Peñarrieta, J.M. Salluca, T. Tejeda, L. Alvarado, J.A. and Bergenståhl, B. (2011). Changes in Phenolic Antioxidants During Chuño Production (Traditional Andean Freeze and Sun-dried Potato). *Journal of Food Composition and Analysis*, 24, 580-587.

[8] Peñarrieta, J.M., Alvarado, J.A. Bergenståhl, B. and B. Åkesson.2009. Total Antioxidant Capacity and Content of Phenolic Compounds in Wild Strawberries (Fragaria vesca) collected in Bolivia. *International Journal of Fruit Science,* 9, 344-359.

[9] Middleton, E.M. and Teramura A. H. (1993). The role of flavonol glycosides and carotenoids in protecting soybean from ultraviolet-B damage. *Plant Physiology*, 103, 741-752.

[10] Atkinson, C.J. Dodds, P.A.A. Ford, Y.Y. Le Mière, J. Taylor, J.M. Blake, P.S. and Paul, N. (2006). Effects of cultivar, fruit and reflected photosynthetically active radiation on Fragaria X ananassa productivity and fruit ellagic acid and ascorbic acid concentrations. *Annals of Botany*, 97, 429-441.

[11] Vila J.(2006). Bioactive Secondary Metabolites Doctoral Thesis, Lund Sweden.

[12] Flores, N. Jiménez, I.A. Giménez, A. Ruiz, G. Gutiérrez, D. Bourdy, G. and Bazzocchi I.L. (2009). Antiparasitic activity of prenylated benzoic acid derivatives from *Piper* species. *Phytochemistry*, Volume 70, Issue 5, March 2009, 621-627.

[13] Valdivia, C.(2006). Bioactive Secondary Metabolites Doctoral Thesis, Lund Sweden.

[14] Sharma, O. P Bath, T. K. and Singh, B.(1998). Thin-layer chromatography of gallic acid, methyl gallate, pyrogallol, phloroglucinol, catechol, resorcinol, hydroquinone, catechin, epicatechin, cinnamic acid, *p*-coumaric acid, ferulic acid and tannic acid. *Journal of Chromatography A,* 822, 167-171.

[15] Gallic acid, [10-02-2013] Available from :URL http://taninos.tripod.com /acidogalico.htm, access date 2/10/2013.

[16] Pretsch, E. Simon, W. Seibl, J. and Clerc, T.(1989). Tables of Spectral Data for Structural Determination of Organic Compounds, 13C-NMR, 1H-NMR, IR, MS, UV/VIS, In Fresenius, W. Huber, J. F. K. Pungor, E. Rechnitz, G.A. Simon, W. and West, T. H.(Eds.), *Chemical Laboratory Practice*. Berlin, Heilderberg: Springer-Verlag.

[17] Friebolin, H.(1993). Basic One- and Two-Dimensional NMR Spectroscopy, Second enlarged edition, Weinheim, VCH Verlagsgesellschaft.

[18] Silverstein, R. M. Bassler, G. C. and Morill, T. C.(1991). Spectrometric Identification of Organic Compounds, , New York, John Wiley and Sons, Inc.

[19] Eldahshan, O.A. (2011). Isolation and Structure Elucidation of Phenolic Compounds of Carob Leaves Grown in Egypt. *Current Research Journal of Biological Sciences* 3, 52-55.

[20] Bravo. J. A. and Sauvain, M. (2002). Chemical education: elemicin, a phenylpropanoid isolated from *Sparattanthelium amazonum*, as a didactical example of structural elucidation. *Bolivian Journal of Chemistry*, 19, 25-33.

INDEX

B

C

E

F

N

Q

R

S

T

U

V

W

Y

Z